1992

Microbial Cell-Cell Interactions

Microbial Cell-Cell Interactions

Editor

Martin Dworkin

Department of Microbiology
University of Minnesota
Minneapolis, Minnesota

American Society for Microbiology
Washington, D.C.

Library of Congress Cataloging-in-Publication Data

Microbial cell-cell interactions/editor, Martin Dworkin.
    p. cm.
  Includes bibliographical references and index.
  ISBN 1-55581-037-3
  1. Cell interaction. 2. Microorganisms—Physiology. I. Dworkin, Martin.
QR96.5.M49 1991
576′.11—dc20
                                                                    91-20748
                                                                         CIP

Cover figures: See chapter 7, Fig. 12.

Contents

v

Predator-Prey Interactions

Authors

Paul V. Dunlap
Biology Department, Woods Hole Oceanographic Institution, Woods Hole, MA 02543

Gary M. Dunny
Bioprocess Technology Institute and Department of Microbiology, University of Minnesota, St. Paul, MN 55108

Martin Dworkin
Department of Microbiology, University of Minnesota, Minneapolis, MN 55455

Ursula Goodenough
Department of Biology, Washington University, St. Louis, MO 63130

Kendall M. Gray
Department of Microbiology, University of Iowa College of Medicine, Iowa City, IA 52242

E. P. Greenberg
Department of Microbiology, University of Iowa, Iowa City, IA 52242

Karin Ippen-Ihler
Department of Medical Microbiology and Immunology, Texas A&M University, College Station, TX 77840

Paul E. Kolenbrander
Laboratory of Microbial Ecology, National Institute of Dental Research, Bethesda, MD 20892

Janet Kurjan
Department of Microbiology and Molecular Genetics, College of Medicine and College of Agriculture and Life Sciences, University of Vermont, Burlington, VT 05405

Sumit Maneewannakul
Department of Medical Microbiology and Immunology, Texas A&M University, College Station, TX 77840

L. Evans Roth
Department of Zoology, University of Tennessee, Knoxville, TN 37996

Edward G. Ruby
Department of Biological Sciences, University of Southern California, University Park Campus, Los Angeles, CA 90089

Pauline Schaap
Cell Biology and Genetics Unit, Zoological Laboratory, University of Leiden, Leiden NL 2311 GP, The Netherlands

Gary Stacey
Center for Legume Research, Department of Microbiology, University of Tennessee, Knoxville, TN 37996

Microbial Cell-Cell Interactions
Edited by Martin Dworkin
© 1991 American Society for Microbiology, Washington, DC 20005

Chapter 1

Introduction

Martin Dworkin

Multicellular organisms are subject to two levels of hierarchical control. There is a flow of information that travels in a vertical as well as in a horizontal direction. In the vertical hierarchy, information travels from the gene to the cell, and it is this direction of control, gene expression, that has received the greatest amount of attention from biologists interested in understanding how an organism controls its development and behavior. However, there is also a flow of information in a horizontal sense as cells interact with each other, tissues communicate with each other, and organs affect each other's activity. It has been a challenge of biology to understand how these horizontal interactions take place and also to understand the regulatory interplay between the vertical and horizontal modes of control.

Microbes are distinguished from the cells of a multicellular organism by their ability to exist for at least part of their normal life cycle in an isolated, unicellular state. It is this unicellular aspect of the life of the microbe that has made it such an attractive experimental model system for examining a wide variety of biological, biochemical, genetic, and molecular questions. However, it is a goal of this book to emphasize an additional feature of the life of the microbe that is often overlooked. Some microbes also manifest the horizontal control referred to above; cells may interact and communicate with other cells or may also exist for part of their life cycle in a multicellular state. The study of microbes can thus contribute to our understanding of the nature of those cell-cell interactions that are an integral feature of multicellularity in higher organisms. In addition, the experimental tractability of microbial systems has made it possible to begin to understand the regulatory interplay that takes place between gene expression and cell-cell interactions.

These horizontal interactions must take place whenever one or more of the following three things has to be accomplished.

(i) When a multicellular structure has to be established. The most striking example of multicellular structures among the microbes is the formation of fruiting bodies by fungi, cellular slime molds, and myxobacteria. In all of these cases this developmental activity is an alternative to the unicellular phase of the organism's life cycle. A less obvious reflection of microbial multicellularity is a chain of cyanobacterial cells differentiated into vegetative cells, heterocysts, and akinetes. Unfortunately, there has not been enough recent work in the area of cyanobac-

1

terial morphogenesis for it to have been included in this volume. The least highly organized microbial multicellular structure is represented by the formation of a biofilm. It has become increasingly clear that, in nature, microbes tend to adhere to surfaces, and they do so often as part of a self-constructed matrix (Costerton et al., 1981a, 1981b). The nature of the cell-cell interactions that occur within these biofilms is, at the moment, a complete mystery.

(ii) When a multicellular activity has to be accomplished. The aggregation and social motility of *Myxococcus xanthus* and of *Dictyostelium discoideum* are excellent examples of the ability of some microbes to behave in a coordinated fashion, and these are described in the chapters on the myxobacteria (chapter 7) and *D. discoideum* (chapter 6). Another illustration of interactive microbial behavior is the regulation of bacterial luminescence as a function of the cell density of the population, described in chapter 8.

(iii) When information has to be distributed among the population. The most obvious and intensively examined example of the distribution of information is, of course, microbial mating, as a result of which genetic changes originating in a single cell are made available to the population as a whole. This aspect of cell-cell interactions is described in the chapters on mating in enteric bacteria (chapters 2 and 3), *Chlamydomonas* (chapter 4), and *Saccharomyces cerevisiae* (chapter 5).

The concept of multicellularity among the eukaryotic microbes is, of course, not new. *Dictyostelium mucoroides* was isolated by Oskar Brefeld in 1869, and since then the cellular slime molds have played a key role in studies of microbial cell-cell interactions (Schaap, this volume). In addition, the mating interactions that occur between opposite mating types of yeasts (Manney and Meade, 1977; Kurjan, this volume), the mucoraceous fungi (Sutter, 1977), *Chlamydomonas* (chapter 4), and ciliates such as *Paramecium, Tetrahymena*, and *Blepharisma* (Nanney, 1977), while not reflecting true structural multicellularity, have also provided valuable insights into the nature of microbial cell-cell interactions.

The situation with regard to the prokaryotes has been somewhat different. Traditionally, bacteria have been thought of as solitary organisms that did not offer the possibility of serving as experimental models for studying cell-to-cell communication.

In my beginning bacteriology course, I was taught that the properties of a population of bacteria were the sum of the properties of the individual cells, nothing more or less. This view of the bacterial organism as a single cell, essentially free of associations or interactions with surrounding or neighboring organisms, was the keystone of the Robert Koch concept of bacteria: the assumption that all of the relevant properties of the bacterium could be deduced from the behavior of the single isolated cell.

The entire cloning/colony technology of bacteriology, based on the spatial separation of single bacterial cells, presupposed that there were no cell-to-cell interactions occurring within the colony itself, that it is a physiologically and genetically homogeneous population of cells. Indeed, the fact that bacteria could be grown as colonies on an agar surface was considered to reflect an artificial, experimental circumstance, and the notion that, within the colony, the cells were interacting with each other in any systematic way had not been seriously considered. The work of Shapiro and Hsu (1989) has shown that a variety of complex interactions do indeed occur within the colony. Furthermore, the conventional technology for the liquid cultivation of bacteria would obscure any cell-cell interactions that might be taking place. In most liquid cultures, the cells are likely to be separated from each other by relatively vast distances. For example, one can calculate that for a cell the size of *Escherichia coli*, at a cell density of 2×10^8 cells per ml there is an average distance of about 25 μm between cell centers. Thus, it is highly unlikely that, under these conditions of laboratory cultivation, the cells would manifest any of their normal cell-cell interactions. Furthermore, if any of these interactions were to require actual physical contact, then the probability of such an interaction occurring in liquid culture is almost nil.

However, those of us who have studied the development and behavior of the myxobacteria have focused our attention for many years on the cell-cell interactions that are characteristic of both the growth and the development of these organisms (Dworkin, 1973; Dworkin and Kaiser, 1985). Thus, the notion that at least some bacteria are able to manifest a rudimentary multicellular behavior has been an integral part of the thinking and the technology associated with these organisms. Nevertheless, their social behavior was for many years considered an exception to the essential nature of bacteria.

The recent ASM Conference on Multicellular Behavior of Bacteria was a dramatic illustration of the fact that the time has come to abandon the traditional notion that cell-cell interactions among the bacteria are limited to a small group of unusual organisms. In view of the fact that the vast majority of bacteria in nature are living in dense association with each other in biofilms or as colonies on surfaces (Costerton et al., 1981a, 1981b), it is becoming increasingly clear that if one wishes to be able to understand these interactions, conventional experimental techniques for examining bacteria are inadequate; more importantly, the conventional ways of thinking about bacteria are inadequate. The conference revealed that a new paradigm of bacteriology has crept into the canon. Microbiologists concerned with bacterial colonization of biological and nonbiological surfaces, with mating, with development, with biodegradation, and with host-parasite relationships are being forced to consider the mechanisms whereby bacterial cells may communicate with each other. Thus, the bacteria, as a group, now join the eukaryotic microbes as appropriate experimental systems for the examination of microbial cell-cell interactions.

One may categorize microbial cell-cell interactions in yet another way. On the one hand, there are those that are dynamic and involve the exchange of information between the cells; on the other hand, there are those whose function is to establish a structural, multicellular matrix. Most of the information on the nature of microbial cell-cell interactions has focused on the dynamic interactions. Thus, information is available about mating interactions, developmental signal exchange, and host-parasite interactions. It is generally the case that microbial cell-cell interactions involved in the actual generation of multicellular structure, i.e., literal morphogenesis, have received considerably less attention. The substance of the chapters in this book reflects that imbalance.

The book is divided into four sections reflecting a superficial taxonomy of microbial cell-cell interactions, as follows.

(i) Mating interactions. The exchange of genetic information between microbes has been studied more intensively than any other cell interaction. Mating interactions in the gram-negative enteric bacteria represent the progenitor of such studies in bacteria, and the chapter by Ippen-Ihler and Maneewannakul describes the sophisticated genetic strategy that has been used. A fascinating new aspect of such prokaryotic mating interactions was revealed by the studies of Dunny and Clewell. They showed that the mating of certain enterococci was preceded by an elaborate exchange of mating signals which they likened to the pheromones of higher organisms (Dunny et al., 1978). Gary Dunny's chapter describes the nature, regulation, and perception of these mating signals and the cell surface changes that they induce.

Among studies of mating in the eukaryotic microbes, yeasts and *Chlamydomonas* have received much if not most of the attention. It is clear from Ursula Goodenough's chapter on the mating dance and sexual interaction between cells of *Chlamydomonas* that the biochemistry of this process is understood in greater detail than is any other contact-mediated interaction in microbes.

It has been clear for many years that yeast cells represented an ideal model system for studing mating between eukaryotic cells. Their versatile genetics and the ease of experimental manipulation of yeasts resulted in the accumulation of a great deal of information on the process. The work of Ira Herskowitz a number of years ago, on mating types in *Saccharomyces cerevisiae*, catalyzed a renewed interest in the subject, and recent advances in our understanding of the role of the cell surface in the courtship and mating process are described in the chapter by Janet Kurjan.

(ii) Developmental interactions. It is appropriate that *Dictyostelium discoideum* and the myxobacteria are presented side by side as the eukaryotic and prokaryotic systems, respectively, that exemplify the study of cell-cell interactions in the context of microbial development. As early as 1955, John Tyler Bonner was preaching the doctrine of multicellularity among the microbes and emphasized the behavioral similarity between the myxobacteria and the cellar slime molds (Bonner, 1955). This similarity has recently been emphasized by Dale Kaiser (1986) in a review comparing the two systems. Kaiser's review points out the behavioral similarities between the two organisms in terms of initiation, aggregation, and intercellular signalling during development, but makes it obvious that the two groups of organisms have evolved different mechanistic strategies for carrying out these processes. Pauline Schaap's chapter on *D. discoideum* and my own on the myxobacteria bring the reader up to date on cell-cell interactions in these organisms.

(iii) Ecological/colonization interactions. It is now clear that, despite the preference by microbiologists that their microbes in the laboratory remain unattached to each other or to surfaces, the microbes themselves show the opposite preference. This fact was appreciated by Henrici over 50 years ago (1933) and is only recently being rediscovered. The majority of bacteria in nature exist as mixed cultures in biofilms attached to surfaces (Costerton et al., 1986). The chapter by

Paul Kolenbrander on coaggregation in the oral ecosystem reflects the beginning attempts to understand what the rules are that determine how the mixed coaggregate is established. This is only the first step toward understanding what sorts of genetic and biochemical interactions occur within the coaggregate and between the mixed members of the population.

There now seems to be general agreement that the function of bacterial luminescence is an ecological one rather than a physiological or biochemical one (chapter 8). With regard to cell-cell interactions, two questions emerge: first, how is an essentially pure culture of the luminescent organism maintained in the light organelle of the host fish? and second, what are the mechanisms of the autoinduction process? This second question is a reflection of the fact that the bacterial luminescence will occur only when the bacterial population is above a certain critical cell density. Thus, the question becomes ''How does a bacterial cell know what the cell density of its population is?'' The cells solve this in a simple and cunning way, and the biochemical and molecular aspects of this cell density phenomenon are discussed in the chapter by Dunlap and Greenberg on the *Vibrio fischeri*-monocentrid fish symbiosis.

The bacterial colonization phenomenon that has received more detailed attention than any other has been legume nodulation by *Rhizobium* spp. It is a model system for examining the invasion of a eukaryotic cell by a prokaryote and for the subsequent establishment of an intracellular symbiosis between them. The chapter by Roth and Stacey presents a description of the morphological, biochemical, and genetic events that lead to the establishment of this endosymbiotic relationship.

(iv) Predator-prey interactions. *Bdellovibrio* is a bacterium that reproduces by entering the periplasmic space of another bacterial cell, growing and dividing there, and ultimately killing the host cell when the reproductive cycle is completed. During the cycle a series of signals are exchanged between the cells. In their chapter on intercellular signalling in the *Bdellovibrio* developmental cycle, Gray and Ruby present evidence for an interesting model. They propose that the signals that coordinate the *Bdellovibrio* development are generated by the prey organism, but only the *Bdellovibrio* survives the interaction.

There are a number of microbial systems that engage in cell-cell interactions that for a number of reasons have not been included in this volume. For example, it is now abundantly clear that during the process of endospore formation by *Bacillus subtilis*, there is a complex interaction taking place between the genome of the mother cell and that of the developing forespore. Regulatory signals in the form of sigma factors that control the developmental process are being synthesized in both cells, and other gene products are being exchanged between the cells (Kunkel et al., 1989). During the generation of the swarm cell by the stalked mother cell of *Caulobacter*, positional information that determines the placement of the swarm cell flagellum and eventually its stalk is passed from one cell to the other (Huguenel and Newton, 1982). And, as alluded to earlier, the one-dimensional pattern formation involved in the spacing of heterocysts along the filament of *Anabaena* cells is an area of cell interactions awaiting some brilliant ideas. I am hopeful that a future volume on microbial cell-cell interactions will include these systems.

The goal of this book is not to illustrate any common theme of microbial cell-cell interactions; there is none. Rather, its intention is to point out the variety of such interactions and to emphasize the different strategies that microbes have evolved as alternatives to their fundamentally unicellular nature.

REFERENCES

Bonner, J. T. 1955. *Cells and Societies*. Princeton University Press, Princeton, N.J.

Brefeld, O. 1869. *Dictyostelium mucoroides*. Ein neuer Organismus und der Verwandschaft der Myxomyceten. *Abh. Seckenberg Naturforsch. Ges.* **7**:85–107.

Costerton, J. W., R. T. Irvin, and K.-J. Cheng. 1981a. The bacterial glycocalyx in nature and disease. *Annu. Rev. Microbiol.* **35**:299–324.

Costerton, J. W., R. T. Irvin, and K.-J. Cheng. 1981b. Role of bacterial surface structures in pathogenesis. *Crit. Rev. Microbiol.* **8**:303–338.

Costerton, J. W., J. C. Nickel, and T. I. Ladd. 1986. Suitable methods for the comparative study of free-living and surface-associated bacterial populations, p. 49–84. *In* J. S. Poindexter and E. R. Leadbetter (ed.), *Bacteria in Nature*, vol. 2. Plenum Publishing Corp., New York.

Dunny, G. M., B. L. Brown, and D. B. Clewell. 1978. Induced cell aggregation and mating in *Streptococcus faecalis*: evidence for a bacterial sex pheromone. *Proc. Natl. Acad. Sci. USA* **75**:3479–3483.

Dworkin, M. 1973. Cell-cell interactions in the myxobacteria, p. 125–142. *In* J. M. Ashworth and J. E. Smith (ed.), *Microbial Differentiation*. Cambridge University Press, Cambridge.

Dworkin, M., and D. Kaiser. 1985. Cell interactions in myxobacterial growth and development. *Science* **230**:18–24.

Henrici, A. T. 1933. Studies of freshwater bacteria. I. A direct microscopic technique. *J. Bacteriol.* **25**:277–286.

Huguenel, E. D., and A. Newton. 1982. Localization of surface structures during procaryotic differentiation: role of cell division in *Caulobacter crescentus*. *Differentiation* **21**:71–78.

Kaiser, D. 1986. Control of multicellular development: *Dictyostelium* and *Myxococcus*. *Annu. Rev. Genet.* **20**:539–566.

Kunkel, B., L. Kroos, H. Poth, P. Youngman, and R. Losick. 1989. Temporal and spatial control of the mother-cell regulatory gene *spoIIID* of *Bacillus subtilis*. *Genes Dev.* **3**:1735–1744.

Manney, T. R., and J. H. Meade. 1977. Cell-cell interactions during mating in *Saccharomyces cerevisiae*, p. 281–321. *In* J. L. Reissig (ed.), *Microbial Interactions*. Chapman and Hall Ltd., London.

Nanney, D. L. 1977. Cell-cell interactions in ciliates: evolutionary and genetic constraints, p. 351–397. *In* J. L. Reissig (ed.), *Microbial Interactions*. Chapman and Hall Ltd., London.

Shapiro, J. A., and C. Hsu. 1989. *Escherichia coli* K-12 cell-cell interactions seen by time-lapse video. *J. Bacteriol.* **171**:5963–5974.

Sutter, R. P. 1977. Regulation of the first stage of sexual development in *Phycomyces blakesleeanus* and in other mucoraceous fungi. *In* D. H. O'Day and P. A. Horgen (ed.), *Eukaryotic Microbes as Model Developmental Systems*. Marcel Dekker, Inc., New York.

Mating Interactions

Microbial Cell-Cell Interactions
Edited by Martin Dworkin
© 1991 American Society for Microbiology, Washington, DC 20005

Chapter 2

Mating Interactions in Gram-Positive Bacteria

Gary M. Dunny

INTRODUCTION

When used in describing a biological activity of procaryotes, the term "mating" generally refers to bacterial conjugation, i.e., the horizontal transfer of genetic material between cells via a direct cell-to-cell contact (Lederberg, 1986). Because of the prominent position that the classical studies of unidirectional conjugal DNA transfer mediated by the *Escherichia coli* sex factor F (Jacob and Wollman, 1961) hold in the history of modern molecular genetics, "bacterial conjugation" often connotes a process involving attachment of cells via sex pili and subsequent transfer of a single-stranded F-plasmid-like DNA molecule from donor to recipient. Indeed, with the exception of experimental evidence for conjugation in *Streptomyces* spp. (reviewed in Hopwood et al., 1985), it was not generally believed that this form of genetic exchange took place in gram-positive bacteria until the mid-1970s. However, it is now recognized that numerous plasmids and transposable elements in many gram-positive microbes encode conjugal gene transfer systems and that the structural and genetic properties of these systems differ in many important ways from those of gram-negative organisms. In this review, I will describe some of the conjugation systems that have been found in gram-positive bacteria, focusing primarily on the cell-cell interactions and the genetic features of the more thoroughly studied gram-positive mating systems of streptococci and related genera. The approach taken will be selective rather than comprehensive (this is almost mandatory, since reports of new conjugative elements in these organisms appear in the literature virtually every week), but will focus on several important themes that will provide a general framework for the consideration of any bacterial conjugation system. One goal of this discourse is to convince the reader of the many unique features of the conjugation systems of gram-positive bacteria which distinguish them from one another and from those of *E. coli* and other gram-negative bacteria.

This chapter will assume a basic knowledge of the key features of conjugation in gram-negative bacteria, and the reader is referred to earlier reviews (Clark and Warren, 1979; Willets and Wilkins, 1984) and to chapter 3 in this volume for detailed information in this area. I will begin with a discussion of the experimental criteria for the demonstration that a given transfer event occurs via conjugation, followed by a summary of the common steps involved in all forms of bacterial conjugation, highlighting some of the important differences and similarities in the way that various gram-positive bacteria carry out these processes. The latter section will include discussion of the potential role of various gene transfer systems in bacterial evolution and interaction. The remainder of the chapter will be devoted to a description and analysis of pheromone-inducible transfer systems and conjugative transposons that have been studied in enterococci, followed by a compilation of some of the conjugal transfer systems which have been described for other gram-positive bacteria. In terms of cell-cell interactions, these other systems have generally been much less intensively studied, and one purpose of mentioning them in the context of this review is to point out some of the interesting and exciting work that remains to be done with these systems.

EXPERIMENTAL CRITERIA FOR DEMONSTRATION THAT A GENE TRANSFER SYSTEM IS CONJUGATIVE

A rigorous experimental demonstration for conjugal gene transfer is not always easy to obtain. The generally accepted proof that a newly discovered genetic transfer system is conjugative consists of experiments demonstrating (i) that transfer requires direct cell-to-cell contact, (ii) that transfer is insensitive to DNases, and (iii) that bacteriophages are not involved in the transfer event. Although these types of experiments are simple conceptually, their execution and the interpretation of the results obtained are not always straightforward. To address the first and third criteria, either U-tube experiments (Lederberg, 1986) or the addition of sterile culture filtrates from the donor strains to recipient cells is employed to demonstrate that the transfer activity does not reside within phage particles or in free DNA released by the donor. In addition, careful investigators examine their strains for the presence of temperate phages which could display transducing activity. Typically, transformation is ruled out by the incorporation of DNase I into mating mixtures. In addition, it is desirable to show that free DNA, containing genes capable of being maintained and expressed in the recipient cells, cannot be transformed into the recipient by incorporating it into mating mixtures. Unfortunately, most of these experiments are compelling only when they yield results indicating that conjugation is not the mechanism of transfer for a given experiment. While the requirement for direct contact between donor and recipient cells can be demonstrated readily, it is much more difficult to rule out a role for bacteriophages or free DNA in a given transfer event. Transducing phages may be defective or an appropriate bacterial indicator strain may be unavailable, making the phage particles difficult to detect. Perhaps the most convincing evidence against phage involvement is a careful electron microscopic examination of donor cell cultures, which is often not carried out in many laboratories (and even this may be inconclusive if the transducing particles are in low titer). Moreover, eliminating transformation as a mechanism of transfer can, in practice, be difficult. Incorporation of 10 to 50 μg of DNase per ml into mating mixtures, the concentration range often used to examine the role of transformation in a newly discovered transfer system, may not remove transforming DNA completely from the medium, especially in the case of matings carried out for prolonged periods or at elevated temperatures. For example, we found that the transfer of conjugative transposon Tn925 between *Enterococcus faecalis* and *Bacillus subtilis* occurred in liquid or on solid surfaces in the presence of 50 μg of DNase per ml (Christie et al., 1987). However, when either a nonleaky competence mutant, *com-56* (Albano et al., 1989), was used as a *Bacillus* recipient or the DNase concentration was increased 10-fold, transfer in liquid was eliminated, but transfer on solid surfaces was not affected (Torres et al., 1991, unpublished observations). Thus, Tn925 can apparently transfer to *B. subtilis* by either transformation (in liquid) or conjugation (on solid surfaces), and it is highly likely that the same situation holds true for other transferable elements. Thus, reports of "conjugal" transfer into certain organisms under conditions where transformation might be occurring should be interpreted with caution. Further complicating the issue are transfer events that involve both free DNA and direct contact (Berghash and Dunny, 1985; Buu-Hoi et al., 1985), which may represent a completely novel class of procaryotic gene transfer. Once a given transferable plasmid or transposon has been thoroughly characterized physically and genetically and transferred to several different bacterial hosts with different genetic backgrounds, it becomes possible to be more confident about the mechanism of transfer. However, in the absence of these data, the commonly used physical experiments to determine transfer mechanisms should be viewed as tenuous when new transferable elements are analyzed or when existing elements are transferred to hosts for which little genetic information is available.

STAGES OF CONJUGATION

Vitually all bacterial conjugal transfer events feature the following steps: (i) initial contact of the donor and recipient cells; (ii) formation of an "effective mating pair (or aggregate)," including a channel between the cells that allows for DNA transfer; (iii) transfer of genetic material between cells; and (iv) establishment of recombinant genomes and resolution of the progeny produced by the mating event. There is considerable variability in different mating systems with regard to the way in which each of these steps is carried out. This has important implications for the type of genetic exchange that results, as well as for the physical and physiological parameters that affect the process. For example, extremely efficient mating systems that have a narrow host range may promote the rapid spread of resistance or virulence traits through populations of a given species, but have relatively little impact on evolutionary events resulting from interspecies genetic transfer. On the other hand, conjugative elements showing a broad host range may have a considerable influence on evolutionary events that result from genetic exchange between unrelated organisms (Clewell and Gawron-Burke, 1986), particularly in the presence of selection for genes encoded by or readily mobilized by the transferable element. Of course, the dissemination of any conjugative element may be enhanced if the transfer-related phenotypes (cell surface or membrane changes) also confer additional selective advantages (or are linked to genes conferring such advantages), e.g., virulence. Obviously, the nature of the initial contact, as well as that of the channel for DNA transfer, is critical in determining the nature and number of genes that are transferred, as well as the conditions under which transfer can take place. As will become evident as the chapter progresses, the type of effective mating pair formed during conjugation can conceivably affect the cellular, as well as the genetic, composition of the progeny of a mating event. In the sections that follow, these points will be further illustrated as the mating systems of selected gram-positive microbes are discussed in the context of the universal steps in conjugal transfer.

PHEROMONE-INDUCIBLE CONJUGATION IN *E. FAECALIS*

The first convincing pieces of evidence for conjugal transfer in nonfilamentous gram-positive bacteria were reports of the transfer of hemolysin/bacteriocin production and antibiotic resistance in *E.* (formerly *Streptococcus*) *faecalis* by Tomura and coworkers (1973) and Jacob and Hobbs (1974). These workers showed that efficient transfer of plasmid-encoded genes could occur in liquid matings between suitably marked *E. faecalis* strains. Several similar plasmids were soon found in other strains (Dunny and Clewell, 1975; Tomich et al., 1979). The transfer process functionally resembled conjugation in *E. coli*. Despite this resemblance and the high frequencies of transfer that characterized this form of mating, the *E. faecalis* donor strains harboring these conjugative plasmids did not appear to produce any surface appendages resembling sex pili (Dunny et al., 1978). Subsequent analyses in the laboratory of D. Clewell (Dunny et al., 1978, 1979) demonstrated that this type of conjugation was an inducible process characterized by the formation of large mating aggregates in mixed cultures of donor and recipient cells. It was found that the recipient cells excreted low-molecular-weight sex pheromones (termed clumping-inducing agents) capable of inducing the synthesis of cell surface adhesins on the donor cell (promoting aggregate formation), as well as additional transfer functions distinct from aggregation (Dunny et al., 1979; Clewell and Brown, 1980; Suzuki et al., 1984; Mori et al., 1984, 1986, 1988).

Further studies showed that *E. faecalis* strains typically excrete a number of low-molecular-weight hydrophobic peptides during normal growth (Dunny et al., 1978, 1979; Suzuki et al., 1984) and that various plasmids have evolved mechanisms that enable their host cell to detect the presence of these compounds (each plasmid, or group of related plasmids, appears to encode a response to one particular pheromone) as a signal of the presence of recipient cells (Dunny et al., 1979; Mori et al., 1986; Clewell et al., 1987). Each plasmid appears to encode one or more mechanisms for preventing the production of the chromosomally encoded pheromone to which it determines a response by its bacterial host cell, but the cell continues to produce other pheromones capable of inducing the transfer of unrelated plasmids (Clewell and Weaver, 1989). In some cases, pheromone-inducible plasmids encode an inhibitor peptide which competitively inhibits pheromone binding but fails to induce a mating response (Clewell et al., 1987). Although a number of pheromone-inducible plasmids, each encoding a response to a different peptide pheromone, have been identified (Dunny et al, 1979; Clewell and Weaver, 1989), most of the detailed studies of this form of mating have utilized the hemolysin plasmid pAD1 (Tomich et al., 1979) or the tetracycline resistance plasmid pCF10 (Dunny et al., 1991). These two plasmids, like most of the others examined, are about 60 kb in size (Clewell and Weaver, 1989; Dunny, 1990). A nomenclature system for the pheromones and inhibitors based on the name of the plasmid which determines a response to a particular pheromone (cAD1 induces the transfer of pAD1, cCF10 induces the transfer of pCF10, etc.) has been established (Clewell and Weaver, 1989), and a compilation of the enterococcal pheromones that have been purified and sequenced at the University of Tokyo in the laboratory of A. Suzuki (Suzuki et al., 1984; Mori et al., 1984, 1986, 1988) appears in Table 1. Two recent reviews (Clewell and Weaver, 1989; Dunny, 1990) of pheromone-inducible plasmid transfer have appeared, emphasizing the gene organization and molecular regulatory mechanisms controlling expression of the transfer systems encoded by pAD1 (Clewell and Weaver, 1989) and pCF10 (Dunny, 1990). In this chapter, I will focus on the aspects of pheromone-inducible conjugation related to cell-cell interactions and refer the interested reader to these

TABLE 1
Amino acid sequences of enterococcal phermones and inhibitors[a]

Pheromone or inhibitor	Plasmid encoding:		Amino acid sequence
	Response	Inhibitor	
Pheromones			
cPD1	pPD1		H-Phe-Leu-Val-Met-Phe-Leu-Ser-Gly-OH
cAD1	pAD1		H-Leu-Phe-Ser-Leu-Val-Leu-Ala-Gly-OH
cAM373	pAM373		H-Ala-Ile-Phe-Ile-Leu-Ala-Ser-OH
cCF10	pCF10		H-Leu-Val-Thr-Leu-Val-Phe-Val-OH
Inhibitors			
iPD1		pPD1	H-Ala-Leu-Ile-Leu-Thr-Leu-Val-Ser-OH
iAD1		pAD1	H-Leu-Phe-Val-Val-Thr-Leu-Val-Gly-OH

[a] See text for futher discussion and references.

previous reviews for further discussions of the genetics, regulation, and molecular biology of these systems.

Signalling Interactions in Pheromone-Inducible Conjugation

Although several pheromones have been purified, characterized chemically, and synthesized, very little is known about the molecular structures and mechanisms involved in the detection and transduction of this chemical signal by the responding cell. One feature of the signal transduction process that is well established is that the responder cells have an extremely sensitive detection system for the pheromone. Producer cells excrete very small amounts, and analyses with pure synthetic pheromones have shown that concentrations required for biological activity are in the range of 10^{-11} to 10^{-12} M (Mori et al., 1984, 1988). In the case of the pCF10 system, there is good evidence (Mori et al., 1988) that as few as one to five molecules of the heptapeptide cCF10 (Table 1) are sufficient to induce a mating response in an *E. faecalis* donor cell. It is also known that the binding is highly specific, with various hydrophobic peptides of similar sequence showing very little stimulation of cells carrying heterologous plasmids (Mori et al., 1984, 1988; Clewell and Weaver, 1989) and with the available evidence indicating that the amino-terminal portion of the peptide may interact directly with the pheromone receptor on the donor cell, as discussed by Clewell and Weaver (1989).

To date there have been no attempts to directly identify the subcellular components of the donor cells that mediate pheromone binding and signal transduction. We recently developed a quantitative assay for removal of pheromone activity from culture medium by *E. faecalis* cells. We have used this assay to determine the pheromone binding ability of strains carrying either pCF10 or pCF10::Tn917 derivatives and to identify regions of the plasmid encoding genes involved in pheromone binding. Our preliminary results (Leeds and Dunny, unpublished data) showed that each wild-type donor cell bound about 100 molecules of pheromone (cells lacking the plasmid do not bind pheromone nonspecifically). Genetic material encoding functions necessary for pheromone binding is located between two regions of pCF10, previously designated tra2 and tra3, that are known to be involved in negative regulation of transfer functions in the absence of pheromone (Christie and Dunny, 1986). Similar types of experiments have been carried out with the pAD1 system (Weaver and Clewell, 1988). The simplest model for the receptor would be a plasmid-encoded membrane protein which, when bound to pheromone, initiated a signal transduction pathway, ultimately leading to the expression of transfer functions. However, there is no direct evidence relating to the biochemical nature of the receptor or the number of plasmid and chromosomal gene products required for binding activity. The region of pCF10 that seems to be required for binding appears to be too small to encode more than one or two genes, and we hope that current attempts to subclone and sequence this region will elucidate the nature of the receptor. Although the size and amino acid sequences of the pheromones (Table 1) preclude labeling by any of several conventional methods, we are attempting to synthesize radiolabeled cCF10 to use in biochemical studies of the kinetics of binding and for identification of the receptor.

These labeling studies, in conjunction with ongoing genetic and molecular analyses of the regulatory genes controlling the pheromone response (reviewed in Dunny, 1990), should also provide a good foundation for detailed biochemical analysis steps in the signal transduction process that follow binding to the receptor. It seems quite reasonable to assume that one or more of the enzymatic reactions, such as phosphorylation or methylation, known to be involved in other procaryotic signalling systems (Boyd and Simon, 1982; Miller et al., 1989) could be involved in pheromone-mediated signal transduction. Supporting this notion are the results of our recent analysis (Christie et al., 1988; Dunny, 1990, unpublished data) of some of the positive regulatory genes involved in the pCF10-encoded pheromone response. Although expression of some of the genes required to activate the pheromone response is regulated at the level of transcription (Chung and Dunny, unpublished data), this is not the case for all of these genes. Our data indicate that the *prgX* regulatory gene encoded by the pCF10 transfer region is required for activation of transcription of one or more structural genes encoding transfer functions. In cells carrying pCF10, this gene is transcribed at a relatively low, constitutive level in the presence or absence of pheromone. This suggests that the *prgX* gene product is probably present at a relatively constant level regardless of whether the cell is responding to a pheromone signal and that the biological activity of this gene product may be affected by the signal transduction pathway. One simple model to explain these results is that the pheromone-induced signal causes a posttranslational modification of the *prgX* gene product, resulting in a change in the biological activity of the protein, and we intend to test this model experimentally.

One other experimental observation which may be relevant to the biochemistry and physiology of the pheromone response comes from studies of the cell surface distribution of pheromone-inducible surface antigens on donor cells. Wirth and coworkers used immunogold labeling for high-resolution study of the pheromone-inducible antigens encoded by pAD1 (Galli et al., 1989; Wanner et al., 1989). They found that these antigens (the nature and function of these antigens will be discussed below) were not uniformly distributed on the cell surface (Fig. 1), but rather were found away from the region where new cell wall synthesis is localized in enterococci (Higgins et al., 1989). One explanation for these results was that the antigens are only inserted in preexisting, or "old," cell wall. An alternative hypothesis, which is also consistent with these observations, is that the antigens are inserted at the time of cell wall synthesis and that addition of pheromone to a donor culture results in a transient "burst" of antigen synthesis caused by the response of the cells to a sudden increase in exogenous pheromone concentration. This would be followed by adaptation to the new external concentration and return to the uninduced state. This type of transient physiological response to pheromone would produce the same type of antigen distribution as was observed. The pheromone-induced antigen produced during the burst would presumably move away from the region of cell wall synthesis, along with the portion of the wall to which it was attached. This model would also be consistent with our preliminary observations (Olmsted and Dunny, unpublished observations) indicating a uniform distribution of pCF10-encoded antigens in cells carrying

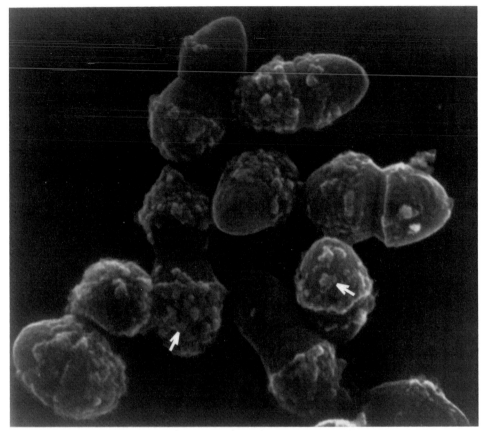

FIGURE 1. Scanning electron micrograph of pheromone-induced *E. faecalis* cells mixed with colloidal gold-labeled antibody against the AS Asa1 (Galli et al., 1989; Wanner et al., 1989). The arrows point to some of the colloidal gold particles bound to the cell surface via their attachment to polyclonal Asa1-specific antibody. Note that the bound gold particles are predominantly located at a considerable distance from the septa of the cells, where new cell wall synthesis is occurring. (Micrograph courtesy of R. Wirth and G. Wanner.)

recombinant plasmids conferring constitutive expression of the antigens. This type of response would also be reminiscent of the adaptation to change in attractant concentration that has been analyzed extensively in several bacterial chemotaxis systems (Boyd and Simon, 1982). In any event, it is obvious that a great deal of further work is necessary to achieve a clear understanding of the biochemistry of pheromone signal transduction, but the experimental tools are now in hand and much more information should be forthcoming in the next few years.

Interactions between Cells in Mating Aggregates

The most striking feature of this mating system, which initally led to its discovery, is the formation in liquid mating mixtures of large bacterial aggregates often visible to the naked eye (Dunny et al., 1978). Aggregation also can be induced in pure cultures of donor cells by the addition of pheromone (Dunny et al., 1978, 1979). These observations led to the original model for pheromone-induced aggregation (Dunny et al., 1979), which is illustrated in Fig. 2. According to this model, all *E. faecalis* cells produce a chromosomally encoded binding substance (BS) which can serve as a receptor for the plasmid-encoded aggregation substance (AS), a surface adhesin whose synthesis is induced by exposure of donor cells to pheromone. Induced donors can either bind to one another or to recipient cells since both produce BS. The AS-BS-mediated binding would allow mating pairs to come into close contact and facilitate efficient plasmid transfer in liquid.

Considerable genetic and biochemical information about both the AS and BS components of the aggregation phenomenon has been obtained. BS must be chromosomally encoded, since it is expressed by plasmid-free cells. Ehrenfeld et al. (1986) showed that the addition of lipoteichoic acid (LTA) purified from *E. faecalis* cells could inhibit aggregation of pheromone-induced donor cells, and they suggested that LTA could be BS. One possible problem with this type of experiment is that LTA is a potent chelator, due to the presence of its polyglycerol phosphate backbone, and the observed inhibition could be an indirect effect. However, genetic analysis of BS in the pCF10 system (Trotter and Dunny, 1990) also suggested the possibility that LTA could be involved in BS, since chromosomal mutants defective in BS showed a concurrent loss of recipient ability, aggregation ability, and alterations in the fatty acid portion of their LTA. The same studies also revealed that the mutants were also missing a 110-kDa protein antigen, and further analyses are required to determine the role of these components in BS activity. It was also shown that the ASs of both pCF10 and pAD1 bind to the same BS and that these plasmids can transfer normally from a BS⁻ donor. These strains do not self-aggregate upon exposure to pheromone, indicating that aggregation does not occur via AS-AS binding. Recent studies from both the pCF10 and the pAD1 systems have clearly identified proteins (now called Asa1 and Asc10, re-

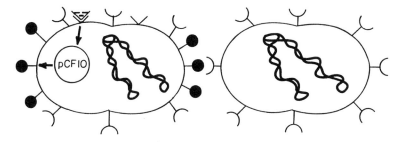

FIGURE 2. Model for pheromone-induced synthesis of a proteinaceous surface adhesin, AS (—●), in the pCF10 system. The cCF10 pheromone (⩡) excreted by recipient cells binds to a receptor (⋎) encoded at least in part by the plasmid. This binding triggers a signal which induces the expression of plasmid-encoded genes, including *prgB*, encoding AS. This adhesin initiates mating pair formation by binding to BS (—⊂) on the recipient cell. See text for further discussion.

spectively) of about 150 kDa that serve as the AS for these plasmids. The conclusion that these proteins mediate aggregation is based on the fact that their expression in *E. faecalis* is sufficient and necessary for aggregation (Christie et al., 1988; Mori et al., 1988; Galli et al., 1989) and that monospecific antibodies against these proteins, and Fab' fragments thereof, inhibit aggregation and mating (Christie et al., 1988; Galli et al., 1989; Dunny et al., unpublished data), as well as recent data from our laboratory to be discussed below. The genes encoding these proteins have been cloned and sequenced (Christie et al., 1988; Galli et al., 1990; Olmsted et al., unpublished data) and have been shown to be highly related at the DNA and protein level (Wirth et al., 1991). Computer analysis of the predicted structures of these proteins indicates that they are globular, secreted surface proteins that may span the cell membrane several times (Wirth et al., 1991). Interestingly, both proteins contain ''RGD'' sequences (Table 2), known to function as motifs for binding to the integrin class of eucaryotic extracellular proteins (Hynes, 1987), suggesting the possibility that these proteins could mediate attachment to host cells in opportunistic infections associated with enterococci, such as bladder infections and endocarditis. The availability of the sequence of these genes, as well as isogenic bacterial strains differing only in the expression of these proteins, should pave the way for detailed structure/function analysis, molecular analysis of AS-BS interactions, and a determination of the role of the AS proteins, if any, in virulence.

Although AS-BS binding is required, it is clearly not sufficient to ensure efficient transfer, but rather represents an initial step in the transfer process, as indicated by several lines of experimental evidence. Clewell and Brown (1980) used a series of ''male-male'' matings between cells carrying derivatives of pAD1 differentially marked with transposons to demonstrate the induction of transfer functions distinct from aggregation by pheromone cAD1. They found that in matings between strains carrying pAD1 derivatives, where only one of the two strains was induced prior to mating, transfer usually occurred from the induced to the uninduced cell, suggesting the possibility of a plasmid-encoded surface (entry)

TABLE 2
RGD-containing sequences in AS proteins
Asa1 and Asc10[a]

Motif	Sequence
1 (residues 261–270)	NVK**RGD**SLQY
2 (residues 593–602)	KVA**RGD**VLSY

[a] Partial amino acid sequences (in one-letter code) of the pheromone-inducible AS proteins Asa1, encoded by pAD1, and Asc10, encoded by pCF10, are shown based on the results of Galli et al. (1990) and Wirth et al. (1991). These sequences each contain an RGD motif (in boldface print) which can serve as a receptor for the integrin superfamily of eucaryotic cell surface receptors, as reviewed by Hynes (1987).

exclusion system that would prevent a donor cell from acquiring a second copy of the same plasmid via conjugation. Ehrenfeld and Clewell (1987) also showed that if a donor strain carries two plasmids, each determining a response to a different pheromone, induction of the strain with one pheromone results in increased transfer of only the plasmid determining a response to that pheromone. Thus, the aggregation of cells does not appear to promote efficient, random transfer of DNA from donor cells, but provides a contact that allows for a subsequent transfer process involving inducible gene products that act preferentially on the pheromone-inducible plasmid itself. This implies that the apparent ability of pheromone-inducible plasmids to promote the low-frequency transfer of chromosomal genes (Franke et al., 1978) and certain nonconjugative plasmids (Dunny and Clewell, 1975) would be dependent on either recombination and cointegrate formation (Clark and Warren, 1979) or specific mobilization genes (Projan and Archer, 1989) and that the access to the channel for DNA transfer in this system is generally restricted to the pheromone-inducible plasmid.

Analysis of pheromone-inducible surface exclusion has also shed some light on the mating interactions involved in this transfer system. In the case of the pCF10 system (Dunny et al., 1985), surface exclusion has been shown to be pheromone inducible, and a 130-kDa antigen, previously called Tra130 and recently renamed Sec10 (Wirth et al., 1991), was shown to be involved. Thus, pure cultures of induced, wild-type donor cells can self-aggregate because they produce both BS and AS, but surface exclusion apparently interferes with a step in DNA transfer occurring after aggregation. Recently, we carried out a series of experiments (Olmstead et al., submitted) utilizing the BS$^-$ mutants described above (Trotter and Dunny, 1990), as well as chimeric plasmids containing cloned fragments of pCF10 (Christie et al., 1988) conferring constitutive expression of various combinations of Asc10 or Sec10 in an otherwise isogenic host. Thus, we were able to examine transfer between cells expressing different combinations of surface proteins. These results showed that transfer in liquid required the interaction between AS on one of the mating partners and BS on the opposite partner. However, mating could occur quite readily when the receptors were reversed from the normal situation, i.e., when the AS on the recipient cell bound to the BS on the donor cell. Thus, it appears that AS-BS binding simply brings cells into sufficiently close proximity for the formation of a mating channel composed of distinct and as yet unidentified components. The experimental data just described have been incorporated in the model shown in Fig. 3 (part I), where the mating channel and the AS-BS complex are shown separated. The channel is depicted as interacting specifically with the plasmid DNA in the donor cell, with transfer from a fixed origin via a single-stranded DNA molecule similar to the F system (Willets and Wilkins, 1984) of E. coli, although there is no experimental evidence for this form of DNA transfer in enterococci as yet. At the completion of the transfer event, it can be imagined that the pheromone-induced AS would gradually be diluted on the cell surface during normal growth, such that the cells would eventually separate, completing the final step in this form of microbial interaction.

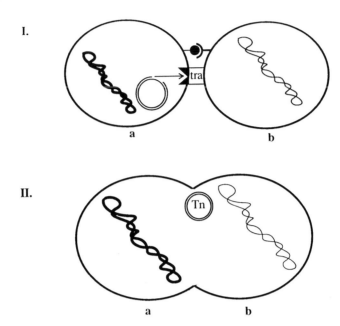

FIGURE 3. Models for the formation of mating channels in pheromone-inducible conjugation (I) and conjugation mediated by conjugative transposons (II). In the first model, mating pair formation is initiated by binding of the plasmid-encoded AS to the BS receptor on the recipient surface. The formation of a mating channel is mediated by additional products, designated tra, which result in the transfer of plasmid DNA. In the second model, a conjugative transposition event is depicted, characterized by the formation of a circular intermediate form of the transposon and a mating involving a fusion between the cells of the mating pair. This fusion allows for extensive recombination between the two genomes, as well as transfer of the transposon.

CONJUGATIVE TRANSPOSONS

Conjugative transposons are genetic elements that are capable of both intermolecular translocation between DNA molecules and intercellular transfer to a wide variety of bacterial hosts via a self-encoded conjugation mechanism. These elements were first identified in enterococci (Franke and Clewell, 1981; Gawron-Burke and Clewell, 1982) and pneumococci (Vijayakumar et al., 1986; Courvalin and Carlier, 1986) and later in a number of other streptococci (Fitzgerald and Clewell, 1985; Hartley et al., 1984). Phenotypically similar elements have also been identified and analyzed in gram-negative *Bacteroides* strains (Odelson et al., 1987), although there have been no reports of homology between the two types of element at the DNA level. Most of the conjugative transposons studied in gram-positive bacteria thus far are at least 15 kb in size, and some very large composite "transposon within a transposon" elements greater than 50 kb have been identified (Vijayakumar et al., 1986; Inamine and Burdett, 1985). A number of antibiotic resistance determinants have been found to be carried by these elements, but the most commonly associated determinant is the *tetM* gene (see Clewell and Gawron-Burke [1986] and Clewell [1990] for reviews). Also, their novel genetic properties make them extremely attractive as research subjects for molecular biologists. Finally, their broad host range means that they have great potential for use in genetic manipulation of a number of medically and industrially important gram-positive bacteria which lack well-developed genetic systems (Bertram et al., 1990; Strätz et al., 1990).

Transposition Mechanism

In recent years, most of the detailed studies of conjugative transposons have focused on Tn916, identified originally (Franke and Clewell, 1981) in *E. faecalis*, and Tn1545, from *Streptococcus pneumoniae* (Courvalin and Carlier, 1986). Much attention has been directed toward molecular and genetic analysis of the mechanism of transposition employed by these elements, with relatively little work on the conjugal transfer mechanism having been carried out. Models for the transposition mechanism have been proposed by several groups (Clewell and Gawron-Burke, 1986; Poyart-Salmeron et al., 1989; Caparon and Scott, 1989). These differ in some details, but they all postulate that transposition is initiated by an excision of the transposon from the donor DNA molecule (which in at least some cases may regenerate the sequence originally disrupted by insertion into that site), followed by the formation of a nonreplicating circular intermediate. In some cases, conjugal transfer of the intermediate may occur between the excision event and the reinsertion of the element into a new target site, although conjugation is not required for transposition. A number of lines of experimental evidence support this type of model, including the isolation of a circular form of the transposon (Scott et al., 1988), the identification of genes encoding enzymes resembling integration and excision enzymes encoded by temperature bacteriophages of *E. coli* (Poyart-Salmeron et al., 1989), and the analysis of the DNA sequences of target sites, excision sites, and the junctions between transposon and target sequences (Caillaud and Courvalin, 1987; Clewell et al., 1988; Caparon and Scott, 1989).

Mechanism of Intercellular Transfer

In contrast to the extensive analysis of the transposition of these elements that has been reported, much less attention has been paid to the means by which conjugative transposons promote intercellular DNA transfer. However, a number of the features of the biology of the conjugal event mediated by these transposons are very different from those of the pheromone-inducible transfer systems described above. One major difference between the two systems is that intercellular transfer of conjugative transposons generally requires that the matings be carried out on solid surfaces, where the cells are forced into close contact for prolonged periods. Even under these conditions, the frequencies of transfer are generally in the range of 10^{-5} to 10^{-8} (Clewell and Gawron-Burke, 1986), as opposed to frequencies of $>10^{-1}$ in the pheromone-inducible systems (Clewell and Weaver, 1989). However, even though the efficiency of transfer of conjugative transposons is quite low, the mechanism of transfer is such that virtually any gram-positive organism, including mycoplasma (Dybvig and Cassell, 1987), and perhaps gram-negative organism (Trieu-Cuot et al., 1988) can act as recipients. Furthermore, there appears to be no surface exclusion function or any other mechanism to prevent a cell from acquiring additional conjugative transposons via multiple mating events. Thus, it might be predicted that the mechanism of donor/recipient attachment, formation of a mating channel, and perhaps the molecular nature of the channel associated with conjugative transposons would be fundamentally different from those of the pheromone-inducible transfer system.

Most of the available data on conjugative transposons indicate that the transfer event usually involves only genes contained within the transposon. For example, in the case of Tn916, mobilization of the nonconjugative plasmid pAD2 by the transposon in matings between *E. faecalis* strains was not detected (Franke and Clewell, 1981). However, Naglich and Andrews (1988) found that transfer for the nonconjugative plasmid pUB110 was promoted by Tn916 in bacilli. In collaboration with S. Zahler and R. Korman, we found that the pheromone-inducible plasmid pCF10 contains a second conjugation system, encoded by a conjugative transposon called Tn925 (Christie et al., 1987). This transposon transfers freely between various gram-positive bacterial genera (Christie et al., 1987; Strätz et al., 1990) in the absence of detectable plasmid DNA, and it shares extensive homology with Tn916 (Christie and Dunny, 1986). In a series of experiments designed to examine the mobilization of various genes in enterococci and bacilli by Tn925, we found that this transposon facilitated the transfer of both chromosomal markers (Torres et al., 1991) and nonconjugative plasmids (Dunny et al., 1991) under conditions where transformation was strictly prohibited. Interestingly, we also found that tetracycline enhances the transfer of Tn925 in *E. faecalis* (Torres et al., 1991).

These results led us to suggest a model for formation of the conjugal DNA transfer associated with Tn925, which is depicted in Fig. 3, part II. We postulate that conjugal transfer of conjugative transposons results in the formation of a mating channel generating a functional diploid cell that may result in extensive interaction and recombination between donor and recipient genomes. This model is not in conflict with the previously discussed excision/insertion model for transposition (Clewell and Gawron-Burke, 1986), as indicated by the depiction of the circular transposition intermediate in Fig. 3. It simply postulates that the mating event is the genetic equivalent of a protoplast fusion, allowing for a great deal of genetic exchange in addition to the relocation of the transposon. In fact, some of our preliminary unpublished results of Tn925-mediated matings in *B. subtilis* suggest the formation of stable diploid strains similar to those resulting from polyethylene glycol-mediated protoplast fusions with this organism (Hotchkiss and Gabor, 1980). In the case of matings between heterogeneous bacterial genera, it might be predicted that any diploids formed would be highly unstable. Furthermore, experiments by Clewell and coworkers (personal communication) indicate that with some strains of *E. faecalis*, the frequency of mobilization of unlinked genes and plasmids by conjugative transposons may be much lower than reported (Torres et al., 1991), indicating that the extent of fusion could vary with different bacterial hosts. Nonetheless, even a transient mixing of genomes could have important evolutionary implications. It should be very interesting to carry out further genetic analysis of this novel form of gene transfer in order to confirm or reject this model and to further elucidate this phenomenon. An alternative explanation for the observed chromosomal transfer mediated by Tn925 is that the transposon somehow causes excision of portions of the chromosome and that these putative plasmidlike forms (PLFs), as well as nonconjugative plasmids, are mobilized by the transposon. Such a phenomenon has been observed (Stevens et al., 1990) with conjugative tetracycline resistance transposons in *Bacteroides* species. In the *Bacteroides* system, tetracycline induces both transfer of the transposon and PLF formation. However, the PLFs are apparently generated from the excision of specific regions of the chromosome, whereas our data indicate that transfer of chromosomal genes is fairly random. However, further work is needed to rule out the involvement of PLFs in Tn925-mediated chromosomal mobilization.

Another extremely attractive area for future study which could shed light on the mechanism discussed above is the analysis of the genes and gene products mediating intercellular gene transfer. In this regard, Clewell and coworkers (Yamamoto et al., 1987; Senghas et al., 1988) have made an extensive set of Tn5 insertions in Tn916 and analyzed the phenotypes of these insertional mutations. Their results indicate that a major portion of Tn916 encodes transfer-related gene products and also provide a firm foundation for the identification of these products. Interestingly, during the course of our analysis of the chromosomal BS determinants (Trotter and Dunny, 1990) involved in pheromone-inducible conjugation in *E. faecalis*, we identified an alteration in the LTA, unrelated to BS, that appeared to be encoded by Tn916. This alteration could possibly be involved in conjugal transfer of Tn916, and it could be very fruitful to examine various insertional mutations in Tn916 for this LTA alteration. With the genetic and biochemical tools now available for the analysis of this system, it should be possible to make great strides in identification of the transfer gene products and their functions during the next few years. However, even with the relatively small amount of currently available data on the mating interactions mediated by conjugative transposons, it is clear that there is a tremendous difference between these elements and the pheromone-inducible plasmids.

BROAD-HOST-RANGE CONJUGATIVE PLASMIDS

Conjugative plasmids capable of transfer to a variety of gram-positive bacterial hosts, including streptococci and staphylococci (Engel et al., 1980), lactobacilli (Gibson et al., 1979), pediococci (Gonzalez and Kunka, 1983), clostridia (Clewell, 1990), and *Listeria* spp. (Buu-Hoi et al., 1984), have been identified in a variety of streptococci and enterococci, as reviewed in Schaberg and Zervos (1986) and Clewell (1990). It appears that the natural hosts for these plasmids may be the enterococci, and in the presence of selective pressure for antibiotic resistance, the enterococci may serve as a reservoir for the dissemination of resistance plasmids (and probably conjugative transposons as well) into the more pathogenic organisms. A recent example of this process is the disturbing report (Poyart-Salmeron et al., 1990) of a clinical isolate of *Listeria monocytogenes* carrying a conjugative R plasmid homologous with several previously identified plasmids from streptococci and enterococci. Such plasmids had been shown previously to be capable of transfer to *Listeria* spp. in the laboratory (Buu-Hoi et al., 1984), but R-plasmid-containing isolates of this organism had not been isolated from patients previously.

These broad-host-range plasmids typically carry the MLS resistance determinant, mediating macrolide, lincosamide, and streptogramin resistance (Horaud et al., 1985), sometimes in combination with other resistance genes (Horaud et al., 1985; Schaberg and Zervos, 1986), and are usually about 30 kb in size. The mating process itself resembles that of the conjugative transposons, requiring prolonged contact or filters or other solid surfaces and not apparently involving any type of mating pheromone. The transfer frequencies are generally in the range of 10^{-4} to 10^{-8}, although frequencies of as high as 10^{-2} have been reported in some instances (Trieu-Cuot et al., 1988). The two best-studied examples of this type of plasmid are pAMβ1 from *E. faecalis* (Clewell et al., 1974) and pIP501 from *Streptococcus agalactiae* (Evans and Macrina, 1983), with most of the experimental attention having been focused on plasmid genes involved in replication (LeBlanc and Lee, 1984) rather than conjugal transfer.

As in the case of the conjugative transposons, virtually nothing is known about the cell-cell interactions or the plasmid-specified gene products that might be involved in donor/recipient attachment, mating channel formation, or DNA transfer. However, some efforts along these lines have been reported recently by Krah and Macrina (1989), who determined that the conjugal transfer genes of pIP501 were contained within adjacent 8.8- and 7.5-kb regions of the plasmid. They used insertional mutagenesis with Tn917 (Youngman et al., 1983) and Tn917-*lac* (Perkins and Youngman, 1986) to define the transfer region and to get an indication of the transcriptional organization. Subcloning of the 7.5-kb region in *E. coli* and analysis of expression in minicells resulted in the identification of several proteins involved in conjugal transfer. Although the functions of these proteins are not yet known, this work represents an important beginning in the analysis of the transfer mechanism of a group of plasmids that are of major significance in relation to evolution and gene transfer in gram-positive microorganisms.

CONJUGATION IN LACTOCOCCI

Lactococci are of great importance to the food and dairy industry as starter cultures for the production of fermented dairy products. Interestingly, most of their industrially important traits, including the ability to ferment lacose and other sugars, protease production, production of various flavor compounds, production of bacteriocins and other antibiotics, and resistance to bacteriophage attack, are usually encoded by extrachromosomal DNA, as reviewed in Kondo and McKay (1985) and Klaenhammer (1987). This makes the study of plasmid biology in lactococci a major area of interest for food microbiologists.

The transfer of lactose fermentation ability (Lac$^+$) by conjugation was first reported by McKay and coworkers (McKay et al., 1980). Later, this group (Walsh and McKay, 1981; Anderson and McKay, 1984) and Gasson and coworkers (Gasson and Davies, 1980; Gasson, 1983) showed that this conjugal transfer process was characterized by novel genetic recombination and DNA rearrangement events, as well as interesting cell-cell interactions. The original transfer event was observed at extremely low frequency and occurred only on solid surfaces, and the mating mixtures grew in uniform suspension when placed in liquid media. However, certain Lac$^+$ transconjugants resulting from these matings were capable of transfer at much higher frequency on solid surfaces or in liquid. These strains also exhibited a novel self-clumping phenotype (Clu$^+$) when the cells were grown in liquid culture. Other transconjugants grew normally in broth and were either unable to serve as conjugal donors for Lac$^+$ or transferred at the same low frequency as the original donor strain. This group also observed a spontaneous and reversible switching from Clu$^+$ to Clu$^-$ in some of the transconjugants. A detailed analysis of these strains and their plasmids by Anderson and McKay (1984) resulted in the notion that a site-specific cointegrate formation between a nonconjugative Lac$^+$ plasmid and a low-copy conjugative plasmid in the donor cell was responsible for the transfer of the Lac$^+$ phenotype. The Clu phenotype switching appeared to be related to the inversion of a segment of DNA on the conjugative plasmid. They were also able to identify a 15- to 20-kb region of the transferable plasmid encoding the clumping and transfer functions. More recently, insertion sequences mediating the formation of cointegrate plasmids have been identified (Polzin and Shimizu-Kadota, 1987). Interestingly, Gasson (1990) recently reported that the conjugal transfer ability in their system resides on a large conjugative transposon, and it seems possible that the low-copy plasmid identified by Anderson and McKay (1984) could also represent a nonreplicating circular intermediate in transposition, similar to that reported for Tn*916* (Caparon and Scott, 1989). The availability of isogenic strains differing in expression of clumping and transfer phenotypes would appear to make this system ideal for analysis using a combined immunological and genetic approach similar to that employed in studies of pheromone-inducible transfer in *E. faecalis* (Dunny, 1990). It is possible that the donor/recipient cell attachment and mating channel formation mechanisms could be similar in the two systems. However, the gene organization and regulatory mechanisms are probably very different, since in the lactococcal system, expression of Clu$^+$ seems to be controlled by the inversion of DNA, while the regulation of the other transfer functions appears to be independent. In any event, it seems clear that the lactococcal conjugation system represents another interesting system, with experimental tools available that make it ripe for an in-depth study. There are clearly a number of important practical reasons for doing such a study since a better understanding of gene transfer and regulatory mechanisms in these organisms is vital in order to develop improved strains for industrial purposes.

CONJUGATION IN STAPHYLOCOCCI AND BACILLI

There are numerous examples, as described in previous sections, of the conjugal transfer of transposons and plasmids from streptococci or enterococci to numerous gram-positive genera, including staphylococci and bacilli. However, there are relatively few examples of naturally occurring conjugative plasmids in these organisms. In the staphylococci, considerable attention has been focused on the conjugal transfer of plasmids determining resistance to aminoglycosides and sometimes other antibiotics, as well (Archer and Johnston, 1983; Schaberg and Zervos, 1986). These plasmids require solid surfaces for transfer, but can promote the mobilization of some nonconjugative R plasmids (Projan and Archer, 1989). Archer and coworkers have carried out considerable genetic and physical analyses of the transfer genes involved in staphylococcal conjugation (Thomas and Archer, 1989), and recently Projan and Archer (1989) identified a mobilization determinant on a nonconjugative plasmid. The latter study implies that transfer of these plasmids may proceed by a mechanism similar to that of sex factors from gram-negative bacteria and that the transfer process associated with these plasmids might not involve a fusionlike event. However, further work will be necessary to confirm this notion.

Most genetic studies with bacilli employ derivatives of *B. subtilis* 168, for which there is a very good transformation system, and a number of transducing phages (Dedonder et al., 1977). This strain has no natural conjugation system although it has been possible to insert broad-host-range conjugative elements (Christie et al., 1987) into this host. However, conjugative plasmids are naturally found in other bacilli, and these appear to represent at least two distinct classes in terms of their transfer mechanisms. *Bacillus thuringiensis* strains carry a variety of conjugative plasmids that encode insecticidal toxins (Chapman and Carlton, 1985; Gonzalez et al., 1982). The transfer of these plasmids has been studied by Carlton and coworkers (Chapman and Carlton, 1985; Gonzalez et al., 1982) and Thorne and associates (Battisti et al., 1985; Reddy et al., 1987; Green et al., 1989). Interestingly, these plasmids transfer in liquid and the frequencies and kinetics of transfer resemble those of the pheromone-inducible systems (Battisti et al., 1985; Chapman and Carlton, 1985), although attempts to demonstrate pheromone involvement in *Bacillus* conjugation were negative (Chapman and Carlton, 1985). A second type of conjugative plasmid is represented by pLS20, isolated from *B. subtilis* (*natto*), which shows a fairly broad host range and requires a solid surface mating to transfer (Koehler and Thorne, 1987). Both types of plasmid have been shown to mobilize other plasmids, and evidence has been obtained for two different mechanisms of mobilization. Small plasmids such as pBC16 are mobilized by a donation mechanism (Clark and Warren, 1979) not requiring cointegrate formation between the two plasmids (Battisti et al., 1985; Green et al., 1989). In the case of the mobilization of small plasmids, a sequence known as ORF-β has been shown to be important (Selinger et al., 1990). As in the case of most of the conjugation systems mentioned thus far, there is very little direct evidence available regarding the mechanisms of cell-cell contact or mating channel formation, and this represents an extremely attractive area for future research.

CONJUGATION IN *STREPTOMYCES* SPP.

Although the *Streptomyces* spp. are gram-positive bacteria, the unique features of their cellular structure, developmental cycle, and DNA base composition would suggest that the conjugal transfer systems of these organisms might be as different from those discussed in the previous sections as they are from those of gram-negative bacteria (Hopwood et al., 1985). Nonetheless, it is appropriate to consider briefly conjugation in *Streptomyces* spp. since it does represent an important form of gene transfer for these organisms, and the study of conjugative genetic elements in these organisms has been subjected to a considerable amount of experimental investigation. It should also be kept in mind that direct conjugal transfer events from *Streptomyces* spp. to other gram-positive bacteria, e.g., staphylococci, enterococci, and bacilli, could be responsible for the transfer of resistance genes which eventually find their way into pathogenic microbes, as will be discussed in the next section.

Streptomyces conjugative plasmids are capable of efficient transfer and mobilization of chromosomal genes, thus mediating an exchange process genetically similar to that of gram-negative transfer systems (as reviewed in Hopwood et al., 1985). However, the analyses of *Streptomyces* conjugation reported thus far suggest that the mechanistic and regulatory features of this form of conjugation may be unique among procaryotes. The plasmids can have a variety of physical forms and interactions with the chromosome (Kieser et al., 1982; Hopwood et al., 1985). The mating process, at least in the laboratory, generally takes place on solid surfaces between growing mycelia, and the genetic analysis of the products is not usually completed until after a population of spores, representing the terminal stage of the developmental cycle of the organism, is obtained. Thus, some features of this system do not lend themselves to detailed analysis of the initial cell-cell interactions and mating channel formation.

Perhaps the best-studied conjugative plasmid in *Streptomyces* spp. is pIJ101 (Kieser et al., 1982; Hopwood et al., 1985), an 8.9-kb element. When pIJ101-mediated matings are carried out on solid surfaces, transfer of the plasmid can be monitored by the presence of "pocks" on the bacterial lawn (Kieser et al., 1982; Hopwood et al., 1985). Such pocks are also observed during the transfer of other conjugative plasmids in *Streptomyces* spp. (Hopwood et al., 1985). It is believed that the pocks result from transient growth inhibition of transconjugant cells by some sort of lethal zygosis activity associated with the incoming plasmid (Kieser et al., 1982; Hopwood et al., 1985). Recently, Cohen and associates have carried out a detailed molecular and genetic analysis of this plasmid, including the determination of the complete nucleotide sequence (Kendall and Cohen, 1988) and genetic and transcriptional analysis of various cloned regions of the plasmid (Kendall and Cohen, 1987; Stein et al., 1989). This work has shown that the region involved in conjugal transfer is very small and encodes a 77-kDa protein required for transfer, along with three small genes whose products appear to be involved in intramycelial spread. The genetic evidence from this group (Kendall and Cohen, 1987) indicates that the pock formation is probably related to the expression of two loci, *kilA* and *kilB*, whose activities are normally suppressed by regulatory *kor* loci. Although the function of these genes is not completely understood, the data are consistent with the notion that pock formation may be related to a transient zygotic induction of *kil* genetic functions in transconjugants.

Since there is only one gene required for transfer, the mechanism of mating channel formation could be much simpler in *Streptomyces* spp. than in other organisms. One possibility is that spontaneous fusions between mycelia could occur that permit the transfer of plasmid DNA. Alternatively, additional conjugation genes, functional for a number of different plasmids, could be present on the chromosome; thus, it could be fruitful to screen random chromosomal insertion mutations for effects on conjugal donor or recipient ability.

MATINGS BETWEEN *E. COLI* AND GRAM-POSITIVE BACTERIA OR EUCARYOTES

Recently, several groups have reported the conjugal transfer of DNA between very distantly related organisms, including transfer from *E. coli* to enterococci (Trieu-Cuot et al., 1987), yeast cells (Heinemann and Sprague, 1989), or *Streptomyces* spp. (Mazodier et al., 1989; Wohlleben and Pielsticker, 1990). In these systems, the investigators engineered test plasmids capable of replication in both *E. coli* and the heterologous recipient cell and included a mobilization function. These studies are of great interest in illustrating the potential for genetic exchange across great evolutionary barriers, and the results may have great practical significance as genetic tools. However, it seems possible that in nature, organisms such as *E. coli* may be more likely to act as a recipient rather than a donor for foreign genes, particularly genes from gram-positive bacteria encoding antibiotic resistance determinants (Trieu-Cuot et al., 1988). It will also be of great interest to determine the potential for genetic exchange between bacteria and mammalian cells, particularly in the case of intracellular pathogens as noted by Davies (1990). In particular, the role of the broad-host-range conjugative plasmids and transposons of gram-positive bacteria in this type of process will be very interesting to assess.

SUMMARY AND CONCLUSIONS

Gram-positive bacteria display a remarkable variety of mechanisms for the conjugal transfer of genetic information. In the case of the phermone-inducible plasmid transfer system of *E. faecalis*, considerable information is available about the details of this process. Although much less is known about other systems, it is clear that they have unique features and that a great deal of interesting and important work remains to be done in order to elucidate the role of these systems in the evolution of their bacterial hosts and other organisms which may serve as recipients.

ACKNOWLEDGMENTS. Research on conjugation in my laboratory has been supported by Public Health Service grant AI19310 from the National Institutes of Health and by grants from the Cornell University Biotechnology Program.

I thank the following colleagues who provided reprints and other material for this article: D. Clewell, S. Cohen, J. Davies, C. Thorne, T. Horaud, P. Courvalin, F. Macrina, R. Wirth, and W. Wohlleben. I also acknowledge the excellent work and intellectual contributions of my students, technicians, and collaborators at Cornell and at the University of Minnesota, J. Gallo, L. Tortorello, P. Christie, O. Torres, S.-M. Kao, J. Chung, A. Viksnins, S. Olmsted, R. Ruhfel, V. Johncox, B.

Bensing, and S. Zahler. The late Ruth Korman made a tremendous contribution to our understanding of Tn925 transfer and was a beloved and respected colleague who is sorely missed.

REFERENCES

Albano, M., R. Breitling, and D. Dubnau. 1989. Nucleotide sequence and genetic organization of the *Bacillus subtilis comG* operon. *J. Bacteriol.* **171**:5386–5404.

Anderson, D. G., and L. L. McKay. 1984. Genetic and physical characterization of recombinant plasmids associated with cell aggregation and high-frequency conjugal transfer in *Streptococcus lactis* ML3. *J. Bacteriol.* **158**:954–962.

Archer, G. L., and J. L. Johnston. 1983. Self-transmissible plasmids in staphylococci that encode resistance to aminoglycosides. *Antimicrob. Agents Chemother.* **24**:70–77.

Battisti, L., B. D. Green, and C. B. Thorne. 1985. Mating system for transfer of plasmids among *Bacillus anthracis, Bacillus cereus,* and *Bacillus thuringiensis. J. Bacteriol.* **162**:543–550.

Berghash, S. R., and G. M. Dunny. 1985. Emergence of multiple bata-lactam-resistance phenotype in group B streptococci of bovine origin. *J. Infect. Dis.* **151**:494–500.

Bertram, J., A. Kuhn, and P. Dürre. 1990. Tn916-induced mutants of *Clostridium acetobutylicum* defective in regulation of solvent formation. *Arch. Microbiol* **153**:373–377.

Boyd, A., and M. Simon. 1982. Bacterial chemotaxis. *Annu. Rev. Physiol.* **44**:501–517.

Buu-Hoi, A., G. Bieth, and T. Horaud. 1984. Broad host range of streptococcal macrolide resistance plasmids. *Antimicrob. Agents Chemother.* **25**:289–291.

Buu-Hoi, A., G. de Cespedes, and T. Horaud. 1985. Deoxyribonuclease-sensitive transfer of an R plasmid in *Streptococcus pyogenes* (group A). *FEMS Microbiol. Lett.* **30**:407–410.

Caillaud, F., and P. Courvalin. 1987. Nucleotide sequence of the ends of the conjugative shuttle transposon Tn1545. *Mol. Gen. Genet.* **209**:110–115.

Caparon, M. G., and J. R. Scott. 1989. Excision and insertion of the conjugative transposon Tn916 involves a noval recombination mechanism. *Cell* **59**:1027–1034.

Chapman, J. S., and B. C. Carlton. 1985. Conjugal plasmid transfer in Bacillus thuringiensis, p. 453–467. *In* D. R. Helinski, S. N. Cohen, D. B. Clewell, D. A. Jackson, and A. Hollaender (ed.), *Plasmids in Bacteria.* Plenum Press, New York.

Christie, P. J., and G. M. Dunny. 1986. Identification of regions of the *Streptococcus faecalis* plasmid pCF-10 that encode antibiotic resistance and pheromone response functions. *Plasmid* **15**:230–241.

Christie, P. J., S.-M. Kao, J. C. Adsit, and G. M. Dunny. 1988. Cloning and expression of genes encoding pheromone-inducible antigens of *Enterococcus (Streptococcus) faecalis. J. Bacteriol.* **170**:5161–5168.

Christie, P. J., R. A. Korman, S. A. Zahler, J. C. Adsit, and G. M. Dunny. 1987. Two conjugation systems associated with *Streptococcus faecalis* plasmid pCF10: identification of a conjugative transposon that transfers between *S. faecalis* and *Bacillus subtilis. J. Bacteriol.* **169**:2529–2536.

Chung, J., and G. Dunny. Unpublished data.

Clark, A. J., and G. J. Warren. 1979. Conjugal transmission of plasmids. *Annu. Rev. Genet.* **13**:99–125.

Clewell, D. B. 1990. Movable genetic elements and antibiotic resistance in enterococci. *Eur. J. Clin. Microbiol. Infect. Dis.* **9**:90–102.

Clewell, D. B., F. Y. An, M. Mori, Y. Ike, and A. Suzuki. 1987. *Streptococcus faecalis* pheromone cAD1 response: evidence that the peptide inhibitor excreted by pAD1-containing cells may be plasmid determined. *Plasmid* **17**:65–68.

Clewell, D. B., and B. L. Brown. 1980. Sex pheromone cAD1 in *Streptococcus faecalis*: induction of a function related to plasmid transfer. *J. Bacteriol.* **143**:1063–1065.

Clewell, D. B., S. E. Flannagan, Y. Oke, J. M. Jones, and C. Gawron-Burke. 1988. Sequence analysis of the termini of the conjugative transposon Tn916. *J. Bacteriol.* **170**:3046–3052.

Clewell, D. B., and M. C. Gawron-Burke. 1986. Conjugative transposons and the dissemination of antibiotic resistance in streptococci. *Annu. Rev. Microbiol.* **40**:653–659.

Clewell, D. B., and K. E. Weaver. 1989. Sex pheromones and plasmid transfer in *Enterococcus faecalis*: a review. *Plasmid* **21**: 175–184.

Clewell, D. B., Y. Yagi, G. M. Dunny, and S. K. Schultz. 1974. Characterization of three plasmid deoxyribonucleic acid molecules in a strain of *Streptococcus faecalis*: identification of a plasmid determining erythromycin resistance. *J. Bacteriol.* **117**:283–289.

Clewell, D. B., et al. 1990. Personal communciation.

Courvalin, P., and C. Carlier. 1986. Transposable multiple antibiotic resistance in *Streptococcus pneumoniae*. *Mol. Gen. Genet.* **205**:291–297.

Davies, J. 1991. Interspecific gene transfer: where next? *Trends Biotechnol.* **8**:198–203.

Dedonder, R. A., J. A. Lepesant, J. Lepesant-Kejzlarova, A. Billaut, M. Steinmetz, and F. Kunst. 1977. Construction of a kit of reference strains for rapid genetic mapping in *Bacillus subtilis*. *Appl. Environ. Microbiol.* **33**:989–993.

Dunny, G. M. 1990. Genetic functions and cell-cell interactions in the pheromone-inducible plasmid transfer system of *Enterococcus faecalis*. *Mol. Microbiol.* **4**:689–696.

Dunny, G. M., B. L. Brown, and D. B. Clewell. 1978. Induced cell aggregation and mating in *Streptococcus faecalis*: evidence for a bacterial sex pheromone. *Proc. Natl. Acad. Sci. USA* **75**:3470–3483.

Dunny, G. M., J. W. Chung, J. C. Gallo, S.-M. Kao, K. M. Trotter, R. Z. Korman, S. B. Olmsted, R. Ruhfel, O. R. Torres, and S. A. Zahler. 1991. Cell-cell interactions and conjugal gene transfer events mediated by the pheromone-inducible plasmid transfer system and the conjugative transposon encoded by *Enterococcus faecalis* plasmid pCF10, p. 9–15. *In* G. M. Dunny, P. P. Cleary, and L. L. McKay (ed.), *Genetics and Molecular Biology of Streptococci, Lactococci, and Enterococci*. American Society for Microbiology, Washington, D.C.

Dunny, G. M., and D. B. Clewell. 1975. Transmissible toxin (hemolysin) plasmid in *Streptococcus faecalis* and its mobilization of a noninfectious drug resistance plasmid. *J. Bacteriol.* **124**:784–790.

Dunny, G. M., R. A. Craig, R. L. Carron, and D. B. Clewell. 1979. Plasmid transfer in *Streptococcus faecalis*: production of multiple pheromones by recipients. *Plasmid* **2**:454–465.

Dunny, G. M., O. R. Torres, J. C. Gallo, and B. Bensing. Unpublished data.

Dunny, G. M., D. L. Zimmerman, and M. L. Tortorello. 1985. Induction of surface exclusion (entry exclusion) by *Streptococcus faecalis* sex pheromones: use of monoclonal antibodies to identify an inducible surface antigen involved in the exclusion process. *Proc. Natl. Acad. Sci. USA* **82**:8582–8586.

Dybvig, K., and G. H. Cassell. 1987. Transposition of gram-positive transposon Tn*916* in *Acholeplasma laidlawii* and *Mycoplasma pulmonis*. *Science* **235**:1392–1394.

Ehrenfeld, E. E., and D. B. Clewell. 1987. Transfer functions of the *Streptococcus faecalis* plasmid pAD1: organization of plasmid DNA encoding response to sex pheromone. *J. Bacteriol.* **169**:3473–3481.

Ehrenfeld, E. E., R. E. Kessler, and D. B. Clewell. 1986. Identification of pheromone-induced surface proteins in *Streptococcus faecalis* and evidence of a role for lipoteichoic acid in the formation of mating aggregates. *J. Bacteriol.* **168**:6–12.

Engel, H. W. B., N. Soedirman, J. A. Rost, W. J. van Leeuwen, and J. D. A. van Embden. 1980. Transferability of macrolide, lincomycin, and streptogramin resistances between group A, B, and D streptococci, *Streptococcus pneumoniae*, and *Staphylococcus aureus*. *J. Bacteriol.* **142**:407–413.

Evans, R. P., and F. L. Macrina. 1983. Streptococcal R plasmid pIP501: endonuclease site map, resistance determinant location, and construction of novel derivatives. *J. Bacteriol.* **154**:1347–1355.

Fitzgerald, G. F., and D. B. Clewell. 1985. A conjugative transposon (Tn*919)) in *Streptococcus sanguis*. *Infect. Immun.* **47**:415–420.

Franke, A. E., and D. B. Clewell. 1981. Evidence for a chromosome-borne resistance transposon (Tn*916*) in *Streptococcus faecalis* that is capable of "conjugal" transfer in the absence of a conjugative plasmid. *J. Bacteriol.* **145**:494–502.

Franke, A. E., G. M. Dunny, B. L. Brown, F. An, D. R. Oliver, S. P. Damle, and D. B. Clewell. 1978. Gene transfer in *Streptococcus faecalis*: evidence for the mobilization of chromosomal determinants by transmissible plasmids, p. 45–47. *In* D. Schlessinger (ed.), *Microbiology—1978*. American Society for Microbiology, Washington, D.C.

Galli, D., F. Lottspeich, and R. Wirth. 1990. Sequence analysis of Enterococcus faecalis aggregation substance encoded by the sex pheromone plasmid pAD1. *Mol. Microbiol.* **4**:895–904.

Galli, D., R. Wirth, and G. Wanner. 1989. Identification of aggregation substances of *Enterococcus faecalis* cells after induction by sex pheromones. *Arch. Microbiol.* **151**:486–490.

Gasson, M. J. 1983. Genetic transfer systems in lactic acid bacteria. *Antonie van Leeuwenhoek J. Microbiol. Serol.* **49:**275–282.

Gasson, M. J. 1990. 3rd Int. ASM Conf. Streptococcal Genet., abstr. 3.

Gasson, J. J., and F. L. Davies. 1980. High-frequency conjugation associated with *Streptococcus lactis* donor cell aggregation. *J. Bacteriol.* **143:**1260–1264.

Gawron-Burke, M. C., and D. B. Clewell. 1982. A transposon in *Streptococcus faecalis* with fertility properties. *Nature* (London) **300:**281–284.

Gibson, M. J., N. M. Chase, S. B. London, and J. London. 1979. Transfer of plasmid-mediated antibiotic resistance from streptococci to lactobacilli. *J. Bacteriol.* **137:**614–619.

Gonzalez, C. F., and B. S. Kunka. 1983. Plasmid transfer in *Pediococcus* spp.: intergeneric and intrageneric transfer of pIP501. *Appl. Environ. Microbiol.* **46:**81–89.

Gonzalez, J. M., Jr., B. J. Brown, and B. C. Carlton. 1982. Transfer of *Bacillus thuringiensis* plasmids coding for delta-endotoxin among strains of *B. thuringiensis* and *B. cereus. Proc. Natl. Acad. Sci. USA* **79:**6951–6955.

Green, B. D., L. Battisti, and C. B. Thorne. 1989. Involvement of Tn*4430* in transfer of *Bacillus anthracis* plasmids mediated by *Bacillus thuringiensis* plasmid pXO12. *J. Bacteriol.* **171:**104–113.

Hartley, D. L., K. R. Jones, J. A. Tobian, D. J. LeBlanc, and F. L. Macrina. 1984. Disseminated tetracycline resistance in oral streptococci: implication of a conjugative transposon. *Infect. Immun.* **45:**13–17.

Heinemann, J. A., and G. F. Sprague, Jr. 1989. Bacterial conjugative plasmids mobilize DNA transfer between bacteria and yeast. *Nature* (London) **340:**205–209.

Higgins, M. L., D. Glaser, D. T. Dicker, and E. T. Zito. 1989. Chromosome and cell wall segregation in *Streptococcus faecium* ATCC 9790. *J. Bacteriol.* **171:**349–352.

Hopwood, D. A., D. J. Lydiate, F. Malpartida, and H. M. Wright. 1985. Conjugative sex plasmids of *Streptomyces*, p. 615–634. *In* D. R. Helinski, S. N. Cohen, D. B. Clewell, D. A. Jackson, and A. Hollaender (ed.), *Plasmids in Bacteria.* Plenum Press, New York.

Horaud, T., C. Le Bouguenec, and K. Pepper. 1985. Molecular genetics of resistance to macrolides, lincosamides and streptogramin B (MLS) in streptococci. *J. Antimicrob. Chemother.* **16:**111–135.

Hotchkiss, R. D., and M. H. Gabor. 1980. Biparental products of bacterial protoplast fusion showing unequal parental chromosome expression. *Proc. Natl. Acad. Sci. USA* **77:**3553–3557.

Hynes, R. O. 1987. Integrins: a family of cell surface receptors. *Cell* **48:**549–554.

Inamine, J., and V. Burdett. 1985. Structural organization of a 67-kilobase streptococcal conjugative element mediating multiple antibiotic resistance. *J. Bacteriol.* **161:**620–626.

Jacob, A. E., and S. J. Hobbs. 1974. Conjugal transfer of plasmid-borne multiple antibiotic resistance in *Streptococcus faecalis* var. *zymogenes. J. Bacteriol.* **117:**360–372.

Jacob, F., and E. L. Wollman. 1961. *Sexuality and Genetics of Bacteria.* Academic Press, Inc., New York.

Kendall, K. J., and S. N. Cohen. 1987. Plasmid transfer in *Streptomyces lividans*: identification of a *kil-kor* system associated with the transfer region of pIJ101. *J. Bacteriol.* **169:**4177–4183.

Kendall, K. J., and S. N. Cohen. 1988. Complete nucleotide sequence of the *Streptomyces lividans* plasmid pIJ101 and correlation of the sequence with genetic properties. *J. Bacteriol.* **170:**4634–4651.

Kieser, T., D. A. Hopwood, H. M. Wright, and C. J. Thompson. 1982. PIJ101, a multicopy broad host-range *Streptomyces* plasmid: functional analysis and development of DNA cloning vectors. *Mol. Gen. Genet.* **185:**223–238.

Klaenhammer, T. R. 1987. Plasmid-directed mechanisms for bacteriophage defense in lactic streptococci. *FEMS Microbiol. Rev.* **46:**313–325.

Koehler, T. M., and C. B. Thorne. 1987. *Bacillus subtilis* (*natto*) plasmid pLS20 mediates interspecies plasmid transfer. *J. Bacteriol.* **169:**5271–5278.

Kondo, J. K., and L. L. McKay. 1985. Gene transfer systems and molecular cloning in Group N streptococci. *J. Dairy Sci.* **68:**2143–2159.

Krah, E. R., III, and F. L. Macrina. 1989. Genetic analysis of the conjugal transfer determinants encoded by the streptococcal broad-host-range plasmid pIP501. *J. Bacteriol.* **171:**6005–6012.

LeBlanc, D. J., and L. N. Lee. 1984. Physical and genetic analyses of streptococcal plasmid pAMβ1 and cloning of its replication region. *J. Bacteriol.* **157:**445–453.

Lederberg, J. 1986. Forty years of genetic recombination in bacteria: a fortieth anniversary reminiscence. *Nature* (London) **324:**627–628.

Leeds, J., and G. Dunny. Unpublished data.

Mazodier, P., R. Petter, and C. Thompson. 1989. Intergeneric conjugation between *Escherichia coli* and *Streptomyces* species. *J. Bacteriol.* **171:**3583–3585.

McKay, L. L., K. A. Baldwin, and P. M. Walsh. 1980. Conjugal transfer of genetic information in group N streptococci. *Appl. Environ. Microbiol.* **40:**84–91.

Miller, J. F., J. J. Mekalanos, and S. Falkow. 1989. Coordinate regulation and sensory transduction in the control of bacterial virulence. *Science* **243:**916–922.

Mori, M., A. Isogai, Y. Sakagami, M. Fujino, C. Kitada, D. B. Clewell, and A. Suzuki. 1986. Isolation and structure of the *Streptococcus faecalis* sex pheromone inhibitor, iAD1, that is excreted by the donor strain harboring plasmid pAD1. *Agric. Biol. Chem.* **50:**539–541.

Mori, M., Y. Sakagami, Y. Ishii, A. Isogai, C. Kitada, M. Fujino, J. C. Adsit, G. M. Dunny, and A. Suzuki. 1988. Structure of cCF10, a peptide sex pheromone which induces conjugative transfer of the *Streptococcus faecalis* tetracycline resistance plasmid, pCF10. *J. Biol. Chem.* 263:14574–14578.

Mori, M., Y. Sakagami, M. Narita, A. Isogai, M. Fujino, C. Kitada, R. A. Craig, D. B. Clewell, and A. Suzuki. 1984: Isolation and structure of the bacterial sex pheromone, cAD1, that induces plasmid transfer in *Streptococcus faecalis.* *FEBS Lett.* **178:**97–100.

Naglich, J. G., and R. E. Andrews, Jr. 1988. Tn*916*-dependent conjugal transfer of PC194 and PUB110 from *Bacillus subtilis* into *Bacillus thuringiensis* subsp. *israelensis.* *Plasmid* **20:**113–126.

Odelson, D. A., J. L. Rasmussen, C. J. Smith, and F. L. Macrina. 1987. Extrachromosomal systems and gene transmission in anaerobic bacteria. *Plasmid* **17:**87–109.

Olmsted, S., and G. Dunny. Unpublished observations.

Olmsted, S., S.-M. Kao, L. J. van Putte, J. C. Gallo, and G. M. Dunny. Submitted for publication.

Perkins, J. B., and P. J. Youngman. 1986. Construction and properties of Tn917-lac, a transposon derivative that mediates transcriptional gene fusions in *Bacillus subtilis.* *Proc. Natl. Acad. Sci. USA* **83:**140–144.

Polzin, K. M., and M. Shimizu-Kadota. 1987. Identification of a new insertion element, similar to gram-negative IS*26,* on the lactose plasmid of *Streptococcus lactis* ML3. *J. Bacteriol.* **169:**5481–5488.

Poyart-Salmeron, C., C. Carlier, P. Trieu-Cuot, A.-L. Courtieu, and P. Courvalin. 1990. Transferable plasmid-mediated antibiotic resistance in *Listeria monocytogenes.* *Lancet* **335:1422–1426.**

Poyart-Salmeron, C., P. Trieu-Cuot, C. Carlier, and P. Courvalin. 1989. Molecular characterization of two proteins involved in the excision of the conjugative transposon Tn*1545:* homologies with other site-specific recombinases. *EMBO J.* **8:**2425–2433.

Projan, S. J., and G. L. Archer. 1989. Mobilization of the relaxable *Staphylococcus aureus* plasmid pC221 by the conjugative plasmid pGO1 involves three pC221 loci. *J. Bacteriol.* **171:**1841–1845.

Reddy, A., L. Battisti, and C. B. Thorne. 1987. Identification of self-transmissible plasmids in four *Bacillus thuringiensis* subspecies. *J. Bacteriol.* **169:**5263–5270.

Schaberg, D. R., and M. J. Zervos. 1986. Intergeneric and interspecies gene exchange in gram-positive cocci. *Antimicrob. Agents Chemother.* **30:**817–822.

Scott, J. R., P. A. Kirchman, and M. G. Caparon. 1988. An intermediate in the transposition of the conjugative transposon Tn*916.* *Proc. Natl. Acad. Sci. USA* **85:**4809–4813.

Selinger, L. B., N. F. McGregor, G. C. Khachatourians, and M. F. Hynes. 1990. Mobilization of closely related plasmids pUB110 and pBC16 by *Bacillus* plasmid pXO503 requires *trans*-acting open reading frame β. *J. Bacteriol.* **172:**3290–3297.

Senghas, E., J. M. Jones, M. Yamamoto, C. Gawron-Burke, and D. B. Clewell. 1988. Genetic organization of the bacterial conjugative transposon Tn*916.* *J. Bacteriol.* **170:**245–249.

Stein, D. S., K. J. Kendall, and S. N. Cohen. 1989. Identification and analysis of transcriptional regulatory signals for the *kil* and *kor* loci of *Streptomyces* plasmid pIJ101. *J. Bacteriol.* **171:**5768–5775.

Stevens, A. M., N. B. Shoemaker, and A. A. Salyers. 1990. The region of a *Bacteroides* conjugal chromosomal tetracycline resistance element which is responsible for production of plasmidlike forms from unlinked chromosomal DNA might also be involved in transfer of the element. *J. Bacteriol.* **172:**4271–4279.

Strätz, M., G. Gottschalk, and P. Dürre. 1990. Transfer and expression of the tetracycline resistance transposon Tn*925* in *Acetobacterium woodii.* *FEMS Microbiol. Lett.* **68:**171–176.

Suzuki, A., M. Mori, Y. Sakagami, A. Isogai, M. Fujino, C. Kitaga, R. A. Craig, and D. B. Clewell. 1984. Isolation and structure of the bacterial sex pheromone, cPD1. *Science* **226**:849–850.

Thomas, W. D., Jr., and G. L. Archer. 1989. Identification and cloning of the conjugative transfer region of *staphylococcus aureus* plasmid pGO1. *J. Bacteriol.* **171**:684–691.

Tomich, P. K., F. Y. An, S. P. Damle, and D. B. Clewell. 1979. Plasmid-related transmissibility and multiple drug resistance in *Streptococcus faecalis* subsp. *zymogenes* strain DS16. *Antimicrob. Agents Chemother.* **15**:828–830.

Tomura, T., T. Hirano, T. Ito, and M. Yoshioka. 1973. Transmission of bacteriocinogenicity by conjugation in group D streptococci. *Jpn. J. Microbiol.* **17**:445–452.

Torres, O. R., R. Z. Korman, S. A. Zahler, and G. M. Dunny. 1991. The conjugative transposon Tn*925*: enhancement of conjugal transfer by tetracycline in *Enterococcus faecalis* and mobilization of chromosomal genes in *Bacillus subtilis* and *E. faecalis*. *Mol. Gen. Genet.* **225**:395–400.

Trieu-Cuot, P., C. Carlier, and P. Courvalin. 1988. Conjugative plasmid transfer from *Enterococcus faecalis* to *Escherichia coli*. *J. Bacteriol.* **170**:4388–4391.

Trieu-Cuot, P., C. Carlier, P. Martin, and P. Courvalin. 1987. Plasmid transfer by conjugation from *Escherichia coli* to gram-positive bacteria. *FEMS Microbiol. Lett.* **48**:289–294.

Trotter, K. M., and G. M. Dunny. 1990. Mutants of *Enterococcus faecalus* deficient as recipients in mating with donors carrying pheromone-inducible plasmids. *Plasmid* **24**:57–67.

Vijayakumar, M. N., S. D. Priebe, and W. R. Guild. 1986. Structure of a conjugative element in *Streptococcus pneumoniae*. *J. Bacteriol.* **166**:978–984.

Walsh, P. M., and L. L. McKay. 1981. Recombinant plasmid associated with cell aggregation and high-frequency conjugation of *Streptococcus lactis* ML3. *J. Bacteriol.* **146**:937–944.

Wanner, G., H. Formanek, D. Galli, and R. Wirth. 1989. Localization of aggregation substances of *Enterococcus faecalis* after induction by sex pheromones. *Arch. Microbiol.* **151**:491–497.

Weaver, K. E., and D. B. Clewell. 1988. Regulation of the pAD1 sex pheromone response in *Enterococcus faecalis*: construction and characterization of *lacZ* transcriptonal fusions in a key control region of the plasmid. *J. Bacteriol.* **170**:4343–4352.

Willets, N., and B. Wilkins. 1984. Processing of plasmid DNA during bacterial conjugation. *Microbiol. Rev.* **48**:24–41.

Wirth, R., S. Olmsted, D. Galli, and G. Dunny. 1991. Comparative analysis of cAD1- and cCF10-induced aggregation substances of *Enterococcus faecalis*, p. 34–38. *In* G. M. Dunny, P. P. Cleary, and L. L. McKay (ed.), *Genetics and Molecular Biology of Streptococci, Lactococci, and Enterococci*. American Society for Microbiology, Washington, D.C.

Wohlleben, W., and A. Pielsticker. 1990. *Investigation of Plasmid Transfer between Escherichia coli and Streptomyces lividans*, 3rd ed., p. 301–305. Volag Chemie, Wenheim, Germany.

Yamamoto, M., J. M. Jones, E. Senghas, C. Gawron-Burke, and D. B. Clewell. 1987. Generation of Tn*5* insertions in streptococcal conjugative transposon Tn*916*. *Appl. Environ. Microbiol.* **53**:1069–1072.

Youngman, P. J., J. B. Perkins and R. Losick. 1983. Genetic transposition and insertional mutagenesis in *Bacillus subtilis* with *Streptococcus faecalis* transposon Tn*917*. *Proc. Natl. Acad. Sci. USA* **80**:2305–2309.

Microbial Cell-Cell Interactions
Edited by Martin Dworkin
© 1991 American Society for Microbiology, Washington, DC 20005

Chapter 3

Conjugation among Enteric Bacteria: Mating Systems Dependent on Expression of Pili

Karin Ippen-Ihler and Sumit Maneewannakul

INTRODUCTION

Conjugation differs from other mechanisms of bacterial DNA transmission in that it is dependent upon contact between the donor and recipient cells. The first bacterial conjugation system to be discovered was described by Lederberg and Tatum (1946), who found that genetic recombinants could be selected from mixtures of certain *Escherichia coli* K-12 strains. As we know today, this phenomenon was dependent upon the presence of a conjugative plasmid, the fertility factor F. The F plasmid was found to be a 100-kb self-transmissible replicon that encodes the functions necessary to transfer its own DNA sequence from the "male" (F$^+$) donor cell to "female" (F$^-$) recipients. Since DNA sequences that become associated with F also become transferable, the F plasmid provided one of the most powerful early tools for studies of *E. coli* genetics: Hfr strains, which carry chromosomal integrations of F, provided unidirectional transfer of long lengths of chromosomal DNA; F$'$ plasmids, which have acquired chromosomal segments, made possible the transmission and independent maintenance of gene sequences needed for merodiploid analyses.

The first evidence that donor cells possessed a unique cell surface component appeared in 1960, when Ørskov and Ørskov (1960) reported that all *E. coli* strains carrying the F fertility factor possessed a specific f^+ antigen. In the same year, Loeb (1960) reported isolating a male-specific bacteriophage which proved to contain RNA (Loeb and Zinder, 1961). The possibility that RNA bacteriophage infection might involve a distinguishable site or appendage expressed by the F plasmid was soon investigated by Crawford and Gesteland (1964), whose electron microscope observations showed that F-specific RNA phages adsorbed along the length of a small number of fimbrial appendages extending from the cell surface. Brinton et al. (1964) confirmed these findings, naming the phage-adsorbing filaments "F-pili" to distinguish them from other types of fimbriae synthesized by both male and female strains. Subsequently, a second type of F-specific bacteriophage, which was filamentous in structure and contained single-stranded DNA (Loeb, 1960), was shown to adsorb to the tips of F-pili (Caro and Schnös, 1966), and the relationship between the f^+ antigen and the F-pilus filament was confirmed (Ishibashi, 1967).

Meanwhile, with the discovery and study of antibiotic resistance and colicinogenic factors in gram-negative bacterial isolates, it also became clear that the conjugative properties of F were not unique. These factors were also frequently found to be transmissible and, in some cases, to exhibit other F-like characteristics. The realization that pilus production was a feature of many conjugation systems stemmed from the studies of Meynell and Datta (Meynell and Datta, 1965, 1966; Meynell et al., 1968), who found that strains carrying certain antibiotic "resistance transfer factors" or colicinogenic (Col) factors were sensitive to the same male-specific phages that infect F^+ cells and that these strains also elaborated phage-adsorbing fimbriae that were morphologically similar to F-pili. Although there were non-F-like transmissible plasmids that did not express male phage receptors, additional bacteriophages, exhibiting specificity for hosts carrying these various non-F-like plasmids, could be isolated. Such phages were also found to utilize plasmid-specific fimbrial receptors. As well as RNA and single-stranded DNA phages, the bacteriophages known to utilize conjugative pilus receptors include double-stranded DNA phages such as J, a short-tailed hexagonal-capsid phage that adsorbs to the side of certain pilus types, and PR4, a lipid-containing phage that adsorbs to pilus tips. Tests for adsorption or infection by such phages have provided one of the most convenient ways of detecting and classifying the transfer systems of plasmids isolated from enteric or *Pseudomonas* hosts. Exploration of morphological and antigenic differences and similarities exhibited by conjugative pili provided additional clues to their degree of relatedness.

Like other plasmids isolated from various members of the family *Enterobacteriaceae* and from various *Pseudomonas* species, many transmissible plasmids have been classified in *E. coli* or *Pseudomonas* incompatibility (Inc) groups (Bukhari et al., 1977). Transmissible plasmids representative of each different Inc group have been investigated, and in each case expression of a conjugative pilus has been detected (Bradley, 1980a, 1980b, 1983, 1984, 1985). Typically, transmissible plasmids within the same Inc group have been found to encode morphologically and antigenically related pili. Additionally, similarities between the pili expressed by compatible plasmids in different Inc groups have also often been evident, and these are sometimes reflected in the Inc group nomenclature used. Incompatibility groups containing plasmids that express an F-like transfer system are considered as an IncF family "complex" which includes at least four Inc groups (IncFI, IncFII, etc.). The pili expressed by $IncI_1$, $IncI_5$, IncB, IncK, and IncZ plasmids are also closely related, so these are considered to be members of the $IncI_1$ complex (Bradley, 1980b, 1984). In contrast, IncH group subclasses reflect differences in transfer properties as well as the results of DNA homology analyses.

Thus, as all of the self-transmissible enteric and *Pseudomonas* plasmids examined have been found to elaborate a fimbrial structure, at present pilus synthesis is considered to be a fundamental characteristic associated with plasmid transfer among these bacteria. Current models presume that, in these organisms, extension of a pilus filament provides the mechanism for initiating intercellular contacts required for conjugation. This article will focus on the known properties of transfer systems that specify pili and, in particular, on the cell surface interactions that appear to be involved in this type of conjugation. Other perspectives and additional details concerning properties that are rather briefly summarized in this chapter are available in other review articles (Bradley, 1980b; Guiney and Lanka, 1989; Ippen-Ihler, 1989; Ippen-Ihler and Minkley, 1986; Iyer, 1989; Paranchych and Frost, 1988; Sukupolvi and O'Connor, 1990; Valentine and Kado, 1989; Willetts and Skurray, 1980, 1987; Willetts and Wilkins, 1984). Pilus-independent conjugation mechanisms, discovered more recently through studies of antibiotic resistance gene transmission among gram-positive bacteria, are discussed in chapter 2 in this text.

THE MECHANISM OF CONJUGATION IN PILUS-DEPENDENT SYSTEMS

The transfer systems expressed by F-like plasmids remain the most well characterized, in terms of the gene functions that they specify, and provide a useful framework for discussion of pilus-dependent conjugation systems in general. As diagrammed in Fig. 1, there is evidence that F plasmid transmission involves at least five different stages, as follows. (i) The tip of an F-pilus extending from a donor cell makes contact with the recipient cell surface. (ii) Depolymerization of F-pilin subunits brings cell surfaces together in "unstable" aggregates which are easily disrupted. (iii) Stabilization of the intercellular contacts occurs. (iv) Starting from a nick at the origin of transfer (*oriT*), a single strand of DNA is unwound in the $5' \rightarrow 3'$ direction and transported into the recipient cell. (v) Cell aggregates disburse once DNA transfer is complete.

DNA transfer is usually accompanied by replacement strand synthesis in the donor and by complementary strand synthesis in the recipient. Transfer of the complete F sequence also permits recircularization of the DNA in the recipient. These events, then, permit the donor bacterium to remain F^+, and the recipient also to become an F^+ donor.

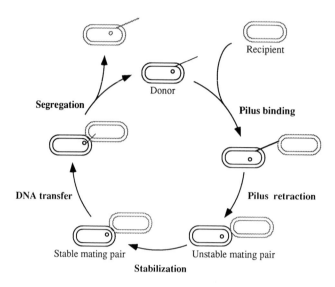

FIGURE 1. Stages in F plasmid conjugative transfer.

As it is the presence of the *oriT* region sequence that determines the site at which DNA displacement and transfer will occur, this is the only feature that must be present in *cis* on a DNA element in order for its transfer to be accomplished. Thus, an F plasmid will "mobilize" and transfer any coresident plasmid that contains a cloned F *oriT* sequence. F plasmid *tra* products can also effect the transfer of other coresident "mobilizable" plasmids which are not, by themselves, self-transmissible. These carry alternate, plasmid-specific *oriT* sequences and encode a small number of *mob* protein products that permit DNA displacement and transport to be initiated at their own *oriT* site. Thus, the conjugative DNA metabolism required for DNA mobilization can often be examined separately from the stages of conjugation that are required for cell contact. Mobilizable plasmids are able to use, for their own transfer, the intercellular contacts established by the products certain families of self-transmissible plasmids express.

Whether stages that precisely correspond to all of the steps indicated in Fig. 1 also occur during conjugation mediated by all pilus-encoding plasmids is not known. The overall effect of plasmid transfer is, certainly, the same: because each recipient becomes a donor, rapid, infectious spread of the plasmid sequences can occur throughout cell populations into which donor cells are introduced. The pervasive association between conjugal pilus expression and transfer among these enteric and pseudomonal plasmids also suggests a basic similarity in the mechanism through which cell contacts are formed, although many differences are also evident. Parallels in the DNA mobilization steps are even stronger. So far, in all cases examined in sufficient detail, unidirectional mobilization of single-stranded DNA has been found to occur after introduction of a nick into a particular, plasmid-specific *oriT* site, and DNA containing an appropriate *oriT* site has been mobilizable in *trans*.

One interesting difference in conjugative DNA metabolism that has emerged is that, while F-like plasmids appear to utilize host functions for donor and recipient complementary strand synthesis, other conjugative plasmids encode their own primase activities. Primase activities have been attributed to plasmids in the IncP and IncI complexes and in groups IncC, IncM, and IncU. Indeed, most plasmids in these Inc groups (except for IncM) can partially suppress the phenotype of a temperature-sensitive *E. coli dnaG* mutant (Guiney and Lanka, 1989). The two best-characterized primases, those expressed by ColIb-P9 (*sog*) and RP4 (*pri*), are not immunologically related to each other. However, *sog* and *pri* are both large genes that specify two sequence-related polypeptide products. Their primase activities appear to be capable of priming complementary strand synthesis in both the donor and recipient (Guiney and Lanka, 1989; Wilkins et al., 1985; Yakobson et al., 1990). During conjugation, plasmid-specified *pri* and *sog* products synthesized in the donor are actually transferred to the recipient cytoplasm (Merryweather et al., 1986b; Rees and Wilkins, 1989, 1990). RP4 *pri* mutants transfer with a lower efficiency into some bacterial species, as the degree in which a recipient cell can compensate for a defect in the donor plasmid *pri* activity is host specific (Merryweather et al., 1986a).

THE ORGANIZATION OF TRANSFER GENES

F-Like Plasmids

As shown in Fig. 2, the genes encoding the F plasmid conjugation system comprise about one third of the 100-kb plasmid and extend over an approximately 33.3-kb region distal to the F origin of transfer (*oriT*) at F coordinate 66.7 (Willetts and Skurray, 1987). Transfer initiated by a nick at F *oriT* results in the processive displacement and transport of DNA sequences from the 5' end such that the "leading region" (to the left of *oriT* in Fig. 2) is first to enter the recipient, and entry of genes in the "transfer region" (coordinates 66.7–100) occurs last.

The transfer systems of other F-like IncF complex plasmids exhibit a close relationship to that of F. The fact that natural isolates of plasmids such as R100 and R1 typically transfer at a much lower level than F is due to the intact transfer gene regulatory system they express; at some time prior to its discovery, the F plasmid acquired an IS*3* insertion that inactivated one of its regulatory genes (see below). Derepressed F-like plasmid derivatives such as R100-1 and R1-19 transfer as efficiently as F. DNA hybridization and sequence analyses of such plasmids have shown that a majority of their transfer genes are highly homologous to those on F and organized in a similar manner. They can also be used to complement most transfer-defective, pilus-deficient F mutants, although some transfer region products do exhibit genetic specificities (Willetts and Maule, 1985). A more remote, but nevertheless evident relationship between the transfer systems of IncF and IncS plasmids has also been revealed by studies of EDP208, a derepressed derivative of F_0 *lac*. The latter, originally found in *Salmonella typhi,* is one of several IncS plasmids that express serologically related pili (Coetzee et al., 1986). F_0 *lac* shows little DNA homology with F, but the organization, regulation, and

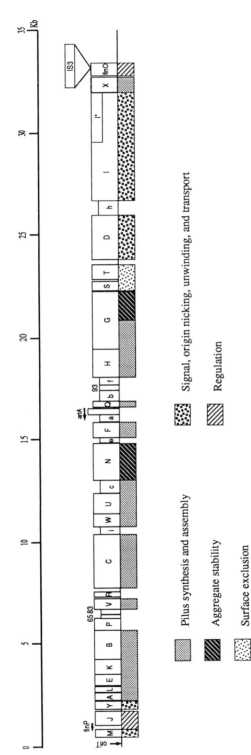

FIGURE 2. Map of the F plasmid transfer region. Boxes indicate the position and size of F transfer region genes, transcribed from left to right. Capital letters indicate *tra* genes (tall boxes); small letters indicate *trb* genes (shorter boxes). Numbers above short unlabeled boxes indicate the positions of additional open reading frames. The *finP* RNA product and *artA* product are encoded by the opposite strand. Where known, functional groups are indicated by the character of the shaded areas below the map. The line above the map indicates distance (kilobases) from the site of the *oriT* nick at F coordinate 66.7 (Thompson et al., 1989). Drawn according to compiled DNA sequence data for *traM-traJ-traY* (Fowler et al., 1983; Inamoto et al., 1988; Thompson and Taylor, 1982), *traA-traL-traE* (Frost et al., 1984), *traK-traB-traP-orf65* (Frost, personal communication), *orf83-traV-traR* (Skurray, personal communication), *traC* (Schandel et al., 1990), *trbI-traW-trbC-traN-trbE* (Maneewannakul, 1990; Maneewannakul et al., 1991), *traU* (Moore et al., 1990), *traF-trbA-artA-traQ-trbB* (Wu and Ippen-Ihler, 1989; Wu et al., 1988), *trbF-traH-traG* (Firth and Skurray, personal communication; Ham et al., 1989b), *traS-traT-traD* (Jalajakumari et al., 1987; Jalajakumari and Manning, 1989), *trbH-traI* (Bradshaw et al., 1990), *traX* (Cram et al., in press), and *traD → finO* (Yoshioka et al., 1987; Yoshioka et al., 1990).

Legend:

- Pilus synthesis and assembly
- Aggregate stability
- Surface exclusion
- Signal, origin nicking, unwinding, and transport
- Regulation

gene content of the EDP208 transfer region display a striking resemblance (Finlay et al., 1986b; Finlay and Paranchych, 1986).

F transfer region genes were named *tra* until the alphabet was exhausted; the designation *trb* was then used for subsequently discovered loci and *art* (*tra* spelled backwards) designates loci encoded on the opposite strand. Thus far, transfer genes on other F-like plasmids have been given parallel designations; e.g., the *traM, traJ, traY,* and *traA* genes of F-like plasmids correspond in function and relative position to the F *traM, traJ, traY,* and *traA* genes. Early experiments with F suggested that there were three main transcription units in the transfer region: individual gene transcripts for both *traM* and *traJ*, and a single long transcript for expression of genes *traY* to *traX*. Each of these promoters has been characterized (Dempsey, 1989; Ippen-Ihler and Minkley, 1986; Silverman et al., 1991; Silverman et al., in press). In addition, evidence for additional, constitutive promoters that may contribute to expression of distal *tra* genes has accumulated, and promoters on the opposite strand, for *finP* and *artA*, have been discovered (Gaffney et al., 1983; Ham et al., 1989a; Ham et al., 1989b; Jalajakumari et al., 1987; Mullineaux and Willetts, 1985; Wu and Ippen-Ihler, 1989). However, it is clear that expression of the majority of F *tra* genes is regulated at the promoter located just prior to *traY* (p_{YX}). This promoter requires both the plasmid positive control product, TraJ, and the host SfrA protein for full activity (Silverman et al., 1991; Silverman et al., in press). Regulation is effected through repression of TraJ synthesis when both of two F-inhibitory genes, *finP* and *finO*, are active. The *finP* gene expresses an antisense RNA that can interact with the *traJ* mRNA leader sequence (Fee and Dempsey, 1986; Finlay et al., 1986c; Frost et al., 1989; Koraimann et al., 1991). Since the F *finO* gene was inactivated by the natural occurrence of an IS*3* insertion at F coordinate 0/100, F transfer genes are constitutively expressed unless a *finO* product is expressed by a coresident F-like plasmid (Cheah and Skurray, 1986; McIntire and Dempsey, 1987a; Yoshioka et al., 1987).

The functions of individual F *tra* genes were originally explored by analysis of the phenotypes and properties of various transfer-deficient mutants. More recently, the nucleotide sequence of the entire F transfer region has been determined, and the products and activities associated with cloned DNA segments have been explored. As summarized in Fig. 2 and Table 1, analysis of F mutant phenotypes and protein activities has permitted the many F *tra* products to be grouped as follows. (i) The regulatory TraJ product is required for transfer operon expression and is, in turn, regulated by *finP* RNA and FinO. (ii) The products of a large number of genes are required for the synthesis and assembly of F-pilus filaments, and all of these products are required for initiation of mating aggregate formation. (iii) Mutants with lesions in *traN* or the distal portion of *traG* express pili but are transfer deficient and do not accumulate stable mating aggregates. (iv) Mutants carrying lesions in *traM, traY, traI,* and *traD* do form stable mating aggregates. TraY and TraI are known to be required for introduction of the *oriT* specific nick, and the TraI product, a multifunctional protein also known as helicase I, can then unwind the DNA in the 5'-to-3' direction. It is thought that, in response to a "mating signal," the TraM protein may be required to mediate

TABLE 1
Products of the F transfer region

Product	$M_r{}^a$ (10^3)	Comment,[b] references
Regulatory		
FinP		A regulatory RNA transcribed from strand opposite the *traJ* mRNA leader (Finlay et al., 1986c; Frost et al., 1989)
FinO	(21.2)	F *finO* has been inactivated by IS*3* (Yoshioka et al., 1987; Yoshioka et al., 1990)
TraJ	27.1	Cytoplasmic, positive control protein (Cuozzo and Silverman, 1986; Cuozzo et al., 1984; Thompson and Taylor, 1982)
Pilus synthesis and assembly		
TraA	12.8	Processed to 7.2-kDa F-pilin subunit; inner membrane protein (Frost et al., 1984; Ippen-Ihler et al., 1984)
TraL	10.4	Predicted to be associated with cytoplasmic membrane (Frost et al., 1984; Frost et al., 1991)
TraE	21.2	Inner membrane protein (Achtman et al., 1979; Frost et al., 1984; Ippen-Ihler et al., 1984; Laine et al., 1985)
TraK	25.6	Predicted to be processed to 23.3-kDa periplasmic protein (Frost, personal communication; Frost et al., 1991)
TraB	50.5	Predicted to be associated with the cytoplasmic membrane (Frost, personal communication; Frost et al., 1991)
TraV	18.6	Predicted to be processed; possible lipoprotein (Frost et al., 1991; Skurray, personal communication)
TraC	99.2	Peripheral membrane protein (Maneewannakul et al., 1987; Schandel et al., 1990; Schandel et al., personal communication)
TraW	23.6	Processed to a 21.7-kDa periplasmic protein (Maneewannakul, 1990; Maneewannakul et al., 1987)
TraU	36.8	Processed to a 34.3-kDa periplasmic protein; *traU* mutants make pili, but in reduced numbers (Moore et al., 1990)
TrbC	23.4	Processed to a 21.2-kDa periplasmic protein (Maneewannakul et al., 1991)
TraF	28.0	Processed to a 25.9-kDa periplasmic protein (Wu et al., 1988)
TraQ	10.9	Inner membrane protein; required for TraA product processing (Wu and Ippen-Ihler, 1989; Wu et al., 1987)
TraH	50.2	Predicted to be processed to a 47.8-kDa periplasmic protein (Ham et al., 1989b)

(Continued on next page)

TABLE 1—*Continued*

Product	M_r^a (10^3)	Comment,[b] references
TraG	102.4	Inner membrane protein; deletion of C terminus does not affect piliation (Firth and Skurray, personal communication; Manning et al., 1981)
TraX	27.5	Required for pilin subunit modification, but not for filament assembly; predicted to be inner membrane protein (Frost et al., 1991; K. Maneewannakul et al., unpublished; Cram et al., in press)
Stabilization of mating aggregates		
TraN	65.7	Processed to 63.8-kDa outer membrane protein (Maneewannkul, 1990; Maneewannakul et al., unpublished)
TraG	102.4	C-terminal portion of protein (TraG*) required for formation of stable aggregates (Firth and Skurray, personal communication; Manning et al., 1981)
Signal, origin nicking, unwinding, and transport		
TraM	14.5	F-specific binding near *oriT*; cytoplasmic protein; required for mating signal? (Di Laurenzio et al., in press; Schwab et al., 1991; Thompson and Taylor, 1982)
TraY	15.2	F-specific soluble protein required for *oriT* nicking; F *traY* may result from duplication of smaller *traY* gene on F-like plasmids (Everett and Willetts, 1980; Frost et al., 1984; Inamoto and Ohtsubo, 1990; Inamoto et al., 1988; Lahue and Matson, 1990)
TraI	192.0	Helicase I; also required for *oriT* nicking (Bradshaw et al., 1990; Traxler and Minkley, 1988)
TraD	81.7	Inner membrane protein, required for DNA transport (Jalajakumari and Manning, 1989; Panicker and Minkley, 1985; Yoshioka et al., 1990)
Surface exclusion		
TraS	16.9	Inner membrane protein (Jalajakumari et al., 1987)
TraT	26.0	Processed to 23.8-kDa outer membrane lipoprotein (Jalajakumari et al., 1987; Perumal and Minkley, 1984)
Function unknown		
TraP	22.0	Predicted to be an inner membrane protein (Frost, personal communication; Frost et al., 1991; Moore et al., 1987)
Orf65	7.1	Unidentified product; predicted to be cytoplasmic (Frost, personal communication; Frost et al., 1991)

(*Continued on next page*)

TABLE 1—*Continued*

Product	M_r^a (10^3)	Comment,[b] references
Orf83	9.1	Unidentified product; predicted to be cytoplasmic (Frost et al., 1991; Skurray, personal communication)
TraR	8.3	Predicted to be cytoplasmic (Moore et al., 1987; Skurray, personal communication)
TrbI	14.1	Intrinsic inner membrane protein (Maneewannakul, 1990)
TrbE	9.9	Intrinsic inner membrane protein (Maneewannakul, 1990)
TrbA	12.9	Inner membrane protein (Wu and Ippen-Ihler, 1989; Wu et al., 1987)
TrbB	19.5	Processed to 17.4-kDa periplasmic protein (Wu and Ippen-Ihler, 1989)
Orf93	10.2	Unidentified product (Wu and Ippen-Ihler, 1989)
TrbF	14.5	Unidentified product; predicted to be inner membrane associated (Ham et al., 1989b)
TrbH	26.2	Unidentified product; predicted to be in inner membrane (Bradshaw et al., 1990)
ArtA	12.1	Unidentified product, predicted to be an inner membrane protein; encoded by the anti-*tra* DNA strand (Wu and Ippen-Ihler, 1989)

[a] Size based on gene product predicted from DNA sequence data (see Fig. 2).

[b] Early experiments suggesting outer membrane (or in some cases inner membrane) locations for *tra* region proteins appear to have been subject to artifacts associated with overproduction of proteins. Therefore, the suggested location of a number of products is based on more recent work and differs from that presented elsewhere (Achtman et al., 1979; Willetts and Skurray, 1987).

changes in the *oriT* protein complex that allow DNA displacement to occur (see below). As DNA displacement (assayed by replacement strand synthesis) can occur in mutants with a *traD* defect, the inner membrane TraD protein is thought to be required for the actual transport of DNA to the recipient. (v) The products of *traT* and *traS* are not required for transfer; these "surface exclusion" proteins reduce the capacity of F^+ cells to act as recipients to other F donors. Surface (or entry) exclusion is highly plasmid specific and reduces the frequency of unproductive mating between F donors about 100-fold. As recipient cells that are subjected to an overwhelming number of F donors may die (lethal zygosis), exclusion genes may be an essential means of avoiding such self-destruction in high-density donor cell populations (Ou, 1980). (vi) The functions associated with an additional group of gene products, discovered by cloning and sequence analysis, remain to be characterized. Mutational analyses in progress have suggested that some of these products are not required for F transfer from *E. coli* under standard mating conditions (Kathir and Ippen-Ihler, in press; Maneewannakul, 1990; Manneewannakul, unpublished). However, as the properties of a number of these proteins suggest they may interact with other *tra* region products, a more subtle or conditional role for these products may eventually be revealed.

The *oriT* sequences of a number of F-like plasmids are known, including those for F (Thompson et al., 1984), R1 (Ostermann et al., 1984), R100 (McIntire and Dempsey, 1987b), ColB4 (Finlay et al., 1986a), and P307 (Göldner et al., 1987). While this chapter will focus on cell surface interactions that occur during conjugation, it is noteworthy that there has been considerable recent progress toward characterizing *oriT* properties (Carter and Porter, 1991; Fu et al., 1991; Thompson et al., 1989; Tsai et al., 1990) and toward characterizing the proteins that interact at the *oriT* site (Bradshaw et al., 1990; Dempsey and Fee, 1990; Fee and Dempsey, 1986; Finlay et al., 1986a, 1986b; Graus-Göldner et al., 1990; Inamoto and Ohtsubo, 1990; Lahue and Matson, 1990; Traxler and Minkley, 1988; Yoshioka et al., 1990).

The Wide-Host-Range IncP, IncN, and IncW Plasmids

The organization of the transfer genes of wide-host-range plasmids representative of groups IncN, IncP, and IncW has also been investigated in some detail. These plasmids are of particular interest since they can both transfer to and replicate in a wide variety of bacteria (see Thomas, 1989; Thomas and Smith, 1987). As for F, the transfer systems they carry include genes for pilus expression, for nicking a specific *oriT* site and mobilizing a single strand of DNA, and for excluding entry of an identical plasmid. Genes required for transfer of these plasmids are also designated *tra*.

The RK2/RP1 IncP transfer system is encoded in two separate regions termed Tra1 (15 kb) and Tra2-3 (10.4 kb). The phenotypes of mutants bearing transposon insertions in Tra2-3 suggest this region expresses gene products involved in P-pilus expression or in surface exclusion (Palombo et al., 1989), although these products are not yet well characterized. However, the structure and products of the Tra1 region, which encodes at least 16 proteins in several operons, have recently been analyzed in considerable detail (see Guiney and Lanka, 1989). The plasmid *oriT* site and plasmid-specific *tra* genes required for the conjugative functions involved in DNA mobilization are contained in a 2.7-kb segment, known as the Tra1 core region, which is located near one end of Tra1 (Guiney et al., 1989; Pansegrau et al., 1988). This region was defined by the minimal RK2 fragment that permitted clones to be mobilized at full transfer efficiency by the heterologous helper plasmid R751. The Tra1 region also apparently includes genes involved in P-pilus synthesis, as well as the *pri* (primase) gene locus which is situated near the furthest end of the Tra1 region from *oriT*.

The *oriT* sequences of the IncP plasmids RK2, RP4, and R751 have been determined (Fürste et al., 1989; Guiney and Yacobson, 1983). Interestingly, characterization of the nick site of RK2/RP4 has revealed a sequence which is found to be highly conserved not only in related plasmids, but also in both border junctions of the transfer DNA of *Agrobacterium tumefaciens* tumor-inducing plasmids (Waters et al., 1991). Recently, considerable insight has been gained into the nature of *oriT*-related events on these plasmids. The RP4 TraI, TraJ, and TraH proteins, which interact at *oriT*, have been purified and assembled in vitro to form an active nucleoprotein complex (the relaxosome; Pansegrau et al., 1990a). After nicking, the RP4 TraI protein remains attached to the 5' end of the nicked DNA strand (Pansegrau et al., 1990b).

On the IncN plasmids pKM101 (R46) and pCU1 and on the IncW plasmid R388 (IncW), all genes necessary for transfer appear to be encoded within a single *tra* region of DNA 14 to 19 kb in length. In each case, clones carrying one end of the *tra* region express conjugal pili and can be used in *trans* to effect the transfer of clones containing the other end of the *tra* region where the *oriT* site and DNA mobilization genes are clustered (Bolland et al., 1990; Iyer, 1989; Thatte et al., 1985; Winans and Walker, 1985a, 1985b, 1985c, 1985d). In the case of pKM101, the mobilizable and piliation *tra* gene segments are known to be separated by loci that encode a nuclease, *nuc*, and the P plasmid inhibition gene, *fip*, while a locus encoding entry exclusion (*eex*) interrupts the set of genes required for pilus expression. Similarly, clones carrying the R388 pili genes also express entry exclusion.

The *oriT* sequences of both R46 (IncN) and R388 (IncW) have been determined and aligned (Coupland et al., 1987; Llosa et al., 1991). Although the gene products that determine conjugal pilus production from these plasmids have not yet been well characterized, progress in this direction should prove very interesting, as these plasmids apparently require a relatively small number of products to assemble the short, rigid pili they express.

IncI Plasmids

Conjugative plasmids in the IncI complex (IncI$_1$, IncB, IncK, IncI$_2$) differ from all other transmissible plasmids in that they encode two types of pili: short, thick rigid filaments that appear to be essential for DNA transfer, and long, thin pili that contribute to the formation of mating aggregates in liquid. Pilus expression and plasmid transmission are normally repressed. Using transposon mutagenesis, Rees et al. (1987) located three blocks of *tra* genes in ColIb-P9, an IncI$_1$ plasmid thought to specify the same conjugation system as R64 and R144. Loci required for thin pilus synthesis fell within an 11-kb region, Tra1. A second region, Tra2, extending from the plasmid primase gene (*sog*) to the surface exclusion locus (*exc*), was predicted to contain thick pilus functions. Region Tra3 included *oriT* and possibly other *tra* functions. ColIb-P9 transfer is known to be unidirectional and oriented such that transfer region genes enter the recipient last (Howland and Wilkins, 1988). Mutations affecting regulation of transfer functions were located between the Tra1 and Tra2 regions, and a locus expressing a nuclease activity similar to that encoded by the IncN plasmid, pKM101, was also identified in this area. In recent analyses of the IncI$_1$ plasmid R64, the smallest self-transmissible clone spanned a 54.7-kb segment on which *oriT* was located at one end and the other included an approximately 19-kb region (Tra1) encoding functions involved in thin pilus formation (Komano et al., 1988; Komano et al., 1990). Interestingly, these studies also indicated that the C-terminal segments of one such gene, *pilV*, are subject to shufflon rearrangements that seem likely to modify the structure of thin pilus filaments.

IncH Plasmids

Plasmids of the IncH group are characteristically large (>100 MDa) and have been placed in two groups on the basis of pilus morphology and antigen specificity. Transfer of plasmids in the IncHI group displays the unusual property of being proficient at 26 to 30°C, but not at 37°C (Anderson, 1975; Smith, 1974). Plasmids in group IncHII differ in that their transfer is not temperature sensitive and H pilus production is constitutive (Bradley et al., 1982b). Three subclasses of the IncHI group have also been distinguished on the basis of phenotypic and incompatibility patterns. The IncHI3 subgroup plasmids exhibit the most marked difference in their transfer-related phenotype, since phage pilHα does not form plaques on strains carrying the representative IncHI3 plasmid, MIP233, and MIP233 pili differ morphologically and serologically from those of other plasmids in the IncH complex (Bradley, 1986; Coetzee et al., 1985).

Recent studies of the IncHI1 plasmid R27 have shown that its transfer genes are located in at least two regions (Taylor et al., 1985). Transfer genes carried by the IncHII plasmid, pHH1508a, have been found to be widely dispersed over the plasmid (Yan and Taylor, 1987). Recent results of complementation analysis between transfer-deficient mutants of R27 and cloned segments of pHH1508a suggest that the transfer mechanisms of these plasmids are related (Newnham and Taylor, 1990).

Leading Region Genes

Genes encoded in "leading region" DNA sequences, which are first to enter the recipient, seem to be dispensable to conjugative transfer from *E. coli*. However, among F-like plasmids, leading region sequences are rather highly conserved, suggesting that such loci may usually contribute to conjugation (Loh et al., 1989). An *ssb* gene, which encodes a single-stranded DNA-binding protein capable of complementing defects in *E. coli ssb*, is found in the leading regions of both F (Kolodkin et al., 1983) and ColIb-P9 (Howland et al., 1989). In both cases, high-level expression of the *ssb* loci requires that transfer genes be expressed at derepressed levels. Homology to F *ssb* has been detected on transmissible plasmids in numerous, diverse Inc groups (Golub and Low, 1986). In some cases, homology to additional F leader region DNA has also been observed. An additional locus, *psiB*, which interferes with induction of the cellular SOS response, was originally characterized on the F-like plasmids R100 and R6-5. Sequences homologous to *psiB* also appear to be conserved in the leader regions of both IncF complex and IncI₁ plasmids (Golub et al., 1988; Howland et al., 1989). Expression of *ssb* and *psiB* would be expected to be advantageous after conjugal entry of single-stranded DNA into recipient cells.

THE PROPERTIES OF CONJUGATIVE PILI

Detection of Pili

Pili can be detected antigenically or through the pilus-specific phage sensitivities they confer. Like other kinds of fimbriae, conjugative pili can be visualized only in the electron microscope. Since the number of conjugative pili expressed per plasmid host cell is usually very low in comparison with the numerous adhesive fimbriae (e.g., type 1 pili) expressed by enteric bacteria, preadsorption with plasmid-specific phages originally provided the most ready means of differentiating conjugative structures from other appendages. Currently, the availability of "bald" *E. coli* host strains (Fim⁻ Fla⁻) has made elaboration of even infrequently expressed, short pili easier to discern.

An additional barrier to detection of many kinds of conjugative pili has been that since plasmid transfer genes are often repressed in natural isolates, elaboration of conjugal pili is a rare event in these strains. This problem has been circumvented by taking advantage of the fact that transfer functions are temporarily derepressed immediately after transfer to a new recipient (Bradley, 1980a). Alternatively, derepressed mutant derivatives, sometimes producing very large numbers of conjugative pili, have been isolated. Variations in the host, growth phase, or growth media have also been found to increase the numbers of conjugative pili expressed on the cell surface (Bradley, 1980a; Curtiss, 1969; Frost et al., 1985; Meynell, 1978).

Morphology and Structure

Pili that are required for conjugation exhibit two distinct morphological types. "Long-flexible" filaments, extending 1 to 2 μm from the cell surface, are produced by a number of transmissible plasmids, including those of the F-like family (Fig. 3). In contrast, "short-rigid" pili, approximately 1/10 this length, are produced by other transfer systems such as those of the wide-host-range plasmids in the IncP, IncW, and IncN groups and the IncI complex plasmids. Both of these types of pili are relatively thick (8 to 11 nm in diameter), and an axial hole can often be distinguished in negatively stained preparations. A third morphological type of pili, the "thin-flexible" pili expressed by IncI complex plasmids, contributes to the efficiency of mating contacts initiated by donors carrying plasmids that express short-rigid pili (see below). Detached pili of all three morphological groups frequently appear to have bleblike "knobs" at one end. Although the complete composition of such knobs is not clear, those found at the base of F-pili are known to contain disorganized F-pilin subunits (Frost et al., 1986; Meynell, 1978; Worobec et al., 1986).

Within morphological classes, variations in the appearance, antigenicity, and phage sensitivity patterns have established differences between the pili expressed by plasmids of each Inc group. The morphology of pili produced by plasmids representative of different Inc groups, together with some of the bacteriophage sensitivities associated with such filaments, are summarized in Table 2. Additional details are summarized elsewhere (Bradley, 1980b, 1981; Ippen-Ihler, 1989; Paranchych and Frost, 1988). At present, it is unclear to what degree the *tra* systems associated with different Inc classes that make morphologically similar pili will prove to be related. As protein product and DNA sequence comparisons become possible, it seems likely that similarities as well as differences will be disclosed and a relatively small number of transfer system families will be revealed. At

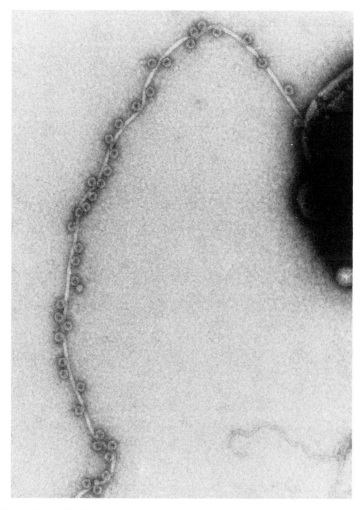

FIGURE 3. An F-pilus with R17 RNA bacteriophages adsorbed along its length.

present, the only pilin gene sequences available are those encoded by IncF and IncS plasmids.

F-pili are formed by assembly of the 70-amino-acid subunit, F-pilin. Although the possibility that additional minor proteins might be present in the F-pilus filament has not been excluded, no other components have yet been detected. The F-pilus filament is a cylinder, 8 nm in diameter with a 2-nm hollow central axis, that appears to extend from the cell through membrane adhesion zones (Bayer, 1975); in the pilus, F-pilin subunits are related by a fivefold rotation axis around the helix axis (Marvin and Folkhard, 1986). The pili expressed by F-like plasmids are essentially similar, but display differences in pilus serology and phage adsorption (Lawn and Meynell, 1970; Paranchych, 1975). Frost and co-workers

TABLE 2
Characteristics of pili produced by conjugative plasmids[a]

Inc group	Representative plasmid	Ratio, transfer plate/broth[b]	Bacteriophage sensitivity		
			Single-stranded DNA	Double-stranded DNA	RNA
Plasmids expressing long-flexible pili					
F complex	R100	0.7	f1		f2, Qβ
S	F_0 *lac*	0.8	f1		F_0 Lac
D	R711G	180	f1	J	D
C	RA1	45	C-2	J	C-1
J	R391	0.9		J	
HI1	R27	5.5			pilHα
HI2	R478	0.3			pilHα
HII	pHH1457	66			pilHα
V	R753	0.4			
T	pIN25[c]	7.0	tf-1		t
X	R6K	250	X, X-2		
P-9[d]	TOL	18		PR4	
Plasmids expressing short-rigid pili					
HI3	MIP233	0.4			
U	pAr-32	3×10^5	X		
W	Sa	4×10^4	·X	PR4	PRD1
P	RP1	2×10^3	X, Ike, Pf3, I_2	PR4	PRD1
N	N3	1×10^4	X, Ike, I_2	PR4	PRD1
M	R831G	6×10^3	X		M
P-10[d]	R91.5	47	X	PR4	
Plasmids expressing short-rigid + thin-flexible pili[e]					
I_2[f]	R721	14	X, Ike, I_2	PR4	
I_1 + B[f]	R144	1.1	If1		$I_α$

[a] Data compiled from Bradley (1980a, 1980b, 1981, 1983, 1984, 1985, 1986), Bradley and co-workers (1980, 1981, 1982a, 1982b, 1982c, 1983), and Coetzee and co-workers (1982, 1983, 1985, 1986). A more detailed compilation of pilus properties, including additional phages, was tabulated by Paranchych and Frost (1988).
[b] The ratio of the transfer frequency observed on an agar surface versus transfer frequency observed in broth. A high ratio indicates a surface preference.
[c] The prototype plasmid Rts1 is not typical as it transfers at lower frequencies and determines few T pili.
[d] *Pseudomonas* plasmid incompatibility classification. The *E. coli* IncP and IncC groups correspond to *Pseudomonas* classes P-1 and P-3, respectively.
[e] As MIP233 pili are very short, they could be a short form of a "flexible" pilus. This would account for their universal mating phenotype (Bradley, 1986).
[f] The pilus types determined by IncI₁ complex plasmids (includes groups I_1, I_5, B, K, and Z) and IncI₂ plasmids differ. Phages IKe and PR4 can use rigid I_2 pili as receptors. Phages $I_α$ and If1 adsorb to the thin I_1 pili.

(Frost et al., 1984; Frost et al., 1985) have characterized both the pilin proteins and pilin structural genes of F (IncFI) and 11 other IncFI and IncFII plasmids, including R100-1, R1, and ColB2. The amino acid sequences of these pilin subunits all exhibit a high degree of homology. These studies also revealed that serological group differences associated with the pilins examined reflect amino acid sequence variations at the pilin amino terminus. The major antigen of F-like pili is the *N*-acetylated pilin amino terminus (Finlay et al., 1985; Worobec et al., 1986). Monoclonal antibodies specific for epitopes at and near the amino terminus have been found to bind to basal knobs, but not to the sides or tips of pili, suggesting that the amino-terminal end of F-pilin is not exposed on the surface of pili (Frost et al., 1986). Two other domains, at residues 14–17 and at the carboxy terminus, appear to be exposed along the length of the F-pilus filament, since missense mutations that affect RNA phage attachment alter residues in these positions (Frost and Paranchych, 1988).

The pilin subunit of the IncS plasmid EDP208 is chemically quite distinct from F-pilin, and these proteins appear to share only 27 conserved amino acid residues (Finlay et al., 1986b). However, the EDP208 pilin amino terminus is also acetylated (Worobec et al., 1985), and the predicted structure of both pilins is very similar. In addition, EDP208, like F, confers sensitivity to bacteriophages of the f1 (M13, fd) type.

Although there has been significant recent progress in identifying the products and sequences of *tra* genes carried by plasmids that express short-rigid pili, pilin proteins and pilin structural genes of these types have not yet been identified, and it is not yet clear to what extent a commonality in ancestry will be revealed by such studies. Certainly, relationships are suggested by overlapping phage sensitivities. The organization of IncN and IncW plasmids is similar, and clones carrying an *oriT* region and mobilization genes derived from an IncW plasmid can be transferred if a coresident IncN plasmid is present (Bolland et al., 1990; Iyer, 1989). Although the assembly mechanism for short-rigid pili also remains to be investigated, the relatively small number of plasmid gene products required for their expression suggests this pathway may be less complex than the F-like plasmid systems described below.

Assembly and Retraction of F-Pili

The 70-amino-acid unassembled F-pilin subunit is found in the cytoplasmic membrane (Moore et al., 1981). This subunit is derived by processing and modification of the 121-amino-acid product of the F *traA* gene (Frost et al., 1984; Ippen-Ihler et al., 1984; Laine et al., 1985). Synthesis of mature pilin is dependent on two other F gene products expressed by the widely separated genes *traQ* (Kathir and Ippen-Ihler, in press; Wu and Ippen-Ihler, 1989) and *traX* (Moore et al., unpublished). The membrane-pilin pool appears to be substantial, since pili can be quickly regenerated even in the presence of protein synthesis inhibitors.

Assembly of an extended filament further depends upon expression of at least 11 additional F gene products: amber mutations in *tra* genes *traL, E, K, B, V, C, W, F, H,* and *G* (see Ippen-Ihler and Minkley, 1986) and insertion mutations in *trbC* (Maneewannakul et al., 1991) all appear to block expression of F-pili. The *traU* gene may also be involved in assembly since, although *traU* mutants have recently been shown to express pili, mutations in *traU* also increase resistance to pilus-specific phages and result in reduced numbers of pili (Moore et al., 1990). The products of all of these genes are assumed to interact with each other. DNA sequence analyses and protein localization studies indicate that the pilin assembly gene products include periplasmic proteins as well as intrinsic and, possibly, peripheral membrane proteins (Table 1; Ham et al., 1989b; Maneewannakul et al., 1991; Moore et al., 1990; Wu and Ippen-Ihler, 1989; Wu et al., 1988). Although no individual enzymatic activities have yet been attributed to these proteins, their concerted activities are thought both to assemble pilin subunits into an extended F-pilus and to permit the depolymerization and reentry of these subunits into the membrane.

The possibility that, once assembled, an extended F-pilus might disassemble back into the donor membrane again was first raised as an explanation for the disappearance of F-pili during filamentous phage infection; that is, that a phage attached to the tip of a retracting F-pilus might "ride down" to the cell surface (Marvin and Hohn, 1969). Indeed, the length of F-pili with attached filamentous phage was observed to shorten with time (Jacobson, 1972). Subsequently, other conditions that caused the disappearance of F-pili were also found. While F-pili remain on cells chilled quickly to 0°C, they vanished from cells cooled to 25°C, exposed briefly to 50°C, or exposed to cyanide, dinitrophenol, or arsenate (Burke et al., 1979; Novotny and Fives-Taylor, 1974, 1978a, 1978b; Novotny et al., 1972; Novotny and Lavin, 1971; O'Callaghan et al., 1973; O'Callaghan et al., 1978). As none of these conditions led to an increase in unattached pili, these findings were most easily explained by F-pilus retraction. The fact that adsorption of RNA phages or antibody to the sides of F-pili could block the NaCN- or 50°C-induced disappearance of pili, presumably by interfering with the depolymerization of pilin, is also consistent with this model. As the ability to adsorb M13 phage is not affected by cyanide poisoning, the pilus tip may still be exposed at the surface of NaCN-treated cells (Novotny and Fives-Taylor, 1974). Similar results were obtained following the addition of arsenate, provided the cells were grown in glycerol minimal medium (O'Callaghan et al., 1973; O'Callaghan et al., 1978). Thus, current models suggest that subunit polymerization and extension of an F-pilus is an energy-dependent and reversible process. The molecular factors that normally regulate it are still unclear, although various growth conditions and mutations are known to affect both the number and the length of pili seen on the cell surface.

It is worth noting that, although the term "retraction" suggests an active process that might need to be "triggered" to depolymerize a pilus that has been extended, it is also quite possible that F-pili are normally relatively transient structures. Depending on an equilibrium between assembly and disassembly of subunits, F-pili might extend and then depolymerize again whether or not they have come in contact with recipient cells or phages. Since subunits appear to be conserved unless pili are removed by blending, depolymerization may usually occur at the base (Sowa et al., 1983), but the possibility that mating contacts might result in depolymerization from the tip into the recipient cell's envelope is also viable. Similarly, addition of subunits could occur at either the inner membrane anchor or at the pilus tip. In the latter case, pilus length might be determined by the rate at which pilin subunits were added to the tip relative to the rate of their removal from the base; a phage or recipient cell interaction at the tip would block assembly and automatically result in depolymerization of the filament (Silverman, 1985).

It is to be hoped that we will soon learn more about the mechanism of this interesting reaction. Until recently most F-pilus assembly mutants studied carried amber mutations that blocked pilus synthesis, and very few F-encoded proteins could be detected under Tra$^+$ conditions. Since the DNA sequence of the complete F *tra* region is now known, however, and most genes and gene products have been identified, these products can be more readily purified from overproducing clones and used to generate specific sera for investigation of assembly protein interactions. In addition, analysis of the complementing activities expressed from mutagenized clones that produce less aberrant proteins should be useful in probing the effect of more subtle changes in protein structure. Some F-like plasmids differ from F in their sensitivity to NaCN (Frost et al., 1985), and RNA-phage-resistant, transfer-proficient F mutants that retain extended pili even after NaCN or heat treatment have been isolated (Burke et al., 1979). Current technology should permit the molecular variations that can result in these altered retraction responses to be identified. Recently, deletion mutations affecting *trbC* have been found to result in transfer-deficient cells which retain some sensitivity to filamentous DNA phage infection despite an apparent lack of F-pili. One *traC* missense mutation has also been found to exhibit this phenotype (Schandel et al., 1987; Schandel et al., 1990). If in these cases the pilus tip is at the cell surface, such mutants may also be useful in identifying an intermediate stage in pilus assembly.

DONOR-RECIPIENT INTERACTIONS

The Role of Pili

In general, plasmids that express a long-flexible type of pilus tend to exhibit a "universal" mating phenotype (Bradley, 1980b). As with the F factor, transfer from donors containing these plasmids occurs at approximately the same efficiency in matings performed either in liquid or on solid surfaces. Data accumulated for F-directed conjugation indicate that contact with the recipient is initiated by the tip of the F-pilus. Such evidence includes light microscope observation of donor-recipient pairs, connected invisibly at a distance of up to one or two cell lengths (Brinton, 1965; Curtiss, 1969); electron microscopic pictures of donor and recipient cells connected by an extended pilus (Brinton, 1971); and experiments showing that binding of F-specific DNA phages (Ippen and Valentine, 1967) or Zn^{2+} ions (Ou, 1978; Ou and Anderson, 1972) to pilus tips blocks pair formation. The tips of purified F-pili also appear to bind to recipient cells (Helmuth and Achtman, 1978). In contrast, pilus-deficient mutants or donors grown under con-

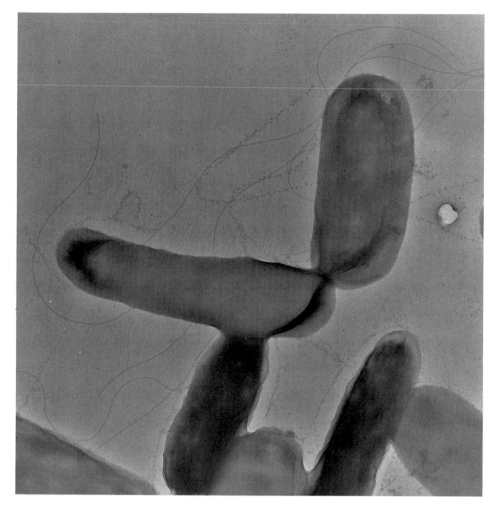

FIGURE 4. Aggregated cells in a mating mixture. Donors that express F-pili, flagella, and fimbriae were mixed with recipients that express no filaments. RNA phages were added to distinguish the F-pilus filaments. The central cell, lying in a horizontal position, is a donor.

ditions that block F-pilus expression (Novotny and Lavin, 1971; Walmsley, 1976), or from which F-pili are removed by blending (Novotny et al., 1969) or sodium dodecyl sulfate (SDS) treatment (Achtman et al., 1978a), are unable to form contacts with recipient cells.

Examination of Hfr and F′ mating mixtures has further shown that donor-recipient cell mixtures contain more aggregated cells than control mixtures and that multicellular aggregates are more typical in mating mixtures than are "mating pairs" (Achtman, 1975; Fig. 4). Although pilus connections can be seen between cells in these aggregates, most cells exhibit more intimate surface contacts. Thus it is thought that the tips of F and other long-thick-flexible conjugative pili serve both to initiate contact with recipient cells and (through depolymerization of pilin subunits) to draw donor and recipient cell surfaces together.

In contrast, the transfer systems of many non-F-like plasmids exhibit a strong surface preference and transfer very inefficiently in liquid media in comparison to their transfer on filter or agar surfaces. Typically, such plasmids express only short-rigid pili (Table 2; Bradley, 1980b). An interesting variation is that, while DNA transfer of $IncI_1$ and $IncI_2$ plasmids is dependent on short-rigid pili, IncI plasmids exhibit a universal mating type because they concomitantly express long-thin (~6-nm-diameter) I-pili. It has been shown that other surface-obligatory transfer systems also become efficient in liquid matings if either the donor or the recipient cell can express these thin filaments (Bradley, 1984).

Therefore, long pili appear to play a role that short-rigid pili do not usually confer: pili with a long-flexible or long-thin morphology are able to initiate more frequent and/or more persistent surface contacts with disbursed recipients than can pili of the short-rigid type. Presumably, donors expressing long pili can interact with more distant recipients. The inefficient transfer in liquid observed for IncT plasmids at 37°C correlates with the shorter length of the pili they appear to express at this temperature (Bradley and Whelan, 1985). Possibly, long-flexible pili may also have more adherent properties than short-rigid pili and can maintain contacts that would otherwise be interrupted by movement in a liquid environment. If, for example, mixtures of motile F-donors and nonmotile F^- cells are examined by light microscopy, donor cells have been observed to tow recipient cells by an "invisible thread" for a considerable length of time (Brinton, 1971). Donor and recipient host factors may also contribute to the persistence of cell surface interactions: F transfer exhibits a surface requirement if recipients are defective in the outer membrane protein OmpA, and the correlation between pilus length and surface preference is not as obvious when *Pseudomonas* hosts are used as donors (Bradley, 1983).

An additional role for conjugative pili is implied by the phenotype of pilus-deficient mutants. Characterization of all transmissible plasmids that express conjugative pili of either the long-thick-flexible or short-rigid type has confirmed that mutations which affect production of these pili are inevitably accompanied by a drastic decrease in the plasmid transfer frequency. These deficiencies are apparent even when cells are held together on a filter. Although most pilus-deficient F mutants retain the capacity to transfer at a low level even in liquid (e.g., 10^{-5} to 10^{-7}), this is usually assumed to reflect reversion or leaky expression of the mutation involved. No case is known in which the transfer defect of an F-pilus assembly mutant is suppressed by surface mating. Thus, it would seem that the role of conjugative pili is not limited to simply establishing contact between cell surfaces.

It follows then that, while extension of an adherent, retractable filament may be an ideal mechanism for bringing cell surfaces together, the type of contact conjugative pili make must also be important. One can imagine that the correct or "functional" mating contact for either a liquid- or a surface-preferred conjugation system might require that particular donor and recipient components be aligned before DNA transport can occur. A reasonable hypothesis is that such a functional contact is made when the pilus tip interacts with a particular recipient receptor (Ippen-Ihler and Minkley, 1986). As F-pilin subunits are channeled to and from the inner membrane, this contact would bring donor components at the base of the F-pilus into apposition with the recipient receptor, possibly at a membrane adhesion zone. Whether pilin subunits and/or pilus assembly products then play an additional role in conjugation remains controversial. Since the first contact between donor and recipient cells relies on an intact set of all gene products required for pilus synthesis and assembly, it has been difficult to assess the contribution of such proteins to conjugal DNA transport or to other observable events. These include both the stabilization of mating aggregates and transmission of a mating signal that allows DNA displacement to occur in the donor and the transport of DNA to begin. The order and timing of the latter events are unknown.

Factors Affecting Aggregate Formation and Stability

For liquid mating systems, the degree of aggregate formation can be assayed by light or electron microscope, by Coulter Counter analysis, or by determining the transfer frequency in cell suspensions that have been diluted after they have been given an opportunity to form. As indicated above, studies of mixtures containing F donors indicate that these form aggregates in which donors and recipient cells in surface contact are prevalent. Although F-pilus connections are also visible in aggregates, there is evidence that extended pili need not be present to maintain mating aggregates and that DNA transfer can occur in the absence of visible filaments. Achtman et al. (1978a) found that addition of a low concentration (0.01%) of SDS to disrupt F-pilus filaments prevented aggregate formation. However, once a mixture of Hfr donors and recipient cells had been incubated for 19 min, addition of SDS did not disrupt aggregates and caused only a twofold reduction in the residual transfer frequency. While F' donor-recipient aggregates were found to be much more sensitive to SDS, this was attributed to the short time required for F' transfer; Achtman et al. (1978a) suggested that completion of transfer resulted in an active disaggregation of cells and that initiation of new mating contacts was required for maintenance of F' aggregates. Additional results obtained with a temperature-sensitive F *lac traD* mutant donor have been consistent with the suggestion that extended pili need not be present at the DNA transport stage. Though *traD* mutants are transfer deficient, they do express pili. Panicker and Minkley (1985) found that when mating aggregates had been allowed to form at the nonpermissive temperature (TraD⁻), the presence of 0.01% SDS did not inhibit DNA transfer during a subsequent incubation at the permissive temperature. Such experiments establish, then, that the distant donor-recipient contacts initiated by an extended F-pilus can progress to a stable stage in which there are cell surface associations that cannot be disrupted by SDS. Additional factors affecting aggregate formation, accumulation, or stability have been identified by analysis of other piliated donor strains and altered recipient strains which exhibit reduced conjugation proficiency.

Conjugation-deficient recipients

Although various mutants that are conjugation-deficient recipients for F (ConF⁻) have been isolated, identification of the recipient cell receptor with which the F-pilus tip interacts has proven difficult (see Achtman et al., 1978a; Ippen-Ihler and Minkley, 1986; Willetts and Skurray, 1980, 1987). Possibly the F-pilus interacts with more than one surface component prior to reaching an ultimate receptor. However, it is clear that defects either in the major outer membrane protein, OmpA, or in lipopolysaccharide (LPS) can affect mating efficiencies. Many ConF⁻ strains exhibit *ompA* defects (see Manoil and Rosenbusch, 1982; Ried and Henning, 1987), and heptose-deficient LPS mutants are also poor recipients for F transfer to either *E. coli* or *Salmonella typhimurium*. The side chains of complete LPS have also been reported to inhibit transfer of F to *S. typhimurium* (Havekes et al., 1976; Sanderson et al., 1981). While addition of small quantities of purified LPS has been reported to inhibit transfer of IncI plasmids, LPS does not appear to inhibit F transfer unless added in combination with OmpA, however. In examining a set of various ConF⁻ mutants, Achtman et al. (1978b) found that membrane preparations from all of the LPS mutants they tested appeared somewhat deficient in OmpA. They also found that all of the OmpA⁻ or heptose-deficient LPS mutants they tested were defective recipients only in liquid matings. Since mating occurred at a normal frequency on agar surfaces, it would appear the mutants did not lack a receptor essential for F transfer. Interestingly, mutation of *ompA* apparently also has no effect on either liquid or surface matings with other closely related F-like plasmids such as R100-1.

However, the analysis of ConF⁻ mutants did offer further insight into the components that normally stabilize F mating aggregates. Achtman et al. (1978b) found that purified F-pili could still bind to ConF⁻ mutants with OmpA or LPS defects. When they examined mixtures of F donors and an *ompA* mutant recipient by electron microscopy, they detected aggregates at about the same frequency as in a wild-type mating mixture, yet they could not detect mating aggregates in any ConF⁻ mixture if they used Coulter Counter analysis. Wild-type mating mixtures to which purified OmpA (dissolved in LPS) had been added displayed the same phenotype. Since, in the Coulter Counter, aggregates are subjected to additional shear forces, they suggested that OmpA is required after F-pili have adsorbed to the recipient and brought cells into surface contact; that is, at the stage during which mating aggregates become stabilized. The stabilization step, then, could involve a direct interaction between OmpA and donor surface components that is not essential to transfer. Alternatively, the presence of OmpA might normally assist in prolonging donor-recipient interactions until stabilization is effected by other, possibly essential, components. In the latter case, a preliminary, adherent interaction between OmpA and the F-pilus is not precluded. However, it would seem that neither OmpA nor LPS can be the only surface element with which the F-pilus tip interacts. Similarly, it is clear that OmpA cannot be required for "functional" pairing and that its absence permits both unstable aggregate formation and DNA transfer.

The surface exclusion protein, TraT

A *traT* product is expressed as a major outer membrane component on the surface of cells carrying IncF and IncS transmissible plasmids. In addition to being one of two *tra* region products that interfere with the entry of an identical plasmid into such donors, the oligomeric TraT lipoprotein enhances the ability of pathogenic bacteria to survive in serum. DNA sequence and product relationships have been explored for a number of *traT* genes, and these, together with details of TraT protein properties, have recently been summarized elsewhere (Sukupolvi and O'Connor, 1990). The TraT protein appears to interfere with an early stage in the formation and stabilization of mating aggregates. If recipients that express a cloned F *traT* gene are used, the number of aggregates detectable in mixtures with F donors is depressed, although transfer levels are only 10- to 20-fold reduced even when high levels of TraT are expressed (Achtman et al., 1977; Minkley and Willetts, 1984). Minkley and Willetts (1984) have suggested that TraT acts as a competitive inhibitor for binding of the pilus tip to the recipient cell surface; it would then reduce the rate at which the conjugal pilus would make a more productive contact with its appropriate receptor and also reduce the rate of functional pairing. TraT effects are stoichiometric and display exclusion specificities that correlate well with differences in the amino terminus of the pilin subunits expressed by different F-like plasmids (Frost et al., 1985; Willetts and Maule, 1984). In addition, the inhibition of transfer effected by purified TraT protein retains this specificity which, apparently, can depend upon a single, conservative amino acid change in the TraT molecule (Minkley and Willetts, 1984; Sukupolvi and O'Connor, 1990). Alternatively, Riede and Eschbach (1986) have suggested that TraT masks a region of the OmpA protein that would normally be exposed on recipient cell surfaces. This, presumably, would also interfere with the stabilization of mating aggregates, while permitting functional pairing and DNA transfer.

The donor components TraG and TraN

Transfer-deficient F *lac* mutants that retain the capacity to express F-pili have also been examined for aggregation defects (Manning et al., 1981). Coulter Counter analysis detected no effect on the aggregation properties of F *lac* derivatives carrying mutations in *traM, traD,* or *traI*. However, piliate F *lac traN* or F *lac traG* strains behaved differently; when recipients were mixed with these mutants, no increase in aggregate formation could be detected by either electron microscopy or Coulter Counter analysis. Although F *lac traN* transfer was somewhat improved in surface matings, *traG* mutant transfer was not enhanced by this procedure. Thus, the *traN* and *traG* products are required at an early stage of aggregation, and, unlike the recipient components discussed above, these proteins appear to be essential for DNA transfer. Manning et al. (1981) have suggested that TraN and TraG may act to stabilize mating aggregates and that, in their absence, aggregates form but continually fall apart again. The *traN* product has recently been shown to be a 66-kDa outer membrane protein at least partially exposed on the cell surface (Maneewannakul, 1990). Thus TraN may be in a position to interact directly with recipient surface components. In contrast, the product of *traG* is a 102-kDa bifunctional protein detectable in inner membrane fractions. While deletions that affect only the carboxy-terminal region of the *traG* product do not affect piliation, mutations affecting the remainder of the molecule result in loss of F-pilus production. Recent work suggests that TraG processing releases a C-terminal polypeptide (TraG*) into the periplasm, where it could interact with internal regions of the TraN protein (Firth and Skurray, personal communication).

THE MATING SIGNAL

Although DNA synthesis is not required for F-mediated DNA transfer, the replacement strand synthesis that can be detected in donor cells provides an indication of whether displacement of the donor DNA strand (from *oriT*) has occurred. Kingsman and Willetts (1978) found that such donor conjugal DNA synthesis was detected only in mating mixtures containing F donors that expressed F-pili. Mutation of *traD* reduced replacement strand synthesis, a finding consistent with the DNA transport role proposed for TraD (Panicker and Minkley, 1985). Replacement strand synthesis did not occur in F *lac traI* or *traM* mutants. TraI (helicase I) would be expected to be essential, since it is now known to function both in *oriT* nicking and DNA unwinding (Traxler and Minkley, 1988). In contrast, TraM was not required in an *oriT* nicking assay (Everett and Willetts, 1980). TraM is essential for transfer and, as it binds to the *oriT* region (Di Laurenzio et al., in press; Schwab et al., 1991) and is a likely component of the F *oriT* nucleoprotein complex, it is thought that this protein may respond to the mating signal and trigger changes in the *oriT* complex that permit strand displacement to occur. Interestingly, Kingsman and Willetts (1978) also found that triggering of donor conjugal DNA synthesis occurred normally in F *lac traN* or *traG* (piliated) mutants.

These results have been seen to have a number of further implications (Kingsman and Willetts, 1978; Manning et al., 1981). First, it is clear that unless the F-pilus tip contacts the recipient cell surface, the mating signal needed to trigger DNA displacement cannot occur. Second, it appears that this signal can be transmitted whether or not mating aggregates have been stabilized by TraN and TraG. Third, since *traI, traM,* and *traD* mutant donors do form stable mating aggregates, triggering of DNA displacement cannot be a requirement for aggregate stabilization. Possibly, signal transmission and aggregate stabilization are independent events that can occur in any order. Alternatively, the data available can be resolved by models in which (i) OmpA participates in a nonessential interaction that can prolong F donor-recipient cell contacts before and/or after functional pairing occurs; (ii) the presence of TraT reduces the frequency of functional pairing by diverting pili from forming more stable contacts that lead to functional pairing; (iii) functional pairing between the F-pilus tip and a recipient component leads to transmission of the mating signal; (iv) donor TraN and TraG components are required after functional pairing and signal transmission to form or maintain a connection that allows DNA transport; and (v) failure to form or maintain this connection causes aggregates to disburse.

While it is evident that the signal must be transmitted at some point after the F-pilus first forms an unstable connection with the recipient and that this signal triggers donor DNA displacement reactions, neither the nature of the signal nor its mechanism of transmission can be specified at present. Possibilities that have been proposed include interactions between the pilus tip and a recipient inner membrane component, changes in the tip that cause pilin to depolymerize into the donor or recipient envelopes, changes in the pilus assembly protein complex, and/or an alteration of the membrane potential caused when one of these events leads to the opening of a channel between the donor and recipient membranes. One clue that a recipient inner membrane component may be involved at this stage is that recipients which express the F surface exclusion proteins also fail to trigger replacement strand synthesis in donor cells (Kingsman and Willetts, 1978). F DNA entry into TraS$^+$ recipients is reduced 100- to 200-fold, although the presence of this protein has very little effect on mating aggregate formation (Achtman et al., 1977). Thus, it has been suggested that entry exclusion effects attributable to the inner membrane protein TraS may reflect TraS interference with mating signal transmission (Manning et al., 1981).

THE MATING CHANNEL

Since the 2-nm core of an F-pilus cylinder should be large enough to accommodate DNA, the discovery of F-pili led immediately to models suggesting that these filaments also provided the passageway for conjugal DNA transport. Since then, the question of whether DNA can pass through an extended F-pilus has remained controversial. Two reports have offered some support for the hypothesis. Using a light microscope and a micromanipulator, Ou and Anderson (1970) removed and grew recipient cells from donor-recipient pairs for which they had observed no wall-to-wall contact. They concluded that, in many cases, donor markers had been inherited. More recently, genetic transfer from donor cells through a 6-μm-thick filter containing straight-through 0.01- to 0.1-μm-diameter pores has been reported (Harrington and Rogerson, 1990). As indicated in the previous sections, however, mating aggregates typically contain cells in close contact, and 0.01% SDS does not disrupt preformed aggregates or prevent DNA transfer. Since extended F-pilus filaments cannot be seen on cells subjected to this treatment, conjugating cells must ordinarily be connected in closer, SDS-resistant contacts. This connection has been imagined to be a localized fusion of the two cellular membranes to form an intercytoplasmic membrane pore through which DNA is passed (see Willetts and Skurray, 1987; Willetts and Wilkins, 1984). Since *tra* proteins involved in pilus expression are required at an earlier stage, their subsequent role in such events is unclear. Thus the question of whether pilin subunits might participate in channeling DNA from one membrane to the other, and occasionally (presumably in an unstable contact) convey DNA through an extended pilus, remains unresolved. At present, there is no evidence that such proteins interact with DNA.

However, DNA transport has been demonstrated to be dependent upon TraD, and this protein has been suggested to form or be part of the DNA export "pore" (Panicker and Minkley, 1985). This large inner membrane protein is required both for F DNA export and for RNA entry during infection by the f2-MS2-R17 family of pilus-specific phages. Interestingly, it is also required for F mobilization of the

ColE1 plasmid, but not for CloDF13 mobilization nor for infection by the RNA phage Qβ (Willetts and Wilkins, 1984). A recent study also suggested that TraD may accompany DNA into the recipient, since a protein of approximately its size appears to be transferred to the recipient membrane during conjugation (Rees and Wilkins, 1990). Although the primase proteins of RP4 and ColIb-P9 move to the recipient cytoplasm during conjugation with these plasmids, F DNA transfer to the cytoplasm was not accompanied by transfer of any cytoplasmic *tra* product or by *ssb* (Rees and Wilkins, 1989, 1990).

SUMMARY

Many of the plasmids prevalent among enteric bacteria and pseudomonads encode conjugation systems. These elements are infectious and capable of transmitting DNA sequences to bacteria with which their host comes in contact. Each of these plasmid conjugation systems carries a set of genes capable of elaborating a filamentous structure on the surface of their bacterial host. In some cases, as in the F system, the pili expressed appear able to initiate progressively adherent intercellular contacts that permit mating to occur in donor-recipient cell aggregates suspended in liquid. The conjugal pili expressed by other plasmids, such as those in the IncP, IncN, and IncW groups, do not seem to confer this property, since transfer of these plasmids occurs preferentially on agar surfaces. Nevertheless, pilus expression appears to be required for conjugation even when cells are held together; thus specific, pilus-related reactions seem to be essential for effecting the intercellular connections required for DNA transport. Pilus retraction may then lead to reactions between other donor and recipient envelope components.

It is clear that the nature of the mating contact is not yet understood at the molecular level. In recent years, research has focused on the systematic identification and characterization of the transfer genes and products encoded by plasmid conjugation systems. The entire sequence of the F transfer region is now known, and the sequences of a number of the related *tra* alleles encoded by F-like plasmids have also been characterized. Most F *tra* region products have been identified, and in many cases their intracellular location has been established. There has also been considerable recent progress toward defining the genes and gene functions encoded by several plasmids in other Inc groups. In addition, experimental systems for analyzing the functions of transfer gene proteins that interact with donor DNA metabolism have been developed and exploited; a number of these cytoplasmic proteins have been purified, and significant advances toward characterizing the DNA binding sites, protein associations, and activities of components involved in *oriT* recognition, nicking, and strand displacement have been made. Until recently, however, few transfer products involved in envelope interactions had been purified. The only conjugal pili and pilin subunits that had been chemically characterized were those in the F-like family, and the only protein that had been purified from donor membranes was the abundant F surface exclusion protein TraT. As this chapter has been written, however, the purification of other transfer-related membrane proteins has progressed, and antisera raised against additional proteins have become available. Thus, while characterization of the activities and functional interactions of individual proteins involved in the cell contact process of conjugation is still at an initial stage, both genetic and biochemical research can now become more focused in this direction.

Many basic questions still need to be addressed. What proteins and factors control pilus extension and retraction? How and where are pilin subunits added and removed, and what determines the length and number of these filaments? What are the actual pilus-recipient and donor-recipient surface interactions that form and stabilize mating contacts, and what biochemical signal causes DNA strand displacement to begin? What is the actual connection through which DNA is passaged, and how is this macromolecular transport event accomplished? Do pilin subunits and/or pilus assembly gene products participate in this process? These are only a few of the intriguing aspects of conjugation which remain to be elucidated.

REFERENCES

Achtman, M. 1975. Mating aggregates in *Escherichia coli* conjugation. *J. Bacteriol.* **123**:505–515.

Achtman, M., N. Kennedy, and R. Skurray. 1977. Cell-cell interactions in conjugating *Escherichia coli:* role of *traT* protein in surface exclusion. *Proc. Natl. Acad. Sci. USA* **74**:5104–5108.

Achtman, M., P. A. Manning, C. Edelbluth, and P. Herrlich. 1979. Export without proteolytic processing of inner and outer membrane proteins encoded by F sex factor *tra* cistrons in *E. coli* minicells. *Proc. Natl. Acad. Sci. USA* **76**:4837–4841.

Achtman, M., G. Morelli, and S. Schwuchow. 1978a. Cell-cell interactions in conjugating *Escherichia coli:* role of F pili and fate of mating aggregates. *J. Bacteriol.* **135**:1053–1061.

Achtman, M., S. Schwuchow, R. Helmuth, G. Morelli, and P. A. Manning. 1978b. Cell-cell interactions in conjugating *Escherichia coli:* Con⁻ mutants and stabilization of mating aggregates. *Mol. Gen. Genet.* **164**:171–183.

Anderson, E. S. 1975. The problem and implications of chloramphenicol resistance in the typhoid bacillus. *J. Hyg.* **74**:289–299.

Bayer, M. E. 1975. Role of adhesion zones in bacterial cell-surface function and biogenesis, p. 393–427. *In* A. Tzagaloff (ed.), *Membrane Biogenesis.* Plenum Press, New York.

Bolland, S., M. Llosa, P. Avila, and F. de la Cruz. 1990. General organization of the conjugal transfer genes of the IncW plasmid R388 and interactions between R388 and IncN and IncP plasmids. *J. Bacteriol.* **172**:5795–5802.

Bradley, D. E. 1980a. Determination of pili by conjugative bacterial drug resistance plasmids of incompatibility groups B, C, H, J, K, M, V, and X. *J. Bacteriol.* **141**:828–837.

Bradley, D. E. 1980b. Morphological and serological relationships of conjugative pili. *Plasmid* **4**:155–169.

Bradley, D. E. 1981. Conjugative pili of plasmids in *Escherichia coli* and *Pseudomonas* species, p. 217–226. *In* S. H. Levy, R. C. Clowes, and E. L. Koenig (ed.), *Molecular Biology, Pathogenicity and Ecology of Bacterial Plasmids.* Plenum Press, New York.

Bradley, D. E. 1983. Specification of the conjugative pili and surface mating systems of *Pseudomonas* plasmids. *J. Gen. Microbiol.* **129**:2545–2556.

Bradley, D. E. 1984. Characteristics and function of thick and thin conjugative pili determined by transfer derepressed plasmids of incompatibility groups I_1, I_2, I_5, B, K, and Z. *J. Gen. Microbiol.* **130**:1489–1502.

Bradley, D. E. 1985. Conjugation systems of IncT plasmids. *J. Gen. Microbiol.* **131**:2665–2671.

Bradley, D. E. 1986. The unique conjugation system of IncHI3 plasmid MIP233. *Plasmid* **16**:63–71.

Bradley, D. E., T. Aoki, T. Kitao, T. Arai, and H. Tschape. 1982a. Specification of characteristics for the classification of plasmids in incompatibility group U. *Plasmid* **8**:89–93.

Bradley, D. E., J. N. Coetzee, T. Bothma, and R. W. Hedges. 1981. Phage X: a plasimd-dependent, broad host-range, filamentous bacterial virus. *J. Gen. Microbiol.* **126**:389–396.

Bradley, D. E., J. N. Coetzee, and R. W. Hedges. 1983. IncI$_2$ plasmids specify sensitivity of filamentous bacteriophage IKe. *J. Bacteriol.* **154**:505–507.

Bradley, D. E., V. M. Hughes, H. Richards, and N. Datta. 1982b. R plasmids of a new incompatibility group determined constitutive production of H pili. *Plasmid* **7**:230–238.

Bradley, D. E., F. A. Sirgel, J. N. Coetzee, R. W. Hedges, and W. F. Coetzee. 1982c. Phages C-2 and J: IncC and IncJ plasmid-dependent phages, respectively. *J. Gen. Microbiol.* **128**:2485–2489.

Bradley, D. E., D. E. Taylor, and D. R. Cohen. 1980. Specification of surface mating systems among conjugative drug resistance plasmids in *Escherichia coli. J. Bacteriol.* **143**:1466–1470.

Bradley, D. E., and J. Whelan. 1985. Conjugation systems of IncT plasmids. *J. Gen. Microbiol.* **131**:2665–2671.

Bradshaw, H. D., Jr., B. A. Traxler, E. G. Minkley, Jr., E. W. Nester, and M. P. Gordon. 1990. Nucleotide sequence of the *traI* (helicase I) gene from the sex factor F. *J. Bacteriol.* **172**:4127–4131.

Brinton, C. C., Jr. 1965. The structure, function, synthesis and genetic control of bacterial pili and a molecular model for DNA and RNA transport in gram negative bacteria. *Trans. N.Y. Acad. Sci.* **27**:1003–1054.

Brinton, C. C., Jr. 1971. The properties of sex pili, the viral nature of "conjugal" genetic transfer systems, and some possible approaches to the control of bacterial drug resistance. *Crit. Rev. Microbiol.* **1**:105–160.

Brinton, C. C., Jr., P. Gemski, Jr., and J. Carnahan. 1964. A new type of bacterial pilus genetically controlled by the fertility factor of *E. coli* K-12 and its role in chromosome transfer. *Proc. Natl. Acad. Sci. USA* **52**:776–783.

Bukhari, A. I., J. A. Shapiro, and S. L. Adhya (ed.). 1977. Appendix B, p. 601–671. *In DNA Insertion Elements, Plasmids, and Episomes.* Cold Spring Harbor Laboratory, Cold Spring Harbor, N.Y.

Burke, J. M., C. P. Novotny, and P. Fives-Taylor. 1979. Defective F pili and other characteristics of F*lac* and Hfr *Escherichia coli* mutants resistant to bacteriophage R17. *J. Bacteriol.* **140**:525–531.

Caro, L. G., and M. Schnös. 1966. The attachment of male-specific bacteriophage f1 to sensitive strains of *Escherichia coli. Proc. Natl. Acad. Sci. USA* **56**:126–132.

Carter, J. R., and R. D. Porter. 1991. *traY* and *traI* are required for *oriT*-dependent enhanced recombination between *lac*-containing plasmids and λp*lac*5. *J. Bacteriol.* **173**:1027–1034.

Cheah, K.-C., and R. Skurray. 1986. The F plasmid carries an IS3 insertion within *finO. J. Gen. Microbiol.* **132**:3269–3275.

Coetzee, J. N., D. E. Bradley, J. Fleming, L. duToit, V. Hughes, and R. W. Hedges. 1985. Phage pilHα: a phage which adsorbs to IncHI and IncHIII plasmid coded pili. *J. Gen. Microbiol.* **131**:1115–1121.

Coetzee, J. N., D. E. Bradley, and R. W. Hedges. 1982. Phages Iα and I2-2: IncI plasmid-dependent bacteriophages. *J. Gen. Microbiol.* **128**:2797–2804.

Coetzee, J. N., D. E. Bradley, R. W. Hedges, J. Fleming, and G. Lecatsas. 1983. Bacteriophage M: an incompatibility group M plasmid-specific phage. *J. Gen. Microbiol.* **129**:2271–2276.

Coetzee, J. N., D. E. Bradley, R. W. Hedges, V. M. Hughes, M. M. McConnell, L. DuToit, and M. Tweehuysen. 1986. Bacteriophages F$_0$lac h, SR, SF: phages which adsorb to pili encoded by plasmids of the S complex. *J. Gen. Microbiol.* **132**:2907–2917.

Coupland, G. M., A. M. Brown, and N. S. Willetts. 1987. The origin of transfer (*oriT*) of the conjugative plasmid R46: characterization by deletion analysis and DNA sequencing. *Mol. Gen. Genet.* **208**:219–225.

Cram, D. S., S. M. Loh, K.-C. Cheah, and R. A. Skurray. *Gene,* in press.

Crawford, E. M., and R. F. Gesteland. 1964. The adsorption of bacteriophage R-17. *Virology* **22**:165–167.

Cuozzo, M., and P. M. Silverman. 1986. Characterization of the F plasmid TraJ protein synthesized in F' and Hfr strains of *Escherichia coli* K12. *J. Biol. Chem.* **261**:5175–5179.

Cuozzo, M., P. M. Silverman, and E. G. Minkley, Jr. 1984. Overproduction in *Escherichia coli* K12 and purification of the TraJ protein encoded by the conjugative plasmid F. *J. Biol. Chem.* **259**:6659–6666.

Curtiss, R. 1969. Bacterial conjugation. *Annu. Rev. Microbiol.* **23**:69–136.

Dempsey, W. B. 1989. Sense and antisense transcripts of *traM,* a conjugal transfer gene of the antibiotic resistance plasmid R100. *Mol. Microbiol.* **3**:561–570.

Dempsey, W. B., and B. E. Fee. 1990. Integration host factor affects expression of two genes at the conjugal transfer origin of plasmid R100. *Mol. Microbiol.* **4:**1019–1028.

Di Laurenzio, L., L. S. Frost, B. B. Finlay, and W. Paranchych. *Mol. Microbiol.*, in press.

Everett, R., and N. Willetts. 1980. Characterization of an *in vivo* system for nicking at the origin of conjugal DNA transfer of the sex factor F. *J. Mol. Biol.* **136:**129–150.

Fee, B. E., and W. B. Dempsey. 1986. Cloning, mapping, and sequencing of plasmid R100 *traM* and *finP* genes. *J. Bacteriol.* **167:**336–345.

Finlay, B. B., L. S. Frost, and W. Paranchych. 1986a. Origin of transfer of IncF plasmids and nucleotide sequences of the type II *oriT*, *traM*, and *traY* alleles from ColB4-K98 and the type IV *traY* allele from R100-1. *J. Bacteriol.* **168:**132–139.

Finlay, B. B., L. S. Frost, and W. Paranchych. 1986b. Nucleotide sequence of the *traYALE* region from IncFV plasmid pED208. *J. Bacteriol.* **168:**990–998.

Finlay, B. B., L. S. Frost, W. Paranchych, J. M. R. Parker, and R. S. Hodges. 1985. Major antigenic determinants of F and ColB2 pili. *J. Bacteriol.* **163:**331–335.

Finlay, B. B., L. S. Frost, W. Paranchych, and N. S. Willetts. 1986c. Nucleotide sequences of five IncF plasmid *finP* alleles. *J. Bacteriol.* **167:**754–757.

Finlay, B. B., and W. Paranchych. 1986. Nucleotide sequence of the surface exclusion genes *traS* and *traT* from the IncF$_0$ *lac* plasmid pED208. *J. Bacteriol.* **166:**713–721.

Firth, N., and R. Skurray. Personal communication.

Fowler, T., L. Taylor, and R. Thompson. 1983. The control region of the F plasmid transfer operon: DNA sequence of the *traJ* and *traY* genes and characterization of the *traY-Z* promoter. *Gene* **26:**79–89.

Frost, L. Personal communication.

Frost, L., S. Lee, N. Yanchar, and W. Paranchych. 1989. *finP* and *fisO* mutations in FinP anti-sense RNA suggest a model for FinOP action in the repression of bacterial conjugation by the F*lac* plasmid JCFLO. *Mol. Gen. Genet.* **218:**152–160.

Frost, L., K. Usher, and W. Paranchych. 1991. Computer analysis of the F transfer region. *Plasmid* **25:**226.

Frost, L. S., B. B. Finlay, A. Opgenorth, W. Paranchych, and J. S. Lee. 1985. Characterization and sequence analysis of pilin from F-like plasmids. *J. Bacteriol.* **164:**1238–1247.

Frost, L. S., J. S. Lee, D. G. Scraba, and W. Paranchych. 1986. Two monoclonal antibodies specific for different epitopes within the amino-terminal region of F pilin. *J. Bacteriol.* **168:**192–198.

Frost, L. S., and W. Paranchych. 1988. DNA sequence analysis of point mutations in *traA*, the F pilin gene, reveal two domains involved in F-specific bacteriophage attachment. *Mol. Gen. Genet.* **213:**134–139.

Frost, L. S., W. Paranchych, and N. Willetts. 1984. DNA sequence of the F *traALE* region that includes the gene for F-pilin. *J. Bacteriol.* **160:**395–401.

Fu, Y.-H., M.-M. Tsai, Y. Luo, and R. C. Deonier. 1991. Deletion analysis of the F plasmid *oriT* locus. *J. Bacteriol.* **173:**1012–1020.

Fürste, J. P., W. Pansegrau, G. Ziegelin, M. Kroger, and E. Lanka. 1989. Conjugative transfer of promiscuous IncP plasmids: interaction of plasmid encoded products with the transfer origin. *Proc. Natl. Acad. Sci. USA* **86:**1771–1775.

Gaffney, D., R. Skurray, and N. Willetts. 1983. Regulation of the F conjugation genes studied by hybridization and *tra-lac* fusion. *J. Mol. Biol.* **168:**103–122.

Göldner, A., H. Graus, and G. Högenauer. 1987. The origin of transfer of P307. *Plasmid* **18:**76–83.

Golub, E., A. Bailone, and R. Devoret. 1988. A gene encoding an SOS inhibitor is present in different conjugative plasmids. *J. Bacteriol.* **170:**4392–4394.

Golub, E. I., and K. B. Low. 1986. Unrelated conjugative plasmids have sequences which are homologous to the leading region of the F factor. *J. Bacteriol.* **166:**670–672.

Graus-Göldner, A., H. Graus, T. Schlacher, and G. Högenauer. 1990. The sequences of genes bordering *oriT* in the enterotoxin plasmid P307: comparison with the sequences of plasmids F and R1. *Plasmid* **24:**119–131.

Guiney, D. G., C. Deiss, V. Simnad, L. Yee, N. Pansegrau, and E. Lanka. 1989. Mutagenesis of the Tra1 core region of RK2 by using Tn5: identification of plasmid-specific transfer genes. *J. Bacteriol.* **171:**4100–4103.

Guiney, D. G., and E. Lanka. 1989. Conjugative transfer of IncP plasmids, p. 27–56. *In* C. M. Thomas (ed.), *Promiscuous Plasmids of Gram-Negative Bacteria.* Academic Press, Inc. (London), Ltd., London.

Guiney, D. G., and E. Yacobson. 1983. Location and nucleotide sequence of the transfer origin of the broad host range plasmid RK2. *Proc. Natl. Acad. Sci. USA* **80:**3595–3598.

Ham, L. M., D. Cram, and R. Skurray. 1989a. Transcriptional analysis of the F plasmid surface exclusion region: mapping of *traS, traT,* and *traD* transcripts. *Plasmid* **21:**1–8.

Ham, L. M., N. Firth, and R. Skurray. 1989b. Nucleotide sequence of the F plasmid transfer gene, *traH:* identification of a new gene and a promoter within the transfer operon. *Gene* **75:**157–165.

Harrington, L. C., and A. C. Rogerson. 1990. The F pilus of *Escherichia coli* appears to support stable DNA transfer in the absence of wall-to-wall contact between cells. *J. Bacteriol.* **172:**7263–7264.

Havekes, L. M., B. J. J. Lugtenberg, and W. P. M. Hoekstra. 1976. Conjugation deficient *E. coli* K-12 F⁻ mutants with heptose-less lipopolysaccharide. *Mol. Gen. Genet.* **146:**43–50.

Helmuth, R., and M. Achtman. 1978. Cell-cell interactions in conjugating *Escherichia coli:* purification of F pili with biological activity. *Proc. Natl. Acad. Sci. USA* **75:**1237–1241.

Howland, C. J., E. E. Rees, P. T. Barth, and B. M. Wilkins. 1989. The *ssb* gene of plasmid Collb-P9. *J. Bacteriol.* **171:**2466–2473.

Howland, C. J., and B. M. Wilkins. 1988. Direction of conjugative transfer of IncI1 plasmid Collb-P9. *J. Bacteriol.* **170:**4958–4959.

Inamoto, S., and E. Ohtsubo. 1990. Specific binding of the TraY protein to *oriT* and the promoter region for the *traY* gene of plasmid R100. *J. Biol. Chem.* **265:**6461–6466.

Inamoto, S., Y. Yoshioka, and E. Ohtsubo. 1988. Identification and characterization of the products from the *traJ* and *traY* genes of plasmid R100. *J. Bacteriol.* **170:**2749–2757.

Ippen, K. A., and R. C. Valentine. 1967. The sex hair of *E. coli* as sensory fiber, conjugation tube, or mating arm? *Biochem. Biophys. Res. Commun.* **27:**674–680.

Ippen-Ihler, K., D. Moore, S. Laine, D. A. Johnson, and N. Willetts. 1984. Synthesis of F-pilin polypeptide in the absence of F *traJ* product. *Plasmid* **11:**116–129.

Ippen-Ihler, K. A. 1989. Bacterial conjugation, p. 33–72. *In* S. B. Levy and R. V. Miller (ed.), *Gene Transfer in the Environment.* McGraw-Hill Publishing Co., New York.

Ippen-Ihler, K. A., and E. G. Minkley, Jr. 1986. The conjugation system of F, the fertility factor of *Escherichia coli. Annu. Rev. Genet.* **20:**593–624.

Ishibashi, M. 1967. F pilus as f⁺ antigen. *J. Bacteriol.* **93:**379–389.

Iyer, V. N. 1989. IncN group plasmids and their genetic systems, p. 165–183. *In* C. M. Thomas (ed.), *Promiscuous Plasmids of Gram-Negative Bacteria.* Academic Press, Inc. (London), Ltd., London.

Jacobson, A. 1972. Role of F pili in the penetration of bacteriophage fl. *J. Virol.* **10:**835–843.

Jalajakumari, M. B., A. Guidolin, H. J. Buhk, P. A. Manning, L. M. Ham, A. L. M. Hodgson, K. C. Cheah, and R. A. Skurray. 1987. Surface exclusion genes *traS* and *traT* of the F sex factor of *Escherichia coli* K12: determination of the nucleotide sequence and promoter and terminator activities. *J. Mol. Biol.* **198:**1–11.

Jalajakumari, M. B., and P. A. Manning. 1989. Nucleotide sequence of the *traD* region in the *Escherichia coli* F sex factor. *Gene* **81:**195–202.

Kathir, P., and K. Ippen-Ihler. *Plasmid,* in press.

Kingsman, A., and N. Willetts. 1978. The requirements for conjugal DNA synthesis in the donor strain during F*lac* transfer. *J. Mol. Biol.* **122:**287–300.

Kolodkin, A. L., M. A. Capage, E. I. Golub, and K. B. Low. 1983. F sex factor of *Escherichia coli* K-12 codes for a single stranded DNA binding protein. *Proc. Natl. Acad. Sci. USA* **80:**4422–4426.

Komano, T., N. Funayama, S.-R. Kim, and T. Nisioka. 1990. Transfer region of IncI1 plasmid R64 and role of shufflon in R64 transfer. *J. Bacteriol.* **172:**2230–2235.

Komano, T., A. Toyoshima, K. Morita, and T. Nisioka. 1988. Cloning and nucleotide sequence of the *oriT* region of the IncI1 plasmid R64. *J. Bacteriol.* **170:**4385–4387.

Koraimann, G., C. Koraimann, V. Koronakis, S. Schlager, and G. Högenauer. 1991. Repression and derepression of conjugation of plasmid R1 by wild-type and mutated *finP* antisense RNA. *Mol. Microbiol.* **5:**77–87.

Lahue, E. E., and S. W. Matson. 1990. Purified *Escherichia coli* F-factor TraY protein binds *oriT. J. Bacteriol.* **172:**1385–1391.

Laine, S., D. Moore, P. Kathir, and K. Ippen-Ihler. 1985. Genes and gene products involved in the synthesis of F-pili, p. 535–553. *In* D. R. Helinski, S. N. Cohen, D. B. Clewell, D. A. Jackson, and A. Hollaender (ed.), *Plasmids in Bacteria.* Plenum Press, New York.

Lawn, A. M., and E. E. Meynell. 1970. Serotypes of sex pili. *J. Hyg.* **68:**683–694.

Lederberg, J., and E. L. Tatum. 1946. Gene recombination in *E. coli. Nature* (London) **158:**558.

Llosa, M., S. Bolland, and F. de la Cruz. 1991. Structural and functional analysis of the broad host range IncW plasmid R388 and comparison with the related IncN plasmid R46. *Mol. Gen. Genet.* **226:**473–483.

Loeb, T. 1960. The isolation of a bacteriophage specific for the F$^+$ and Hfr mating types of *Escherichia coli* K-12. *Science* **131:**932–933.

Loeb, T., and N. Zinder. 1961. A bacteriophage containing RNA. *Proc. Natl. Acad. Sci. USA* **47:**282–289.

Loh, S., D. Cram, and R. Skurray. 1989. Nucleotide sequence of the leading region adjacent to the origin of transfer on plasmid F and its conservation among conjugative plasmids. *Mol. Gen. Genet.* **219:**177–186.

Maneewannakul, K. Unpublished data.

Maneewannakul, K., D. Moore, C. Hamilton, and K. Ippen-Ihler. Unpublished data.

Maneewannakul, S. 1990. Ph.D. thesis. Texas A&M University, College Station.

Maneewannakul, S., P. Kathir, and K. Ippen-Ihler. Unpublished data.

Maneewannakul, S., P. Kathir, D. Moore, L.-A. Le, J. H. Wu, and K. Ippen-Ihler. 1987. Location of F plasmid transfer operon genes *traC* and *traW* and identification of the *traW* product. *J. Bacteriol.* **169:**5119–5124.

Maneewannakul, S., K. Maneewannakul, and K. Ippen-Ihler. 1991. Characterization of *trbC*, a new F plasmid *tra* operon gene that is essential to conjugative transfer. *J. Bacteriol.* **173:**3872–3878.

Manning, P. A., G. Morelli, and M. Achtman. 1981. *traG* protein of the F sex factor of *Escherichia coli* K-12 and its role in conjugation. *Proc. Natl. Acad. Sci. USA* **78:**7487–7491.

Manoil, C., and J. P. Rosenbusch. 1982. Conjugation-deficient mutants of *Escherichia coli* distinguish classes of functions of the outer membrane OmpA protein. *Mol. Gen. Genet.* **187:**148–156.

Marvin, D. A., and W. Folkhard. 1986. Structure of F-pili: reassessment of the symmetry. *J. Mol. Biol.* **191:**299–300.

Marvin, D. A., and B. Hohn. 1969. Filamentous bacterial viruses. *Bacteriol. Rev.* **33:**172–209.

McIntire, S., and W. B. Dempsey. 1987a. Fertility inhibition gene of plasmid R100. *Nucleic Acids Res.* **15:**2029–2041.

McIntire, S., and W. B. Dempsey. 1987b. *oriT* sequence of the antibiotic resistance plasmid R100. *J. Bacteriol.* **169:**3829–3832.

Merryweather, A., P. T. Barth, and B. M. Wilkins. 1986a. Role and specificity of plasmid RP4-encoded DNA primase in bacterial conjugation. *J. Bacteriol.* **167:**12–17.

Merryweather, A., C. E. D. Rees, N. M. Smith, and B. M. Wilkins. 1986b. Role of *sog* polypeptides specified by plasmid ColIb-P9 and their transfer between conjugating bacteria. *EMBO J.* **5:**3007–3012.

Meynell, E. 1978. Experiments with sex pili: an investigation of the characters and function of F-like and I-like sex pili based on their reactions with antibody and phage, p. 207–233. *In* D. E. Bradley, E. Raizen, P. Fives-Taylor, and J. Ou (ed.), *Pili.* International Conferences on Pili, Washington, D.C.

Meynell, E., and N. Datta. 1965. Functional homology of the sex factor and resistance transfer factor. *Nature* (London) **207:**884–885.

Meynell, E., and N. Datta. 1966. The nature and incidence of conjugation factors in *Escherichia coli. Genet. Res.* **7:**141–148.

Meynell, E., G. G. Meynell, and N. Datta. 1968. Phylogenetic relationships of drug-resistance factors and other transmissible bacterial plasmids. *Bacteriol. Rev.* **32:**55–83.

Minkley, E. G., Jr., and N. S. Willetts. 1984. Overproduction, purification and characterization of the F *traT* protein. *Mol. Gen. Genet.* **196:**225–235.

Moore, D., K. Maneewannakul, C. Hamilton, and K. Ippen-Ihler. Unpublished data.

Moore, D., K. Maneewannakul, S. Maneewannakul, J. Wu, K. Ippen-Ihler, and D. E. Bradley. 1990. Characterization of the F plasmid conjugative transfer gene *traU*. *J. Bacteriol.* **172:**4263–4270.

Moore, D., B. A. Sowa, and K. Ippen-Ihler. 1981. Location of an F-pilin pool in the inner membrane. *J. Bacteriol.* **146**:251–259.

Moore, D., J. H. Wu, P. Kathir, C. M. Hamilton, and K. Ippen-Ihler. 1987. Analysis of transfer genes and gene products within the *traB-traC* region of the *Escherichia coli* fertility factor F. *J. Bacteriol.* **169**:3994–4002.

Mullineaux, P., and N. Willetts. 1985. Promoters in the transfer region of plasmid F, p. 605–614. *In* D. Helinski, S. Cohen, D. Clewell, and A. Hollaender (ed.), *Plasmids in Bacteria*. Plenum Press, New York.

Newnham, P. J., and D. E. Taylor. 1990. Genetic analysis of transfer and incompatibility functions within the IncH1 plasmid R27. *Plasmid* **23**:107–118.

Novotny, C. P., and P. Fives-Taylor. 1974. Retraction of F pili. *J. Bacteriol.* **117**:1306–1311.

Novotny, C. P., and P. Fives-Taylor. 1978a. Retraction of F pili at 50°C; its effect on mating and donor-specific phage infection; and detachment of F pili in Zn^{++}, p. 259–278. *In* D. E. Bradley, E. Raizen, P. Fives-Taylor, and J. Ou (ed.), *Pili*. International Conferences on Pili, Washington, D.C.

Novotny, C. P., and P. Fives-Taylor. 1978b. Effects of high temperature on *Escherichia coli* F-pili. *J. Bacteriol.* **133**:459–464.

Novotny, C. P., P. Fives-Taylor, and K. Lavin. 1972. Effects of growth inhibitors and ultraviolet irradiation on F-pili. *J. Bacteriol.* **112**:1083–1089.

Novotny, C. P., and K. Lavin. 1971. Some effects of temperature on the growth of F pili. *J. Bacteriol.* **98**:1307–1319.

Novotny, C. P., E. Raizen, W. S. Knight, and C. C. Brinton, Jr. 1969. Functions of F pili in mating pair formation and male bacteriophage infection studied by blending spectra and reappearance kinetics. *J. Bacteriol.* **98**:1307–1319.

O'Callaghan, R., J. Coward, and W. Paranchych. 1978. The loss of F-pili by arsenate poisoning, p. 279–289. *In* D. E. Bradley, E. Raizen, P. Fives-Taylor, and J. Ou (eds.), *Pili*. International Conferences on Pili, Washington, D.C.

O'Callaghan, R. J., L. Bundy, R. Bradley, and W. Paranchych. 1973. Unusual arsenate poisoning of the F pili of *Escherichia coli*. *J. Bacteriol.* **115**:76–81.

Ørskov, I., and F. Ørskov. 1960. An antigen termed f$^+$ occurring in F$^+$ coli strains. *Acta Pathol. Microbiol. Scand.* **48**:37–46.

Ostermann, E., F. Kricek, and G. Högenauer. 1984. Cloning the origin of transfer region of the resistance plasmid R1. *EMBO J.* **3**:1731–1735.

Ou, J. T. 1978. Kinetics of maturation of preliminary mating pairs into effective mating pairs in bacterial conjugation in *E. coli* K12, p. 245–247. *In* D. E. Bradley, E. Raizen, P. Fives-Taylor, and J. Ou (ed.), *Pili*. International Conferences on Pili, Washington, D.C.

Ou, J. T. 1980. Mating due to loss of surface exclusion as a cause for thermosensitive growth of bacteria containing the Rts1 plasmid. *Mol. Gen. Genet.* **180**:501–510.

Ou, J. T., and T. F. Anderson. 1970. Role of pili in bacterial conjugation. *J. Bacteriol.* **111**:177–185.

Ou, J. T., and T. F. Anderson. 1972. Effect of Zn^{2+} on bacterial conjugation. *J. Bacteriol.* **111**:177–185.

Palombo, E. A., K. Yusoff, V. A. Stanisich, V. Krishnapillai, and N. S. Willetts. 1989. Cloning and genetic analysis of *tra* cistrons of the Tra2/Tra3 region of plasmid RP1. *Plasmid* **22**:59–69.

Panicker, M. M., and E. G. Minkley, Jr. 1985. DNA transfer occurs during a cell surface contact stage of F sex factor-mediated bacterial conjugation. *J. Bacteriol.* **162**:584–590.

Pansegrau, W., D. Balzer, V. Kruft, R. Lurz, and E. Lanka. 1990a. In vitro assembly of relaxosomes at the transfer origin of plasmid RP4. *Proc. Natl. Acad. Sci. USA* **87**:6555–6559.

Pansegrau, W., G. Ziegelin, and E. Lanka. 1988. The origin of conjugative IncP plasmid transfer: interaction with plasmid encoded products and the nucleotide sequence at the relaxation site. *Biochim. Biophys. Acta* **951**:365–374.

Pansegrau, W., G. Ziegelin, and E. Lanka. 1990b. Covalent association of the *traI* product of plasmid RP4 with the 5′ terminal nucleotide at the relaxation nick site. *J. Biol. Chem.* **265**:10637–10644.

Paranchych, W. 1975. Attachment, ejection and penetration stages of the RNA phage infectious process, p. 85–111. *In* N. Zinder (ed.), *RNA Phages*. Cold Spring Harbor Laboratory, Cold Spring Harbor, N.Y.

Paranchych, W., and L. S. Frost. 1988. The physiology and biochemistry of pili. *Adv. Microbiol. Physiol.* **29**:53–114.

Perumal, N. B., and E. G. Minkley, Jr. 1984. The product of the F sex factor *traT* surface exclusion gene in a lipoprotein. *J. Biol. Chem.* **259**:5357–5360.

Rees, C. E. D., D. E. Bradley, and B. M. Wilkins. 1987. Organization and regulation of the conjugation genes of IncI₁ plasmid ColIb-P9. *Plasmid* **18**:223–236.

Rees, C. E. D., and B. M. Wilkins. 1989. Transfer of *tra* proteins into the recipient cell during bacterial conjugation mediated by plasmid ColIb-P9. *J. Bacteriol.* **171**:3152–3157.

Rees, C. E. D., and B. M. Wilkins. 1990. Protein transfer into the recipient cell during bacterial conjugation: studies with F and RP4. *Mol. Microbiol.* **4**:1199–1205.

Ried, G., and U. Henning. 1987. A unique amino acid substitution in the outer membrane protein OmpA causes conjugational deficiency in *Escherichia coli* K-12. *FEBS Lett.* **223**:387–390.

Riede, I., and M. L. Eschbach. 1986. Evidence that TraT interacts with OmpA of *Escherichia coli*. *FEBS Lett.* **205**:241–245.

Sanderson, K. E., J. Janzer, and J. Head. 1981. Influence of lipopolysaccharide and protein in the cell envelope on recipient capacity in conjugation of *Salmonella typhimurium*. *J. Bacteriol.* **148**:283–293.

Schandel, K. A., S. Maneewannakul, K. Ippen-Ihler, and R. E. Webster. 1987. A *traC* mutant that retains sensitivity to f1 bacteriophage but lacks F-pili. *J. Bacteriol.* **169**:3151–3159.

Schandel, K. A., S. Maneewannakul, R. A. Vonder Haar, K. Ippen-Ihler, and R. E. Webster. 1990. Nucleotide sequence of the F plasmid gene *traC* and identification of its product. *Gene* **96**:137–140.

Schandel, K. A., M. Mueller, and R. E. Webster. Personal communication.

Schwab, M., H. Gruber, and G. Högenauer. 1991. The TraM protein of plasmid R1 is a DNA binding protein. *Mol. Microbiol.* **5**:439–446.

Silverman, P. M. 1985. Host cell-plasmid interactions in the expression of DNA donor activity by F⁺ strains of *Escherichia coli* K-12. *BioEssays* **2**:254–259.

Silverman, P. M., E. Wickersham, and R. Harris. 1991. Regulation of the F plasmid traY promoter in *E. coli* by host and plasmid factors. *J. Mol. Biol.* **218**:119–128.

Silverman, P. M., E. Wickersham, S. Rainwater, and R. Harris. *J. Mol. Biol.*, in press.

Skurray, R. Personal communication.

Smith, W. H. 1974. Thermosensitive transfer factors in chloramphenicol-resistant strains of *Salmonella typhi. Lancet* **2**:281–282.

Sowa, B. A., D. Moore, and K. Ippen-Ihler. 1983. Physiology of F-pilin synthesis and utilization. *J. Bacteriol.* **153**:962–968.

Sukupolvi, S., and C. D. O'Connor. 1990. TraT lipoprotein, a plasmid-specified mediator of interactions between gram-negative bacteria and their environment. *Microbiol. Rev.* **54**:331–341.

Taylor, D. E., E. C. Brose, S. Kwan, and W. Yan. 1985. Mapping of transfer regions within incompatibility group HI plasmid R27. *J. Bacteriol.* **162**:1221–1226.

Thatte, V., D. E. Bradley, and V. N. Iyer. 1985. N conjugative transfer system of plasmid pCU1. *J. Bacteriol.* **163**:1229–1236.

Thomas, C. M. (ed.). 1989. *Promiscuous Plasmids of Gram-Negative Bacteria*. Academic Press, Inc. (London), Ltd., London.

Thomas, C. M., and C. A. Smith. 1987. Incompatibility group P plasmids: genetics evolution and use in genetic manipulation. *Annu. Rev. Microbiol.* **41**:77–101.

Thompson, R., and L. Taylor. 1982. Promoter mapping and DNA sequencing of the F plasmid transfer genes *traM* and *traJ. Mol. Gen. Genet.* **188**:513–518.

Thompson, R., L. Taylor, K. Kelly, R. Everett, and N. Willetts. 1984. The F plasmid origin of transfer: DNA sequence of wild-type and mutant origins and location of origin-specific nicks. *EMBO J.* **3**:1175–1180.

Thompson, T. L., M. B. Centola, and R. C. Deonier. 1989. Location of the nick at *oriT* of the F plasmid. *J. Mol. Biol.* **207**:505–512.

Traxler, B. A., and E. G. Minkley, Jr. 1988. Evidence that DNA helicase I and *oriT* site-specific nicking are both functions of the F TraI protein. *J. Mol. Biol.* **204**:205–209.

Tsai, M.-M., Y.-H. F. Fu, and R. Deonier. 1990. Intrinsic bends and integration host factor binding at F plasmid *oriT. J. Bacteriol.* **172**:4603–4609.

Valentine, C. R. I., and C. I. Kado. 1989. Molecular genetics of IncW plasmids, p. 125–163. *In* C. M. Thomas (ed.), *Promiscuous Plasmids of Gram-Negative Bacteria.* Academic Press, Inc. (London), Ltd., London.

Walmsley, R. H. 1976. Temperature dependence of mating-pair formation in *Escherichia coli. J. Bacteriol.* **126:**222–224.

Waters, V. L., K. H. Hirata, W. Pansegrau, E. Lanka, and D. G. Guiney. 1991. Sequence identity in the nick regions of INCP plasmid transfer origins and T-DNA borders of *Agrobacterium* Ti plasmids. *Proc. Natl. Acad. Sci. USA* **88:**1456–1460.

Wilkins, B. M., L. K. Chatfield, C. C. Wymbs, and A. Merryweather. 1985. Plasmid DNA primases and their role in bacterial conjugation, p. 585–603. *In* D. R. Helinski, S. N. Cohen, D. B. Clewell, D. A. Jackson, and A. Hollaender (ed.), *Plasmids in Bacteria.* Plenum Press, New York.

Willetts, N., and J. Maule. 1984. Interactions between the surface exclusion systems of some F-like plasmids. *Genet. Res.* **24:**81–89.

Willetts, N., and J. Maule. 1985. Specificities of IncF plasmid conjugation genes. *Genet. Res.* **47:**1–11.

Willetts, N., and R. Skurray. 1980. The conjugation system of F-like plasmids. *Annu. Rev. Genet.* **14:**41–76.

Willetts, N., and R. Skurray. 1987. Structure and function of the F factor and mechanism of conjugation, p. 1110–1133. *In* F. C. Neidhardt, J. L. Ingraham, K. B. Low, B. Magasanik, M. Schaechter, and H. E. Umbarger (ed.), *Escherichia coli and Salmonella typhimurium: Cellular and Molecular Biology.* American Society for Microbiology, Washington, D.C.

Willetts, N., and B. Wilkins. 1984. Processing of plasmid DNA during bacterial conjugation. *Microbiol. Rev.* **48:**24–41.

Winans, S. C., and G. C. Walker. 1985a. Conjugal transfer system of the IncN plasmid pKM101. *J. Bacteriol.* **161:**402–410.

Winans, S. C., and G. C. Walker. 1985b. Entry exclusion determinants of IncN plasmid pKM101. *J. Bacteriol.* **161:**411–416.

Winans, S. C., and G. C. Walker. 1985c. Identification of pKM101-encoded loci specifying potentially lethal gene products. *J. Bacteriol.* **161:**417–424.

Winans, S. C., and G. C. Walker. 1985d. Fertility inhibition of RP1 by IncN plasmid pKM101. *J. Bacteriol.* **161:**425–427.

Worobec, E. A., L. S. Frost, P. Pieroni, G. D. Armstrong, R. S. Hodges, J. M. R. Parker, B. B. Finlay, and W. Paranchych. 1986. Location of the antigenic determinants of conjugative F-like pili. *J. Bacteriol.* **167:**660–665.

Worobec, E. A., W. Paranchych, J. M. R. Parker, A. K. Taneja, and R. S. Hodges. 1985. Antigen-antibody interaction: the immunodominant region of EDP208 pili. *J. Biol. Chem.* **260:**938–943.

Wu, J. H., and K. Ippen-Ihler. 1989. Nucleotide sequence of *traQ* and adjacent loci in the *Escherichia coli* K-12 F plasmid transfer operon. *J. Bacteriol.* **171:**213–221.

Wu, J. H., P. Kathir, and K. Ippen-Ihler. 1988. The product of the F plasmid transfer operon gene, *traF*, is a periplasmic protein. *J. Bacteriol.* **170:**3633–3639.

Wu, J. H., D. Moore, T. Lee, and K. Ippen-Ihler. 1987. Analysis of *E. coli* K12 F factor transfer genes: *traQ, trbA* and *trbB. Plasmid* **18:**54–69.

Yakobson, E., C. Deiss, K. Hirata, and D. G. Guiney. 1990. Initiation of DNA synthesis in the transfer origin region of RK2 by the plasmid-encoded primase: detection using defective M13 phage. *Plasmid* **23:**80–84.

Yan, W., and D. E. Taylor. 1987. Characterization of transfer regions within the HII incompatibility group plasmid pHH1508a. *J. Bacteriol.* **169:**2866–2868.

Yoshioka, Y., Y. Fujita, and E. Ohtsubo. 1990. Nucleotide sequence of the promoter-distal region of the *tra* operon of plasmid R100, including *traI* (DNA helicase I) and *traD* genes. *J. Mol. Biol.* **214:**39–53.

Yoshioka, Y., H. Ohtsubo, and E. Ohtsubo. 1987. Repressor gene *finO* in plasmids R100 and F: constitutive transfer of plasmid F is caused by insertion of IS3 into F *finO. J. Bacteriol.* **169:**619–623.

Microbial Cell-Cell Interactions
Edited by Martin Dworkin
© 1991 American Society for Microbiology, Washington, DC 20005

Chapter 4

Chlamydomonas Mating Interactions

Ursula W. Goodenough

INTRODUCTION

Facets of the *Chlamydomonas* mating reaction have been reviewed regularly (Goodenough, 1977; Goodenough et al., 1980; Goodenough, 1985; Snell, 1985; Goodenough and Ferris, 1987; Goodenough and Adair, 1989; van den Ende et al., 1988, 1990). In this review I attempt to cover the entire mating reaction from the perspective of cell biology/biochemistry, focusing particular attention on the sexual adhesion molecules, on mechanisms of signal transduction, and on mechanisms of receptor mobility in the plane of the flagellar membrane.

EVOLUTIONARY PERSPECTIVES

Chlamydomonas reinhardtii is a unicellular biflagellate green alga. Its phylogenetic position has recently been analyzed using the conserved sequence elements present in small-subunit rRNA genes (Gunderson et al., 1987). These studies indicate that the eukaryotic protists do not represent a single "lower" taxon but rather have diverged as several discrete lineages so that, for example, *C. reinhardtii* is more closely related to the vertebrates than is either *Dictyostelium* or *Euglena*. The basic split between the "higher" plants and animals appears to have occurred roughly midway during eukaryotic evolution, and *C. reinhardtii* appears to have evolved from the same ancestor that gave rise to corn and wheat. Moreover, of the eukaryotic microorganisms analyzed to date by this technique, *C. reinhardtii* emerges as the most closely related to the hypothetical protistan cell that gave rise to both the higher plants and animals, with a slightly earlier branch giving rise to the fungi. Therefore, the mating reaction of modern *Chlamydomonas* is of considerable evolutionary interest, for many of its molecular strategies may have operated in the hypothetical protistan ancestor and hence may have been transmitted to the "higher" lineages that evolved from this ancestor. Certain of these strategies would of course be expected to have undergone modification or elimination in subsequent lineages, but our present picture of eukaryotic evolution indicates that the unicellular ancestor of higher plants and animals was a green flagellate that may have mated in much the same fashion as *Chlamydomonas* (see also Janssens, 1988).

The genus *Chlamydomonas* offers the opportunity for further evolutionary insights in that it includes two quite different groups, one represented by *C. reinhardtii* and the other by *C. eugametos*. The two groups have distinctive cell walls (Roberts, 1974; Goodenough and Heuser, 1988a, 1988b) and distinctive mating systems (Lewin, 1954b; Wiese, 1969). Recent studies (Jupe et al., 1988) indicate that *C. reinhardtii* and *C. eugametos* diverged from one another early in the *Chlamydomonas* lineage. It follows that the traits they share in common are likely to have been possessed by their common ancestor—that is, by the flagellated unicell postulated to sit at the base of the "higher" eukaryotic tree—while their unique traits likely reflect adaptations made by one or both lineages. In this review, careful consideration will be given to similarities versus differences between the *C. reinhardtii* and *C. eugametos* mating systems in an attempt to better understand the phenotype of our "base-line ancestor."

Yet another opportunity for evolutionary analysis is provided by the order Volvocales, which includes *Chlamydomonas* unicells, various colonial forms (*Gonium, Pandorina, Eudorina,* and *Pleodorina*) of increasing complexity, and finally the truly multicellular *Volvox*. Since *C. reinhardtii* is clearly more closely related to *Volvox* than is *C. eugametos*, both in phenotype (Adair et al., 1987; Goodenough and Heuser, 1988a) and at the level of DNA sequences (Rausch et al., 1989; Larson et al., in press), the radiation leading to *Volvox* presumably initiated with a *C. reinhardtii*-like organism and occurred more recently (~50 million years ago) than the *eugametos-reinhardtii* schism. Again it is possible to argue that traits shared by the *C. reinhardtii* and *Volvox* mating systems were present as well in their common ancestor, thereby shedding light on conserved features of the volvocine lineage.

Finally, it is of interest to consider the class Chlorophyta (green algae) as a whole (Pickett-Heaps, 1975). This group has retained the use of motile unicellular flagellated gametes even as it has evolved into the broad-leafed *Ulva mirabilis* (sea lettuce) and the highly differentiated *Nitella,* and certain features of their mating interactions have been shown to be surprisingly similar to those utilized by *Chlamydomonas,* as detailed in later sections of this review.

The Chlorophytes have come to occupy virtually every aquatic niche on the planet. It has been argued (e.g., Stanley, 1975) that successful eukaryotic lineages are those that have the capacity to diverge rapidly from parental species; that is, a key attribute is the frequent formation of new species (Wiese and Wiese, 1977). By definition, each speciation event requires both sexual reproduction and the ability to develop sexual isolation mechanisms. It follows, then, that a successful mating system should be one with established mechanisms for frequent change, as considered again in a later section.

GAMETOGENESIS

Chlamydomonas cells are of two mating types, *plus* (mt^+) and *minus* (mt^-), with mating type controlled by a single complex locus on linkage group VI (Smith and Regnery, 1950; Ebersold et al., 1962; Goodenough and Ferris, 1987) in *C. reinhardtii.* Cells growing vegetatively (mitotically) give no evidence of expressing any of the traits known to be encoded in, or controlled by, the *mt* locus, although this may only reflect our ignorance of such traits. In any case, when nitrogen is depleted (Sager and Granick, 1954), the vegetative cells undergo two critical differentiations: they reorganize their metabolism such that they enter into a stable G_0 phase of their cell cycle, in which they can survive for months if not years as long as they remain hydrated (Martin and Goodenough, 1975), and they also express gamete-specific traits, described in detail below, which permit them to recognize, and fuse with, a gamete of opposite *mt* should one be encountered (Fig. 1). The resultant diploid zygote is capable of expressing yet another set of novel genes (Ferris and Goodenough, 1987; Woessner and Goodenough, 1989; Matters and Goodenough, in press) which permits the construction of a thick wall, thereby preventing desiccation of the cell. Thus the G_0 gamete can survive nitrogen starvation whereas the zygote can survive both nitrogen starvation and desiccation, the latter presumably a common stress confronted by this predominantly soil organism.

Several differences have been reported between gametogenesis in *C. reinhardtii* and that in *C. eugametos* (van den Ende et al., 1988). In *C. reinhardtii,* novel gene expression requires both nitrogen starvation and light (Treier et al., 1989); in *C. eugametos,* nutrient stresses other than nitrogen depletion can also induce gametogenesis (Tomson et al., 1985), and light is required only for agglutinin activation in certain strains (Kooijman et al., 1988; see below). In *C. reinhardtii,* mitosis accompanies gametogenesis under some but not all circumstances (Matsuda et al., 1990), and differentiation is not completed until cells are suspended from plates into liquid medium (Saito et al., 1988); these facets of differentiation have not yet been explored for *C. eugametos.*

Gametic differentiation in all known *Chlamydomonas* species is "reversible"; i.e., gametes lose their mating abilities and resume mitotic cycling when am-

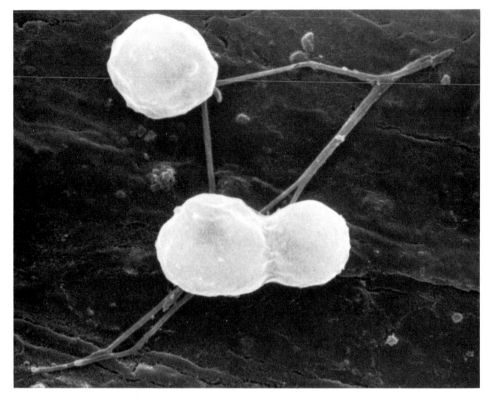

FIGURE 1. Mating gametes of *C. reinhardtii*. The two lower cells have just fused to form a zygote; the third cell displays a slightly curved fertilization tubule extending between its two flagella. Adhesive foci are localized to the flagellar tips at upper right. At lower left, the disadhesion which follows cell fusion has commenced. The unmated mt^+ gamete does not lose its adhesiveness whereas the young zygote does; this allows the zygote to swim freely and the unmated gamete to swim off and find a new mt^- partner. Magnification, ×7,900.

monium (or nitrate) is restored to the medium. Thus there is no separate germ line and no obligate death of the soma. Interestingly, *Volvox* has evolved a separate germ line: at each generation, certain cells are set aside which, if sexually induced, go on to form either sperm in "males" or eggs in "females" (reviewed by Kirk and Harper, 1987). If not fertilized, the eggs can revert to vegetative growth but the sperm do not survive.

A second unique feature of *Volvox* gametogenesis is that a hormone is involved. Metabolic stress, including nitrogen starvation but also heat shock (Kirk and Kirk, 1986), causes males to secrete a glycoprotein which induces the formation of sperm and also, in females, the formation of eggs. The cDNA sequence of this protein has recently been obtained (Mages et al., 1988), but its induction mechanism is not understood.

Nothing is known, in any of these systems, about how metabolic stress induces the expression of gametic traits or, in *Volvox,* the expression of the hormone. There are, however, well-characterized prokaryotic programs wherein nitrogen starvation induces the formation of a dormant asexual spore whose cell wall is different from the vegetative wall (Losick et al., 1986; Cutting et al., 1990). Regulation in these cases involves a highly conserved group of genes (Ronson et al., 1987). Therefore, as discussed more fully elsewhere (Goodenough, 1985), it is possible that features of a rather ancient program of cellular differentiation in response to metabolic stress are still operative in the Volvocales, with the requirement for sexual fusion interposed between perception of the stress and construction of a novel cell type.

Saito and Matsuda (1991) have recently identified two gene mutations, *dif-1* and *dif-2,* which render *C. reinhardtii* cells temperature sensitive for gametic differentiation: at restrictive temperature, they fail to acquire any gametic traits in response to nitrogen starvation. The two gene loci are not linked to *mt* nor to one another, and further study of these strains promises to yield important insights into the early events of gametogenesis.

FLAGELLAR AGGLUTININS

A key gametic trait is the acquisition of flagellar agglutinability, which permits sexual recognition. Thus when fully differentiated *plus* gametes are mixed with *minus* gametes, they instantly stick together by their flagellar surfaces (Fig. 1), and all subsequent events in the mating reaction are triggered by this initial adhesion. Since vegetative *plus* and *minus* cells are nonadhesive, either to themselves or to gametes, the gametic flagella clearly acquire novel molecules which have come to be called agglutinins (Wiese, 1965). Three gene loci have been identified that are necessary for the biosynthesis of active agglutinins (Goodenough et al., 1978; Hwang et al., 1981), but it is not yet known whether they encode the agglutinins per se.

We have recently written a detailed account of the experiments leading to the purification of the *C. reinhardtii* agglutinins (Goodenough and Adair, 1989), and a parallel effort in the Musgrave–van den Ende laboratory has led to agglutinin purification from *C. eugametos* and the closely related *C. moewusii* (reviewed by van den Ende et al., 1988, 1990). Figure 2 shows the *plus* molecules from *C. reinhardtii;* the *minus* proteins are only subtly different in morphology (Goodenough et al., 1985). Figure 3 shows *plus* and *minus* proteins from *C. eugametos.* By visual inspection alone, all are clearly homologous, with long fibrous domains that bend or kink at abrupt angles. Biochemically, all have been found to be hydroxyproline-rich glycoproteins (HRGPs) (Cooper et al., 1983; Collin-Osdoby and Adair, 1985; Samson et al., 1987b), with arabinose and galactose representing the major sugars (Klis et al., 1989). Returning to our opening discussion, then, the fact that *C. reinhardtii* and *C. eugametos* use the same type of glycoproteins to effect sexual adhesion suggests that their common ancestor did so also.

A particularly interesting feature of the agglutinins is that they are also homologous to the proteins of the *C. reinhardtii* and *C. eugametos* cell wall (Cooper et al., 1983; Adair, 1985; Adair and Snell, 1990). Figures 4 and 5 show the units that make up the so-called W2 layer (Roberts et al., 1972; Goodenough and Heu-

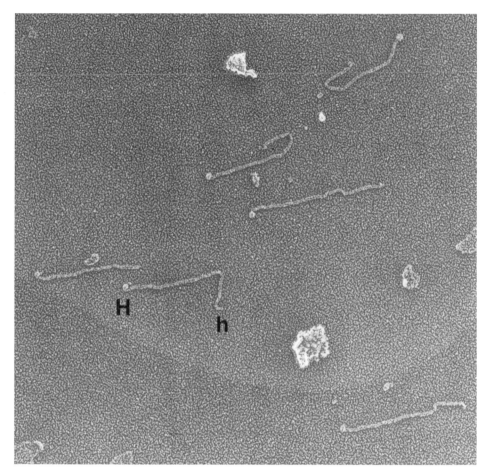

FIGURE 2. Agglutinin glycoproteins purified from mt^+ gametes of *C. reinhardtii*, adsorbed to a mica surface. and quick-freeze deep-etched (Heuser, 1983) (this and all subsequent deep-etch micrographs were prepared by John Heuser, Washington University Medical School, St. Louis, Mo.). Each molecule displays a head (H) at one end and a hook (h) at the other. Magnification, ×166,600.

ser, 1985) of this wall: the long fibrous proteins in this case associate with a central "spine" to form "fishbone units," but their basic structural homology with the agglutinins is evident, and they are also rich in hydroxyproline (Hills et al., 1975). Other examples of wall HRGPs include components of the *C. reinhardtii* W6 layer (Fig. 6) and components of the *C. reinhardtii* zygote cell wall (Grief et al., 1987; Woessner and Goodenough, 1989). HRGPs have, moreover, continued to be used as the major structural protein component of higher-plant cell walls (Showalter and Rumeau, 1990). It seems reasonable to speculate, therefore, that the baseline ancestor of the plant kingdom, and possibly the common ancestor of both plants and animals, utilized HRGPs in its cell wall, and that some common ancestor of the Chlorophytes came up with the notion of displaying these on the flagellar surface for sexual purposes.

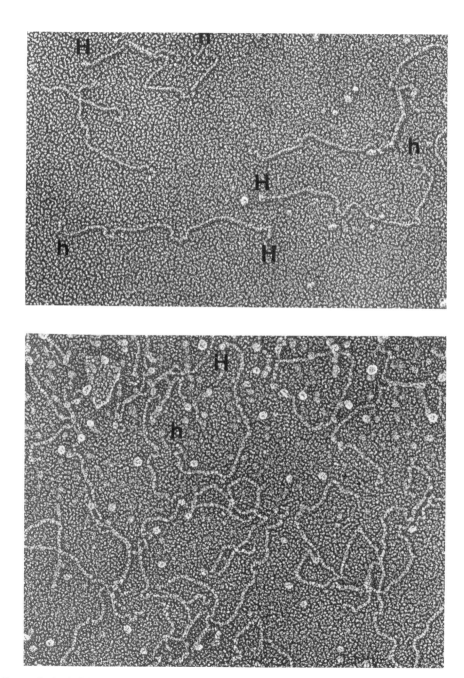

FIGURE 3. Agglutinin glycoproteins purified from mt^+ (top) and mt^- (bottom) gametes of *C. eugametos* and provided by Franz Klis (University of Amsterdam, Amsterdam, The Netherlands). Head (H) and hook (h) domains can be recognized although they are difficult to discern in the mt^- material. Magnification, ×190,000.

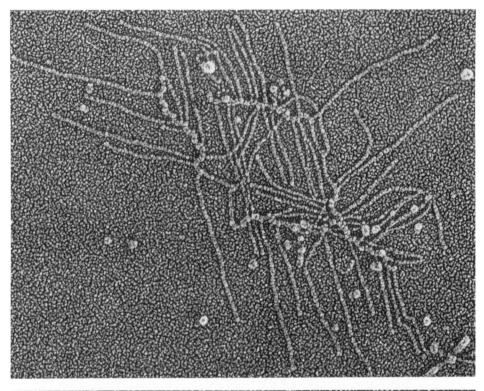

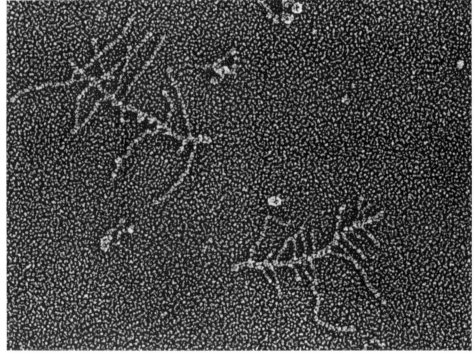

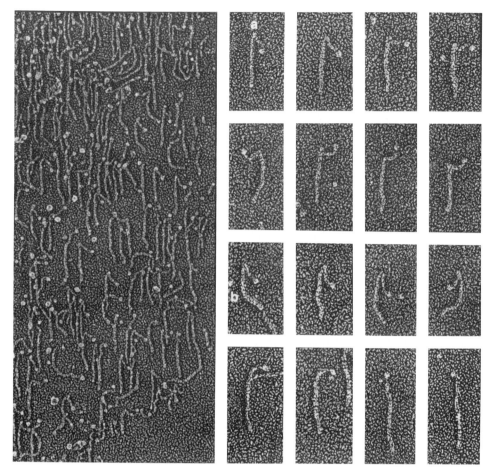

FIGURE 6. GP1 glycoproteins purified from *C. reinhardtii* cell wall extracts (from Goodenough et al., 1986). Magnification: left, ×139,200; right, ×217,500.

The agglutinins of other chlorophyte groups have not, unfortunately, been isolated or characterized as yet. However, mating in *Volvox* (Kirk and Harper, 1987), in *Ulva* (Bråten, 1971), and in *Acetabularia* (Hämmerling, 1934) is initiated by flagellar adhesion involving biflagellate gametes, and it seems likely that HRGPs will be found to be utilized in these interactions as well. Granted this presumed homology, the agglutinins emerge as key players in the chlorophyte speciation process since, for every speciation event, new agglutinins must be

FIGURE 4 [top]. Fishbone units generated from the W2 layer of the *C. reinhardtii* cell wall by exposure to lysin. Magnification, ×257,600.

FIGURE 5 [bottom]. Fishbone units from the cell wall of *C. eugametos*. Magnification, ×257,600.

formed that continue to carry *plus/minus*-type specificity for one another but no longer adhere to the agglutinins of sibling species. Granted the postulated adaptiveness of rapid speciation, it becomes important to learn how the agglutinins might be endowed with the ability to change their recognition/adhesion properties.

To address this question, we must first consider agglutinin structure in more detail. (i) When *C. eugametos* agglutinins (Fig. 3) are compared with those of *C. moewusii* (Samson et al., 1987a; Samson et al., 1987b; Crabbendam et al., 1986), their overall homology is striking, but the mt^- molecules differ from their mt^+ counterparts in that they are more flexible and do not carry a prominent globular domain ("head") at one end. (ii) In *C. reinhardtii,* the mt^+ agglutinin is again a rigid protein with a head (Fig. 2). Strikingly, however, the mt^- agglutinin of *C. reinhardtii* is also a rigid protein with a head, the *plus/minus* differences being confined to minor differences in head size and head/shaft connections rather than the major differences observed in the *C. eugametos* group.

These data can be integrated by postulating that the original agglutinins in the common *reinhardtii/eugametos* ancestor were both rigid proteins like those in the present-day *reinhardtii* species and that at some point following the *reinhardtii/eugametos* schism, the *minus* member of the dyad underwent a series of alterations to attain its more flexible form. During the relatively recent split between *C. eugametos* and *C. moewusii,* then, the dimorphism was conserved.

The common theme that emerges in comparing these proteins is that they have undergone changes in topology. One such strategy is to change length: the rigid molecules of *C. reinhardtii* are both 225 nm; the rigid mt^+ protein of *C. moewusii* is 210 nm and that of *C. eugametos* is 276 nm; and the flexible mt^- proteins are an extraordinary 336 nm in length. Similarly, in the wall proteins, the "ribs" of the fishbone units illustrated in Fig. 4 and 5 are 30 and 90 nm for *C. eugametos* and 100 and 200 nm for *C. reinhardtii.* A second motif has been a change in the position of the kinks that occur along the fibrous shafts. Thus the mt^+ agglutinin of *C. reinhardtii* tends to bend at two positions whereas the mt^- protein bends at only one (Goodenough et al., 1985), while the *C. eugametos* and *C. moewusii* mt^- agglutinins have kinks at four set positions (Crabbendam et al., 1986; Samson et al., 1987b). Perspective on these differences emerges when we consider what we would expect if we compared any other set of *Chlamydomonas* proteins: their tubulins, clathrins, or dyneins might differ by one or several amino acids, but we would be startled indeed by a 50% increase in the length of the polypeptide chains or by major changes in protein conformation or amino acid composition.

What is it about the HRGPs that allows them to change so rapidly? Two speculations have been set forward (Goodenough, 1985) and can be summarized briefly here. The first notes that since all known HRGP sequences are internally repetitive, the HRGP genes may be prone to mispairing and hence to undergoing deletions and additions that could radically alter the lengths of the protein products. The second proposes that HRGPs may be hybrid molecules, spliced together posttranslationally to generate the final products. If this is the case, and the evidence is still largely circumstantial (Goodenough et al., 1985), then deletion, addition, or rearrangement of splice sites could quickly generate the very different sets of proteins encountered.

An intriguing and as yet mysterious property of certain *C. eugametos* and *C. moewusii* agglutinins is that their agglutinability requires light and is rapidly lost when cells are returned to darkness (Kooijman et al., 1988). It is also remarkable that all forms of agglutinin remain fully active after glutaraldehyde or osmium fixation (Wiese and Jones, 1963; Goodenough and Weiss, 1975; Mesland and van den Ende, 1979; Goodenough, 1986), suggesting that adhesiveness does not require major changes in protein conformation. A third puzzling feature of agglutinin biochemistry pertains to its amino acid composition. The extended domains of HRGPs are thought to be formed by polyproline II helical configurations (van Holst and Varner, 1984). Given the extensive fibrous domains of the agglutinins (Fig. 2 and 3), their hydroxyproline content ($\sim$10 to 12%; summarized by van den Ende et al., 1990) is surprisingly low; in contrast, the similar GP1 wall protein (Fig. 6) is 33% hydroxyproline (Goodenough et al., 1986) and a typical extensin is 30 to 40% hydroxyproline (Showalter and Rumeau, 1990). No explanation can as yet be given for this anomaly.

An extensive literature has accumulated on the effects of various proteases, glycosidases, and metabolic inhibitors on agglutinin activity, from which the following can be inferred. (i) The adhesivity of gametes is much more sensitive to proteolysis than is the adhesivity of isolated agglutinin glycoproteins (Goodenough et al., 1980; Saito and Matsuda, 1984b; Collin-Osdoby and Adair, 1985), possibly because the in vivo targets are in fact protease-sensitive membrane components with which the agglutinins associate (van den Ende et al., 1990; see below). (ii) A variety of differential drug/enzyme sensitivities have been found when comparing *plus* and *minus* agglutinin proteins (Matsuda et al., 1982; Wiese et al., 1983; Saito and Matsuda, 1984b; Collin-Osdoby and Adair, 1985; Dutcher and Gibbons, 1988; van den Ende et al., 1990), but as is always the case with such studies, it is not possible to say whether the targets are the adhesive sites per se or domains whose disruption leads to a loss of adhesivity elsewhere in the molecule (e.g., due to a denaturation of the protein). (iii) Somewhat more informative is the observation that while both the *plus* and *minus* agglutinins of *C. reinhardtii*, but apparently not *C. eugametos/moewusii* (Wiese and Jones, 1963), are sensitive to disulfide reducing agents, the *plus* protein is much more sensitive than the *minus* (Collin-Osdoby and Adair, 1985; Saito and Matsuda, 1986; Goodenough, 1986). High concentrations of reducing agent cause the globular head domain of the *plus* protein (Fig. 2) to straighten into a linear configuration (Goodenough et al., 1985); it will be important to determine whether this occurs differentially at lower concentrations.

Several laboratories have obtained monoclonal antibodies (MAbs) which recognize epitopes common to both *plus* and *minus* agglutinins, with some recognizing other flagellar proteins as well (Adair, 1985; Homan et al., 1988). Of particular interest is MAb 66.3, which binds to both the *plus* and *minus* agglutinins of *C. eugametos,* but inactivates only the *minus* adhesivity (Homan et al., 1988). The site of mAB 66.3 binding to the two agglutinins has not yet been determined.

AGGLUTININ INTERACTIONS

When the agglutinin proteins of *C. reinhardtii* are visualized in association with the flagellar surface, they are all oriented such that their globular heads face outward and their terminal "hooks" (Fig. 2 and 3) anchor into the flagellar membrane (Fig. 7). The association with the membrane is extrinsic—agglutinins can be released using chelating agents (Adair et al., 1982) or chaotropes (Musgrave et al., 1981)—which suggests the presence of intrinsic agglutinin-binding proteins (ABPs) in the flagellar membrane (the targets of protease sensitivity postulated

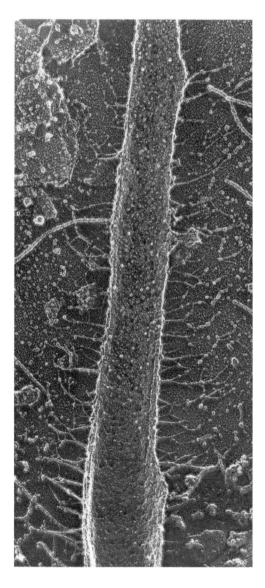

FIGURE 7. Agglutinins extending from both sides of an *mt*$^+$ gametic flagellum. Magnification, ×71,400.

above). Candidate ABPs include two high-molecular-weight polypeptides whose synthesis is induced during adhesion in *C. reinhardtii* (Snell et al., 1983) and a 125-kDa polypeptide which copurifies with the *C. eugametos* agglutinin during immunoaffinity chromatography using mAb 66.3 (Kalshoven et al., 1990). The agglutinins are apparently aligned in rows along the flagellar surface (Goodenough et al., 1985; Tomson et al., 1990b). This implies a linkage between ABPs and the underlying axoneme, meaning that ABPs may be formally analogous to the integrins which anchor both to the actin cytoskeleton and to extrinsic surface molecules (Buck and Horwitz, 1987; Marcantonio and Hynes, 1988; Springer, 1990).

The disposition of agglutinins illustrated in Fig. 7 suggests two models for how the agglutinins might associate during sexual adhesion. In the first model, the globular heads effect recognition/adhesion, and the fibrous shafts simply serve as "extension rods" to ensure that the heads project far enough out into the environment. In the second model, the heads may mediate some initial recognition event, but a necessary if not sufficient feature of the adhesion reaction is an interaction between the agglutinin shafts.

That the head is important for adhesivity is indicated by the observation that several biochemical treatments that destroy agglutinin activity in *C. reinhardtii* also destroy the integrity of the head (Goodenough et al., 1985) and by the observation that repeated freeze-thawing of the *mt*⁺ *C. eugametos* proteins causes them to both aggregate by their heads and lose activity (Crabbendam et al., 1986). On the other hand, the fact that the fishbone units (Fig. 4 and 5) interact along their shafts to form the W2 layers of the cell wall (Goodenough and Heuser, 1985, 1988b) argues for the homologous occurrence of shaft-to-shaft interactions between agglutinin fibrils.

Direct visualization of the sexual adhesion reaction by quick-freeze, deep-etch transmission electron microscopy has affirmed that by the time stable adhesions have formed between *C. reinhardtii* gametes, their agglutinin fibrils have interacted shaft-to-shaft to form a meshwork of branching fibers that interconnect the two flagellar membranes (Fig. 8). Whether the heads have participated in setting up this interaction remains an open question, but the importance of fibrillar associations is clear. Our interest in the length, kinks, and other topological features of the shafts has therefore not been misplaced: if the shafts were simply extension rods, these features would be relatively uninteresting, but since the shafts appear to be participating in the adhesion reaction, their unique structures are potentially critical to the process.

There is not enough regularity in the agglutinin meshworks to yield a detailed picture of how the shafts interact, but several other HRGP arrays in *Chlamydomonas* signal the importance of shaft topology. The simplest, illustrated in Fig. 9, is a flagellar-associated structure, of unknown function, whose shape and open weave are reminiscent of a hammock. The hammock monomer is 55 nm long and has one kink. It associates head-to-tail to form a long zig-zag chain with the kinks all facing in one direction. A second chain then aligns laterally with the first such that its heads bind to the kinks, with additional chains adding on to complete the unit.

A more complex interaction is displayed by GP1 monomers when they form

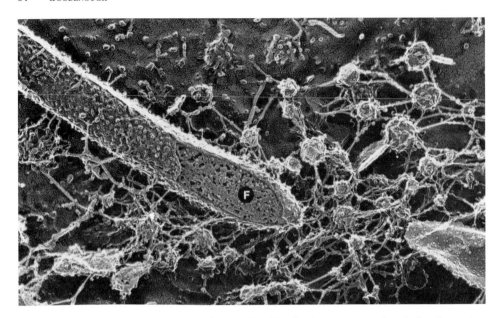

FIGURE 8. Two flagella (one labeled F) of *C. reinhardtii* adhering to one another during the mating reaction. A collection of flagellar membrane vesicles accumulates in the adherent region; these carry agglutinin proteins associated in a complex meshwork. Magnification, ×69,000.

the W6B layer of the *C. reinhardtii* cell wall (Fig. 10). GP1 resembles the hammock monomer but is 100 nm long and carries two kinks (Fig. 6). Here the shafts associate laterally to form dimeric fibers, but the associations are staggered such that the head of one protein contacts the kink of the next (Goodenough and Heuser, 1988a), the result being a crystalline array of double-stranded hexagons.

The last example is given by the W2 layer of the *Chlamydomonas* wall, whose fishbone units have already been considered (Fig. 4 and 5). The ribs of the fishbones interact in an antiparallel fashion, in contrast to the foregoing examples but like the agglutinins, to form a branching three-dimensional meshwork that is very similar to the agglutinin meshworks (Goodenough and Heuser, 1985, 1988b). A key unanswered question is whether this rib-to-rib interaction occurs only between long and short ribs to produce heterodimeric fibers, or whether homodimers (long-long and short-short) also form. If pairing will only occur between long and short ribs, the process is formally equivalent to the obligate pairing of *plus* and *minus* agglutinins.

Taken together, the studies of agglutinin interaction and cell wall assembly in *Chlamydomonas* indicate that the two are equivalent processes, that is, that sexual adhesion occurs by HRGP interactions that were probably first worked out during the evolution of cell-wall assembly. The use of extracellular matrix (ECM) recognition mechanisms to effect fertilization has been conserved throughout the eukaryotes, even as the ECM components themselves have undergone enormous diversification: cell wall glycoproteins are used by yeast gametes (Pierce and Ballou, 1983; Lipke et al., 1989), long-chain carbohydrates by echi-

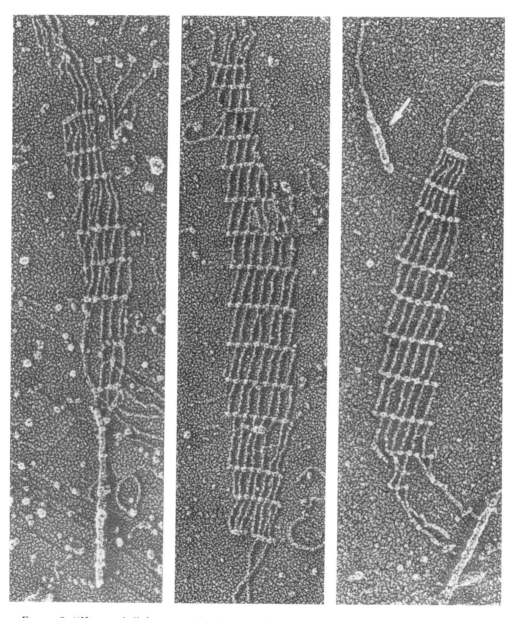

FIGURE 9. "Hammocks" from gametic *C. reinhardtii*. The structure on the left suggests that the hammock rolls up into a thick fiber, possibly forming a mastigonemelike flagellar appendage. Arrow, right panel, points to a fractured portion of a "regular" mastigoneme. Magnification, ×221,625.

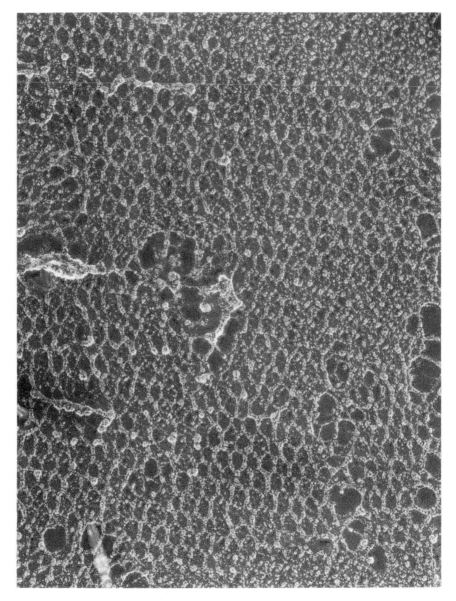

FIGURE 10. W6B layer of the *C. reinhardtii* cell wall displaying a sheet of hexagonal units formed by the GP1 monomer (Fig. 6). Magnification, ×200,000.

noderms (Glabe et al., 1982), arabinogalactan proteins by higher plants (Gell et al., 1986; Pennell and Roberts, 1990), and zona pellucida fibers by mammals (Wasserman et al., 1985; Leyton and Saling, 1989a). This diversification takes us full circle. We began with the premise that sexual recognition molecules should be capable of rapid diversification. We established that *Chlamydomonas* utilizes

ECM molecules to effect sexual recognition, albeit ECM molecules that bind to the cell membrane. And we are now reminded that ECM molecules are, as a class, highly diverse, usually internally repetitive, and usually fibrillar and glycosylated.

The association of agglutinin proteins with ABPs in the flagellar membrane endows them with an additional feature: they acquire the ability to move in the plane of the membrane. As detailed in later sections, the aggregation of agglutinins is thought to play a key role in sexual signalling. Here we can cite a recent paper (Tomson et al., 1990a) indicating that agglutinin aggregation in *C. eugametos* also enhances the sexual binding capacity of the flagellum. While this effect, called "contact activation," is interpreted to occur as a consequence of signalling, the data seem also consistent with the occurrence of some sort of cooperativity: relatively weak binding sites may acquire greater affinity for their (multivalent) ligand when they are clustered together.

SIGNALLING

In most eukaryotes, the initial sexual recognition/adhesion event is spatially and temporally separated from the fusion of gametes to form a zygote (e.g., in plants, pollen-stigma interactions are followed by pollen tube growth; in marine invertebrates, sperm/jelly-coat interactions are followed by the acrosome reaction and jelly penetration). In most eukaryotes also, both sexes participate in the fertilization process with a series of mating responses: e.g., pollen tube growth is either permitted or blocked by activities of the female gametophyte (Nasrallah and Nasrallah, 1986; Cornish et al., 1987), and the marine egg, once fertilized, sets up several blocks to polyspermy (Iwao and Jaffe, 1989).

In *Chlamydomonas,* both of these features of fertilization are operative. Adhesion occurs via flagella, but gamete fusion occurs between specialized domains on the plasmalemma (Fig. 1); therefore, adhesion is always a prelude to fusion, and two distinct parts of the cells are involved. Moreover, both gamete types react to adhesion by a series of mating responses which culminate in cell fusion, responses to be described in detail in later sections.

It became apparent by 1975 (Goodenough and Weiss, 1975; Goodenough, 1977) that a signal must be generated by flagellar adhesion which triggers the cell-body mating responses, a signal that might in theory travel down the flagellar membrane, down the flagellar microtubules, and/or via the soluble flagellar matrix which is presumably continuous with the cytoplasm. Solter and Gibor (1977) developed this concept as well, making the analogy between the *Chlamydomonas* flagella and the sensory cilia found throughout the metazoa.

During the past decade, enormous interest has flourished in cell signalling mechanisms, and two primary effectors have emerged: cyclic nucleotides and Ca^{2+}, the latter often associated with the products of inositol phospholipid metabolism. These subjects are reviewed constitutively (e.g., Levitzki, 1988; Berridge and Irvine, 1989; Ullrich and Schlessinger, 1990), and the role of these effectors in lower-eukaryote signalling is considered in other chapters in this volume. Of particular interest for the *Chlamydomonas* system are recent studies on the sensory cilia of visual, olfactory, and mechanosensory cells (e.g., Wood et al., 1989; Boekhoff et al., 1990; Breer et al., 1990; Dhallan et al., 1990). It is becoming clear that invertebrate photoreceptors transmit light signals via inositol triphosphate (IP_3) whereas vertebrates utilize cyclic GMP (cGMP); moreover, some olfactory stimuli in rat cilia elicit IP_3 production while others elicit cAMP.

Not surprisingly, therefore, research on the nature of the *Chlamydomonas* sexual signal has focused on two notions: the signal might be mediated by Ca^{2+} fluxes or by cAMP fluxes. In addition, much attention has been given to the role of receptor aggregation in the flagellar membrane. The available data, reviewed below, suggest that a Ca^{2+} flux is not serving as a "second messenger"; that cAMP fluxes are central to the process; that membrane protein aggregation induces the cAMP fluxes; and that Ca^{2+} or Ca^{2+}-binding proteins or both are important for coupling aggregation to cAMP generation.

The Calcium Flux Hypothesis

In 1980, Claes reported that the calcium ionophore A23187 induced a key mating response, cell wall lysis, in *C. reinhardtii* gametes, and proposed that Ca^{2+} fluxes might trigger mating responses. A number of investigators have been unable to confirm this observation, and mating responses are not elicited by a second Ca^{2+} ionophore, ionomycin (Pasquale and Goodenough, 1987). Nonetheless, this report stimulated several subsequent studies. Bloodgood and Levin (1983) observed a dramatic increase in ^{45}Ca in the medium when ^{45}Ca-loaded gametes were mated, and suggested that Ca^{2+} sequestered in internal compartments might be released into the cytoplasm during mating and then spill out into the medium. An alternative interpretation of their results, however, is that ^{45}Ca bound to components in the cell wall is released during the wall lysis that accompanies mating. Kaska et al. (1985) produced scanning electron microscopic images showing a redistribution of the internal Ca^{2+} signal during mating, but it is not clear that cellular ultrastructure was consistently preserved in this study.

The major difficulty with the calcium flux hypothesis, first pointed out by Snell et al. (1982), is that Ca^{2+} fluxes are already used by gametes to signal phototactic behavior and flagellar excision. Thus the positive phototactic response is induced by Ca^{2+} influx over the concentration range 10^{-9} to 10^{-7} M (Kamiya and Witman, 1984); the photophobic response is induced by 10^{-6} M Ca^{2+} (Hyams and Borisy, 1978); and flagella detach from the cell at $>10^{-6}$ M Ca^{2+} (Salisbury et al., 1987; Sanders and Salisbury, 1989). In other words, it would appear that general fluxes in $[Ca^{2+}]_i$ are already used by gametes to signal other responses and hence are not "available" to mediate mating-signal transduction. Support for this deduction comes from unpublished experiments in collaboration with R. Crain. Crain and his colleagues have detected a marked increase in IP_3 generation when flagellar excision is induced (Quarmby et al., submitted), but we detect no IP_3 changes during the course of the mating reaction. Experiments with Ca-indicator dyes such as fura-2 are needed to settle this question, but for the present it appears that IP_3-mediated cytoplasmic Ca^{2+} fluxes are not mediators of the agglutination signal in *Chlamydomonas*.

This reasoning has general applicability for microorganisms: a single cell in a particular state of differentiation can have but one pattern of response to a given local concentration of a signalling molecule. An event such as Ca^{2+} influx might well affect a cycling cell differently than a noncycling cell, or a G_1 cell differently than a G_2 cell, but the enormous multiplicity of responses to Ca^{2+} influx displayed by multicellular organisms is founded on the enormous multiplicity of different cell types in these organisms.

In addition to wholesale changes in free Ca^{2+}, cells can also compartmentalize Ca^{2+} by local sequestering, and such ''bound Ca^{2+}'' may well define important cellular activities. In a later section we consider experiments which suggest that perturbations of flagellar Ca^{2+} can result in an inhibition of signalling and hence that Ca^{2+} may play an important role in maintaining a signalling-competent flagellum.

The cAMP Hypothesis

In 1984, an important paper by Pijst et al. (1984b) documented that adhesion between *C. eugametos* gametes resulted in a dramatic 30-fold increase in intracellular levels of cAMP. Moreover, isolated flagella from one mating type elicited the response when adhered to gametes of opposite mating type. We confirmed these observations for *C. reinhardtii* (Pasquale and Goodenough, 1987). Neither set of data, however, proved a direct role for cAMP in sexual signalling, because the cAMP rise could well be some secondary response. In sea urchin sperm, for example, contact with egg jelly elicits a sharp rise in cAMP levels (Garbers and Kopf, 1980), but no role for cAMP in fertilization has yet been demonstrated, and the key signalling event appears instead to be a rise in cytoplasmic pH.

That cAMP indeed functions as the primary sexual signal in *Chlamydomonas* was demonstrated directly by showing that agents that elevate cAMP levels in gametes of a single mating type (e.g., exogenous dibutyryl [db-] cAMP) are able to elicit all known mating responses (Pasquale and Goodenough, 1987; Goodenough, 1989; Snell et al., 1989). Such gametes are not undergoing adhesion; therefore, adhesion must function to generate elevated cAMP levels. In unpublished experiments, D. Kirk has similarly shown that db-cAMP elicits an immediate disintegration of the wall surrounding the sperm packet of *Volvox carteri;* since this is the initial in vivo response when the sperm undergo flagellar agglutination with female spheroids (Kirk and Harper, 1987), it would appear that *Volvox* also uses cAMP as a primary signal. Recently, Kooijman et al. (1990) have shown that a subset of mating responses in *C. eugametos* can be induced by db-cAMP, although they argue for a more complex mode of regulation in this system.

In a subsequent section we consider how elevated cAMP might elicit mating responses. Here we focus on the relationship between adhesion and cAMP generation.

The Receptor-Aggregation Hypothesis

In the 1970s there developed considerable interest in the concept that the cross-linking of surface receptors in lymphocytes, and the resultant ''patching,'' signalled the lymphocytes to differentiate (Schreiner and Unanue, 1976). Since then, numerous reports have documented the importance of ligand-induced membrane rearrangements in cell signalling (Isenberg et al., 1987; Leyton and Saling, 1989b; Ullrich and Schlessinger, 1990), although I am aware of only one report that this occurs for receptors associated with adenylate cyclase (Couraud et al., 1981).

The importance of receptor aggregation for signalling in *Chlamydomonas* was first proposed for *C. reinhardtii* by Claes (1977) and has been extensively expanded (Goodenough and Jurivich, 1978; Mesland and van den Ende, 1978; Mesland et al., 1980; Goodenough et al., 1980; Musgrave and van den Ende, 1987; Kooijman et al., 1989; Kooijman et al., 1990; Pasquale, 1989; van den Ende et al., 1990; Tomson et al., 1990a). The following principles have emerged from these studies. (i) Cross-linking of agglutinins in gametes of a single mating type, using either polyclonal antibodies or MAbs, generates a burst of cAMP and elicits mating responses. (ii) Monovalent (Fab) preparations have neither effect. (iii) *C. reinhardtii* mutant strains lacking agglutinins (Adair et al., 1983; Saito et al., 1985) can be induced to generate cAMP and undergo mating responses when isoagglutinated by polyclonal antisera directed against the flagellar surface; hence the key effect of agglutination is apparently to rearrange and/or bring together the ABPs discussed in an earlier section. (iv) In *C. eugametos* (but not in *C. reinhardtii*) the ABPs can be selectively cross-linked with wheat germ agglutinin, and the cAMP/mating response is again elicited.

Receptor aggregation in *Chlamydomonas* can be analogized to patching in lymphocytes or to hormone-induced receptor oligomerization. These local rearrangements are normally followed by a migration of the aggregated foci to the flagellar tips (Fig. 1 and 11), a phenomenon designated "tipping" (Goodenough and Jurivich, 1978) with obvious analogies to lymphocyte "capping." Agents that block tipping, such as colchicine, vinblastine (Mesland et al., 1980; Homan et al., 1988), and trifluoperazine (TFP) (Detmers and Condeelis, 1986), also block cAMP generation and subsequent mating responses (Pasquale and Goodenough, 1987, and unpublished observation from my laboratory). This could be interpreted to indicate that tipping is necessary for the cAMP response, but it appears instead that tipping requires both receptor aggregation and cAMP generation and these agents inhibit tipping by blocking cAMP generation. Hence, tipping appears to be a response to signal generation rather than an active participant.

The mechanism of tipping is an interesting subject and will be considered separately in a later section. First we consider what is known about coupling receptor aggregation to cAMP generation.

Receptor-Cyclase Coupling

Gametic flagellar membranes possess an adenylate cyclase of high specific activity (Pasquale and Goodenough, 1987), and it is our working hypothesis, adopted in the following discussion, that this cyclase is stimulated in response to ABP aggregation. Direct evidence for this stimulation has recently been obtained in my laboratory (Saito et al., 1991). A second alternative is that most of the stimulation affects an adenylate cyclase located in the cell bodies, in which case

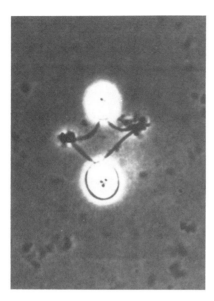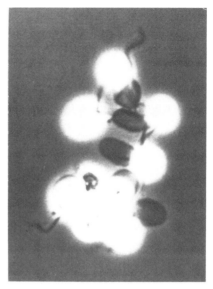

FIGURE 11 [left]. Pair of mt^+ gametes of *C. reinhardtii* isoagglutinated by a monoclonal antibody directly against a flagellar surface carbohydrate epitope (Adair, 1985). The antibody complexes have migrated to the flagellar tips ("tipping"). Magnification, ×1,000.

FIGURE 12 [right]. Group of mt^+ gametes of *C. reinhardtii* isoagglutinated as in Fig. 11. As a consequence of signal transduction, cell walls (dark oblongs) are released from the cell bodies. Magnification, ×1,000.

coupling would necessitate more indirect interactions than those considered here; again, we have recently found that the cell body enzyme is stimulated during mating (Saito et al., 1991). And finally, agglutination might inhibit cAMP phosphodiesterase activity in the flagellum and/or the cell bodies, as suggested by some preliminary experiments (Pasquale, 1989).

Three general mechanisms have been described in the metazoa for coupling ligand binding to cyclase activation. The first entails the mediation of G_s proteins, which also operate in lower eukaryotes (Levitski, 1988; Kumagai et al., 1989; Kurjan, 1990). While there is molecular evidence for G-protein homologs in *Chlamydomonas* (Schloss, 1990) and in higher plants (Ma et al., 1990), no G-protein-mediated signal transduction has yet been demonstrated in green organisms, and we have as yet found no evidence for a G-protein modulation of flagellar adenylate cyclase activity (Pasquale and Goodenough, 1987).

The second known activator of adenylate cyclase is Ca-calmodulin, which stimulates several enzymes in the metazoa (MacNeil et al., 1985; Coussen et al., 1985; Grost et al., 1987) but has not been shown to be directly active in protozoa (Janssens, 1988) despite an observed Ca^{2+} stimulation in certain systems (Gustin and Nelson, 1987). We have thus far failed to detect any Ca^{2+} effects on the *Chlamydomonas* flagellar adenylate cyclase in vitro.

In the third mode of coupling, the receptor itself is a cyclase, and ligand binding induces a conformational change which stimulates enzyme activity directly (Singh et al., 1988; Chinkers and Garbers, 1989; Chinkers et al., 1989). Such receptors have thus far been guanylate cyclases and have been found only in metazoa.

With no information about coupling mechanisms in *Chlamydomonas*, we can nonetheless list a number of "uncoupling reagents" which have no effect on adhesion, which inhibit signalling and cAMP generation in vivo, and which fail to inhibit the flagellar adenylate cyclase in vitro. These include lidocaine (Snell et al., 1982), TFP and W-7 (Detmers and Condeelis, 1986; Pasquale and Goodenough, 1987), colchicine and vinblastine (Mesland et al., 1980; Homan et al., 1988), and cadmium and lanthanum (unpublished observations from this laboratory). The effects of these inhibitors can be reversed by supplying gametes with db-cAMP, indicating that they are indeed affecting some coupling event(s). Moreover, the lidocaine, colchicine, and vinblastine effects can be reversed by exogenous Ca^{2+} (Snell et al., 1982; unpublished data). Since TFP and W-7 bind to calmodulin and calmodulin is abundant in the flagellum (Gitelman and Witman, 1980), and since Cd^{2+} and La^{3+} are calcium antagonists, these results collectively suggest a need for some Ca-dependent component in the coupling event. The recent discovery of stretch-activated ion channels in diverse cell types, and their possible regulation of voltage-sensitive Ca^{2+} channels (Saimi et al., 1988; Morris and Sigurdson, 1989; Martinac et al., 1990), suggests the appealing notion that adhesive cross-links might activate such channels in the flagellar membrane, but there is as yet no evidence for this. Therefore, at this stage we can only state definitively that it is possible to experimentally uncouple adhesion from cAMP generation in this system by a variety of reagents and that identification of the coupling mechanism(s) represents a high research priority in this laboratory.

PRIMARY RESPONSE TO cAMP ELEVATION

There are two basic eukaryotic responses to elevated cAMP levels: activation of cAMP-dependent protein kinases (Hoppe, 1985) and the opening of cAMP- (or cGMP-) gated monovalent cation channels (Cobbs and Pugh, 1985; Nakamura and Gold, 1987; Yau and Baylor, 1989). Since the mating reaction can occur in distilled water and shows no dependency on external monovalent cations, the channel hypothesis is not obviously applicable. In contrast, the mating reaction is inhibited by H-8, an inhibitor of cAMP-dependent protein kinase activity (Hidaka et al., 1984; Pasquale and Goodenough, 1987). Direct evidence for a specific stimulation of cAMP-dependent protein phosphorylation is currently being sought in this laboratory. Functional identification of any mating-induced phosphoproteins will then be the next goal. They may include such signal-generating substrates as the ABP (compare Benovic et al., 1985) or the cAMP phosphodiesterase (compare Grant et al., 1988), or they may represent proteins that participate directly in the mating responses. These responses, i.e., tipping and flagellar tip activation, cell wall lysis, mating structure activation and cell fusion, and flagellar disadhesion, are considered in the next four sections of this review.

TIPPING AND FLAGELLAR TIP ACTIVATION

Tipping (Goodenough and Jurivich, 1978) was introduced in the preceding section as the migration of cross-linked agglutinins (Mesland and van den Ende, 1979; Goodenough et al., 1980; Homan et al., 1987) and ABPs (Kooijman et al., 1989) to the flagellar tips during the mating reaction (Fig. 11). Several additional kinds of flagellar surface translocations have also been described for *Chlamydomonas* (reviewed by Bloodgood, 1990). In this section an attempt is made to integrate these phenomena.

Surface translocations along the *Chlamydomonas* flagellum occur in three fashions: retrograde (towards the basal body), bidirectional, and anterograde (towards the tip).

Retrograde Motility

Retrograde motility occurs during gliding (Lewin, 1954a): a gliding cell attaches to the substrate "nose down," such that its flagella are oriented 180° from one another, and then moves either to the right or to the left, its lead flagellum always making the force-generating contact with the cell surface (Bloodgood, 1981, 1990). Since the two flagella are mirror-symmetric (Hoops and Witman, 1983), this means that gliding always occurs over doublets 9, 1, and 2 of the axoneme and the direction of force is always retrograde. Gliding rates are rapid, ~2 μm/s.

Bidirectional Motility

Bidirectional motility is observed when large particles, such as polystyrene beads or bacteria, associate with the flagellar surface (Bloodgood, 1977). The particles move past one another as if on tracks, tracks that are assumed but not proven to be defined by the underlying axonemal doublets, and the rates (~2 μm/s) are comparable to gliding rates. A particle may transverse the entire flagellar length (10 to 12 μm) in a single smooth motion but more often will stop partway and then continue in either the same or reverse direction. Each attached particle moves independent of all others. Moreover, a gliding cell engaged in retrograde movement can simultaneously support the bidirectional movement of particles attached to the upper surfaces of its flagella.

Anterograde Motility

Anterograde motility occurs when flagellar surface molecules are cross-bridged. When these molecules are the gametic agglutinins or their postulated ABPs, their tipping is "exclusive": other membrane proteins are not "co-tipped" (Homan et al., 1987; Kooijman et al., 1989). It is possible, on the other hand, to induce the tipping of other membrane proteins by cross-linking them with antibodies or lectins, in which case they too will move to the tips (Bloodgood et al., 1986). In all cases involving antibodies or lectins, monovalent derivatives fail to migrate to the tips. Hence it would seem that the stimulus for anterograde movement is a local, intermolecular cross-linking, and that such cross-linking is not induced by the more global surface association of polystyrene particles.

Two occasions have been described wherein polystyrene beads are nonetheless tipped: the first is during the natural mating reaction (Hoffman and Goodenough, 1980); the second is when cells of one mating type are treated with db-cAMP (Goodenough, 1989). In both cases the cAMP levels are elevated; hence the notion arises that the same motors that mediate bidirectional translocation might be converted to "anterograde-only" motors by a cAMP-dependent phosphorylation event. An analogous phenomenon has been described for motors in fish chromatophore cells: cAMP-dependent phosphorylation transforms the pigment-translocating motors into anterograde-only motors (Rozdzial and Haimo, 1986; Lynch et al., 1986) such that the pigment disperses. Another interpretation for such phenomena must be offered as well, however: cAMP might cause motors and/or their cargo to become trapped at the flagellar tips or at the chromatophore periphery when they arrive, rather than inducing their biochemical modification to an anterograde state.

Whether by trapping or by exclusive anterograde movement, another event associated with gametic tipping is known as flagellar tip activation (FTA) (Mesland et al., 1980; Crabbendam et al., 1984; Snell, 1981), wherein a dense material accumulates beneath the flagellar membrane and the A-microtubules elongate and fill with an electron-opaque material. FTA occurs in response to flagellar agglutination via agglutinins, concanavalin A, polyclonal antisera, and, in *C. eugametos*, wheat germ agglutinin (Kooijman et al., 1989). Curiously, it does not occur in response to the tipping of the MAb 66.3 antibody specific for the *mt⁻* agglutinin of *C. eugametos* (Homan et al., 1988), although no thin sections were analyzed in this study. Whether FTA occurs as a consequence of antibody tipping by vegetative cells is an important unanswered question. Of particular interest is the observation that an FTA response is elicited when gametes of one mating type are treated with db-cAMP (Pasquale and Goodenough, 1987), even though there is no evidence of agglutinin migration or tipping (Goodenough, 1989; Kooijman et al., 1990).

To integrate these observations, we propose that flagella surface phenomena occur in three stages: (i) ligand attachment induces (ii) motor association with attached domains and hence (iii) translocation of attached domains. If ligand attachment does not involve close intermolecular cross-links, then a cAMP rise is not induced and the attached motors support bidirectional translocation. If the ligand instead effects cross-linking and hence adenylate cyclase activation, then the attached domains are subject to anterograde influences and move to, or are trapped at, the tips. If db-cAMP is administered without any cross-linking of surface domains, motors would be expected to accumulate in the tips even though "empty handed," and FTA might be the morphological consequence of such accumulation.

The notion that flagellar surface phenomena entail three discrete steps is useful in analyzing the effects of various reagents on the system. Thus Bloodgood and Salomonsky (1990) find that micromolar levels of extracellular Ca^{2+} are required for anterograde movement of concanavalin A or antibody by vegetative flagella and that this movement is prevented by TFP, W-7, Ba^{2+}, lidocaine, and calcium-channel blockers. Several of these agents, as noted earlier, also prevent cAMP elevation during mating, suggesting that they might in both cases interfere with the mechanism which couples cross-linking to cAMP elevation. Supporting this is the observation that db-cAMP reverses the TFP inhibition of antibody tipping (Bloodgood, 1990). The kinase inhibitor H-8 also disrupts both gametic and vegetative-cell tipping, and as expected, this is not reversed by db-cAMP (Pasquale and Goodenough, 1987; Bloodgood, 1990). Thus we are once again brought back to the concept of some "upstream" Ca-associated event(s) at the flagellar surface as important for modulating external cross-linking signals, and a "downstream" elevation of cAMP levels as a consequence of effective signal transduction. Since Ca depletion, lidocaine, and TFP also inhibit the bidirectional translocation of polystyrene microspheres (Bloodgood et al., 1979; Snell et al., 1982; Detmers and Condeelis, 1986), these agents may additionally interfere with the association of motors with occupied surface domains.

Finally, we note that colchicine and vinblastine have been observed to inhibit the contact activation of agglutinins during mating (Tomson et al., 1990a), an effect interpreted as a block in signal transduction mechanisms. In fact, these agents may also have the effect of preventing surface cross-linking, which would explain their inhibition of cAMP generation, tipping, and FTA in gametic cells (Mesland et al., 1980; Homan et al., 1988; and unpublished results from this laboratory). Consistent with this reasoning, neither colchicine nor vinblastine inhibits bidirectional microsphere movement (Bloodgood et al., 1979; Hoffman and Goodenough, 1980), which, we propose, occurs independently of intermolecular cross-linking. Inconsistent with this reasoning, however, is the observation that colchicine fails to inhibit anterograde antibody redistribution in vegetative cells (Bloodgood, personal communication).

In summary, then, we postulate the following two sequences. (i) Polystyrene bead association induces the attachment of bidirectional motors to the large object, an attachment that requires external Ca^{2+} and is inhibited by lidocaine, TFP, and chymotrypsin digestion (Hoffman and Goodenough, 1980). (ii) Surface cross-linking induces the attachment of the same motors, which therefore also requires Ca^{2+} and is inhibited by lidocaine, TFP, and trypsin. Some aspect of surface cross-linking may also be sensitive to colchicine and vinblastine, explaining their inhibition of subsequent events. The key subsequent event is the activation of adenylate cyclase; the resultant rise in cAMP generates motors that move anterograde only. It is important to keep in mind that some of the drugs employed in these studies may well inhibit at several points: for example, TFP, which is known to bind to a number of molecules besides calmodulin (reviewed by Bloodgood, 1990), might inhibit both motor attachment and cyclase activation, possibly but not necessarily by binding to the same targets. It is also important to stress that whereas it is attractive to imagine that cAMP converts bidirectional motors to anterograde-only motors in this system, the two types of motors may in fact be unrelated, with cAMP stimulating a motor which is otherwise inactive. Clearly none of these questions can be settled until the relevant proteins are identified and characterized.

THE WALL-LYSIS MATING RESPONSE

The first visible cell body response to flagellar adhesion or cAMP presentation is the lysis of the cell wall (reviewed by Adair and Snell, 1990). In *C. reinhardtii,* this response is readily monitored by phase or differential interference contrast (DIC) microscopy: within seconds after *plus* and *minus* gametes adhere, cell walls are seen to slip off the gametes and float free in the medium (Fig. 12). The enzyme activity responsible for this effect—originally termed autolysin (Claes, 1971) and more recently shortened to lysin (Snell, 1985)—is sensitive to EDTA (Matsuda et al., 1984), so the process of wall dissolution can be temporally resolved by adding EDTA at various times.

There are clearly two primary targets of lysin activity. The first is the fibrous matrix of the W1 layer, which anchors the wall to the plasmalemma (Roberts et al., 1972; Goodenough and Heuser, 1985); the second is the association between the wall layers and the flagellar collars, two cylindrical structures that form tunnels within the wall through which the flagella pass (Imam and Snell, 1988; Goodenough and Heuser, 1988a). Once the plasmalemma/wall and the collar/wall connections are severed, the collars slip up and off the flagellar shafts while the wall proper slips back and off the cell body. Following these two primary events, the shed cell walls continue to disassemble, the key event being the breakdown of the insoluble W2 layer (Matsuda et al., 1985; Goodenough and Heuser, 1985; Imam and Snell, 1988) to generate "fishbones" (Fig. 4 and 5), structures considered earlier in a different context. All of the events of wall lysis can apparently be affected by a single metalloenzyme with a subunit molecular size of 60 kDa (Matsuda et al., 1984; Matsuda et al., 1985). Therefore, either all three targets (W1/membrane, collar/wall, and W2) are biochemically equivalent, or else the lysin enzyme carries out more than one catalytic activity (see Jaenicke et al., 1987; Waffenschmidt et al., 1988).

In *C. eugametos,* only one of these three targets is a substrate for its lysin. Sexual adhesion causes an apical opening to develop in the walls of mating gametes, but the W1/plasmalemma connectives remain patent, the W2 layer remains intact, and the bulk of the wall continues to surround the mating cells (Brown et al., 1968; Triemer and Brown, 1975; vanWinkle-Swift et al., 1987). Applying our rule about interpreting *C. eugametos/C. reinhardtii* differences, then, it can be argued that their common ancestor produced a lysin targeted for the wall apex; whether the ancestor proceeded with total wall dissolution (subsequently lost by *C. eugametos*) or whether the *C. reinhardtii* line "invented" total wall dissolution cannot be decided.

There is an interesting homology between the cell apex of *C. eugametos* and that of *C. reinhardtii:* both possess flagellar collars that are very similar in their crystalline structure (Goodenough and Heuser, 1988b). The evolutionary thread is not a simple one, however, because whereas this collar is clearly a target of the *C. reinhardtii* lysin (Imam and Snell, 1988), the *C. eugametos* lysin digests the so-called papilla region between the collars, leaving the collars in place (Brown et al., 1968; vanWinkle-Swift et al., 1987). In any case, divergence has occurred to the extent that the *C. reinhardtii* lysin has no lytic effect on the *C. eugametos* apex (Schlösser, 1981) and is immunologically unrelated to any *C. eugametos* proteins (Waffenschmidt et al., 1989).

The question of whether the papilla alone or the entire wall is released in response to sexual adhesion may appear to be a rather baroque point of concern, but in fact it highlights a key distinction between the *C. reinhardtii* and *C. eugametos* groups: not only are the *C. reinhardtii* gametes "naked," but the resultant quadriflagellated zygotes remain naked for about an hour before they secrete their zygotic cell walls (Cavalier-Smith, 1976). In contrast, the *C. eugametos* line never makes this leap, instead retaining its gametic cell wall until the primary zygotic cell wall is laid down beneath it (Brown et al., 1968). While it seems maladaptive for a soil organism to have an obligate naked stage in its life cycle, a useful consequence for biological research is that *C. reinhardtii* protoplasts are fully viable, and cell-wall-less mutants are readily obtained (Davies and Plaskitt, 1971); by contrast, *C. eugametos* seems inexorably embedded in its chainmail-like wall. Interestingly, two aquatic green algae, *Hydrodictyon* (Marchant and Pickett-Heaps, 1972) and *Ulva* (Bråten, 1971; Melkonian, 1980), also generate naked gametes.

Returning now to the main thread of this review, we can ask what is known about the release of lysin in response to the primary cAMP signal. Recent experiments by Snell and co-workers (Buchanan et al., 1989; Snell et al., 1989) have revealed an interesting mechanism for *C. reinhardtii*. An antibody specific for lysin recognizes two forms of the enzyme, an inactive prolysin of 62 kDa and an active lysin of 60 kDa. The prolysin localizes to the periplasmic (W1) layer of the wall (Milliken and Weiss, 1984; Matsuda et al., 1987). Mating releases an activity called *p*-lysinase which effects the 62-to-60-kDa clip, activating the lysin and allowing it to go into solution. Release of *p*-lysinase can also be induced by incubating cells with db-cAMP. Whether *p*-lysinase is actually secreted from the cell by exocytosis or is instead also a periplasmic enzyme subjected to yet another level of activation remains to be established. In any case, the chain of activation presumably initiates within the cell in response to elevated cAMP, and this system appears ideally suited to working out the biochemistry of a cAMP-mediated effect.

Lysin activity is also stimulated in response to the cAMP signal in *C. eugametos* (Kooijman et al., 1990) and in *Volvox* (Kirk, unpublished; see Waffenschmidt et al., 1990), but nothing is yet known about the mechanisms involved.

THE MATING STRUCTURE ACTIVATION RESPONSE

The second visible cell body response to flagellar adhesion or db-cAMP presentation is mating structure activation. The activation events are clearly homologous in *C. reinhardtii* and *C. eugametos*, but more detailed information is available for *C. reinhardtii* (Friedmann et al., 1968; Goodenough and Weiss, 1975; Weiss et al., 1977; Forest et al., 1978; Goodenough and Jurivich, 1978; Goodenough et al., 1982; Detmers et al., 1983; Forest, 1983, 1987; Weiss, 1983).

The mating structures in *C. reinhardtii* are differentiated domains of the plasmalemma that are found only in gametes. The cytoplasmic face of the mating structure membrane carries a narrow, electron-dense material called the membrane zone; the external face of the membrane carries a thick carbohydrate coat called the fringe. The organelle is circular in shape, resembling a small saucer (~400 nm in diameter), and the saucer center carries a cluster of intramembranous particles believed to be involved in fusion between gametes.

The mating structures occupy fixed locations in the *C. reinhardtii* cell. The mt^- organelle is located at the cell apex on the same side of the flagellar basal bodies as is the eyespot (the so-called *cis* half of the cell), whereas the mt^+ organelle is located in the opposite, *trans* position (Holmes and Dutcher, 1989). Both positions are presumably defined by the microtubule "roots" which radiate out from the basal apparatus. One set of roots contacts an edge of the mating structure via several fibrous connectives (Goodenough and Weiss, 1978). Therefore, the mt^+ locus apparently specifies that the mt^+ mating structure will form in association with a *trans* root whereas the mt^- locus specifies the *cis* root. Holmes and Dutcher (1989) speculate that this arrangement ensures that the gametes fuse with a fixed polarity such that their two eyespots wind up on the same side of the zygote and phototactic steering mechanisms (Kamiya and Witman, 1984) are not compromised.

The mt^+ mating structure possesses a so-called doublet zone beneath its membrane zone which is not found associated with the mt^- mating structure. In response to mating or elevated cAMP levels, the doublet zone serves as a nucleating site for actin polymerization, the result being the rapid erection of a microvillar fertilization tubule (Fig. 1) with striking structural similarity to the acrosome of marine sperm (Tilney and Kallenbach, 1979). The tip of the fertilization tubule continues to carry the fringe and the presumptive fusogenic particles. Growth of the fertilization tubule normally ceases when its tip contacts, and then fuses with, the mt^- mating structure. If no mt^- gametes are present (e.g., if mt^+ gametes are activated with db-cAMP), the fertilization tubule grows out for several micrometers and becomes progressively tapered at its tip as if the pool of actin monomer were limiting. Membrane for tubule outgrowth is possibly provided by coated vesicles that accumulate in the region (Weiss, 1983).

The mt^- mating structure also undergoes structural changes in response to mating/elevated cAMP, albeit less dramatic. The region puffs out to form a small bulbous protrusion, with cytoplasm filling the outpocketing rather than actin. If no mt^+ gametes are present, these protrusions become enlarged, but they remain difficult to resolve by light microscopy (fertilization tubules can be readily seen using DIC optics [Pasquale and Goodenough, 1987]).

Two sets of mutant strains have permitted dissection of the fusion reaction into two stages, mating-structure adhesion and mating-structure fusion (Forest, 1987). (i) The mt^+ mutants *imp-1* and *imp-11* (Goodenough et al., 1976; Galloway and Goodenough, 1985) carry no detectable fringe on the tips of their fertilization tubules, and neither is able to make any stable cell body contacts with mt^- gametes even though all other mating-related responses appear to occur normally (Goodenough et al., 1982). (ii) The mt^- mutant *gam-10* is temperature sensitive for fusion; at restrictive temperature its mating structures adhere to the tips of fertilization tubules but no fusion ensues, so that the pairs remain tethered even when flagellar adhesion is experimentally disrupted, and fuse only when the temperature is shifted down (Forest, 1983).

The conversion of mating structures from inactive organelles to actively adhesive/fusogenic organelles is dependent on elevated cAMP levels, as perhaps most dramatically documented by the following experiment (Pasquale and Goodenough, 1987). We removed the cell walls from two nonagglutinating mutant strains, *imp-2 mt*⁺ and *imp-12 mt*⁻, mixed them together, and as expected, observed no adhesion and no cell fusion. A second group was first activated with db-cAMP and then mixed. Again no flagellar adhesion took place, but within minutes we observed numerous pairs conjoined at their apical ends and then fully fused as zygotes; within 40 min after mixing, 75% of the cells had fused. This outcome was quite surprising. We had assumed that the correct apposition of activated mating structures required that the gametes be immobilized by flagellar adhesion, but in fact, rapidly swimming gametes can nonetheless make effective adhesive contacts via their mating structures alone.

The mating structure fringe is apparently responsible for the *mt*-specific adhesion (activated *mt*⁺ cells do not fuse with each other, nor do activated *mt*⁻ cells). The fringe material does not look at all like agglutinin by deep-etching (Goodenough and Heuser, unpublished); it is short (~25 nm) and pliant. Moreover, mutations affecting agglutinin production have no effect on fringe and vice versa (Goodenough et al., 1982), and tunicamycin inhibits fusability but not agglutinability in *mt*⁻ gametes (Matsuda et al., 1982). We conclude, therefore, that *Chlamydomonas* has evolved two independent adhesion systems, one involving HRGPs and one involving a material that does not resemble HRGPs morphologically (biochemical information on this material is very much needed).

In the *C. eugametos* group, no mating structures have been identified, although candidate dense domains are evident in the micrographs of Brown et al. (1968). Both mating types produce apical papillae which fuse in a far more symmetrical fashion than is observed with *C. reinhardtii*. Nonetheless, one of the two papillae in mating pairs appears to contain a fibrous material which is lacking in the other (Triemer and Brown, 1975), and one cell in each pair is often observed to send out a papilla before the other (Mesland, 1976), suggesting a process akin to fertilization tubule formation. The actinlike nature of this fibrous material is particularly conspicuous in the micrographs of Brown et al. (1968) and vanWinkle-Swift et al. (1987). The major difference, then, is that the cytoplasmic bridge forms medially rather than to one side of the basal apparatus; the careful specification of mating-structure asymmetry in *C. reinhardtii* described above (Holmes and Dutcher, 1989) would thus appear to be a later evolution.

Associated with the formation of a medial cytoplasmic bridge is the unique formation by the *C. eugametos* group of vis-à-vis pairs: the bridge remains patent for several hours, the *mt*⁻ flagella become paralyzed, and the tandem cells are propelled by the two *mt*⁺ flagella for several hours before they withdraw and zygote maturation proceeds (Wiese, 1965). In contrast, the asymmetric cytoplasmic bridge of *C. reinhardtii* pairs is a transient structure which quickly opens out to allow full cytoplasmic confluence (Fig. 1); both pairs of flagella continue to beat, and the young zygote swims as a quadiflagellated cell for several hours before flagellar withdrawal (Friedmann et al., 1968).

A dramatic example of structural conservation in the green algae is that both *Hydrodictyon* (Marchant and Pickett-Heaps, 1972) and *Ulva* (Melkonian, 1980) gametes possess unambiguous mating structures. The organelles in *Hydrodictyon* are medial, like the putative *C. eugametos* structures (Brown et al., 1968), and extend to form papillae at conjugation; however, fusion generates quadiflagellated cells rather than stable vis-à-vis pairs. In *Ulva* the mating structures are asymmetrically placed over two-membered flagellar roots as in *C. reinhardtii* (Goodenough and Weiss, 1978), and again quadiflagellated cells are formed (vis-à-vis pair formation may well require that the gametes retain their cell walls, which is not the case for either *Hydrodictyon* or *Ulva*).

THE AGGLUTININ INACTIVATION/MOBILIZATION RESPONSE

The final mating response that has been characterized in *Chlamydomonas* involves both the cell body and the flagellum and hence can bring together our consideration of the two. Lewin (1952) was the first to report that once *Chlamydomonas* gametes fuse, their flagella become nonadhesive (Fig. 1). Snell and his colleagues (reviewed by Snell, 1985) went on to show that when mt^+ gametes adhere to isolated mt^- flagella (or the reciprocal), the isolated flagella lose their adhesiveness whereas the gametes retain theirs and can stick to fresh rounds of flagella. Snell and Moore (1980) put these observations together by showing that in matings involving the *imp-1 mt$^+$* mutant that cannot fuse, a replacement of inactivated agglutinins takes place during the prolonged adhesion reaction. Thus the concept developed that during a normal mating, agglutinin inactivation is the normal consequence of adhesion; that inactivated agglutinins are replaced by a cell body pool of agglutinins and, in prolonged *imp-1* matings, by a cycloheximide-sensitive synthesis of new agglutinins; and that when sexual fusion takes place, the replacement process ceases and the flagella ipso facto disadhere. A similar turnover has been described for *C. eugametos* (Wiese et al., 1984; Pijst et al., 1984a), and it also occurs when agglutinins are removed from the flagellar surface by EDTA (Saito and Matsuda, 1984a).

A major gap in our understanding of the *Chlamydomonas* mating reaction is that we do not understand the mechanism of agglutinin inactivation. Inactivated agglutinins from *C. eugametos* comigrate with active agglutinins by sodium dodecyl sulfate-polyacrylamide gel electrophoresis (Pijst et al., 1984a), and no evidence has been found for modifying enzymes or other such activities. Indeed, it is not even clear whether the agglutinins are inactivated or whether they are simply released from the flagellar surface after participating in adhesion for a short period of time. Thus Tomson et al. (1990a) suggest a weakening of agglutinin binding to the membrane as a consequence of adhesion, and Snell et al. (1986) have described a MAb which apparently does not bind to the agglutinins per se but has the interesting property of causing mt^+ but not mt^- agglutinins to release from the flagellar surface. Electron microscopy reveals the formation of abundant flagellar membrane vesicles during adhesion (Fig. 8), but this process cannot be invoked to explain the inactivation of glutaraldehyde-fixed flagella by gametes of the opposite mating type.

Two recent sets of experiments have, on the other hand, enhanced our understanding of the replacement process. (i) Tomson et al. (1986) and Demets et al. (1988) observed mating interactions between *C. eugametos* gametes and concluded that the initial collisions between gametes stimulated the agglutinability of both partners by about four- to eightfold. (ii) Goodenough (1989) found that treating *C. reinhardtii* gametes of one mating type with db-cAMP causes their entire cytoplasmic pool of agglutinins to be mobilized to the flagellar surface, resulting in an eightfold increase in agglutinability, an effect also observed for *C. eugametos* (Kooijman et al., 1990) and confirmed for *C. reinhardtii* (Hunnicutt et al., 1991). The following scenario therefore emerges. Unmated *Chlamydomonas* gametes apparently display relatively few agglutinins on their flagellar surfaces, retaining most of their presynthesized proteins in a cytoplasmic pool. Initial adhesive contacts with the opposite mating type not only enhance the adhesiveness of the resident agglutinins, apparently because of their microaggregation (Tomson et al., 1990a), but also generate a rise in cAMP which induces the mobilization of some of the pool agglutinins to the flagellar surface, thereby permitting more adhesive contacts, more cAMP, and a positive-feedback loop that eventually mobilizes the entire pool. This process takes minutes, and mating is usually completed in minutes. Therefore, the cycle can be viewed as serving several purposes: agglutinins are not wastefully displayed until a partner is encountered; the initial contacts are quickly reinforced with fresh agglutinin; whatever inactivation is built into the system is automatically countered by replacement until fusion is accomplished; and the limited pool size assures that the resultant zygotes are rendered free-swimming.

To this scenario must, however, be added the likelihood that cell fusion per se has some additional effects. When cell fusion is prevented by the *imp-1* mutation and the limited cytoplasmic pool is exhausted, the gametes do not disadhere as the above scenario would predict; instead, they initiate an active synthesis of new agglutinin in a cycloheximide-sensitive fashion. This is clearly an unnatural situation, but it points to the occurrence of an important signal exerted by cell fusion which dictates "stop replacement, time to disadhere."

The cytoplasmic pool has been interpreted to reside either on the cell surface (Pijst et al., 1983; Musgrave et al., 1986) or in the cell interior (Saito et al., 1985). Recent experiments (Hunnicutt et al., 1991) localize at least some of the pool to the plasmalemma: when deflagellated cells are surface labeled and washed, the regenerated flagella carry labeled agglutinins. The cell surface agglutinins are not adhesive, and they are prevented from migrating to the flagellar surface until the flagella population is depleted or a cAMP signal is generated. Why the cell surface agglutinins are nonadhesive is mysterious, but when we consider such unexpected phenomena as the light activation of *C. eugametos* agglutinins (Kooijman et al., 1988), we can quickly conclude that much remains to be understood about these proteins.

THE CELL FUSION SIGNAL

Although perhaps beyond the scope of this volume, brief consideration should be given to the profound consequences of gametic cell fusion (Fig. 1). Not only is disagglutination triggered and, hence, adenylate cyclase activation reversed (Pasquale and Goodenough, 1987); there also occurs, in *C. reinhardtii* at least, an activation of novel gene transcription (Ferris and Goodenough, 1987). The new transcripts are certainly present within 10 min after cells are mated, indicating that they probably start being synthesized as soon as the cells fuse; moreover, they never appear in a mating involving *imp-1,* indicating that cell fusion per se is required. They appear long before the gametic nuclei fuse (this usually occurs after 1 h of zygotic fusion [Cavalier-Smith, 1976; Minami and Goodenough, 1978]), and they quickly become very abundant. We have obtained partial or complete nucleotide sequences for eight of these; one is a clear-cut HRGP, with repeating Pro-Ser motifs, and a second contains such motifs (Woessner and Goodenough, 1989). Our working hypothesis, therefore, is that some of these transcripts encode the zygote cell wall proteins that are quickly secreted around the naked *C. reinhardtii* zygote (Minami and Goodenough, 1978; Grief et al., 1987). Other zygote-specific genes, at least one of which is linked to *mt,* may dictate such events as nuclear fusion and flagellar regression. An obvious future research goal is to determine how the fusion of mt^+ and mt^- cytoplasms can so quickly activate this set of genes.

While cloning experiments have not yet been initiated with *C. eugametos,* cell fusion signals are clearly operative here as well. Fusion also dictates flagellar disadhesion (Mesland and van den Ende, 1978) and the synthesis of a primary, and then a secondary, zygote cell wall (Brown et al., 1968). In the related homothallic species *C. monoica,* vanWinkle-Swift et al. (1987) have shown that a mutation that permits cell fusion but blocks nuclear fusion has no effect on spore wall assembly, documenting that here too, cytoplasmic confluence and not nuclear fusion represents the critical event in turning on the "zygote program."

THE MATING-TYPE LOCI

We come then to the heart of the matter: why do *Chlamydomonas* cells set up and execute this elaborate mating ritual before they can begin synthesizing their highly adaptive zygote cell wall? Surely asexual sporulation in response to stress, as practiced by many prokaryotes, would seem to be the way to go. I have pondered this question elsewhere (Goodenough, 1985) without notable insights. Since it has become useful when presented with evolutionary enigmas (e.g., the origin of introns) to invoke the notion of "selfish DNA," the same kind of thinking can be applied here, although perhaps quite inappropriately.

It is clear in *C. reinhardtii* that the development of the gametic state is directly controlled by the *mt* locus: all identified genes that directly affect gametic phenotypes are either linked to *mt* or are regulated in *trans* by *mt* (Goodenough and Ferris, 1987). Indirect evidence for such a system has been obtained as well for the *C. eugametos*-like *C. monoica* (vanWinkle-Swift and Hahn, 1986). It is also clear for both groups of *Chlamydomonas* that the proper development and germination of the zygote are dependent on *mt*-linked or *mt*-controlled genes and that *mt* heterozygosity is required for the expression of the zygote program (reviewed by Goodenough and Ferris, 1987). The selfish-DNA argument, then, holds that a "selfish" *mt* locus gained control of both gametogenesis genes and spore formation genes and coupled them in series such that mating and heterozygosity had to occur before the adaptive sporulation could take place. While the mechanism of this takeover is quite difficult to imagine, the notion is put forward here if only to concisely summarize, and stress, the key role played by *mt* in setting up all the activities described in this article.

We have become comfortable, in the past few years, with the notion of a few "master genes" in *Drosophila* controlling all of early fly development. An analogous master gene complex clearly governs the key life cycle transitions in *Chlamydomonas*. Whether the analogy can be extended to the level of homology awaits a molecular characterization of the *Chlamydomonas mt* locus, a major project under way in this laboratory.

ACKNOWLEDGMENTS. This review is dedicated, with deep affection, to Lutz Wiese (1921–1989) and W. Steven Adair (1946–1990), two pioneers in the study of *Chlamydomonas* sexual agglutination. Wiese first identified the agglutinin proteins and Adair first purified them; both pursued their science with skill, imagination, and flair.

Research from my laboratory is supported by grants from the National Institutes of Health, National Science Foundation, and U.S. Department of Agriculture. Warm thanks are extended to John Heuser for the electron micrographs that illustrate this review and to Irma Morose for her skillful preparation of the manuscript.

REFERENCES

Adair, W. S. 1985. Characterization of the *Chlamydomonas* sexual agglutinins. *J. Cell Sci.* 2(Suppl.):233–260.

Adair, W. S., C. Hwang, and U. W. Goodenough. 1983. Identification and visualization of the sexual agglutinin from the mating-type plus flagellar membrane of Chlamydomonas. *Cell* 33:183–193.

Adair, W. S., B. C. Monk, R. Cohen, and U. W. Goodenough. 1982. Sexual agglutinins from the *Chlamydomonas* flagellar membrane. Partial purification and characterization. *J. Biol. Chem.* 257:4593–4602.

Adair, W. S., and W. J. Snell. 1990. The *Chlamydomonas reinhardtii* cell wall: structure, biochemistry, and molecular biology, p. 14–85. *In* W. S. Adair and R. P. Mecham (eds.), *Organization and Assembly of Plant and Animal Extracellular Matrix*. Academic Press, Inc., San Diego, Calif.

Adair, W. S., S. A. Steinmetz, D. M. Mattson, U. W. Goodenough, and J. E. Heuser. 1987. Nucleated assembly of *Chlamydomonas* and *Volvox* cell walls. *J. Cell Biol.* 105:2373–2382.

Benovic, J. L., L. J. Pike, R. A. Cerione, C. Staniszewski, T. Yoshimasa, J. Codina, M. G. Caron, and R. J. Lefkowitz. 1985. Phosphorylation of the mammalian β-adrenergic receptor by cyclic AMP-dependent protein kinase. *J. Biol. Chem.* 260:7094–7101.

Berridge, M. J., and R. F. Irvine. 1989. Inositol phosphates and cell signalling. *Nature* (London) 341:197–205.

Bloodgood, R. A. 1977. Rapid motility occurring in association with the *Chlamydomonas* flagellar membrane. *J. Cell Biol.* 75:983–989.

Bloodgood, R. A. 1981. Flagella-dependent gliding motility in *Chlamydomonas*. *Protoplasma* 106:183–192.

Bloodgood, R. A. 1990. Gliding motility and flagellar glycoprotein dynamics in *Chlamydomonas*, p. 91–128. *In* R. A. Bloodgood (ed.), *Ciliary and Flagellar Membranes*. Plenum Press, New York.

Bloodgood, R. A. 1990. Personal communication.

Bloodgood, R. A., E. M. Leffler, and A. T. Bojcsuk. 1979. Reversible inhibition of *Chlamydomonas* surface motility. *J. Cell Biol.* **82**:664–674.

Bloodgood, R. A., and E. N. Levin. 1983. Transient increase in calcium efflux accompanies fertilization in *Chlamydomonas. J. Cell Biol.* **97**:397–404.

Bloodgood, R. A., and N. L. Salomonsky. 1990. Calcium influx regulates antibody-induced glycoprotein movements within the *Chlamydomonas* flagellar membrane. *J. Cell Sci.* **96**:27–33.

Bloodgood, R. A., M. P. Woodward, and N. L. Salomonsky. 1986. Redistribution and shedding of flagellar membrane glycoproteins visualized using an anticarbohydrate monoclonal antibody and concanavalin A. *J. Cell Biol.* **102**:1797–1812.

Boekhoff, I., E. Tareilus, J. Strotmann, and H. Breer. 1990. Rapid activation of alternative second messenger pathways in olfactory cilia from rats by different odorants. *EMBO J.* **9**:2453–2458.

Bråten, T. 1971. The ultrastructure of fertilization and zygote formation in the green alga *Ulva mutabilis* Føyn. *J. Cell Sci.* **9**:621–635.

Breer, H., I. Boekhoff, and E. Tareilus. 1990. Rapid kinetics of second messenger formation in olfactory transduction. *Nature* (London) **345**:65–68.

Brown, R. M., C. Johnson, and H. C. Bold. 1968. Electron and phase-contrast microscopy of sexual reproduction in *Chlamydomonas moewusii. J. Phycol.* **4**:100–140.

Buchanan, M. J., S. H. Imam, W. A. Eskue, and W. J. Snell. 1989. Activation of the cell wall degrading protease, lysin, during sexual signaling in *Chlamydomonas:* the enzyme is stored as an inactive, higher relative molecular mass precursor in the periplasm. *J. Cell Biol.* **108**:199–207.

Buck, C. A., and A. F. Horwitz. 1987. Cell surface receptors for extracellular matrix molecules. *Annu. Rev. Cell Biol.* **3**:179–205.

Cavalier-Smith, T. 1976. Electron microscopy of zygospore formation in *Chlamydomonas reinhardtii. Protoplasma* **87**:297–315.

Chinkers, M., and D. L. Garbers. 1989. The protein kinase domain of the ANP receptor is required for signaling. *Science* **245**:1392–1394.

Chinkers, M., D. L. Garbers, M. Chang, D. G. Lowe, H. Chin, D. V. Goeddel, and S. Schulz. 1989. A membrane form of guanylate cyclase is an atrial natiuretic peptide receptor. *Nature* (London) **338**:78–83.

Claes, H. 1971. Autolyse der Zellwand bei den Gameten von *Chlamydomonas reinhardii. Arch. Mikrobiol.* **78**:180–188.

Claes, H. 1977. Non-specific stimulation of the autolytic system in gametes from *Chlamydomonas reinhardtii. Exp. Cell Res.* **108**:221–229.

Claes, H. 1980. Calcium ionophore-induced stimulation of secretory activity in *Chlamydomonas reinhardi. Arch. Microbiol.* **124**:81–86.

Cobbs, W. H., and E. N. Pugh. 1985. Cyclic GMP can increase rod outer-segment light-sensitive current 10-fold without delay of excitation. *Nature* (London) **313**:585–587.

Collin-Osdoby, P., and W. S. Adair. 1985. Characterization of the purified *Chlamydomonas minus* agglutinin. *J. Cell Biol.* **101**:1144–1152.

Cooper, J. B., W. S. Adair, R. P. Mecham, J. E. Heuser, and U. W. Goodenough. 1983. *Chlamydomonas* agglutinin is a hydroxyproline-rich glycoprotein. *Proc. Natl. Acad. Sci. USA* **80**:5898–5901.

Cornish, E. C., J. M. Pettitt, I. Bonig, and A. E. Clarke. 1987. Developmentally controlled expression of a gene associated with self-incompatibility in *Nicotiana alata. Nature* (London) **326**:99–102.

Couraud, P., C. Delavier-Klutchko, O. Durieu-Trautmann, and A. D. Strosberg. 1981. Antibodies raised against β-adrenergic receptors stimulate adenylate cyclase. *Biochem. Biophys. Res. Commun.* **99**:1295–1302.

Coussen, F., J. Haiech, J. d'Alayer, and A. Momeron. 1985. Identification of the catalytic subunit of brain adenylate cyclase: a calmodulin binding protein of 135 kDa. *Proc. Natl. Acad. Sci. USA* **82**:6736–6740.

Crabbendam, K. J., F. M. Klis, A. Musgrave, and H. van den Ende. 1986. Ultrastructure of the *plus* and *minus* mating-type sexual agglutinins of *Chlamydomonas eugametos*, as visualized by negative staining. *J. Ultrastruct. Mol. Struct. Res.* **96**:151–159.

Crabbendam, K. J., N. Nanninga, A. Musgrave, and H. van den Ende. 1984. Flagellar tip activation in vis-à-vis pairs of *Chlamydomonas eugametos. Arch. Microbiol.* **138**:220–223.

Cutting, S., V. Oke, A. Driks, R. Losick, S. Lu, and L. Kroos. 1990. A forespore checkpoint for mother cell gene expression during development in B. subtilis. *Cell* **62:**239–250.

Davies, D. R., and A. Plaskitt. 1971. Genetical and structural analyses of cell-wall formation in *Chlamydomonas reinhardii. Genet. Res.* **17:**33–43.

Demets, R., A. M. Tomson, W. L. Homan, D. Stegwee, and H. van den Ende. 1988. Cell-cell adhesion in conjugating *Chlamydomonas* gametes: a self-enhancing process. *Protoplasma* **145:**27–36.

Detmers, P. A., and J. Condeelis. 1986. Trifluoperazine and W-7 inhibit mating in *Chlamydomonas* at an early stage of gametic interaction. *Exp. Cell Res.* **163:**317–326.

Detmers, P. A., U. W. Goodenough, and J. Condeelis. 1983. Elongation of the fertilization tubule in *Chlamydomonas:* new observations on the core microfilaments and the effect of transient intracellular signals on their structural integrity. *J. Cell. Biol.* **97:**522–532.

Dhallan, R. S., K.-W. Yau, K. A. Schrader, and R. R. Reed. 1990. Primary structure and functional expression of a cyclic nucleotide-activated channel from olfactory neurons. *Nature* (London) **347:**184–187.

Dutcher, S. K., and W. Gibbons. 1988. Isolation and characterization of dominant tunicamycin resistance mutations in *Chlamydomonas reinhardtii* (Chlorophyceae). *J. Phycol.* **24:**230–236.

Ebersold, W. T., R. P. Levine, E. E. Levine, and M. A. Olmsted. 1962. Linkage maps in *Chlamydomonas reinhardii. Genetics* **47:**531–543.

Ferris, P. J., and U. W. Goodenough. 1987. Transcription of novel genes, including a gene linked to the mating-type locus, induced by *Chlamydomonas* fertilization. *Mol. Cell. Biol.* **7:**2360–2366.

Forest, C. L. 1983. Specific contact between mating structure membranes observed in conditional fusion-defective *Chlamydomonas* mutants. *Exp. Cell. Res.* **148:**143–154.

Forest, C. L. 1987. Genetic control of plasma membrane adhesion and fusion in *Chlamydomonas* gametes. *J. Cell Sci.* **88:**613–621.

Forest, C. L., D. A. Goodenough, and U. W. Goodenough. 1978. Flagellar membrane agglutination and sexual signaling in the conditional *gam*-1 mutant of *Chlamydomonas. J. Cell Biol.* **79:**74–84.

Friedmann, I., A. L. Colwin, and L. H. Colwin. 1968. Fine structural aspects of fertilization in *Chlamydomonas reinhardi. J. Cell Sci.* **3:**115–128.

Galloway, R. E., and U. W. Goodenough. 1985. Genetic analysis of mating locus-linked mutations in *Chlamydomonas reinhardi. Genetics* **111:**447–461.

Garbers, D. L., and G. S. Kopf. 1980. The regulation of spermatozoa by calcium and cyclic nucleotides. *Adv. Cyclic Nucleotide Res.* **13:**251–306.

Gell, A. C., A. Bacic, and A. E. Clarke. 1986. Arabinogalactan-proteins of the female sexual tissue of *Nicotiana alata. Plant Physiol.* **82:**885–889.

Gitelman, S. E., and G. B. Witman. 1980. Purification of calmodulin from *Chlamydomonas:* calmodulin occurs in cell bodies and flagella. *J. Cell Biol.* **98:**764–770.

Glabe, C. G., L. B. Grabel, V. D. Vacquier, and S. D. Rosen. 1982. Carbohydrate specificity of sea urchin sperm binding: a cell surface lectin mediating sperm-egg adhesion. *J. Cell Biol.* **94:**123–125.

Goodenough, U. W. 1977. Mating interactions in *Chlamydomonas,* p. 323–350. *In* J. L. Reissig (ed.), *Microbial Interactions. Receptors and Recognition, Series B.* Chapman and Hall, London.

Goodenough, U. W. 1985. An essay on the origins and evolution of eukaryotic sex, p. 123–140. *In* H. O. Halvorson and A. Monroy (ed.), *The Origin and Evolution of Sex.* Alan R. Liss, New York.

Goodenough, U. W. 1986. Experimental analysis of the adhesion reaction between isolated *Chlamydomonas* flagella. *Exp. Cell Res.* **166:**237–246.

Goodenough, U. W. 1989. Cyclic AMP enhances the sexual agglutinability of *Chlamydomonas* flagella. *J. Cell Biol.* **109:**247–252.

Goodenough, U. W., and W. S. Adair. 1989. Recognition proteins of *Chlamydomonas reinhardii* (Chlorophyceae), p. 171–185. *In* A. W. Coleman, L. J. Goff, and J. R. Stein-Taylor (ed.), *Algae as Experimental Systems.* Alan R. Liss, New York.

Goodenough, U. W., W. S. Adair, E. Caligor, C. L. Forest, J. L. Hoffman, D. A. M. Mesland, and S. Spath. 1980. Membrane-membrane and membrane-ligand interactions in *Chlamydomonas* mating, p. 131–152. *In* N. B. Gilula (ed.), *Membrane-Membrane Interactions.* Raven Press, New York.

Goodenough, U. W., W. S. Adair, P. Collin-Osdoby, and J. E. Heuser. 1985. Structure of the *Chlamydomonas* agglutinin and related flagellar surface proteins *in vitro* and *in situ. J. Cell. Biol.* **101:**924–941.

Goodenough, U. W., P. A. Detmers, and C. Hwang. 1982. Activation for cell fusion in *Chlamydomonas:* analysis of wild-type gametes and nonfusing mutants. *J. Cell Biol.* **92**:378–386.

Goodenough, U. W., and P. J. Ferris. 1987. Genetic regulation of development in *Chlamydomonas,* p. 171–189. *In* W. Loomis (ed.), *Genetic Regulation of Development.* Alan R. Liss, New York.

Goodenough, U. W., B. Gebhart, R. P. Mecham, and J. E. Heuser. 1986. Crystals of the *Chlamydomonas reinhardtii* cell wall: polymerization, depolymerization, and purification of glycoprotein monomers. *J. Cell Biol.* **103**:405–417.

Goodenough, U. W., and J. E. Heuser. 1985. The *Chlamydomonas* cell wall and its constituent glycoproteins analyzed by the quick-freeze deep-etch technique. *J. Cell Biol.* **101**:1550–1568.

Goodenough, U. W., and J. E. Heuser. 1988a. Molecular organization of cell wall crystals from *Chlamydomonas reinhardtii* and *Volvox carteri. J. Cell Sci.* **90**:717–733.

Goodenough, U. W., and J. E. Heuser. 1988b. Molecular organization of the cell wall and cell-wall crystals from *Chlamydomonas eugametos. J. Cell Sci.* **90**:735–750.

Goodenough, U. W., and J. E. Heuser. Unpublished data.

Goodenough, U. W., C. Hwang, and H. Martin. 1976. Isolation and genetic analysis of mutant strains of *Chlamydomonas reinhardi* defective in gametic differentiation. *Genetics* **82**:169–186.

Goodenough, U. W., C. Hwang, and A. J. Warren. 1978. Sex-limited expression of gene loci controlling flagellar membrane agglutination in the *Chlamydomonas* mating reaction. *Genetics* **89**:235–243.

Goodenough, U. W., and D. Jurivich. 1978. Tipping and mating-structure activation induced in *Chlamydomonas* gametes by flagellar membrane antiserum. *J. Cell Biol.* **79**:680–693.

Goodenough, U. W., and R. L. Weiss. 1975. Gametic differentiation in *Chlamydomonas reinhardtii.* III. Cell wall lysis and microfilament-associated mating structure activation in wild-type and mutant strains. *J. Cell Biol.* **67**:623–637.

Goodenough, U. W., and R. L. Weiss. 1978. Interrelationships between microtubules, a striated fiber, and the gametic mating structure of *Chlamydomonas reinhardtii. J. Cell Biol.* **76**:430–438.

Grant, P. G., A. F. Mannarinom, and R. W. Colman. 1988. cAMP-mediated phosphorylation of the low-K_m cAMP phosphodiesterase markedly stimulates its catalytic activity. *Proc. Natl. Acad. Sci. USA* **85**:9071–9075.

Grief, C., M. A. O'Neill, and P. J. Shaw. 1987. The zygote cell wall of *Chlamydomonas reinhardii:* a structural, chemical, and immunological approach. *Planta* **170**:433–445.

Grost, M. K., D. G. Toscano, and W. A. Toscano. 1987. Calmodulin-mediated adenylate cyclase from mammalian sperm. *J. Biol. Chem.* **262**:8672–8676.

Gunderson, J. H., H. Elwood, A. Jugold, K. Kindle, and M. L. Sogin. 1987. Phylogenetic relationships between chlorophytes, chrysophytes, and oomycetes. *Proc. Natl. Acad. Sci. USA* **84**:5823–5827.

Gustin, M. C., and D. L. Nelson. 1987. Regulation of ciliary adenylate cyclase by Ca^{2+} in *Paramecium. Biochem. J.* **246**:337–345.

Hämmerling, J. 1934. Über die Geschlechtsverhältnisse von *Acetabularia mediterranea* und *Acetabularia wettsteinii. Arch Protistenk.* **83**:57–97.

Heuser, J. E. 1983. Procedure for freeze-drying molecules adsorbed to mica flakes. *J. Mol. Biol.* **169**:155–195.

Hidaka, H., M. Inagalei, S. Kawamoto, and Y. Sasaki. 1984. Isoquinoline sulfonamides, novel and potent inhibitors of cyclic nucleotide dependent protein kinase and protein kinase C. *Biochemistry* **23**:5036–5041.

Hills, G. J., J. M. Phillips, M. R. Gay, and K. Roberts. 1975. Self-assembly of a plant cell wall in vitro. *J. Mol. Biol.* **96**:431–441.

Hoffman, J. L., and U. W. Goodenough. 1980. Experimental dissection of flagellar surface motility in *Chlamydomonas. J. Cell Biol.* **86**:656–665.

Holmes, J. A., and S. K. Dutcher. 1989. Cellular asymmetry in *Chlamydomonas reinhardtii. J. Cell Sci.* **94**:273–285.

Homan, W. L., A. Musgrave, H. de Nobel, R. Wagter, A. H. J. Kolk, D. de Wit, and H. van den Ende. 1988. Monoclonal antibodies directed against the sexual binding site of *Chlamydomonas eugametos* gametes. *J. Cell Biol.* **107**:177–189.

Homan, W., C. Sigon, W. van den Briel, R. Wagter, H. de Nobel, D. Mesland, A. Musgrave, and H. van den Ende. 1987. Transport of membrane receptors and the mechanics of sexual cell fusion in *Chlamydomonas eugametos. FEBS Lett.* **215**:323–326.

Hoops, H. J., and G. B. Witman. 1983. Outer doublet heterogeneity reveals structural polarity related to beat direction in *Chlamydomonas* flagella. *J. Cell Biol.* **97**:902–908.

Hoppe, J. 1985. cAMP-dependent protein kinases: conformational changes during activation. *Trends Biochem. Sci.* **10**:29–31.

Hunnicutt, G. R., M. G. Kosfiszer, and W. J. Snell. 1991. Cell body and flagellar agglutinins in *Chlamydomonas reinhardtii*. The cell body plasma membrane is a reservoir for agglutinins whose migration to the flagella is regulated by a functional barrier. *J. Cell Biol.* **111**:1605–1616.

Hwang, C., B. C. Monk, and U. W. Goodenough. 1981. Linkage of mutations affecting *minus* flagellar membrane agglutinability to the *mt⁻* mating-type locus of *Chlamydomonas*. *Genetics* **99**:41–47.

Hyams, J. S., and G. G. Borisy. 1978. Isolated flagellar apparatus of *Chlamydomonas:* characterization of forward swimming and alteration of waveform and reversal of motion by calcium ions *in vitro*. *J. Cell Sci.* **33**:235–253.

Imam, S. H., and W. J. Snell. 1988. The *Chlamydomonas* cell wall degrading enzyme, lysin, acts on two substrates within the framework of the wall. *J. Cell Biol.* **106**:2211–2221.

Isenberg, W. M., R. P. McEver, D. R. Phillips, M. A. Shuman, and D. F. Bainton. 1987. The platelet fibrinogen receptor: an immunogold-surface replica study of agonist-induced ligand binding and receptor clustering. *J. Cell Biol.* **104**:1655–1663.

Iwao, Y., and L. A. Jaffe. 1989. Evidence that the voltage-dependent component in the fertilization process is contributed by the sperm. *Dev. Biol.* **134**:446–451.

Jaenicke, L., W. Kuhne, R. Spessert, U. Wahle, and S. Waffenschmidt. 1987. Cell-wall lytic enzymes (autolysins) of *Chlamydomonas reinhardtii* are (hydroxy)proline-specific proteases. *Eur. J. Biochem.* **170**:485–491.

Janssens, P. M. W. 1988. The evolutionary origin of eukaryotic transmembrane signal transduction. *Comp. Biochem. Physiol.* **90A**:209–223.

Jupe, E. R., R. L. Chapman, and E. A. Zimmer. 1988. Nuclear ribosomal RNA genes and algal phylogeny—the *Chlamydomonas* example. *Biosystems* **21**:223–230.

Kalshoven, H. W., A. Musgrave, and H. van den Ende. 1990. Mating receptor complex in the flagellar membrane of *Chlamydomonas eugametos* gametes. *Sex Plant Reprod.* **3**:77–87.

Kamiya, R., and G. B. Witman. 1984. Submicromolar levels of calcium control the balance of beating between the two flagella in demembranated models of *Chlamydomonas*. *J. Cell Biol.* **98**:97–107.

Kaska, D. D., I. C. Piscopo, and A. Gibor. 1985. Intracellular calcium redistribution during mating in *Chlamydomonas reinhardii*. *Exp. Cell Res.* **160**:371–379.

Kirk, D. Unpublished data.

Kirk, D. L., and J. F. Harper. 1987. Genetic, biochemical, and molecular approaches to *Volvox* development and evolution. *Int. Rev. Cytol.* **99**:217–293.

Kirk, D. L., and M. M. Kirk. 1986. Heat shock elicits production of sexual inducer in *Volvox*. *Science* **231**:51–54.

Klis, F. M., K. Crabbendam, P. van Egmond, and H. van den Ende. 1989. Ultrastructure and properties of the sexual agglutinins of the biflagellate green alga *Chlamydomonas moewusii*. *Sex Plant Reprod.* **2**:213–218.

Kooijman, R., P. de Wildt, S. Beumer, G. van der Uliet, W. Homan, H. Kalshoven, A. Musgrave, and H. van den Ende. 1989. Wheat germ agglutinin induces mating reactions in *Chlamydomonas eugametos* by cross-linking agglutinin-associated glycoproteins in the flagellar membrane. *J. Cell. Biol.* **109**:1677–1687.

Kooijman, R., P. de Wildt, W. L. Homan, A. Musgrave, and H. van den Ende. 1988. Light affects flagellar agglutinability in *Chlamydomonas eugametos* by modification of the agglutinin molecules. *Plant Physiol.* **86**:216–223.

Kooijman, R., P. de Wildt, W. van den Briel, S. Tan, A. Musgrave, and H. van den Ende. 1990. Cyclic AMP is one of the intracellular signals during the mating of *Chlamydomonas eugametos*. *Planta* **181**:529–537.

Kumagai, A., M. Pupillo, R. Gundersen, R. Miake-Lye, P. N. Devreotes, and R. A. Firtel. 1989. Regulation and function of G_α protein subunits in Dictyostelium. *Cell* **57**:265–275.

Kurjan, J. 1990. G proteins in yeast *Saccharomyces cerevisiae*. *In G Proteins*, p. 571–599. Academic Press, Inc., New York.

Larson, A., M. Kirk, and D. Kirk. Molecular phylogeny of the Volvocine flagellates. *Mol. Biol. Evol.*, in press.

Levitzki, A. 1988. From epinephrine to cyclic AMP. *Science* **241**:800–806.

Lewin, R. A. 1952. Studies on the flagella of algae. I. General observations on *Chlamydomonas moewusii* Gerloff. *Biol. Bull.* **103**:74–79.

Lewin, R. A. 1954a. Mutants of *Chlamydomonas moewusii* with impaired motility. *J. Gen. Microbiol.* **11**:358–368.

Lewin, R. A. 1954b. Sex in unicellular algae, p. 100–133. *In* D. H. Wenrich (ed.), *Sex in Microorganisms*. American Association for the Advancement of Science, Washington, D.C.

Leyton, L., and P. Saling. 1989a. 95 kD sperm proteins bind ZP3 and serve as tyrosine kinase substrates in response to zona binding. *Cell* **57**:1123–1130.

Leyton, L., and P. Saling. 1989b. Evidence that aggregation of mouse sperm receptors by ZP3 triggers the acrosome reaction. *J. Cell Biol.* **108**:2163–2168.

Lipke, P. N., D. Wojciechowicz, and J. Kurjan. 1989. AGα1 is the structural gene for the *Saccharomyces cerevisiae* α-agglutinin, a cell surface glycoprotein involved in cell-cell interactions during mating. *Mol. Cell. Biol.* **9**:3155–3165.

Losick, R., P. Youngman, and P. J. Piggot. 1986. Genetics of endospore formation in *Bacillus subtilis*. *Annu. Rev. Genet.* **20**:625–669.

Lynch, T. J., B. Wu, J. D. Taylor, and T. T. Tchen. 1986. Regulation of pigment organelle translocation. *J. Biol. Chem.* **261**:4212–4216.

Ma, H., M. F. Yanofsky, and E. H. Meyerowitz. 1990. Molecular cloning and characterization of GPA1, a G protein α subunit gene from *Arabidopsis thaliana*. *Proc. Natl. Acad. Sci. USA* **87**:3821–3825.

MacNeil, S., T. Lakey, and S. Tomlinson. 1985. Calmodulin regulation of adenylate cyclase activity. *Cell Calcium* **6**:213–226.

Mages, H.-W., H. Tschochner, and M. Sumper. 1988. The sexual inducer of *Volvox carteri:* primary structure deduced from cDNA sequence. *FEBS Lett.* **234**:407–410.

Marcantonio, E. E., and R. O. Hynes. 1988. Antibodies to the conserved cytoplasmic domain of the integrin β_1 subunit react with proteins in vertebrates, invertebrates, and fungi. *J. Cell Biol.* **106**:1763–1772.

Marchant, H. J., and J. D. Pickett-Heaps. 1972. Ultrastructure and differentiation of *Hydrodictyon reticulatum*. IV. Conjugation of gametes and the development of zygospores and azygospores. *Aust. J. Biol. Sci.* **25**:279–291.

Martin, N. C., and U. W. Goodenough. 1975. Gametic differentiation in *Chlamydomonas reinhardtii*. I. Production of gametes and their fine structure. *J. Cell Biol.* **67**:587–605.

Martinac, B., J. Adler, and C. Kung. 1990. Amphipaths activate mechanosensitive ion channels of *Escherichia coli*. *Nature* (London) **348**:261–263.

Matsuda, Y., T. Saito, M. Koseki, and T. Shimada. 1990. The *Chlamydomonas* non-synchronous and synchronous gametogenesis as analyzed by the activity of cell body-agglutinin and cell wall lytic enzyme. *Plant Physiol. (Life Sci. Adv.)* **9**:1–6.

Matsuda, Y., T. Saito, T. Yamaguchi, and H. Kawase. 1985. Cell wall lytic enzyme released by mating gametes of *Chlamydomonas reinhardtii* is a metalloprotease and digests the sodium perchlorate-insoluble component of cell wall. *J. Biol. Chem.* **260**:6373–6377.

Matsuda, Y., T. Saito, T. Yamaguchi, M. Koseki, and K. Hayashi. 1987. Topography of cell wall lytic enzyme in *Chlamydomonas reinhardtii:* form and location of the stored enzyme in vegetative cell and gamete. *J. Cell Biol.* **104**:321–329.

Matsuda, Y., K. Sakamoto, N. Kiuchi, T. Mizouchi, Y. Tsubo, and A. Kobata. 1982. Two tunicamycin-sensitive components involved in agglutination and fusion of *Chlamydomonas reinhardtii* gametes. *Arch. Microbiol.* **131**:87–90.

Matsuda, Y., A. Yamasaki, T. Saito, and T. Yamaguchi. 1984. Purification and characterization of the cell wall lytic enzyme released by mating gametes of *Chlamydomonas reinhardtii*. *FEBS Lett.* **166**:293–297.

Matters, G. L., and U. W. Goodenough. A gene/pseudogene tandem duplication encodes a cysteine-rich protein expressed during zygote development in *Chlamydomonas reinhardtii*. *Mol. Gen. Genet.*, in press.

Melkonian, M. 1980. Flagellar roots, mating structure and gametic fusion in the green alga *Ulva lactuca* (Uvales). *J. Cell Sci.* **46**:149–169.

Mesland, D. A. M. 1976. Mating in *Chlamydomonas eugametos*. A scanning electron microscopical study. *Arch. Microbiol.* **109**:31–35.

Mesland, D. A. M., J. L. Hoffman, E. Caligor, and U. W. Goodenough. 1980. Flagellar tip activation stimulated by membrane adhesions in *Chlamydomonas* gametes. *J. Cell Biol.* **894**:599–617.

Mesland, D. A. M., and H. van den Ende. 1978. The role of flagellar adhesion in sexual activation of *Chlamydomonas eugametos*. *Protoplasma* **98**:115–129.

Milliken, B. E., and R. L. Weiss. 1984. Distribution of concanavalin A binding carbohydrates during mating in *Chlamydomonas*. *J. Cell Sci.* **66**:223–239.

Minami, S. A., and U. W. Goodenough. 1978. Novel glycopolypeptide synthesis induced by gametic cell fusion in *Chlamydomonas reinhardtii*. *J. Cell Biol.* **77**:165–180.

Morris, C. E., and W. J. Sigurdson. 1989. Stretch-inactivated ion channels coexist with stretch-activated ion channels. *Science* **243**:807–809.

Musgrave, A., P. de Wildt, Y. van Etten, H. Pijst, R. Kooijman, W. Homan, and H. van den Ende. 1986. Evidence for a functional membrane barrier in the transition zone between the flagellum and the cell body of *Chlamydomonas eugametos* gametes. *Planta* **167**:544–553.

Musgrave, A., and H. van den Ende. 1987. How *Chlamydomonas* court their partners. *Trends Biochem. Sci.* **12**:470–473.

Musgrave, A., E. van Eyk, R. de Welscher, R. Broekman, P. F. Lens, W. L. Homan, and H. van den Ende. 1981. Sexual agglutination factor from *Chlamydomonas eugametos*. *Planta* **153**:361–369.

Nakamura, T., and G. H. Gold. 1987. A cyclic nucleotide-gated conductance in olfactory receptor cilia. *Nature* (London) **325**:442–444.

Nasrallah, M. E., and J. B. Nasrallah. 1986. Molecular biology of self-incompatibility in plants. *Trends Genet.* **2**:239–244.

Pasquale, S. M. 1989. Contact-induced signaling in gametes of *Chlamydomonas reinhardtii*. Ph.D. thesis. Washington University, St. Louis, Mo.

Pasquale, S. M., and U. W. Goodenough. 1987. Cyclic AMP functions as a primary sexual signal in gametes of *Chlamydomonas reinhardtii*. *J. Cell Biol.* **105**:2279–2293.

Pennell, R. I., and K. Roberts. 1990. Sexual development in the pea is presaged by altered expression of arabinogalactan protein. *Nature* (London) **344**:547–549.

Pickett-Heaps, J. D. 1975. *Green Algae: Structure, Reproduction, and Evolution in Selected Genera.* Sinauer Associates, Sunderland, Mass.

Pierce, M., and C. E. Ballou. 1983. Cell-cell recognition in yeast. Characterization of the sexual agglutination factors from *Saccharomyces kluyveri*. *J. Biol. Chem.* **258**:3576–3582.

Pijst, H. L. A., F. A. Ossendorp, P. van Egmond, A. M. I. E. Kamps, A. Musgrave, and H. van den Ende. 1984a. Sex-specific binding and inactivation of agglutination factor in *Chlamydomonas eugametos*. *Planta* **160**:529–535.

Pijst, H. L. A., R. van Driel, P. M. W. Janssens, A. Musgrave, and H. van den Ende. 1984b. Cyclic AMP is involved in sexual reproduction of *Chlamydomonas eugametos*. *FEBS Lett.* **174**:132–136.

Pijst, H. L. A., R. J. Zilber, A. Musgrave, and H. van den Ende. 1983. Agglutination factor in the cell body of *Chlamydomonas eugametos*. *Planta* **158**:403–409.

Quarmby, L. M., Y. Yeuh, J. L. Cheshire, L. Keller, W. Snell, and R. C. Crain. Submitted for publication.

Rausch, H., N. Larsen, and R. Schmitt. 1989. Phylogenetic relationships of the green alga *Volvox carteri* deduced from small-subunit ribosomal RNA comparisons. *J. Mol. Evol.* **29**:255–265.

Roberts, K. 1974. Crystalline glycoprotein cell walls of algae: their structure, composition and assembly. *Philos. Trans. R. Soc. London Ser. B* **268**:129–146.

Roberts, K., M. Gurney-Smith, and G. J. Hills. 1972. Structure, composition, and morphogenesis of the cell wall of *Chlamydomonas reinhardi*. I. Ultrastructure and preliminary chemical analysis. *J. Ultrastruct. Res.* **40**:599–613.

Ronson, C. W., B. T. Nixon, and F. M. Ausubel. 1987. Conserved domains in bacterial regulatory proteins that respond to environmental stimuli. *Cell* **49**:579–581.

Rozdzial, M. M., and L. T. Haimo. 1986. Bidirectional pigment granule movements of melanophores are regulated by protein phosphorylation and dephosphorylation. *Cell* **47**:1061–1070.

Sager, R., and S. Granick. 1954. Nutritional control of sexuality in *Chlamydomonas reinhardi. J. Gen. Physiol.* **37:**729–742.

Saimi, Y., B. Martinac, M. C. Gustin, M. R. Culbertson, J. Adler, and C. Kung. 1988. Ion channels in *Paramecium*, yeast, and *Escherichia coli. Trends Biochem. Sci.* **13:**304–309.

Saito, T., and Y. Matsuda. 1984a. Sexual agglutination of mating-type minus gametes in *Chlamydomonas reinhardii*. I. Loss and recovery of agglutinability of gametes treated with EDTA. *Exp. Cell Res.* **152:**322–330.

Saito, T., and Y. Matsuda. 1984b. Sexual agglutination of mating-type minus gametes in *Chlamydomonas reinhardtii*. II. Purification and characterization of minus agglutinin and comparison with plus agglutinin. *Arch. Microbiol.* **139:**95–99.

Saito, T., and Y. Matsuda. 1986. The *Chlamydomonas* plus and minus agglutinin: difference in sensitivity to dithiothreitol. *Arch. Microbiol.* **146:**25–29.

Saito, T., and Y. Matsuda. 1991. Isolation and characterization of *Chlamydomonas* temperature-sensitive mutants affecting gametic differentiation under nitrogen-starved conditions. *Curr. Genet.* **19:**65–72.

Saito, T., L. Small, and U. W. Goodenough. Submitted for publication.

Saito, T., Y. Tsubo, and Y. Matsuda. 1985. Synthesis and turnover of cell body-agglutinin as a pool for flagellar surface-agglutinin in *Chlamydomonas reinhardii. Arch. Microbiol.* **142:**107–210.

Saito, T., Y. Tsubo, and Y. Matsuda. 1988. A new assay system to classify non-mating mutants and to distinguish between vegetative cell and gamete in *Chlamydomonas reinhardii. Curr. Genet.* **14:**59–63.

Salisbury, J. L., M. Sanders, and L. Harpst. 1987. Flagellar root contraction and nuclear movement during flagellar regeneration in *Chlamydomonas reinhardtii. J. Cell Biol.* **105:**1799–1805.

Samson, M. R., F. M. Klis, K. J. Crabbendam, P. van Egmond, and H. van den Ende. 1987a. Purification, visualization and characterization of the sexual agglutinins of the green alga *Chlamydomonas moewusii yapensis. J. Gen. Microbiol.* **133:**3183–3191.

Samson, M. R., F. M. Klis, W. L. Homan, P. Van Egmond, A. Musgrave, and H. van den Ende. 1987b. Composition and properties of the sexual agglutinins of the flagellated green algae *Chlamydomonas eugametos. Planta* **170:**314–321.

Sanders, M. A., and J. L. Salisbury. 1989. Centrin-mediated microtubule severing during flagellar excision in *Chlamydomonas reinhardtii. J. Cell Biol.* **108:**1751–1760.

Schloss, J. A. 1990. A *Chlamydomonas* gene encodes a G protein β subunit-like polypeptide. *Mol. Gen. Genet.* **221:**443–452.

Schlösser, U. G. 1981. Algal wall-degrading enzymes—autolysines. Plant carbohydrates II. *Encycl. Plant Physiol. N. Ser.* **138:**333–351.

Schreiner, G., and E. R. Unanue. 1976. Membrane and cytoplasmic changes in B lymphocytes induced by ligand-surface immunoglobulin interaction. *Adv. Immunol.* **24:**37–165.

Showalter, A. M., and D. Rumeau. 1990. Molecular biology of plant cell wall hydroxyproline-rich glycoproteins, p. 247–281. *In* W. S. Adair and R. P. Mecham (ed.), *Organization and Assembly of Plant and Animal Extracellular Matrix.* Academic Press, Inc., San Diego.

Singh, S., D. G. Lowe, D. S. Thorpe, H. Rodriguez, W. Kuang, L. J. Dangott, M. Chinkers, D. V. Goeddel, and D. L. Garbers. 1988. Membrane guanylate cyclase is a cell-surface receptor with homology to protein kinases. *Nature* (London) **334:**708–712.

Smith, G. M., and D. C. Regnery. 1950. Inheritance of sexuality in *Chlamydomonas reinhardi. Proc. Natl. Acad. Sci. USA* **36:**246–248.

Snell, W. J. 1981. Flagellar adhesion and deadhesion in *Chlamydomonas* gametes: effects of tunicamycin and observations on flagellar tip morphology. *J. Supramol. Struct. Cell Biochem.* **16:**371–376.

Snell, W. J. 1985. Cell-cell interactions in *Chlamydomonas. Annu. Rev. Plant Physiol.* **36:**287–315.

Snell, W. J., M. Buchanan, and A. Clausell. 1982. Lidocaine reversibly inhibits fertilization in *Chlamydomonas:* a possible role for calcium in sexual signalling. *J. Cell Biol.* **94:**607–612.

Snell, W. J., A. Clausel, and W. S. Moore. 1983. Flagellar adhesion in *Chlamydomonas* induces synthesis of two high molecular weight cell surface proteins. *J. Cell Biol.* **96:**589–597.

Snell, W. J., W. A. Eskue, and M. J. Buchanan. 1989. Regulated secretion of a serine protease that activates an extracellular matrix-degrading metalloprotease during fertilization in *Chlamydomonas*. *J. Cell Biol.* **109**:1689–1694.

Snell, W. J., M. Kosifszec, A. Clausell, N. Pecillo, S. Iman, and G. Hunnicutt. 1986. A monoclonal antibody that blocks adhesion of *Chlamydomonas mt⁺* gametes. *J. Cell Biol.* **103**:2449–2456.

Snell, W. J., and W. S. Moore. 1980. Aggregation-dependent turnover of flagellar adhesion molecules in *Chlamydomonas* gametes. *J. Cell Biol.* **84**:203–210.

Solter, K. M., and A. Gibor. 1977. Evidence for role of flagella as sensory transducers in mating of *Chlamydomonas reinhardi*. *Nature* (London) **265**:444–445.

Springer, T. A. 1990. Adhesion receptors of the immune system. *Nature* (London) **346**:425–434.

Stanley, S. M. 1975. Clades versus clones in evolution: why we have sex. *Science* **190**:382–383.

Tilney, L. G., and N. Kallenbach. 1979. Polymerization of actin. VI. The polarity of the actin filaments in the acrosomal process and how it might be determined. *J. Cell Biol.* **81**:608–623.

Tomson, A. M., R. Demets, N. P. M. Bakker, D. Stegwee, and H. van den Ende. 1985. Gametogenesis in liquid cultures of *Chlamydomonas eugametos*. *J. Gen. Microbiol.* **131**:1553–1560.

Tomson, A. M., R. Demets, A. Musgrave, R. Kooijman, D. Stegwee, and H. van den Ende. 1990a. Contact activation in *Chlamydomonas* gametes by increased binding capacity of sexual agglutinins. *J. Cell Sci.* **95**:293–301.

Tomson, A. M., R. Demets, C. A. M. Sigon, D. Stegwee, and H. van den Ende. 1986. Cellular interactions during the mating process in *Chlamydomonas eugametos*. *Plant Physiol.* **81**:522–526.

Tomson, A. M., R. Demets, E. A. van Spronsen, G. J. Brackenhoff, D. Stegwee, and H. van den Ende. 1990b. Turnover and transport of agglutinins in conjugating *Chlamydomonas* gametes. *Protoplasma* **155**:200–209.

Treier, U., S. Fuchs, M. Weber, W. W. Wakarchuk, and C. F. Beck. 1989. Gametic differentiation in *Chlamydomonas reinhardtii*: light dependence and gene expression patterns. *Arch. Microbiol.* **152**:572–577.

Triemer, R. E., and R. M. Brown. 1975. The ultrastructure of fertilization in *Chlamydomonas moewusii*. *Protoplasma* **84**:315–325.

Ullrich, A., and J. Schlessinger. 1990. Signal transduction by receptors with tyrosine kinase activity. *Cell* **61**:203–212.

van den Ende, H., F. M. Klis, and A. Musgrave. 1988. The role of flagella in sexual reproduction of *Chlamydomonas eugametos*. *Acta Bot. Neerl.* **37**:327–350.

van den Ende, H., A. Musgrave, and F. M. Klis. 1990. The role of flagella in the sexual reproduction of *Chlamydomonas* gametes, p. 129–147. *In* R. A. Bloodgood (ed.), *Ciliary and Flagellar Membranes*. Plenum Publishing Corp., New York.

van Holst, G. J., and J. E. Varner. 1984. Reinforced polyproline-II conformation in a hydroxyproline-rich cell wall glycoprotein from carrot root. *Plant Physiol.* **74**:247–251.

vanWinkle-Swift, K. P., G. R. Aliaga, and J. C. Pommerville. 1987. Haploid spore formation following arrested cell fusion in *Chlamydomonas* (Chlorophyta). *J. Phycol.* **23**:414–427.

vanWinkle-Swift, K. P., and J.-H. Hahn. 1986. The search for mating type-limited genes in the homothallic alga *Chlamydomonas monoica*. *Genetics* **113**:601–619.

Waffenschmidt, S., M. Knittler, and L. Jaenicke. 1990. Characterization of a sperm lysin of *Volvox carteri*. *Sex Plant Reprod.* **3**:1–6.

Waffenschmidt, S., W. Kuhne, and L. Jaenicke. 1989. Immunological characterization of gamete autolysins in *Chlamydomonas reinhardtii*. *Bot. Acta* **102**:73–79.

Waffenschmidt, S., R. Spessert, and L. Jaenicke. 1988. Oligosaccharide side chains of wall molecules are essential for cell-wall lysis in *Chlamydomonas reinhardtii*. *Planta* **175**:513–519.

Wasserman, P. M., J. D. Bleil, H. M. Florman, J. M. Greve, R. J. Roler, G. S. Salsmann, and F. G. Samuels. 1985. The mouse egg's receptor for sperm: what is it and how does it work? *Cold Spring Harbor Symp. Quant. Biol.* **50**:11–19.

Weiss, R. L. 1983. Coated vesicles in the contractile vacuole/mating structure region of *Chlamydomonas*. *J. Ultrastruct. Res.* **85**:33–44.

Weiss, R. L., D. A. Goodenough, and U. W. Goodenough. 1977. Membrane differentiations at sites specialized for cell fusion. *J. Cell Biol.* **72**:144–160.

Wiese, L. 1965. On sexual agglutination and mating-type substances (gamones) in isogamous heterothallic chlamydomonads. I. Evidence of the identity of the gamones with the surface components responsible for sexual flagellar contact. *J. Phycol.* **1**:46–54.

Wiese, L. 1969. Algae, p. 135–188. *In* C. B. Metz and A. Monroy (ed.), *Fertilization, Comparative Morphology, Biochemistry, and Immunology*. Academic Press, Inc. (London) Ltd., London.

Wiese, L., A. Hardcastle, and M. Stewart. 1984. Inactivation of the (+) gamete during the mating-type reaction in *Chlamydomonas*. *Gam. Res.* **9**:441–449.

Wiese, L., and R. F. Jones. 1963. Studies on gamete copulation in heterothallic chlamydomonads. *J. Cell Comp. Physiol.* **61**:265–274.

Wiese, L., and W. Wiese. 1977. On speciation by evolution of gametic incompatibility: a model case in *Chlamydomonas*. *Am. Nat.* **111**:733–742.

Wiese, L., L. A. Williams, and D. L. Baker. 1983. A general and fundamental molecular bipolarity of the sex cell contact mechanism as revealed by tunicamycin and bacitracin in *Chlamydomonas*. *Am. Nat.* **122**:806–816.

Woessner, J. P., and U. W. Goodenough. 1989. Molecular characterization of a zygote wall protein: an extension-like molecule in *Chlamydomonas reinhardtii*. *Plant Cell* **1**:901–911.

Wood, S. F., E. Z. Szuts, and A. Fein. 1989. Inositol triphosphate production in squid photoreceptors. Activation by light, aluminum fluoride, and guanine nucleotides. *J. Biol. Chem.* **264**:12970–12976.

Yau, K.-W., and D. A. Baylor. 1989. Cyclic GMP-activated conductance of retinal photoreceptor cells. *Annu. Rev. Neurosci.* **12**:289–327.

Microbial Cell-Cell Interactions
Edited by Martin Dworkin
© 1991 American Society for Microbiology, Washington, DC 20005

Chapter 5

Cell-Cell Interactions Involved in Yeast Mating

Janet Kurjan

INTRODUCTION

Mating in the yeast *Saccharomyces cerevisiae* involves fusion of haploid cells of opposite mating type, called a and α, to produce a third cell type, the a/α diploid (Fig. 1; reviewed by Cross et al., 1988; Herskowitz, 1989). Cell type is determined by information present at the mating type locus, *MAT,* which encodes regulatory proteins. These proteins regulate unlinked genes, resulting in the specific expression of different sets of gene products in the three cell types: a-specific genes are expressed only in a cells, α-specific genes only in α cells, haploid-specific genes in both a and α cells, and a/α-specific genes only in a/α cells. Some of these genes are also subject to other types of regulation, such as induction by exposure to peptide pheromones, as described below.

Many of the cell-type-specific proteins are involved in cell-cell interactions that play roles in the mating process. One type of interaction involves the secretion of and response to peptide pheromones. The second type of interaction involves

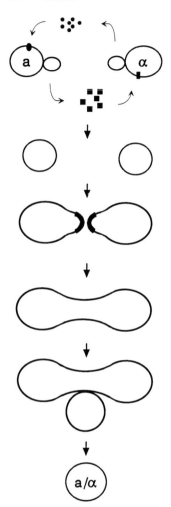

FIGURE 1. Yeast mating. Peptide pheromones, **a**-factor and α-factor, are secreted by **a** and α cells, respectively. These pheromones bind to specific receptors on cells of the opposite mating type and elicit a set of responses. Cells arrest as unbudded cells in the G1 phase of the cell cycle but continue to grow and elongate to form the altered morphology popularly called shmoos. A number of gene products are induced during this response; products involved in direct cell-cell contact and fusion localize to the shmoo tip. The cells fuse to form a dumbbell-shaped zygote, and after nuclear fusion within the zygote, **a**/α buds are produced.

cell surface glycoproteins that interact with one another to mediate cellular adhesion. Some of the components involved in these interactions were identified by the isolation and characterization of mutants defective in mating (*ste* mutants; MacKay and Manney, 1974; Hartwell, 1980). Mutations in other components do not result in a sterile phenotype; therefore, other approaches were used to identify these components.

The peptide pheromones, **a**-factor and α-factor, are secreted by **a** and α cells, respectively, and **a** and α cells respond to the pheromone produced by cells of the opposite mating type (Fig. 1; Cross et al., 1988). Production of α-factor occurs through the classical secretory pathway and has been extensively characterized, including the identification of all of the genes involved in processing of the α-factor precursor (Fuller et al., 1988). Production of the lipopeptide **a**-factor does not occur through the classical secretory pathway but instead involves a novel mechanism (McGrath and Varshavsky, 1989; Kuchler et al., 1989).

Many components involved in response to pheromone have been identified (Cross et al., 1988). Specific receptors bind pheromone to initiate the response; **a** cells express the α-factor receptor, Ste2, and α cells express the **a**-factor receptor, Ste3. The intracellular machinery involved in this response is similar in cells of both mating types. An early intracellular step involves the action of a heterotrimeric G protein that has sequence similarity to mammalian and other G proteins involved in a wide variety of signal transduction systems. Downstream components include protein kinases and a transcription factor involved in gene induction of components involved in pheromone production and response, components that play roles in cell and nuclear fusion, and components involved in desensitization to pheromone.

Adhesion of cells of opposite mating type is mediated by the sexual agglutinins, which are cell surface glycoproteins that are induced by pheromone (Lipke and Kurjan, in press). The agglutinins expressed by **a** and α cells, **a**-agglutinin and α-agglutinin, respectively, interact with one another to promote aggregation or agglutination of mixtures of cells of the two mating types. The agglutinins are only necessary for efficient mating under certain conditions (see below; Lipke et al., 1989; Roy et al., 1991). Other species of budding yeast with similar mating systems express analogous agglutinins. Until recently, biochemical approaches were used to characterize the agglutinins from several yeast species. In the past several years, a molecular approach has been undertaken in *S. cerevisiae*.

Response to pheromone also elicits cell wall and nuclear changes that are thought to be necessary for cell and nuclear fusion. The recent identification of genes involved in these processes should facilitate characterization of the mechanisms of these fusion events. In this review, the term "yeast" will be used to indicate *S. cerevisiae;* in sections where comparisons to other yeast species are made, I will refer to the species by their formal names.

PHEROMONE PRODUCTION

The discovery of the sex pheromones α-factor and **a**-factor involved the observation that cell culture supernatants contain activities that result in cell cycle arrest, morphological alterations, and induction of agglutinability of cells of the opposite mating type (Duntze et al., 1970; Bucking-Throm et al., 1973; Shimoda et al., 1976; Hagiya et al., 1977; Betz et al., 1978). Purification and sequencing of the active component from α cells indicated that these multiple activities are induced by the same peptide pheromone.

Production of α-Factor

The α-factor peptide purified from α cell supernatants is a mixture of 12- and 13-amino-acid peptides that differ by the presence or absence of an N-terminal Trp residue (Sakurai et al., 1976; Stotzler and Duntze, 1976; Stotzler et al., 1976). Synthetic peptides of both lengths are active in inducing all of the responses observed for natural α-factor. Mutants defective in secretion (*sec* mutants) show defects in α-factor secretion, suggesting that α-factor is secreted through the classical secretory pathway (Julius et al., 1984b).

The first α-factor structural gene to be identified (*MFα1*) was cloned by isolation of a high-copy-number plasmid that resulted in overproduction of α-factor in a strain that is defective in α-factor production due to a *MATα* mutation Fig. 2A; Kurjan and Herskowitz, 1982). A second α-factor structural gene (*MFα2*) was cloned by hybridization to *MFα1*, and both *MFα1* and *MFα2* were cloned using a degenerate oligonucleotide corresponding to the α-factor sequence (Singh et al., 1983; Kurjan, 1985). Gene disruption analysis indicates that the majority of α-factor is produced by *MFα1* and that only a small proportion of α-factor is produced by *MFα2* (Kurjan, 1985). Both single mutants retain mating ability; an *mfα2* mutation alone has little or no effect on mating efficiency, and an *mfα1* mutation alone results in only a few-fold decrease in mating. Double mutants (α *mfα1 mfα2*) show no detectable α-factor production or mating. These results indicate that *MFα1* and *MFα2* are the only α-factor structural genes and that at least one *MFα* gene is necessary for mating by α cells. The ability of the single mutants to mate explains why α-factor structural genes were not identified by isolation of α-specific sterile mutants.

The structure of the *MFα1* and *MFα2* genes indicates that α-factor is processed from precursors of 165 and 120 amino acids, respectively (Fig. 2A; Kurjan and Herskowitz, 1982; Singh et al., 1983). Both gene products have a hydrophobic N terminus with features of signal sequences of secreted proteins. The signal sequences are followed by regions (called pro regions) of about 60 amino acids containing three putative sites for N-linked glycosylation. The C termini of the precursors consist of tandem repeats of α-factor or α-factor-like sequences, each preceded by a spacer peptide of six to eight amino acids. The *MFα1* precursor contains four repeats, all of which contain the sequence determined for secreted α-factor. *MFα1* constructs with one to four spacer-α-factor repeats show levels of secreted α-factor proportional to the number of repeats, suggesting that all four repeats are processed to produce mature α-factor (Caplan et al., 1991). The *MFα2* precursor contains two repeats; one encodes α-factor, and the second encodes a peptide that differs from α-factor by two amino acid substitutions, which will be referred to as α-factor'. The α-factor' peptide was not detected among secreted α-factor; however, because *MFα1* is responsible for the majority of α-factor production (Kurjan, 1985), a small proportion of *MFα2*-encoded α-factor' could have been missed. Analysis of precursor processing has focused on the *MFα1* precursor, as described below, and there is no evidence as to whether or not α-factor' is actually produced. Synthetic α-factor' and α-factor have similar activities in induction of cell cycle arrest, morphological changes, and agglutinability (Raths et al., 1986; Kurjan and Lipke, 1986). Also, strains expressing equivalent levels of α-factor or α-factor' mate at similar levels (Caplan and Kurjan, 1991). The activities of α-factor and α-factor' therefore seem to be equivalent.

The original sequence of the *MFα1* precursor (prepro-α-factor) showed features consistent with secretion through the classical secretory pathway and processing by mechanisms similar to those characterized for processing of precursors in higher eukaryotes (Kurjan and Herskowitz, 1982). This processing pathway

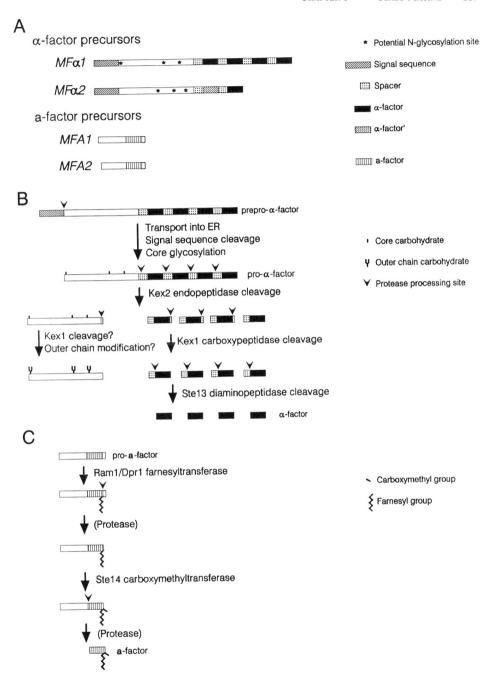

FIGURE 2. Pheromone structural genes and precursor processing. (A) The α-factor and a-factor precursor structures, based on gene sequences, are shown. The pathways for precursor processing to produce α-factor (B) and a-factor (C) are shown. Details are described in the text. The order of the Kex1 and Ste13 steps has not been determined (B). The processed pro region has not been detected; therefore, its fate is unknown. The proteases involved in two of the a-factor processing steps have not been identified (C).

has been extensively characterized, including the identification of the genes encoding the processing enzymes (Fig. 2B). Prepro-α-factor is translocated into the endoplasmic reticulum and a 19-amino-acid signal sequence is cleaved (Waters et al., 1988); an initial report suggesting that the signal sequence was not cleaved was based on aberrant gel mobility of various precursor forms (Julius et al., 1984a; Waters et al., 1988). One unusual property of prepro-α-factor is that it can be translocated across microsomal membranes in vitro posttranslationally as well as cotranslationally (Hansen et al., 1986; Rothblatt and Meyer, 1986; Waters and Blobel, 1986). Processing of the *MFα2* precursor has not been analyzed, but a precursor of the expected size can be synthesized in vitro and translocated into microsomal membranes (Rothblatt et al., 1987).

All three potential N-glycosylation sites in the pro region of pro-α-factor are modified by addition of core N-linked carbohydrate (Fig. 2B; Emter et al., 1983; Julius et al., 1984a). Outer chain modification of pro-α-factor occurs in mutants defective in transport out of the Golgi or in precursor processing (Julius et al., 1984a; Julius et al., 1984b); whether similar modification of the processed pro peptide occurs in wild-type cells has not been determined (Julius et al., 1984b). Elimination of N-glycosylation of pro-α-factor, either by tunicamycin addition or mutation of the glycosylation sites, results in reduced α-factor secretion and intracellular accumulation of pro-α-factor (Julius et al., 1984a; Caplan et al., 1991), indicating that N-glycosylation is not essential for secretion of α-factor but does seem to facilitate transport through the secretory pathway.

Three proteolytic processing steps are involved in production of mature α-factor from the *MFα1* precursor (Fig. 2B). The α-factor production and α-specific mating defects of *kex2* mutants, which were identified by a defect in production of the secreted killer toxin (Wickner and Leibowitz, 1976), suggested that *KEX2* might be involved in pro-α-factor processing. An endopeptidase activity capable of cleaving on the C-terminal side of dibasic sequences is dependent on *KEX2* (Julius et al., 1984b; Achstetter and Wolf, 1985). *KEX2* encodes a membrane-bound, Ca^{2+}-dependent serine protease that cleaves on the C-terminal side of the Lys-Arg residues in the pro-α-factor spacers (Julius et al., 1984b; Achstetter and Wolf, 1985; Bathurst et al., 1987; Wagner and Wolf, 1987; Mizuno et al., 1988; Fuller et al., 1989). Kex2 cleavage occurs in the late Golgi (Julius et al., 1984a; Julius et al., 1984b; Fuller et al., 1989; Payne and Schekman, 1989). A putative aspartyl protease encoded by *YAP3* can substitute for Kex2 cleavage when both *YAP3* and *MFα1* are expressed on multicopy plasmids, but Yap3 does not seem to play a role in pro-α-factor processing under normal conditions (Egel-Mitani et al., 1990).

Kex2 cleavage produces a product with a mature C terminus derived from the fourth *MFα1* repeat, and products with the extra Lys-Arg residues at the C termini derived from the other three repeats (Fig. 2B). A membrane-bound serine carboxypeptidase encoded by *KEX1* results in the trimming of these residues from the C terminus to produce mature α-factor (Dmochowska et al., 1987; Wagner and Wolf, 1987; Cooper and Bussey, 1989). *KEX1* was originally identified among mutants that showed a defect in killer toxin secretion (Wickner and Leibowitz, 1976). Because Kex1 is not required for maturation of the C-terminal α-factor repeat, *kex1* mutations reduce but do not eliminate α-factor production and therefore do not result in mating defects. Dibasic residues are frequently found flanking polypeptide hormone sequences in precursors in higher eukaryotes. Processing by an endopeptidase that recognizes dibasic residues, followed by carboxypeptidase B trimming, is therefore a highly conserved and ubiquitous mechanism (Steiner et al., 1980; Herbert and Uhler, 1982).

Maturation of the N terminus of α-factor requires removal of four or six amino acid spacers, which are made up of Glu/Asp-Ala repeats, remaining after Kex2 cleavage (Fig. 2B). The similarity of these spacer sequences to regions removed from the N terminus of melittin, the major component of honeybee venom, suggested that a similar mechanism might be involved in α-factor processing (Kurjan and Herskowitz, 1982). *STE13* is required for mating by α cells and encodes a membrane-bound diaminopeptidase that removes N-terminal Glu-Ala and Asp-Ala dipeptides (Julius et al., 1983; Fuller et al., 1988). Two to three consecutive Ste13 processing steps are required to remove the four to six amino acid spacers to produce mature α-factor. Similar enzymes involved in N-terminal precursor processing have been identified in insects, frogs, and pigs, indicating that this mechanism is also quite conserved, although it may be a rarer event than processing at dibasic residues (Kreil, 1990).

Production of a-Factor

Determination of the sequence of **a**-factor was more difficult than for α-factor, largely due to the lipophilic modification of **a**-factor. Active **a**-factor is a mixture of two 12-amino-acid peptides that differ by a single amino acid substitution; the C-terminal Cys residue of **a**-factor is modified by *trans, trans-S*-farnesylation and carboxymethylation (Betz et al., 1987; Anderegg et al., 1988). The activity of synthetic farnesylated and carboxymethylated **a**-factor has confirmed this identification (Xue et al., 1989). Both farnesylation and carboxymethylation are necessary for activity.

Two **a**-factor structural genes, *MFA1* and *MFA2*, were identified using degenerate oligonucleotides corresponding to the **a**-factor peptide sequence (Fig. 2A; Brake et al., 1985). Disruption of either *MFA1* or *MFA2* results in about a twofold decrease in **a**-factor production, indicating that each gene contributes equally to **a**-factor production (Michaelis and Herskowitz, 1988). Whereas the single mutants are capable of efficient mating, disruption of both genes eliminates **a**-factor production and **a** mating, indicating that these are the only **a**-factor structural genes and that at least one *MFA* gene is required for mating.

The *MFA1* and *MFA2* genes encode precursors of 36 and 38 amino acids, respectively (Fig. 2A; Brake et al., 1985). Each precursor encodes a single copy of the **a**-factor sequence; a single amino acid difference in the two genes accounts for the two forms of **a**-factor (Betz et al., 1987).

The **a**-factor precursors do not contain hydrophobic regions that resemble signal sequences or canonical sites for protein processing in the secretory pathway, consistent with the lack of a defect in **a**-factor secretion in *sec* mutants. The similar C-terminal C-A-A-X (C = Cys, A = an aliphatic amino acid, and X =

any amino acid) sequences of *MFA1* and *MFA2* and Ras suggested that processing and modification of the **a**-factor precursors might be similar to processing and modification of Ras (Fig. 2C). The Cys residue of mammalian *ras* is modified by lipid addition and carboxymethylation, and the three C-terminal amino acids are not present in the mature protein. The *RAM1/DPR1* gene is required for both Ras activity and **a**-factor production; the latter requirement results in **a**-specific sterility in *ram1/dpr1* mutants (Powers et al., 1986; Fujiyama et al., 1987). *RAM1/DPR1* encodes a component of the farnesyltransferase necessary for farnesylation of **a**-factor (Schafer et al., 1989; Schafer et al., 1990). The *RAM2* gene, which is necessary for wild-type levels of **a**-factor, may encode an additional farnesyltransferase component (Goodman et al., 1990). Another **a**-specific sterile gene, *STE14*, is necessary for carboxymethylation of **a**-factor and may encode the methyltransferase (Hrycyna and Clarke, 1990; Marr et al., 1990). Farnesylation of Cys requires the presence of the C-terminal three amino acids of the precursor, indicating that isoprenylation occurs before proteolytic processing (Schafer et al., 1990). The component(s) responsible for C-terminal proteolytic processing has not been identified. Results from an in vitro processing system confirm that the order of C-terminal events is isoprenylation, proteolytic processing, and carboxymethylation (Marcus et al., 1991). An N-terminal processing step is also necessary to produce mature **a**-factor, but no information is available on the component(s) responsible for this step.

The remaining **a**-specific sterile mutant defective in **a**-factor production (*ste6*) allowed the identification of a protein involved in secretion of the cytoplasmically synthesized **a**-factor. The *STE6* sequence shows similarity to several proteins involved in pumping compounds out of cells (McGrath and Varshavsky, 1989; Kuchler et al., 1989), including the multi-drug resistance (MDR1) gene product in mammalian cells, the chloroquine resistance gene in mosquitos, and the cystic fibrosis transmembrane conductance regulator (Higgins, 1989; Ringe and Petsko, 1990; Davies, 1990). These proteins contain tandem duplications, each of which contains six putative transmembrane-spanning domains and an ATP binding site. Mutations in *ste6* result in intracellular accumulation of pro-**a**-factor and **a**-factor (Kuchler et al., 1989). Overall, these results suggest that the processed **a**-factor lipopeptide associates with the membrane and is then secreted by Ste6 in an ATP-dependent manner.

COURTSHIP

Changing the receptor and pheromone expressed by a cell changes the mating type with which it mates, suggesting that the sending and receiving of pheromone result in directional signals that are important for mating (Bender and Sprague, 1989). In recent elegant experiments, Jackson and Hartwell (1990a, 1990b) have shown that potential mating partners signal one another by the secretion of pheromone in an early step called courtship. This polarized signal allows cells to identify an appropriate mating partner and is proposed to act as a morphogen that promotes the elongation of the shmoo tips of the two potential partners towards one another (Fig. 1). In these experiments, cells were found to choose the partner that secretes the highest level of pheromone.

This ability to discriminate a potential mating partner is eliminated at very high pheromone concentrations (Jackson and Hartwell, 1990b). These high concentrations can result from addition of extremely high levels of exogenous pheromone or from a mutation in one of the mating partners that results in a defect in degradation of the opposite pheromone. Very high pheromone concentrations also reduce mating of wild-type cells (Sena et al., 1973; Kurjan, 1985), indicating that the ability to discriminate potential mating partners is necessary for efficient mating.

The elucidation of the phenomena of courtship helps explain several previous results. Because α-factor is secreted into the extracellular environment, it seemed likely that addition of exogenous α-factor would alleviate the mating defect of α *mfα1 mfα2* mutants. In matings of these double mutants to wild-type **a** strains, however, exogenous α-factor does not alleviate the mating defect (Kurjan, 1985). One possible hypothesis to explain this result was that the *MFα* precursors play a second role in mating separate from α-factor production. If so, some *mfα1* mutations should eliminate mating without eliminating α-factor production. Instead, a close correlation between the level of α-factor production and the level of mating has been seen with mutants containing a wide variety of structural alterations of the *MFα1* precursor (Caplan and Kurjan, 1991). A second hypothesis seems more consistent with the finding that addition of exogenous pheromone to **a** cells leads to results similar to those found for α cells, i.e., exogenous **a**-factor does not alleviate the mating defect of **a** *mfa1 mfa2* mutants (Michaelis and Herskowitz, 1988). This hypothesis is that a reciprocal directionality of the signalling process between cells is necessary for efficient mating, i.e., that cells need to "see" cells of the opposite mating type secrete pheromone.

The need for the directional signal resulting from secretion of pheromone by a potential mating partner is eliminated under conditions where cells have lost the ability to discriminate pheromone levels (Jackson and Hartwell, 1990b). Exogenous α-factor can partially alleviate the α *mfα1 mfα2* mating defect in crosses to **a** *bar1/sst1* strains, which are defective in degradation of α-factor, resulting in very high α-factor concentrations (Kurjan, 1985). Also, an α-factor concentration about 10-fold higher than used in the previous study is able to alleviate the α *mfα1 mfα2* mating defect even in crosses to wild-type cells (Jackson and Hartwell, 1990b). Under both of these conditions, the **a** strain has lost the ability to discriminate due to the very high level of α-factor present. In either mating type, loss of the ability to discriminate pheromone levels due to a desensitization defect resulting from an *sst2* mutation (described below) allows mating to cells unable to produce pheromone; i.e., **a** *mfa1 mfa2* and α *mfα1 mfα2* cells are able to mate with *sst2* cells of the opposite mating type (Kurjan, 1985; Michaelis and Herskowitz, 1988).

Intracellular activation of the pheromone response pathway allows a low level of mating in the absence of receptor (individual examples are described below; Bender and Sprague, 1986; Nakayama et al., 1987; Jahng et al., 1988; Nakayama et al., 1988b; Dolan and Fields, 1990). This intracellular activation can occur by induction of a positive component of the pathway or repression of a negative component. Intracellular activation of the pathway does not provide the directional signal normally provided by exposure to a potential mating partner secreting pheromone, explaining the low level of mating seen under these conditions. Thus, the need for an extracellular signal is partially eliminated by intracellular activation of the pheromone response pathway or by cellular defects resulting in the elimination of the cell's ability to discriminate the signal, as described above.

PHEROMONE RESPONSE PATHWAY

Exposure of haploid cells to the pheromone produced by the opposite mating type results in a similar set of responses in both mating types. Cells arrest as unbudded cells in the G1 phase of the cell cycle but continue to enlarge and form extended forms called shmoos (Fig. 1). Genes that encode components involved in many aspects of pheromone response and mating have been identified.

Pheromone Receptors

α-Factor receptor

One of the original sterile mutants, *ste2*, is an α-specific sterile, is defective in response to α-factor, and shows normal production of **a**-factor, consistent with a defect in the α-factor receptor. Binding of radiolabeled α-factor to temperature-sensitive *ste2* mutants is thermolabile, suggesting that Ste2 is a component of the receptor (Jenness et al., 1983). Expression of *Saccharomyces kluyveri* Ste2 rather than *S. cerevisiae* Ste2 results in a change in specificity; the resulting strain efficiently responds to *S. kluyveri* α-factor (Marsh and Herskowitz, 1988). Truncation of *STE2* results in decreased size of an α-factor-binding protein (Blumer et al., 1988), and *Xenopus* oocytes expressing Ste2 bind labeled α-factor (Yu et al., 1989). Together, these results show that *STE2* encodes the α-factor receptor.

Sequencing of the *STE2* gene showed that it could encode a protein with seven hydrophobic domains that were proposed to represent transmembrane domains (Burkholder and Hartwell, 1985; Nakayama et al., 1985). The resemblance to rhodopsin, the G protein-linked membrane protein involved in vertebrate phototransduction, was striking. The later discovery that other G protein-linked receptors, such as β-adrenergic and muscarinic receptors, showed a similar structural pattern suggested that response to α-factor might be mediated by a G protein (Marsh and Herskowitz, 1987).

a-Factor receptor

Another original sterile mutant, *ste3*, is an **a**-specific sterile, produces α-factor, but is defective in response to **a**-factor, suggesting that *STE3* might encode the **a**-factor receptor (Hagen et al., 1986). A Ste3–β-galactosidase fusion localizes to a membrane fraction. The *STE3* sequence is not similar to the *STE2* sequence, but does contain the same structural pattern of seven hydrophobic, putative transmembrane domains (Nakayama et al., 1985; Hagen et al., 1986). The finding that

Ste3 expression in **a** cells allows induction of aspects of **a**-factor response provides additional evidence that Ste3 is the **a**-factor receptor (Bender and Sprague, 1986).

The Intracellular Pathway Is Similar in a and α Cells

Elegant experiments, in which the pheromone receptors are expressed under the control of a galactose-inducible promoter in various types of haploid cells, indicate that the receptors are the only haploid-specific products that determine the specificity of pheromone response (Bender and Sprague, 1986; Nakayama et al., 1987). *matα1* mutants normally do not express either receptor; expression of Ste3 (**a**-factor receptor) allows response to **a**-factor, and expression of Ste2 (α-factor receptor) allows response to α-factor. Autocrine pheromone response in **a** cells expressing both Ste3 (**a**-factor receptor) and **a**-factor is sufficient to allow a low level of mating to α cells that do not produce α-factor (Bender and Sprague, 1986). The reverse experiment, i.e., expression of both Ste2 (α-factor receptor) and α-factor in α cells that do not express Ste3, allows a low level of mating to **a** cells (Nakayama et al., 1987). These results indicate that the intracellular response pathway is shared by **a** and α cells. Consistent with this conclusion, all intracellular components of the pathway that have been identified (described below) affect pheromone response and mating in both mating types.

G Protein

Three genes with similarity to G protein subunits from mammalian systems have been identified. The *SCG1* (also called *GPA1*), *STE4,* and *STE18* genes show sequence similarity to the α, β, and γ genes, respectively (Dietzel and Kurjan, 1987a; Nakafuku et al., 1987; Whiteway et al., 1989). Mutations in these genes result in phenotypes in **a** and α but not **a**/α cells.

Disruption of *SCG1* leads to constitutive activation of the pheromone response pathway, resulting in growth arrest and changes in cellular morphology, called shmooing (Dietzel and Kurjan, 1987a; Miyajima et al., 1987). Conditional expression of Scg1 indicates that turning off of Scg1 function leads to induction of pheromone-inducible genes and allows a low level of receptor-independent mating (Jahng et al., 1988; Nakayama et al., 1988b). These results indicate that Scg1 plays a negative role in the pheromone response pathway; in the absence of pheromone it acts to keep the pathway off.

Disruption of *STE4* or *STE18* results in defects in pheromone response and mating, indicating that βγ plays a positive role; it activates the pathway after exposure to pheromone (Whiteway et al., 1989). Double mutants (*scg1 ste4* and *scg1 ste18*) show normal growth and morphology and defects in pheromone response and mating; i.e., *ste4* and *ste18* are epistatic to *scg1*. This result shows that βγ acts downstream of α in this pathway (Fig. 3).

In the current model for this pathway, the G protein exists as a heterotrimer with bound GDP in the absence of pheromone (Fig. 3); binding of pheromone to receptor results in guanine nucleotide exchange, dissociation of α-GTP from βγ, and activation of a downstream effector by free βγ (Dietzel and Kurjan, 1987a; Whiteway et al., 1989). The negative role of α is to bind to βγ to prevent it from

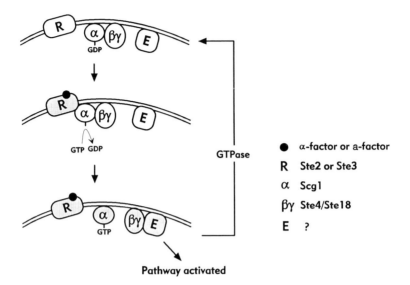

● α-factor or a-factor

R   Ste2 or Ste3

α   Scg1

βγ   Ste4/Ste18

E   ?

FIGURE 3. Model for action of the G protein involved in pheromone response. Activated forms of components are dotted. In the absence of pheromone, the G protein exists as a heterotrimer with bound GDP. Activation of receptor by pheromone binding results in guanine nucleotide exchange and dissociation of α-GTP from βγ. Free βγ activates a downstream effector, which has not been identified. A GTPase activity intrinsic to the α subunit shifts the system back to the basal heterotrimeric state.

activating the effector, and the positive role of βγ is to activate the effector. A GTPase activity intrinsic to the α subunit shifts the system back to the basal heterotrimeric state. This model is similar to models for the best-characterized vertebrate systems, except that in these systems, after dissociation of α-GTP from βγ, α-GTP activates an effector, i.e., α plays a positive role. Direct activation of phospholipase A2 by βγ has been reported in mammalian systems (Jelsema and Axelrod, 1987), so an active role for βγ may not be limited to yeast.

Altered expression levels of the G protein subunit genes provide additional results consistent with this model and indicate that there is a critical balance between the levels of the components. Overexpression of Scg1 results in decreased pheromone response and mating, consistent with its negative role (Dietzel and Kurjan, 1987a; Kang et al., 1990; Cole et al., 1990). Increased expression of Ste4 results in activation of the pathway (Whiteway et al., 1990; Cole et al., 1990; Nomoto et al., 1990); this activation requires Ste18 and is eliminated by increased expression of Scg1. Also, the ability of catecholamine (β-adrenergic agonist) to induce aspects of the pheromone response pathway in an ''in vivo'' reconstitution system in which cells express the β-adrenergic receptor, its associated α subunit, $α_s$, and yeast βγ indicates that βγ is the activating subunit (King et al., 1990).

Several lines of evidence suggest that Scg1 binds guanine nucleotides and associates with the pheromone receptors. The addition of GTP[γ-S], a nonhydrolyzable GTP analog, along with labeled α-factor shifts the affinity of the receptor-pheromone binding to a lower affinity state (Blumer and Thorner, 1990), similar to effects in mammalian systems. This effect is dependent on all three components of the G protein. These results suggest that the yeast G protein homolog binds guanine nucleotides and is involved in the receptor interaction. In addition, mutations in the putative guanine nucleotide binding domains of Scg1 that have been shown to have effects on guanine nucleotide binding or GTPase activity in *ras* or α_s result in phenotypes predicted based on their effect in the other systems (Kurjan et al., 1991). The carboxy termini of mammalian α subunits have been implicated in receptor interactions. C-terminal Scg1 mutations result in mating defects, consistent with a similar role in receptor interactions in the yeast system (Hirsch et al., 1991). Some mutations have more severe effects in one mating type than in the other, suggesting that the specificity for the interactions with the nonhomologous **a**- and α-factor receptors may differ.

Components That Act at the Level of the G Protein

Several components have been identified that act at the level of the G protein and may play roles in modification or synthesis of the G protein subunits. Loss-of-function mutations in gene products necessary for the positive role of $\beta\gamma$ should result in defects in activation of the pheromone response pathway and mating. Loss-of-function mutations in gene products necessary for the negative role of α should result in constitutive activation of the pathway and a resulting growth defect.

A mutation isolated by its ability to suppress the *scg1* growth defect was shown to be allelic to *ram1/dpr1* (Nakayama et al., 1988a), suggesting that isoprenylation of a downstream component of the pheromone response pathway is necessary for its activity. The C terminus of *STE18*, the γ subunit gene, has the C-A-A-X motif (Whiteway et al., 1989) that has been associated with isoprenylation. The mature form of Ste18 is found in a membrane fraction, but Ste18 with a Cys-to-Ser mutation is found in the soluble fraction and is nonfunctional, suggesting that Ram1/Dpr1 is involved in the modification, membrane localization, and function of Ste18 (Finegold et al., 1990). Surprisingly, *ram1/dpr1* mutations have a fairly mild effect on mating of α cells; the more severe effect seen in **a** cells is due to the defect in production of **a**-factor. Some other gene product may therefore be sufficient for adequate modification and localization of Ste18 to provide some function.

The *CDC72* gene was identified by mutations that result in arrest at the same point of the cell cycle as α-factor at the restrictive temperature (Jahng et al., 1988). *CDC72* is allelic to the *NMT1* gene (Duronio et al., 1989; Reed, personal communication), which encodes *N*-myristoyltransferase. *cdc72* mutants are defective in membrane localization of Scg1, indicating that myristoylation of Scg1, which contains a consensus site for this modification, is necessary for proper localization.

cdc36 and *cdc39* mutants also arrest at the same point of the cell cycle as *scg1* mutants, the point of pheromone arrest. Epistasis analyses of *cdc36* and *cdc39* and G protein mutants indicate that the products of these genes act at the level of the G protein (Neiman et al., 1990; de Barros Lopes et al., 1990). Cdc36 and Cdc39 have been suggested to act at the level of the α subunit, possibly by activation of GTPase or stabilization of the GDP-bound form of α (de Barros Lopes et al., 1990). The ability of a C-terminal Scg1 truncation that results in sterility to suppress the growth arrest of a *cdc39* mutant is consistent with these hypotheses (Neiman et al., 1990). The same Scg1 truncation does not suppress *cdc36* growth arrest, suggesting that Cdc36 may act at a different level. It might affect Scg1 expression or be involved in a modification that is necessary for the interaction of Scg1 with Ste4/Ste18.

Another mutant, *srm1*, was identified by its ability to allow mating of receptorless mutants (Clark and Sprague, 1989). It did not suppress a *ste4* mutant, however, suggesting that it might also act at the level of the G protein. More recent results indicate that the mating phenotype is an indirect effect of the mutation. Srm1 is allelic to *prp20,* which was originally identified due to a splicing defect (Aebi et al., 1990; Fleischmann et al., in press; Clark, personal communication). Srm1/Prp20 is a nuclear protein that seems to be involved in nuclear structure; therefore, it is unlikely to directly affect G protein function.

Downstream Components

Several additional components of the pheromone response pathway, most of which were identified as sterile mutants, act downstream of the G protein (Fig. 4; Blinder et al., 1989; Nakayama et al., 1988b). The *STE7, STE11,* and *FUS3* genes show sequence similarity to kinases (Teague et al., 1986; Elion et al., 1990; Rhodes et al., 1990). A conserved residue necessary for kinase function in other kinases is essential for Ste11 function, and a potential Ste11 substrate has been observed (Rhodes et al., 1990). Ste4, Ste5, Ste7, Ste11, and Ste12 are all phosphoproteins (Rhodes et al., 1990; Cole and Reed, 1991) and are therefore possible substrates for the three kinases.

a- and α-specific products such as **a**-factor, the agglutinins, and the barrier protease that degrades α-factor are induced by the opposite pheromone (Strazdis and MacKay, 1983; Manney, 1983; Terrance et al., 1987a; Achstetter, 1989). This induction occurs by increased RNA levels. Many additional genes have been shown to be under similar control, including genes involved in desensitization and cell and nuclear fusion (McCaffrey et al., 1987; Dietzel and Kurjan, 1987b; Trueheart et al., 1987; Meluh and Rose, 1990). A consensus sequence found upstream of pheromone-inducible genes is implicated in this induction (Kronstad et al., 1987; Trueheart et al., 1987). Ste12 binds to this sequence (Dolan et al., 1989; Errede and Ammerer, 1989) and therefore is likely to correspond to the transcription factor necessary for this regulation. Some modification, possibly the phosphorylation of Ste12 that occurs in yeast but not *Escherichia coli,* is necessary for efficient binding (Dolan et al., 1989). Increased Ste12 expression is sufficient to induce several aspects of pheromone response and a low level of mating in *ste* mutants, including *ste7* and *ste11,* indicating that Ste12 acts downstream of these two kinases (Dolan and Fields, 1990).

Recently, *FAR1*, a downstream component necessary for arrest of cells in

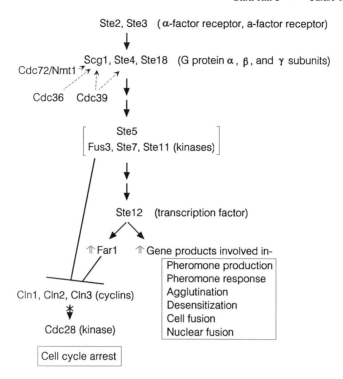

FIGURE 4. Components of the pheromone response pathway. The known components of the pheromone response pathway are indicated. Arrows indicate that a protein plays a positive or activating role, and lines ending in bars indicate that a protein plays a negative or inhibitory role. An arrow with an ⬍ indicates that the step is inhibited by an upstream inhibitory component. The actions of proteins implicated in modification of components of the pathway are indicated by dashed arrows. ⇑ represents increased expression of the indicated products. Double arrows between components indicate that the number of steps between the components is not known. Components in brackets may act at the same or different steps.

the G1 phase of the cell cycle but not for other responses to pheromone, including altered morphogenesis and pheromone induction of genes, has been identified (Fig. 4; Chang and Herskowitz, 1990). The *FAR1* gene is itself pheromone inducible. Together, these results indicate that Far1 acts downstream of Ste12. Mutation of *CLN2,* one of three cyclin genes (Cross, 1988, 1990a; Wittenberg et al., 1990), suppresses the arrest defect of *far1* mutations. The three cyclins are necessary for activity of Cdc28, a kinase necessary for transversing the point of the G1 phase of the cell cycle at which pheromone arrest occurs. It is therefore proposed that after induction by pheromone, Far1 prevents Cln2 activity, thereby reducing Cdc28 activity. All three cyclins must be inactivated to eliminate Cdc28 activity entirely, resulting in cell cycle arrest; therefore, other components must inactivate the other two cyclins. It has been proposed that Fus3 inactivates Cln3, but no candidate for inactivation of Cln1 is known (Chang and Herskowitz, 1990; Elion et al., 1990).

Basal Expression of the Pheromone Response Pathway

Mutations of many of the intracellular components of the pheromone response pathway that result in sterility lead to decreased basal expression of pheromone-inducible products, including the agglutinins, the pheromones, and the pheromone receptors (Hartwell, 1980). Pheromone-inducible genes show reduced basal RNA levels in mutants defective in intracellular components of the pheromone response pathway (Fields and Herskowitz, 1985; Fields et al., 1988). Mutations that increase sensitivity to pheromone show the opposite effect, i.e., increased basal expression of pheromone-inducible genes (Dietzel and Kurjan, 1987b). Interestingly, receptor mutants do not show a similar decrease in pheromone-inducible gene products (Hartwell, 1980). These results indicate that the pathway is in an equilibrium in which it is activated at a low level in the absence of pheromone and that this basal activation does not require receptor. Exposure to pheromone increases the level of activation, and mutation of intracellular components eliminates the basal level of activity. Because sterile mutations in all three G protein subunits show decreased basal expression, this effect is likely to occur at the G protein level; i.e., a low level of the GTP-bound form of the α subunit and of free $\beta\gamma$ is most likely present even in the absence of pheromone (Fig. 3).

Desensitization to Pheromone

After a period of growth arrest upon exposure to pheromone, cells are able to reinitiate growth even in the continuous presence of pheromone (Chan and Otte, 1982a, 1982b; Moore, 1984). This recovery involves both degradation of the opposite pheromone and a desensitization process in which cells are able to bind pheromone, but this binding does not activate the response; i.e., the pathway is uncoupled.

Degradation of pheromone involves specific proteases. **a** cells secrete an activity, called barrier, which degrades α-factor (Tanaka and Kita, 1977; Ciejek and Thorner, 1979). **a**-specific mutants that result in supersensitivity to pheromone (*sst1* or *bar1*) are defective in this activity (Sprague and Herskowitz, 1981; Chan and Otte, 1982a). The sequence of the *BAR1* gene suggests that it encodes a pepsinlike protease (MacKay et al., 1988). Barrier activity and the *BAR1* gene are induced by α-factor (Manney, 1983); therefore, exposure to α-factor increases the level of an activity that reduces α-factor levels, promoting recovery.

An extracellular α-specific activity that degrades **a**-factor has recently been detected (Marcus et al., 1991). This activity is not found in cell-free culture supernatants and is present in the particulate fraction from cell extracts, indicating that it remains associated with the cell. Activity is increased by exposure to **a**-factor and is inhibited by some proteases, suggesting that it is a pheromone-inducible protease. An α-specific mutant (*ssl1*) that is supersensitive to **a**-factor and had been reported to be defective in an activity that inactivates **a**-factor (Steden et al., 1989) shows wild-type degradation of **a**-factor (Marcus et al., 1991). *SSL1* therefore does not encode the activity described above, but could be involved in uptake of **a**-factor.

The ability of cells to recover from α-factor arrest under conditions where no α-factor degradation occurs indicates that there is a recovery process independent of pheromone degradation (Moore, 1984). This desensitization is not due simply to loss of receptors from the cell surface. After exposure to α-factor, receptor is internalized as indicated by loss of cell surface binding sites (Jenness and Spatrick, 1986). α-Factor is internalized along with the receptor and then degraded (Chvatchko et al., 1986). New receptor synthesis results in reappearance of binding sites; however, this reaccumulation occurs at a time when cells are still desensitized.

One component implicated in desensitization is the *SST2* product (Chan and Otte, 1982a; Dietzel and Kurjan, 1987b). *sst2* mutations result in supersensitivity to pheromone and a defect in recovery from arrest after removal of pheromone. The mechanism of Sst2 action has not been determined.

Desensitization may occur at multiple levels of the pheromone response pathway. Truncations of the cytoplasmic tail of the α-factor receptor, Ste2, lead to increased sensitivity to pheromone (Konopka et al., 1988; Reneke et al., 1988). This effect is independent of *BAR1* and *SST2*. The cytoplasmic tail of Ste2 is phosphorylated on Ser and Thr residues, and exposure to pheromone results in a 50 to 75% increase in phosphorylation (Reneke et al., 1988). Phosphorylation of the cytoplasmic C terminus of G protein-linked receptors in vertebrate receptors has been shown to be involved in desensitization. Two recent reports disagree as to whether the *ste2-T326* truncation mutant, which lacks 106 amino acids of the Ste2 C terminus, shows normal recovery after removal of α-factor; one group has shown that recovery is wild type (Jackson and Hartwell, 1990b), and the other group has shown that recovery is delayed (Cole and Reed, 1991). If the former result is correct, the supersensitivity of this mutant may not be due to a defect in desensitization. Without directly testing recovery, therefore, a supersensitive phenotype cannot be interpreted to indicate a defect in desensitization.

It has also been proposed that the GTP-bound form of Scg1 acts to desensitize cells (Miyajima et al., 1989; Stone and Reed, 1990). This suggestion is based mainly on an unusual pattern of inhibition of growth by α-factor on plates (Miyajima et al., 1989; Stone and Reed, 1990; Kurjan et al., 1991) of a mutant (*scg1^{Val50}*) with an alteration shown to result in a GTPase defect in *ras* and α$_s$. Certain aspects of the phenotype of this mutant differ when the mutant gene is present on a plasmid or as a gene replacement (Kurjan et al., 1991). It is difficult to conclude with certainty that the unusual phenotype is due to desensitization. An *scg1^{Ser302}* mutant shows supersensitivity to pheromone, but this supersensitivity is not correlated with a defect in recovery from pheromone (Jackson and Hartwell, 1990b). Further experiments are necessary to determine whether Scg1 plays a role in desensitization.

Recent results indicate that Ste4, the G protein β subunit, shows a low level of phosphorylation of Ser and some Thr residues and that the level of phosphorylation is increased upon exposure to pheromone (Cole and Reed, 1991). Deletion of an internal region not present in mammalian β subunits results in increased sensitivity to pheromone and a defect in recovery, suggesting that phosphorylation of β is involved in desensitization. This effect is independent of *SST2*. The deletion mutant also showed partial activation of the pathway in the absence of pheromone; this effect was suggested to be due to a defect in desensitization to the low basal activity of the pathway described above. In the best-characterized mammalian systems, where α-GTP is the active component, the action of the GTPase and reassociation of α-GDP with free βγ returns the system to the basal unactivated state. In the yeast system, where βγ is proposed to interact with an effector after dissociation from α-GTP, return of α to the GDP-bound form by the intrinsic GTPase (Fig. 3) might not be sufficient to return the system to the basal state. A modification of β that promotes its dissociation from the effector might be necessary to allow reassociation with α-GDP. It therefore is intuitively appealing to have one aspect of desensitization occur at the level of the β subunit in yeast.

Potential for Reconstitution of Other Systems in Yeast

Expression of rat α_s (the Gα subunit that acts to stimulate adenylate cyclase in mammalian cells) in yeast complements the *scg1* growth defect (Dietzel and Kurjan, 1987a). Based on the model for the pheromone response pathway, this result suggests that the α_s subunit can interact with yeast βγ to keep the pathway inactivated (Kang et al., 1990). The inability of the *scg1* strains expressing α_s to mate indicates that this heterologous α subunit cannot interact with the pheromone receptors.

The ability of α_s to complement the *scg1* growth defect suggested that it might be possible to reconstitute mammalian systems in yeast and use genetic approaches to study interactions between components. Functional β-adrenergic receptor has been expressed in yeast as a fusion protein in which the first few amino acids of this receptor are replaced by the N terminus of Ste2 (King et al., 1990). In a mutant lacking Scg1 and expressing α_s as well as β-adrenergic receptor, exposure to catecholamine results in gene induction and morphological changes but not mating. The mating defect is consistent with an inability of exogenous pheromone to allow mating in strains defective in the pheromone structural genes (Kurjan, 1985; Michaelis and Herskowitz, 1988). Unfortunately, the inability of rat α_{i2} (the Gα subunit involved in inhibition of adenylate cyclase in mammalian cells) and α_o (a Gα subunit present in high levels in the brain) to efficiently complement the *scg1* growth defect (Kang et al., 1990) suggests that it will not be straightforward to extend this approach to other Gα-mediated systems, although it may be possible to construct hybrid Gα genes that are useful for such an approach.

SEXUAL AGGLUTININS

Response to pheromone results in induction of the sexual agglutinins, thus promoting adhesion between cells of opposite mating type. The agglutinins from several species of budding yeast have been biochemically characterized (reviewed by Lipke and Kurjan, in press). In each species, the agglutinin expressed by one mating type (α-agglutinin in *S. cerevisiae*) is composed of a single polypeptide that is modified by addition of N-linked and possibly also O-linked carbohydrate

(Fig. 5). The activity of these agglutinins is eliminated by heat but not by treatment with reducing agents or removal of the N-linked sugar by endoglycosidase H. The opposite mating type of each species expresses an agglutinin (a-agglutinin in *S. cerevisiae*) that contains about 90% carbohydrate, most or all of which is O-linked. It is composed of multiple subunits, a core subunit that mediates cell surface attachment and one or more binding subunits (i.e., the subunit that binds the opposite agglutinin) attached to the core by disulfide linkage. This agglutinin is resistant to heat but is altered by reducing agents. Although these biochemical features are shared by the agglutinins of the different species, the actual agglutinin interactions are species specific (Burke et al., 1980; Pierce and Ballou, 1983; Yamaguchi et al., 1984a); e.g., the *S. cerevisiae* a-agglutinin binds to the *S. cerevisiae* α-agglutinin but not to analogs from other yeast species.

Agglutinin structural genes from *S. cerevisiae* have been isolated using genetic and biochemical approaches. The genetic approach involved the isolation of agglutination-defective mutants (Lipke et al., 1989; Roy et al., 1991). Mutations that result in an effect on agglutination only in **a** cells or only in α cells identify putative a-agglutinin and α-agglutinin structural genes, respectively. Complementation groups have been determined and agglutinin structural genes have been isolated by complementation of the agglutination defects. The biochemical approach involved determination of N-terminal amino acid sequences of purified agglutinins (Hauser and Tanner, 1989; Tanner, personal communication). Degenerate oligonucleotides were synthesized and used to isolate structural genes.

The general features of the agglutinins and their genes will be described here, with an emphasis on the agglutinins from *S. cerevisiae*. A more detailed review

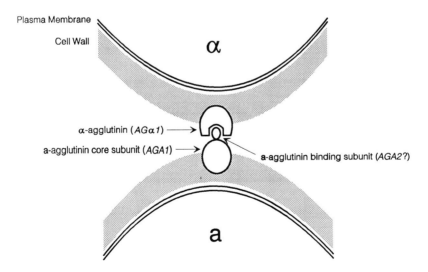

FIGURE 5. Sexual agglutinins. The agglutinins are cell wall glycoproteins that interact with one another to promote cellular adhesion and aggregation. α-Agglutinin is composed of a single subunit. a-Agglutinin is composed of two subunits, a core subunit that mediates cell surface attachment and a binding subunit that binds to α-agglutinin.

of the results that led to this overview is presented elsewhere (Lipke and Kurjan, in press).

Features of α-Agglutinin

Treatment of *S. cerevisiae* α cells with β-glucanase, which hydrolyzes cell wall glucan, yields a large glycoprotein with an estimated molecular size of 200 kDa (Shimoda and Yanagishima, 1975; Yamaguchi et al., 1982; Terrance et al., 1987b; Hauser and Tanner, 1989). Removal of N-linked carbohydrate by endoglycosidase treatment yields multiple species with sizes ranging from 160 to about 50 kDa. All of these species retain the ability to bind the opposite agglutinin. Peptide mapping indicated that these forms are all breakdown products of a single polypeptide, and protein sequencing indicated that they all contain the same N-terminal sequence.

The putative α-agglutinin structural gene, *AGα1,* was identified by both the genetic and biochemical approaches described above and shows features predicted for the α-agglutinin structural gene (Lipke et al., 1989; Hauser and Tanner, 1989). Localization of α-agglutinin to the cell surface involves transport through the secretory pathway (Tohoyama and Yanagishima, 1985, 1987). The amino terminus of the *AGα1* open reading frame is hydrophobic and resembles signal sequences of proteins targeted to the secretory pathway (Lipke et al., 1989). The N-terminal sequence of purified α-agglutinin starts at amino acid 20 of the open reading frame, indicating that a 19-amino-acid signal sequence is removed from the primary translation product (Hauser and Tanner, 1989). *AGα1* could encode a 70.5-kDa protein (including the putative signal sequence and anchor sequence described below) and has 12 putative N-glycosylation sites, consistent with a glycoprotein of about 160 kDa composed of about 50% N-linked carbohydrate. α-Agglutinin may contain O-linked as well as N-linked carbohydrate (Terrance et al., 1987b), and the high proportions of Ser and Thr residues in *AGα1* are potential sites for addition of O-linked sugar. Antibodies against purified α-agglutinin recognize a fusion protein expressed in *E. coli* that includes amino acids 128 to 356 of the *AGα1* coding sequences (Lipke et al., 1989). The N-terminal amino acid sequence of purified α-agglutinin is found in *AGα1* (Hauser and Tanner, 1989). These results indicate that *AGα1* is the α-agglutinin structural gene.

Other features of the *AGα1* coding sequences allowed us to speculate about functional domains of α-agglutinin (Lipke et al., 1989), and we now have some preliminary evidence consistent with these speculations. In addition to the hydrophobic N-terminal signal sequence, *AGα1* has a hydrophobic C terminus that does not have the charged residues characteristic of a transmembrane domain. This feature is similar to the precursors to proteins linked to the cell membrane by glycosyl phosphatidylinositol (GPI) anchors, which have hydrophobic C termini that are removed before attachment to the anchor (Cross, 1990b). We therefore speculated that the hydrophobic C terminus of *AGα1* is a signal sequence for the attachment of a GPI anchor.

If α-agglutinin remained attached to the cell membrane, it would have to be in a highly extended conformation for its binding domain to be exposed on the surface of the cell wall, where it is available to interact with **a**-agglutinin. Also, agglutinins can be released by β-glucanase, suggesting that the agglutinins are attached to the cell wall rather than the membrane. We therefore proposed that the putative GPI anchor acts to transport the agglutinin to the cell membrane, but that it is then transferred to the cell wall by an undetermined mechanism (Lipke et al., 1989). Truncation of *AGα1* to eliminate the hydrophobic C terminus results in secretion of α-agglutinin, indicating that this C-terminal domain does play a role in cell surface anchorage (Wojciechowicz et al., in preparation). Whether this role is mediated by a GPI anchor will require further experiments.

We also speculated that the relatively acidic region between amino acids 200 and 300 of the *AGα1* coding sequence corresponds to the binding domain (Lipke et al., 1989). This proposal was based on the acidic nature of the binding fragments of the analogous agglutinins from *Hansenula wingei* and *Pichia amethionina* (Burke et al., 1980; Mendonca-Previato et al., 1982). A C-terminal truncation of *AGα1* to amino acid 350 results in secretion of a smaller product that retains activity in binding the opposite agglutinin; truncation to amino acid 280 results in lower-level secretion of a protein with very low binding activity (Wojciechowicz et al., in preparation). These results localize the C-terminal boundary of the binding domain to between amino acids 280 and 350. An exciting observation (Wojciechowicz et al., in preparation) is that the region between amino acids 200 and 300 shows sequence and structural similarity to the immunoglobulin fold structure, a conserved feature found in many cell surface adhesion proteins in higher organisms (Williams and Barclay, 1988).

α-Agglutinin is an α-specific product that is induced by exposure to **a**-factor (Terrance et al., 1987a). There are strain differences in the level of basal expression and the degree of induction; however, most strains show detectable basal expression and only moderate induction. Agglutinability is increased only 1.5- to 2.5-fold by pheromone, and protein levels increase about 6-fold (Wojciechowicz and Lipke, 1989). An *AGα1* transcript is present in α but not **a** or **a**/α cells, and its level is increased at least 20-fold in α cells by exposure to **a**-factor, indicating that regulation occurs at the RNA level (Lipke et al., 1989). Sequences similar to consensus sequences identified for α-specific expression and pheromonal induction are present upstream of the *AGα1* coding sequence.

Features of a-Agglutinin

a-Agglutinin and its analogs from other yeasts consist of multiple subunits and have a very high carbohydrate content (80 to 95%), most or all of which is O-linked. The most thorough biochemical analysis of *S. cerevisiae* **a**-agglutinin has been carried out on the small binding subunit, which is released from **a** cells by treatment with reducing agents and is able to bind to the opposite agglutinin (Orlean et al., 1986; Watzele et al., 1988). Although larger forms of **a**-agglutinin from *S. cerevisiae* have been analyzed less extensively, they show properties similar to those of analogous agglutinins from other yeasts (Hagiya et al., 1977; Yamaguchi et al., 1984b; Wagner and Lipke, personal communication). These agglutinins are composed of a highly glycosylated core subunit that mediates cell surface anchorage, with one or more binding subunits with a lower carbohydrate content attached to the core by disulfide linkage. The **a**-agglutinin analogs from *H. wingei* and *P. amethionina* are multivalent; i.e., there are several binding subunits attached to each core subunit (Taylor and Orton, 1967; Yen and Ballou, 1974; Mendonca-Previato et al., 1982). The **a**-agglutinins from *S. cerevisiae* and *S. kluyveri* are monovalent; i.e., a single binding subunit is attached to each core subunit (Pierce and Ballou, 1983; Lipke, personal communication). The *H. wingei* and *P. amethionina* **a**-agglutinin analogs, particularly the core subunits, contain very high proportions of Ser and Thr residues (>50%), and most of these residues are modified by attachment of O-linked carbohydrate.

Most of the **a**-specific agglutination-defective mutants isolated in the genetic screen described above identify the *AGA1* gene (Roy et al., 1991). Features of this gene and of the *aga1* mutants suggest that *AGA1* encodes the **a**-agglutinin core subunit. *AGA1* contains an N-terminal hydrophobic sequence that resembles a signal sequence for transport to the cell surface through the secretory pathway. It also has a hydrophobic C terminus, and a truncation that eliminates this domain allows secretion of **a**-agglutinin, consistent with a role for this sequence in cell surface anchorage, possibly mediated by a GPI anchor. The *aga1* mutants secrete the binding subunit, indicating that *AGA1* is necessary for cell surface anchorage of the binding subunit and suggesting that *AGA1* encodes the **a**-agglutinin core subunit. Other than the hydrophobic N and C termini, the *AGA1* coding sequence is composed of >50% Ser and Thr residues, which is consistent with the amino acid composition of the analogous agglutinins from other yeasts and the high degree of O-linked glycosylation of **a**-agglutinin.

A single mutant, *aga2*, represents a second **a**-specific complementation group (Roy et al., 1991) that may identify an **a**-agglutinin binding subunit structural gene. A portion of the **a**-agglutinin binding subunit gene has been cloned using degenerate oligonucleotides corresponding to the N-terminal sequence of purified **a**-agglutinin binding fragment (Tanner, personal communication). Additional analysis is necessary to determine structural features of the gene and whether or not the cloned gene corresponds to *AGA2*.

The **a**-agglutinin binding fragment is **a** specific and induced by pheromone (Watzele et al., 1988; Terrance and Lipke, 1981). Again, there are strain differences in basal levels and degree of induction, but most strains show very low (often undetectable) basal expression and strong induction by pheromone. Surprisingly, an *AGA1* RNA is expressed at low levels in both **a** and α cells and is induced in both mating types by exposure to the opposite pheromone (Roy et al., 1991). Consistent with this expression pattern, a haploid-specific consensus sequence and several close matches to the pheromone induction sequence are present upstream of the *AGA1* coding sequence. The **a**-agglutinin core subunit may therefore be expressed in both **a** and α cells exposed to pheromone. No phenotype has been observed in α *aga1* mutants; therefore, if Aga1 is expressed in α cells, any role remains undetermined. Because the **a**-agglutinin binding subunit is necessary for binding activity, expression of the core subunit in α cells does not contradict the **a**-specific pattern of expression of active **a**-agglutinin. A strong prediction, however, is that the **a**-agglutinin binding fragment gene is expressed only in **a** cells.

Role of the Agglutinins in Mating

It seemed surprising that agglutinin structural gene mutants were not identified in extensive screens for sterile mutants. The original nonspecific mutants that showed agglutination defects were later shown to be defective in components of the pheromone response pathway, and the agglutination defects resulted from the reduced basal expression of pheromone-inducible genes in these mutants, as described above. Mutations in agglutinin structural genes should show **a**- or α-specific agglutination defects, but normal pheromone production and response. Because such mutants were not identified among the sterile mutants, we undertook the screens for agglutination-defective mutants described above (Lipke et al., 1989; Roy et al., 1991). The agglutinin structural gene mutants show levels of mating similar to those of isogenic wild-type strains in a qualitative plate assay, thus explaining the lack of identification of agglutinin structural gene mutants among sterile mutants. In quantitative assays, these mutants showed only two- to fourfold decreases in mating frequency when tested on solid medium; when assayed in liquid medium, however, the agglutinin mutants showed several orders of magnitude lower mating frequency than the isogenic wild-type strains. These results indicate that the agglutinins are not necessary for mating under conditions that promote cell-cell contact, but that they are necessary under conditions that do not promote cell-cell contact. Because yeast cells are not motile, two cells of opposite mating type that randomly contact one another in liquid medium must remain in contact in order to mate. On solid medium, however, cells are in contact with one another, and under these conditions, the agglutinins seem to have little effect on mating.

CELL AND NUCLEAR FUSION

The conjugation process involves a series of steps that culminate in cell fusion, followed by nuclear migration and fusion to produce the diploid zygote. Response to pheromone is involved in many aspects of this process. Paired cells are thought to elongate towards one another, and the extended shmoo tips fuse, resulting in formation of the zygote bridge (Fig. 1). Exposure to pheromone results in changes in the cell wall, including altered glucan and mannan composition, increased porosity, and decreased resistance to β-glucanase (Lipke et al., 1976; Lipke and Ballou, 1980; de Nobel et al., 1990). As observed in the electron microscope, the cell wall becomes more diffuse and thin at the shmoo tip (Lipke et al., 1976). During zygote formation, the cell wall at the region of contact between mating cells becomes thinner and the cell membranes fragment and then rejoin to form the membrane of the zygote (Osumi et al., 1974). Two genes (*FUS1* and *FUS2*) have been identified that are involved in this process (McCaffrey et al., 1987; Trueheart et al., 1987). In matings of *fus* mutants, stable "prezygotes" with a shape similar to wild-type zygotes are formed, but a partition remains that prevents cytoplasmic fusion. Remnants of cell wall remain at this partition and membrane fusion does not occur.

Mutation of a single *fus* gene in a mating pair has little effect on mating (McCaffrey et al., 1987; Trueheart et al., 1987). As the number of *fus* mutations in the pair of mating cells is increased, however, the level of mating decreases. Expression of one *FUS* gene on a high-copy plasmid partially complements the other mutation, suggesting that the function of the two genes is redundant. Both *FUS* genes are induced by pheromone, and the *FUS1* gene product localizes to the shmoo tip in cells treated with pheromone (Fig. 1) and to the neck of the zygote after fusion. *FUS1* encodes a membrane protein with O-linked carbohydrate (Trueheart and Fink, 1989). These results suggest that the *FUS* products localize to the position of cell fusion upon induction by pheromone and play a role in cell wall breakdown and/or membrane fusion.

The amount of cell wall chitin is higher at the shmoo tip than elsewhere, and enzymes involved in chitin synthesis are induced by pheromone (Schekman and Brawley, 1979; Appeltauer and Achstetter, 1989; Watzele and Tanner, 1989). What, if any, role this change plays in the fusion process is unknown. The presence of vesicles that resemble secretory vesicles in the shmoo tip suggests that secretion occurs at this location. Because many of the proteins involved in pheromone response, cell interactions, and cell fusion are transported through the secretory pathway, secretion at this specialized site may provide a mechanism for concentrating components involved in pheromone response and cell fusion in this region.

The single spindle pole body (SPB) of the nucleus of cells exposed to pheromone shows a morphology similar to cells in late G1 during normal vegetative growth (Byers and Goetsch, 1975). This SPB has an extranuclear satellite from which cytoplasmic microtubules extend. The satellite-bearing SPB is oriented towards the shmoo tip. In zygotes, these microtubules enter the opposite cell and can sometimes be seen to interconnect the SPBs of the two nuclei. The nuclei migrate towards one another, and the SPBs fuse and produce a single contiguous nuclear envelope. Exposure to pheromone is necessary for efficient nuclear fusion (Rose et al., 1986). Two genes (*KAR1* and *KAR3*) that are necessary for SPB structure and microtubule function are also required for nuclear fusion (Rose and Fink, 1987; Meluh and Rose, 1990). Mutations in these genes prevent nuclear fusion, and the heterokaryotic zygotes produce haploid buds. The *KAR3* gene is induced by pheromone (Meluh and Rose, 1990). Thus, pheromone response is necessary for nuclear fusion within the zygote and production of a/α diploid buds.

PERSPECTIVES

Several areas of yeast research have indicated the extensive evolutionary conservation of many fundamental cellular processes. This conservation is found in many aspects of the yeast mating system. The similarity of the α-agglutinin binding domain to the immunoglobulin fold domains found in many cell surface proteins involved in adhesion in mammalian systems suggests that mechanisms of adhesion mediated by glycoprotein-glycoprotein interactions may be highly conserved. Many aspects of pheromone production are also highly conserved, including the mechanism of secretion and processing of α-factor. The similarity of *STE6* to *MDR* and other pumps suggests that a-factor secretion also involves a conserved mechanism. The pheromone receptors show structural features in common with the receptors from G protein-mediated pathways in all systems ranging from other microorganisms to vertebrates. Aspects of structure and mechanism of action of G proteins are also highly conserved. Although some aspects of the interactions involved in yeast mating are likely to be unique to yeast, it seems likely that phenomena elucidated in yeast may facilitate understanding of other systems. In addition, it should be possible to reconstitute interactions from other systems in yeast and use the powerful genetics possible in this system to study interactions between components.

ACKNOWLEDGMENTS. I thank D. Johnson, C. Raper, and S. Horton for comments on the manuscript.

REFERENCES

Achstetter, T. 1989. Regulation of α-factor production in *Saccharomyces cerevisiae*: a-factor pheromone-induced expression of the *MFα1* and *STE13* genes. *Mol. Cell. Biol.* **9:**4507–4514.

Achstetter, T., and D. H. Wolf. 1985. Hormone processing and membrane-bound proteinases in yeast. *EMBO J.* **4:**173–177.

Aebi, M., M. W. Clar, U. Viyayraghavan, and J. Abelson. 1990. A yeast mutant, *PRP20*, altered in mRNA metabolism and maintenance of the nuclear structure, is defective in a gene homologous to the human gene *RCC1* which is involved in the control of chromosome condensation. *Mol. Gen. Genet.* **224:**72–80.

Anderegg, R. J., R. Betz, S. A. Carr, J. W. Crabb, and W. Duntze. 1988. Structure of *Saccharomyces cerevisiae* mating hormone a-factor. *J. Biol. Chem.* **263:**18236–18240.

Appeltauer, U., and T. Achstetter. 1989. Hormone-induced expression of the *CHS1* gene from *Saccharomyces cerevisiae*. *Eur. J. Biochem.* **181:**243–247.

Bathurst, I. C., S. O. Brennan, R. W. Carrell, L. S. Cousens, A. J. Brake, and P. J. Barr. 1987. Yeast KEX2 protease has the properties of a human proalbumin converting enzyme. *Science* **235:**348–350.

Bender, A., and G. F. Sprague, Jr. 1986. Yeast peptide pheromones, a-factor and α-factor, activate a common response mechanism in their target cells. *Cell* **47:**929–937.

Bender, A., and G. F. Sprague, Jr. 1989. Pheromones and pheromone receptors are the primary determinants of mating specificity in the yeast *Saccharomyces cerevisiae*. *Genetics* **121:**463–476.

Betz, R., J. W. Crabb, H. E. Meyer, R. Wittig, and W. Duntze. 1987. Amino acid sequences of a-factor mating peptides from *Saccharomyces cerevisiae*. *J. Biol. Chem.* **262:**546–548.

Betz, R., W. Duntze, and T. R. Manney. 1978. Mating factor-mediated sexual agglutination in *Saccharomyces cerevisiae*. *FEMS Lett.* **4:**107–110.

Blinder, D., S. Bouvier, and D. D. Jenness. 1989. Constitutive mutants in the yeast pheromone response: ordered function of the gene products. *Cell* **56:**479–486.

Blumer, K. J., J. E. Reneke, and J. Thorner. 1988. The *STE2* gene product is the ligand-binding component of the α-factor receptor of *Saccharomyces cerevisiae*. *J. Biol. Chem.* **263:**10836–10842.

Blumer, K. J., and J. Thorner. 1990. β and γ subunits of a yeast guanine nucleotide-binding protein are not essential for membrane association of the α subunits but are required for receptor coupling. *Proc. Natl. Acad. Sci. USA* **87:**4363–4367.

Brake, A., C. Brenner, R. Najarian, P. Laybourn, and J. Merryweather. 1985. Structure of genes encoding precursors of the yeast peptide mating pheromone a-factor, p. 103–108. *In* M. J. Gething (ed.), *Current Communications in Molecular Biology: Protein Transport and Secretion*. Cold Spring Harbor Laboratory, Cold Spring Harbor, N.Y.

Bucking-Throm, E., W. Duntze, L. H. Hartwell, and T. R. Manney. 1973. Reversible arrest of haploid yeast cells at the initiation of DNA synthesis by a diffusible sex factor. *Exp. Cell Res.* **76:**99–110.

Burke, D., L. Mendonca-Previato, and C. E. Ballou. 1980. Cell-cell recognition in yeast: purification of *Hansenula wingei* 21-cell agglutination factor and comparison of the factors from three genera. *Proc. Natl. Acad. Sci. USA* **77**:318–322.

Burkholder, A. C., and L. H. Hartwell. 1985. The yeast α-factor receptor: structural properties deduced from the sequence of the *STE2* gene. *Nucleic Acids Res.* **13**:8463–8475.

Byers, B., and L. Goetsch. 1975. Behavior of spindles and spindle plaques in the cell cycle and conjugation of *Saccharomyces cerevisiae*. *J. Bacteriol.* **124**:511–523.

Caplan, S., R. Green, J. Rocco, and J. Kurjan. 1991. Glycosylation and structure of the yeast *MFα1* α-factor precursor is important for efficient transport through the secretory pathway. *J. Bacteriol.* **173**:627–635.

Caplan, S., and J. Kurjan. 1991. Role of α-factor and the *MFα1* α-factor precursor in yeast mating. *Genetics* **127**:299–307.

Chan, R. K., and C. A. Otte. 1982a. Isolation and genetic analysis of *Saccharomyces cerevisiae* mutants supersensitive to G1 arrest by a-factor and α-factor pheromones. *Mol. Cell Biol.* **2**:11–20.

Chan, R. K., and C. A. Otte. 1982b. Physiological characterization of *Saccharomyces cerevisiae* mutants supersensitive to G1 arrest by a factor and α factor pheromones. *Mol. Cell. Biol.* **2**:21–29.

Chang, F., and I. Herskowitz. 1990. Identification of a gene necessary for cell cycle arrest by a negative growth factor of yeast: FAR1 is an inhibitor of a G1 cyclin, CLN2. *Cell* **63**:999–1011.

Chvatchko, Y., I. Howald, and H. Riezman. 1986. Two yeast mutants defective in endocytosis are defective in pheromone response. *Cell* **46**:355–364.

Ciejek, E., and J. Thorner. 1979. Recovery of *Saccharomyces cerevisiae* a cells from G1 arrest by α factor pheromone requires endopeptidase action. *Cell* **18**:623–635.

Clark, K. L. 1991. Personal communication.

Clark, K. L., and G. F. Sprague, Jr. 1989. Yeast pheromone response pathway: characterization of a suppressor that restores mating to receptorless mutants. *Mol. Cell. Biol.* **9**:2682–2694.

Cole, G. M., and S. I. Reed. 1991. Pheromone-induced phosphorylation of a G protein β subunit in S. cerevisiae is associated with an adaptive response to mating pheromone. *Cell* **64**:703–716.

Cole, G. M., D. E. Stone, and S. I. Reed. 1990. Stoichiometry of G protein subunits affects the *Saccharomyces cerevisiae* mating pheromone signal transduction pathway. *Mol. Cell. Biol.* **10**:510–517.

Cooper, A., and H. Bussey. 1989. Characterization of the yeast *KEX1* gene product: a carboxypeptidase involved in processing secreted precursor proteins. *Mol. Cell. Biol.* **9**:2706–2714.

Cross, F. 1988. *DAF1*, a mutant gene affecting size control, pheromone arrest, and cell cycle kinetics of *Saccharomyces cerevisiae*. *Mol. Cell. Biol.* **8**:4675–4684.

Cross, F. 1990a. Cell cycle arrest caused by *CLN* gene deficiency in *Saccharomyces cerevisiae* resembles START-I arrest and is independent of the mating-pheromone signalling pathway. *Mol. Cell. Biol.* **10**:6482–6490.

Cross, F., L. H. Hartwell, C. Jackson, and J. B. Konopka. 1988. Conjugation in *Saccharomyces cerevisiae*. *Annu. Rev. Cell Biol.* **4**:430–457.

Cross, G. A. M. 1990b. Glycolipid anchoring of plasma membrane proteins. *Annu. Rev. Cell Biol.* **6**:1–39.

Davies, K. 1990. Cystic fibrosis: complementary endeavours. *Nature* (London) **348**:110–111.

de Barros Lopes, M., J.-Y. Ho, and S. I. Reed. 1990. Mutations in cell division cycle genes *CDC36* and *CDC39* activate the *Saccharomyces cerevisiae* mating pheromone response pathway. *Mol. Cell. Biol.* **10**:2966–2972.

de Nobel, J. G., F. M. Klis, J. Prem, T. Munnik, and H. van den Ende. 1990. The glucanase-soluble mannoproteins limit cell wall porosity in *Saccharomyces cerevisiae*. *Yeast* **6**:491–499.

Dietzel, C., and J. Kurjan. 1987a. The yeast *SCG1* gene: a Gα-like protein implicated in the a- and α-factor response pathway. *Cell* **50**:1001–1010.

Dietzel, C., and J. Kurjan. 1987b. Pheromonal regulation and sequence of the *Saccharomyces cerevisiae SST2* gene: a model for desensitization to pheromone. *Mol. Cell. Biol.* **7**:4169–4177.

Dmochowska, A., D. Dignard, D. Henning, D. Y. Thomas, and H. Bussey. 1987. Yeast *KEX1* gene encodes a putative protease with a carboxypeptidase B-like function involved in killer toxin and α-factor precursor processing. *Cell* **50**:573–584.

Dolan, J. W., and S. Fields. 1990. Overproduction of the yeast STE12 protein leads to constitutive transcriptional induction. *Genes Dev.* **4**:492–502.

Dolan, J. W., C. Kirkman, and S. Fields. 1989. The yeast STE12 protein binds to the DNA sequence mediating pheromone induction. *Proc. Natl. Acad. Sci. USA* **86:**5703–5707.

Duntze, W., V. MacKay, and T. R. Manney. 1970. *Saccharomyces cerevisiae;* a diffusible sex factor. *Science* **168:**1472–1473.

Duronio, R. J., D. A. Towler, R. O. Heuckeroth, and J. I. Gordon. 1989. Disruption of the yeast N-myristoyl transferase gene causes recessive lethality in yeast. *Science* **243:**796–800.

Egel-Mitani, M., H. P. Flygenring, and M. T. Hansen. 1990. A novel aspartyl protease allowing *KEX2*-independent *MFα* propheromone processing in yeast. *Yeast* **6:**127–137.

Elion, E. A., P. L. Grisafi, and G. R. Fink. 1990. FUS3 encodes a cdc2 + /CDC28-related kinase required for the transition from mitosis into conjugation. *Cell* **60:**649–664.

Emter, O., B. Mechler, T. Achstetter, H. Muller, and D. H. Wolf. 1983. Yeast hormone α-factor is synthesized as a high molecular weight precursor. *Biochem. Biophys. Res. Commun.* **116:**822–829.

Errede, B., and G. Ammerer. 1989. STE12, a protein involved in cell-type-specific transcription and signal transduction in yeast, is part of protein-DNA complexes. *Genes Dev.* **3:**1349–1361.

Fields, S., D. T. Chaleff, and G. F. Sprague, Jr. 1988. Yeast *STE7, STE11,* and *STE12* genes are required for expression of cell-type-specific genes. *Mol. Cell. Biol.* **8:**551–556.

Fields, S., and I. Herskowitz. 1985. The yeast *STE12* product is required for expression of two sets of cell-type specific genes. *Cell* **42:**923–930.

Finegold, A. A., W. R. Schafer, J. Rine, M. Whiteway, and F. Tamanoi. 1990. Common modification of trimeric G proteins and ras protein: involvement of polyisoprenylation. *Science* **249:**165–169.

Fleischmann, M., M. W. Clark, W. Forrester, M. Wickens, T. Nishimoto, and M. Aebi. Analysis of yeast *prp20* mutations and functional complementation by the human homologue *RCC1,* a protein involved in the control of chromosome consensation. *Mol. Gen. Genet.,* in press.

Fujiyama, A., K. Matsumoto, and F. Tamanoi. 1987. A novel yeast mutant defective in the processing of ras proteins: assessment of the effect of the mutation on processing steps. *EMBO J.* **6:**223–228.

Fuller, R. S., A. Brake, and J. Thorner. 1989. Yeast prohormone processing enzyme (*KEX2* gene product) is a Ca^{2+}-dependent serine protease. *Proc. Natl. Acad. Sci. USA* **86:**1434–1438.

Fuller, R. S., R. E. Sterne, and J. Thorner. 1988. Enzymes required for yeast prohormone processing. *Annu. Rev. Physiol.* **50:**345–362.

Goodman, L. E., S. R. Judd, C. C. Farnsworth, S. Powers, M. H. Gelb, J. A. Glomset, and F. Tamanoi. 1990. Mutants of *Saccharomyces cerevisiae* defective in the farnesylation of Ras proteins. *Proc. Natl. Acad. Sci. USA* **87:**9665–9669.

Hagen, D. C., G. McCaffrey, and G. F. Sprague, Jr. 1986. Evidence the yeast *STE3* gene encodes a receptor for the peptide pheromone a factor: gene sequence and implications for the structure of the presumed receptor. *Proc. Natl. Acad. Sci. USA* **83:**1418–1422.

Hagiya, M., K. Yoshida, and N. Yanagishima. 1977. The release of sex-specific substances responsible for sexual agglutination from haploid cells of *Saccharomyces cerevisiae. Exp. Cell Res.* **104:**263–272.

Hansen, W., P. D. Garcia, and P. Walter. 1986. *In vitro* protein translocation across the yeast endoplasmic reticulum: ATP-dependent post-translational translocation of the prepro-α-factor. *Cell* **45:**397–406.

Hartwell, L. H. 1980. Mutants of *Saccharomyces cerevisiae* unresponsive to cell division control by polypeptide mating hormone. *J. Cell Biol.* **85:**811–822.

Hauser, B. A., and W. Tanner. 1989. Purification of the inducible α-agglutinin and molecular cloning of the gene. *FEBS Lett.* **255:**90–94.

Herbert, E., and M. Uhler. 1982. Biosynthesis of polyprotein precursors to regulatory peptides. *Cell* **30:**1–2.

Herskowitz, I. 1989. A regulatory hierarchy for cell specialization in yeast. *Nature* (London) **342:**749–757.

Higgins, C. 1989. Transport proteins: export-import family expands. *Nature* (London) **340:**342.

Hirsch, J. P., C. Dietzel, and J. Kurjan. 1991. The carboxy terminus of *SCG1,* the yeast Gα protein involved in yeast mating, is implicated in interactions with the pheromone receptors. *Genes Dev.* **5:**467–474.

Hrycyna, C. A., and S. Clarke. 1990. Farnesyl cysteine C-terminal methyltransferase activity is dependent upon the *STE14* gene product in *Saccharomyces cerevisiae. Mol. Cell. Biol.* **10:**5071–5076.

Jackson, C. L., and L. H. Hartwell. 1990a. Courtship in *Saccharomyces cerevisiae:* an early cell-cell interaction during mating. *Mol. Cell. Biol.* **10:**2202–2213.

Jackson, C. L., and L. H. Hartwell. 1990b. Courtship in S. cerevisiae: both cell types choose mating partners by responding to the strongest pheromone signal. *Cell* **63:**1039–1051.

Jahng, K.-Y., J. Ferguson, and S. I. Reed. 1988. Mutations in a gene encoding the α subunit of a *Saccharomyces cerevisiae* G protein indicate a role in mating pheromone signaling. *Mol. Cell. Biol.* **8:**2484–2493.

Jelsema, C. L., and J. Axelrod. 1987. Stimulation of phospholipase A_2 activity in bovine rod outer segments by the βγ subunits of transducin and its inhibition by the α subunit. *Proc. Natl. Acad. Sci. USA* **84:**3623–3627.

Jenness, D. D., A. C. Burkholder, and L. H. Hartwell. 1983. Binding of α-factor pheromone to yeast **a** cells: chemical and genetic evidence for an α-factor receptor. *Cell* **35:**521–529.

Jenness, D. D., and P. Spatrick. 1986. Down regulation of the α-factor pheromone-receptor in *Saccharomyces cerevisiae. Cell* **46:**345–353.

Julius, D., L. Blair, A. Brake, G. Sprague, and J. Thorner. 1983. Yeast α factor is processed from a larger precursor polypeptide: the essential role of a membrane-bound dipeptidyl aminopeptidase. *Cell* **32:**839–852.

Julius, D., A. Brake, L. Blair, R. Kunisawa, and J. Thorner. 1984a. Isolation of the putative structural gene for the lysine-arginine-cleaving endopeptidase required for processing of yeast prepro-α-factor. *Cell* **37:**1075–1089.

Julius, D., R. Schekman, and J. Thorner. 1984b. Glycosylation and processing of prepro-α-factor through the yeast secretory pathway. *Cell* **36:**309–318.

Kang, Y.-S., J. Kane, J. Kurjan, J. M. Stadel, and D. J. Tipper. 1990. Effect of expression of mammalian Gα and hybrid mammalian-yeast Gα proteins on the yeast pheromone response signal transduction pathway. *Mol. Cell. Biol.* **10:**2582–2590.

King, K., H. G. Dolhman, J. Thorner, M. G. Caron, and R. J. Lefkowitz. 1990. Control of yeast mating signal transduction by a mammalian β_2-adrenergic receptor and $G_s\alpha$ subunit. *Science* **250:**121–123.

Konopka, J. B., D. D. Jenness, and L. H. Hartwell. 1988. The C-terminus of the Saccharomyces cerevisiae α-pheromone receptor mediates an adaptive response to pheromone. *Cell* **54:**609–618.

Kreil, G. 1990. Processing of precursors by dipeptidylaminopeptidase: a case of molecular ticketing. *Trends Biochem. Sci.* **15:**23–26.

Kronstad, J. W., J. A. Holly, and V. L. MacKay. 1987. A yeast operator overlaps an upstream activation site. *Cell* **50:**369–377.

Kuchler, K., R. E. Sterne, and J. Thorner. 1989. *Saccharomyces cerevisiae STE6* gene product: a novel pathway for protein export in eukaryotic cells. *EMBO J.* **8:**3973–3984.

Kurjan, J. 1985. α-Factor structural gene mutations in *Saccharomyces cerevisiae:* effects on α-factor production and mating. *Mol. Cell. Biol.* **5:**787–796.

Kurjan, J., and I. Herskowitz. 1982. Structure of a yeast pheromone gene (*MFα*): a putative α-factor precursor contains four tandem copies of mature α-factor. *Cell* **30:**933–943.

Kurjan, J., J. P. Hirsch, and C. Dietzel. 1991. Mutations in the guanine nucleotide binding domains of a yeast Gα protein confer a constitutive or uninducible state to the pheromone response pathway. *Genes Dev.* **5:**475–483.

Kurjan, J., and P. N. Lipke. 1986. Agglutination and mating activity of the *MFα2*-encoded α-factor analog in *Saccharomyces cerevisiae. J. Bacteriol.* **168:**1472–1475.

Lipke, P. 1991. Personal communication.

Lipke, P. N., and C. E. Ballou. 1980. Altered immunochemical reactivity of *Saccharomyces cerevisiae* **a**-cells after α-factor-induced morphogenesis. *J. Bacteriol.* **141:**1170–1177.

Lipke, P. N., and J. Kurjan. *Microbiol. Rev.*, in press.

Lipke, P. N., A. Taylor, and C. E. Ballou. 1976. Morphogenic effects of α-factor on *Saccharomyces cerevisiae* **a** cells. *J. Bacteriol.* **127:**610–618.

Lipke, P. N., D. Wojciechowicz, and J. Kurjan. 1989. *AGα1* is the structural gene for the *Saccharomyces cerevisiae* α-agglutinin, a cell surface glycoprotein involved in cell-cell interactions during mating. *Mol. Cell. Biol.* **9:**3155–3165.

MacKay, V. L., and T. R. Manney. 1974. Mutations affecting sexual conjugation and related processes in *Saccharomyces cerevisiae*. I. Isolation and phenotypic characterization of nonmating mutants. *Genetics* **76:**255–271.

MacKay, V. L., S. K. Welch, M. Y. Insley, T. R. Manney, J. Holly, G. C. Saari, and M. L. Parker. 1988. The *Saccharomyces cerevisiae BAR1* gene encodes an exported protein with homology to pepsin. *Proc. Natl. Acad. Sci. USA* **85:**55–59.

Manney, T. 1983. Expression of the *BAR1* gene in *Saccharomyces cerevisiae:* induction by the α mating pheromone of an activity associated with a secreted protein. *J. Bacteriol.* **155:**291–301.

Marcus, S., C.-B. Xue, F. Naider, and J. M. Becker. 1991. Degradation of a-factor by a *Saccharomyces cerevisiae* α-mating-type-specific endopeptidase: evidence for a role in recovery of cells from G_1 arrest. *Mol. Cell. Biol.* **11:**1030–1039.

Marr, R. S., L. C. Blair, and J. Thorner. 1990. *Saccharomyces cerevisiae STE14* gene is required for COOH-terminal methylation of a-factor mating pheromone. *J. Biol. Chem.* **265:**20057–20060.

Marsh, L., and I. Herskowitz. 1987. Conservation of a receptor/signal transduction system. *Cell* **50:**995–996.

Marsh, L., and I. Herskowitz. 1988. STE2 protein of *Saccharomyces kluyveri* is a member of the rhodopsin/β-adrenergic receptor family and is responsible for recognition of the peptide ligand α factor. *Proc. Natl. Acad. Sci. USA* **85:**3855–3859.

McCaffrey, G., F. J. Clay, K. Kelsay, and G. F. Sprague, Jr. 1987. Identification and regulation of a gene required for cell fusion during mating of the yeast *Saccharomyces cerevisiae*. *Mol. Cell. Biol.* **7:**2680–2690.

McGrath, J. P., and A. Varshavsky. 1989. The yeast *STE6* gene encodes a homologue of the mammalian multidrug resistance P-glycoprotein. *Nature* (London) **340:**400–404.

Meluh, P. B., and M. D. Rose. 1990. KAR3, a kinesin-related gene required for yeast nuclear fusion. *Cell* **60:**1029–1041.

Mendonca-Previato, L., D. Burke, and C. E. Ballou. 1982. Sexual agglutination factor from the yeast *Pichia amethionina*. *J. Cell. Biochem.* **19:**171–178.

Michaelis, S., and I. Herskowitz. 1988. The a-factor pheromone of *Saccharomyces cerevisiae* is essential for mating. *Mol. Cell. Biol.* **8:**1309–1318.

Miyajima, I., K.-I. Arai, and K. Matsumoto. 1989. *GPA1*[Val-50] mutation in the mating-factor signaling pathway in *Saccharomyces cerevisiae*. *Mol. Cell. Biol.* **9:**2289–2297.

Miyajima, I., M. Nakafuku, N. Nakayama, C. Brenner, A. Miyajima, K. Kaibuchi, K. Arai, Y. Kaziro, and K. Matsumoto. 1987. *GPA1*, a haploid-specific essential gene, encodes a yeast homolog of mammalian G protein which may be involved in mating factor signal transduction. *Cell* **50:**1011–1019.

Mizuno, K., T. Nakamura, T. Ohshima, S. Tanaka, and H. Matsuo. 1988. Yeast *KEX2* gene encodes an endopeptidase homologous to subtilisin-like serine proteases. *Biochem. Biophys. Res. Commun.* **156:**246–254.

Moore, S. A. 1984. Yeast cells recover from mating pheromone α factor-induced division arrest by desensitization in the absence of α factor destruction. *J. Biol. Chem.* **259:**1004–1010.

Nakafuku, M., H. Itoh, S. Nakamura, and Y. Kaziro. 1987. Occurrence in *Saccharomyces cerevisiae* of a gene homologous to the cDNA coding for the α subunit of mammalian G proteins. *Proc. Natl. Acad. Sci. USA* **84:**2140–2144.

Nakayama, N., K. Arai, and K. Matsumoto. 1988a. Role of *SCP2*, a suppressor of a *gpa1* mutation, in the mating-factor signaling pathway of *Saccharomyces cerevisiae*. *Mol. Cell. Biol.* **8:**5410–5416.

Nakayama, N., Y. Kaziro, K.-I. Arai, and K. Matsumoto. 1988b. Role of *STE* genes in the mating-factor signaling pathway mediated by *GPA1* in *Saccharomyces cerevisiae*. *Mol. Cell. Biol.* **8:**3777–3783.

Nakayama, N., A. Miyajima, and K. Arai. 1985. Nucleotide sequences of *STE2* and *STE3*, cell type-specific sterile genes from *Saccharomyces cerevisiae*. *EMBO J.* **4:**2643–2648.

Nakayama, N., A. Miyajima, and K. Arai. 1987. Common signal transduction system shared by *STE2* and *STE3* in haploid cells of *Saccharomyces cerevisiae:* autocrine cell-cycle arrest results from forced expression of *STE2*. *EMBO J.* **6:**249–254.

Neiman, A. M., F. Chang, K. Komachi, and I. Herskowitz. 1990. *CDC36* and *CDC39* are negative elements in the signal transduction pathway of yeast. *Cell Reg.* **1:**391–401.

Nomoto, S., N. Nakayama, K.-I. Arai, and K. Matsumoto. 1990. Regulation of the yeast pheromone response pathway by G protein subunits. *EMBO J.* **9**:691–696.

Orlean, P., H. Ammer, M. Watzele, and W. Tanner. 1986. Synthesis of an O-glycosylated cell surface protein induced in yeast by α factor. *Proc. Natl. Acad. Sci. USA* **83**:6263–6266.

Osumi, M., C. Shimoda, and N. Yanagishima. 1974. Mating reaction in *Saccharomyces cerevisiae*. V. Changes in the fine structure during the mating reaction. *Arch. Microbiol.* **97**:27–38.

Payne, G. S., and R. Schekman. 1989. Clathrin: a role in the intracellular retention of a Golgi membrane protein. *Science* **245**:1358–1365.

Pierce, M., and C. E. Ballou. 1983. Cell-cell recognition in yeast: characterization of the sexual agglutination factors from *Saccharomyces kluyveri*. *J. Biol. Chem.* **258**:3476–3482.

Powers, S., S. Michaelis, D. Broek, S. Santa Anna-A, J. Fields, I. Herskowitz, and M. Wigler. 1986. RAM, a gene of yeast required for a functional modification of RAS proteins and for production of mating pheromone a-factor. *Cell* **47**:412–422.

Raths, S., P. Shenbagamurthi, F. Naider, and J. M. Becker. 1986. Biological activity of the Asn-5, Arg-7 tridecapeptide encoded by *MFα2* of *Saccharomyces cerevisiae*. *J. Bacteriol.* **168**:1468–1471.

Reed, S. I. 1991. Personal communication.

Reneke, J. E., K. J. Blumer, W. E. Courchese, and J. Thorner. 1988. The carboxy-terminal segment of the yeast α-factor receptor is a regulatory domain. *Cell* **55**:221–234.

Rhodes, N., L. Connell, and B. Errede. 1990. STE11 is a protein kinase required for cell-type-specific transcription and signal transduction in yeast. *Genes Dev.* **4**:1862–1874.

Ringe, D., and G. A. Petsko. 1990. Cystic fibrosis; a transport problem? *Nature* (London) **346**:312–313.

Rose, M. D., and G. R. Fink. 1987. *KAR1*, a gene required for function of both intranuclear and extranuclear microtubules in yeast. *Cell* **48**:1047–1060.

Rose, M. D., B. R. Price, and G. R. Fink. 1986. *Saccharomyces cerevisiae* nuclear fusion requires prior activation by alpha factor. *Mol. Cell. Biol.* **6**:3490–3497.

Rothblatt, J. A., and D. I. Meyer. 1986. Secretion in yeast: reconstitution of the translocation and glycosylation of α-factor and invertase in a homologous cell-free system. *Cell* **44**:619–628.

Rothblatt, J. A., J. R. Webb, G. Ammerer, and D. I. Meyer. 1987. Secretion in yeast: structural features influencing the post-translational translocation of prepro-α-factor *in vitro*. *EMBO J.* **6**:3455–3463.

Roy, A., C. F. Lu, P. N. Lipke, D. Marykwas, and J. Kurjan. 1991. *AGA1* encodes the *Saccharomyces cerevisiae* cell surface attachment subunit of the cell adhesion glycoprotein a-agglutinin. *Mol. Cell. Biol.* **11**:4196–4206.

Sakurai, A., S. Tamura, N. Yanagishima, and C. Shimoda. 1976. Structure of the peptidyl factor inducing sexual agglutination in *Saccharomyces cerevisiae*. *Agric. Biol. Chem.* **40**:1057–1058.

Schafer, W. R., R. Kim, R. Sterne, J. Thorner, S. H. Kim, and J. Rine. 1989. Genetic and pharmacological suppression of oncogenic mutations in *ras* genes of yeast and humans. *Science* **245**:379–385.

Schafer, W. R., C. E. Trueblood, C.-C. Yang, M. P. Mayer, S. Rosenberg, C. D. Poulter, S.-H. Kim, and J. Rine. 1990. Enzymatic coupling of cholesterol intermediates to a mating pheromone precursor and to the Ras protein. *Science* **249**:1133–1139.

Schekman, R., and V. Brawley. 1979. Localized deposition of chitin on the yeast cell surface in response to mating pheromone. *Proc. Natl. Acad. Sci. USA* **76**:645–649.

Sena, E. P., D. N. Radin, and S. Fogel. 1973. Synchronous mating in yeast. *Proc. Natl. Acad. Sci. USA* **70**:1373–1377.

Shimoda, C., and N. Yanagishima. 1975. Mating reaction in *Saccharomyces cerevisiae*. VIII. Mating type-specific substances responsible for sexual cell agglutination. *Antonie van Leeuwenhoek* **41**:521–532.

Shimoda, C., N. Yanagishima, A. Sakurai, and S. Tamura. 1976. Mating reaction in *Saccharomyces cerevisiae*. IX. Regulation of sexual cell agglutinability of a-type cells by a sex factor produced by alpha type cells. *Arch. Microbiol.* **108**:27–34.

Singh, A., E. Y. Chen, J. M. Lugovoy, C. N. Chang, R. A. Hitzman, and P. H. Seeburg. 1983. *Saccharomyces cerevisiae* contains two discrete genes coding for the α-factor pheromone. *Nucleic Acids Res.* **11**:4049–4063.

Sprague, G. F., Jr., and I. Herskowitz. 1981. Control of yeast cell type by the mating type locus. I. Identification and control of expression of the a-specific gene, *BAR1*. *J. Mol. Biol.* **153:**357–372.

Steden, M., R. Betz, and W. Duntze. 1989. Isolation and characterization of *Saccharomyces cerevisiae* mutants supersensitive to G1 arrest by the mating hormone a-factor. *Mol. Gen. Genet.* **219:**439–444.

Steiner, D. F., P. S. Quinn, S. J. Chan, J. Marsh, and H. S. Trager. 1980. Processing mechanisms in the biosynthesis of proteins. *Ann. N. Y. Acad. Sci.* **343:**1–16.

Stone, D. E., and S. I. Reed. 1990. G protein mutations that alter the pheromone response in *Saccharomyces cerevisiae*. *Mol. Cell. Biol.* **10:**4439–4446.

Stotzler, D., and W. Duntze. 1976. Isolation and characterization of four related peptides exhibiting α-factor activity from *Saccharomyces cerevisiae*. *Eur. J. Biochem.* **65:**257–262.

Stotzler, D., H. Kiltz, and W. Duntze. 1976. Primary structure of α-factor peptides from *Saccharomyces cerevisiae*. *Eur. J. Biochem.* **69:**397–400.

Strazdis, J. R., and V. L. MacKay. 1983. Induction of yeast mating pheromone a-factor by a cells. *Nature* (London) **305:**543–545.

Tanaka, T., and H. Kita. 1977. Degradation of mating factor by α-mating type cells of *Saccharomyces cerevisiae*. *J. Biochem.* **82:**1689–1693.

Tanner, W. 1991. Personal communication.

Taylor, N. W., and W. L. Orton. 1967. Sexual agglutination in yeast. V. Small particles of 5-agglutinin. *Arch. Biochem. Biophys.* **120:**602–608.

Teague, M. A., D. T. Chaleff, and B. Errede. 1986. Nucleotide sequence of the yeast regulatory gene *STE7* predicts a protein homologous to protein kinases. *Proc. Natl. Acad. Sci. USA* **83:**7371–7375.

Terrance, K., P. Heller, and P. N. Lipke. 1987a. Pheromone induction of agglutinability in *Saccharomyces cerevisiae*. *J. Bacteriol.* **169:**4811–4815.

Terrance, K., P. Heller, Y.-S. Wu, and P. N. Lipke. 1987b. Identification of glycoprotein components of α-agglutinin, a cell adhesion protein from *Saccharomyces cerevisiae*. *J. Bacteriol.* **169:**475–482.

Terrance, K., and P. N. Lipke. 1981. Sexual agglutination in *Saccharomyces cerevisiae*. *J. Bacteriol.* **148:**889–896.

Tohoyama, H., and N. Yanagishima. 1985. The sexual agglutination substance is secreted through the yeast secretory pathway in *Saccharomyces cerevisiae*. *Mol. Gen. Genet.* **201:**446–449.

Tohoyama, H., and N. Yanagishima. 1987. Site of pheromone action and secretion pathway of a sexual agglutination substance during its induction by pheromone a in α cells of *S. cerevisiae*. *Curr. Genet.* **12:**271–275.

Trueheart, J., J. D. Boeke, and G. R. Fink. 1987. Two genes required for cell fusion during yeast conjugation: evidence for a pheromone-induced surface protein. *Mol. Cell. Biol.* **7:**2316–2328.

Trueheart, J., and G. R. Fink. 1989. The yeast cell fusion protein FUS1 is O-glycosylated and spans the plasma membrane. *Proc. Natl. Acad. Sci. USA* **86:**9916–9920.

Wagner, J.-C., and D. H. Wolf. 1987. Hormone (pheromone) processing enzymes in yeast: the carboxy-terminal processing enzyme of the mating pheromone α-factor, carboxypeptidase yscα, is absent in α-factor maturation-defective *kex1* mutant cells. *FEBS Lett.* **221:**423–426.

Wagner, N., and P. N. Lipke. 1991. Personal communication.

Waters, M. G., and G. Blobel. 1986. Secretory protein translation in a yeast cell-free system can occur postranslationally and requires ATP hydrolysis. *J. Cell Biol.* **102:**1543–1550.

Waters, M. G., E. A. Evans, and G. Blobel. 1988. Prepro-α-factor has a cleavable signal sequence. *J. Biol. Chem.* **263:**6209–6214.

Watzele, G., and W. Tanner. 1989. Cloning of the glutamine:fructose-6-phosphate amidotransferase gene from yeast: pheromonal regulation of its transcription. *J. Biol. Chem.* **264:**8753–8758.

Watzele, M., F. Klis, and W. Tanner. 1988. Purification and characterization of the inducible a agglutinin of *Saccharomyces cerevisiae*. *EMBO J.* **7:**1483–1488.

Whiteway, M., L. Hougan, D. Dignard, D. Y. Thomas, L. Bell, G. C. Saari, F. J. Grant, P. O'Hara, and V. L. MacKay. 1989. The *STE4* and *STE18* genes of yeast encode potential β and γ subunits of the mating factor receptor-coupled G protein. *Cell* **56:**467–477.

Whiteway, M., L. Hougan, and D. Y. Thomas. 1990. Overexpression of the *STE4* gene leads to mating response in haploid *Saccharomyces cerevisiae*. *Mol. Cell. Biol.* **10:**217–222.

Wickner, R. B., and M. J. Leibowitz. 1976. Two chromosomal genes required for killing expression in killer strains of *Saccharomyces cerevisiae*. *Genetics* **82**:429–442.

Williams, A. F., and A. N. Barclay. 1988. The immunoglobulin superfamily—domains for cell surface recognition. *Annu. Rev. Immunol.* **6**:381–405.

Wittenberg, C., K. Sugimoto, and S. I. Reed. 1990. G1-specific cyclins of S. cerevisiae: cell cycle periodicity, regulation by mating pheromone, and association with the p34^{CDC28} protein kinase. *Cell* **62**:225–237.

Wojciechowicz, D., J. Kurjan, and P. N. Lipke. Manuscript in preparation.

Wojciechowicz, D., and P. N. Lipke. 1989. α-Agglutinin expression in *S. cerevisiae*. *Biochem. Biophys. Res. Commun.* **161**:45–51.

Xue, C. B., G. A. Caldwell, J. M. Becker, and F. Naider. 1989. Total synthesis of the lipopeptide a-mating factor of *Saccharomyces cerevisiae*. *Biochem. Biophys. Res. Commun.* **162**:253–257.

Yamaguchi, M., K. Yoshida, I. Banno, and N. Yanagishima. 1984a. Mating-type differentiation in ascosporogenous yeasts on the basis of mating-type-specific substances responsible for sexual cell-cell recognition. *Mol. Gen. Genet.* **194**:24–30.

Yamaguchi, M., K. Yoshida, and N. Yanagishima. 1982. Isolation and partial characterization of cytoplasmic α agglutination substance in the yeast *Saccharomyces cerevisiae*. *FEBS Lett.* **139**:125–129.

Yamaguchi, M., K. Yoshida, and N. Yanagishima. 1984b. Isolation, and biochemical and biological characterization of an a-mating-type-specific glycoprotein responsible for sexual agglutination from the cytoplasm of a-cells, in the yeast *Saccharomyces cerevisiae*. *Arch. Microbiol.* **140**:113–119.

Yen, P. H., and C. E. Ballou. 1974. Partial characterization of the sexual agglutination factor from *Hansenula wingei* Y-2340 type 5 cells. *Biochemistry* **13**:2428–2437.

Yu, L., K. J. Blumer, N. Davidson, H. A. Lester, and J. Thorner. 1989. Functional expression of the yeast α-factor receptor in *Xenopus* oocytes. *J. Biol. Chem.* **264**:20847–20850.

Developmental Interactions

Microbial Cell-Cell Interactions
Edited by Martin Dworkin
© 1991 American Society for Microbiology, Washington, DC 20005

Chapter 6

Intercellular Interactions during *Dictyostelium* Development

Pauline Schaap

INTRODUCTION

The dictyostelids represent an interesting group of species, which are hovering on the borderline between uni- and multicellularity. The development of these organisms demonstrates how, during the course of evolution, the initial step to multicellularity may have been made. In the unicellular amoebal stage of development, cells locate their prey, bacteria, by sensing bacterial secretion products as pteridines. Upon depletion of the food source, the more simple species start to secrete pteridines themselves and to attract each other, resulting in the formation of multicellular aggregates. More advanced species utilize a separate chemotactic system for mutual attraction.

Dictyostelids are simple eukaryotes whose genome size is about 50,000 kb. The single-copy portion of the genome is about seven times the size of the *Escherichia coli* genome and only 2% of the mammalian genome (Sussman and Rayner, 1971; Firtel and Bonner, 1972). Amoebae can spend many generations as single cells, but they can also perform a quite intricate pattern of morphogenesis and cell differentiation.

The life cycle of all *Dictyostelium* species proceeds along the following general lines. Under appropriate conditions, amoebae emerge from spores in the top layer of the soil and start feeding on bacteria. Upon starvation, cells attract each other and form aggregates. The initially hemispherical cell mound becomes surrounded by a slime sheath and forms a nipple-shaped structure, called the tip, which coordinates cell movement in the remainder of the cell mass (Raper, 1940).

First the aggregate elongates to form a finger-shaped structure, the slug, which in some species migrates freely over the substratum, guided by signals such as light and heat (Bonner et al., 1950). After a period of migration, the slug raises itself erect to start the process of culmination. Cells at the tip move into a central prefabricated cellulose tube and, meanwhile, differentiate into highly vacuolated stalk cells, which become rigid by the formation of a cellulose cell wall. The reverse fountain movement continues until the stalk has reached a certain length and has carried the bulk of the cell mass aloft (Raper and Fennell, 1952). These cells then turn into compact spores and become surrounded by a mucopolysaccharide spore coat.

Besides this asexual life cycle, cellular slime molds have two other modes of survival. In some species, starving cells can round up and encyst as single cells, called microcysts. The cytoplasm of microcysts is less compact than that of spores and also their cell wall is less thick (Hohl et al., 1970). High levels of ammonia produced by the cells are considered to induce microcyst formation (Choi and O'Day, 1982).

Alternatively, many species, such as *Dictyostelium discoideum, D. mucoroides, D. minutum,* and *Polysphondylium,* form so-called macrocysts, which are of great interest for genetic studies because they represent the somewhat crude sex life of the slime mold. During macrocyst formation, starving cells form multicellular aggregates by chemotaxis, which become surrounded by a thick cellulose wall. Two cells of opposite mating type fuse to form a zygote, which subsequently devours all the cells in the aggregate. After some time the nucleus of the giant cell divides to give rise to several smaller nuclei, and after a long period of dormancy the macrocyst germinates and releases a small number of cells (Blaskovics and Raper, 1957; Filosa and Dengler, 1972; Erdos et al., 1972; MacInnes and Francis, 1974; Saga et al., 1983; Szabo et al., 1982). The choice between macrocyst formation or the asexual life cycle is most likely controlled by ethylene and cyclic AMP (cAMP) levels, with ethylene stimulating the macrocyst pathway and cAMP the fruiting body pathway (Amagai, 1984, 1987; Amagai and Filosa, 1984). Macrocyst development has not attracted extensive research efforts, because the germination of these structures takes several weeks and is difficult to control. One hopes that this disadvantage will be overcome by better methods to induce rapid germination (Abe and Maeda, 1986).

Most research in the *Dictyostelium* field has been dedicated to the asexual life cycle of the species *D. discoideum.* Cellular interactions controlling the developmental program of this species will therefore be mainly described, but in order to obtain a more general overview, available data from other cellular slime mold species will be included.

Dictyostelium development consists of two major processes which are both controlled by intercellular communication. First, multicellular structures perform a series of shape changes, which result from the coordinated movement of individual cells. Meanwhile, cells respond to spatiotemporal signals which dictate the expression of different classes of genes. Morphogenetic movement and gene expression are interdependent processes and, as will be shown in the following paragraphs, often utilize the same signals.

CONTROL OF CELL MOVEMENT DURING DEVELOPMENT

Preaggregative Cell Movement

Cell movement in the feeding stage is mainly directed towards finding and eating bacteria. In this stage, amoebae of all investigated species are chemotactically responsive to several pteridines, which are secreted by bacteria (Pan et al., 1972, 1975). Starvation initiates the social phase of development. A few amoebae in the cell population now become sources of chemoattractant themselves. Some species, such as *D. minutum* and *D. lacteum,* use the simple strategy of secreting the attractant of their bacterial food source, i.e., folates or pterins (Van Haastert et al., 1982a; De Wit and Konijn, 1983). Aggregation in these species is a fairly simple process involving the gradual movement of individual cells towards the cell that initiated secretion of the attractant (Gerisch, 1964).

Many species developed a very striking aggregation mechanism, in which the chemoattractant is not identical to that of the food source and is secreted in an oscillatory fashion. The original signal is relayed by surrounding cells, resulting in concentric or spiral patterns of cell movement (Roos et al., 1975; Shaffer, 1975; Tomchik and Devreotes, 1981). In several species that aggregate in an oscillatory fashion, e.g., *D. mexicanum* and *D. vinaceo-fuscum,* the attractant has not yet been identified (see Raper, 1984; Schaap and Wang, 1984). In *D. discoideum* and about six related species, the compound that induces both the chemotactic and the relay response is cAMP (Konijn et al., 1967). In *Polysphondylium,* which also aggregates in an oscillatory manner, the chemoattractant is a modified dipeptide called glorin (Shimamura et al., 1982). Curiously, glorin cannot induce glorin relay (De Wit et al., 1988), which means that in addition to glorin another compound is secreted that controls oscillatory behavior. All investigated chemotactic systems utilize specific extracellular enzymes such as cAMP phosphodiesterase (cAMP-PDE), folate deaminase, or glorinase to reduce background chemoattractant levels (Malchow et al., 1972; Pan and Wurster, 1978; De Wit et al., 1988).

Postaggregative Cell Movement

After aggregation, the cell mass goes through a series of shape changes which ultimately results in the formation of one or more fruiting bodies. This morphogenetic movement is guided by a small group of cells called the tip, which strongly resembles classical embryonic organizers (Spemann, 1938; Raper, 1940; Rubin and Robertson, 1975). The number and time of appearance of initial and accessory tips determine the species-specific maximal size of the fruiting body as well as its specific pattern of side branches. Tips of all cellular slime mold species most likely control development by acting as autonomous cAMP oscillators. Cells in *D. discoideum* slugs show periodic waves of cell movement (Durston and Vork, 1979) and partially retain chemotactic responsiveness to cAMP (Matsukuma and Durston, 1979; Kitami, 1984; Mee et al., 1986) and the ability to relay cAMP signals (Kesbeke et al., 1986; Otte et al., 1986).

The primitive species *D. minutum,* which utilizes continuous folate secretion to control the aggregation process, exhibits oscillations inside completed aggregates initiated by the tip region. Cells move towards the oscillation center and push it upwards as the newly formed tip. This process is disrupted by exogenously applied cAMP, but not by folate. Just before oscillations become visible, *D. minutum* cells accumulate surface cAMP receptors and cAMP-PDE and obtain the ability to relay cAMP pulses (Schaap et al., 1983; Schaap et al., 1984). All investigated species show disruption of multicellular development by cAMP and expression of the cAMP chemotactic machinery after aggregation (George, 1977; Schaap and Wang, 1984), which suggests that in all cellular slime molds, the tip controls morphogenetic movement by secreting cAMP pulses. In the course of evolution some species, such as *D. discoideum,* may have additionally utilized the cAMP signalling system to control the process of aggregation.

Although cAMP oscillations may coordinate cell movement during slug migration and culmination, a single chemotactic gradient may not be sufficient to account for movement of the whole structure. As argued by Odell and Bonner (1986), net movement is zero when cells in the slug crawl in the same direction as similarly crawling cells. They propose that perpendicular to the cAMP gradient, which dictates orientation, a second gradient exists which dictates chemokinesis. This gradient is highest at the center and lowest at the periphery of the slug, causing cells at the center to move at the expense of cells at the periphery. Two possible candidates for the chemokinetic factor are the unidentified factor STF (slug turning factor), described by Fisher et al. (1981), or ammonia, an end product of protein degradation.

An alternative mechanism is suggested by the observation that during migration, slugs leave distinct footprints of matrix proteins behind in the slime trail. Slug migration is proposed to consist of alternative phases in which the slug anterior moves forward, attaches to the substratum, and pulls in the posterior to the place of attachment (Vardy et al., 1986). This model falls somewhat short of explaining how the cells move straight upwards during initial slug formation and culmination, a problem which is circumvented in the first model.

Ammonia has also been implicated in several other aspects of morphogenetic movement. The transition from migration to culmination can be induced by depleting ammonia levels in the slug (Schindler and Sussman, 1977). Culminating structures produce a gas which acts as a repellent and determines the orientation of the fruiting body (Bonner and Dodd, 1962). The gas in question is most likely ammonia (Bonner et al., 1986; Feit and Sollitto, 1987). Culmination can also be initiated by overhead illumination, and the direction of slug migration can in general be guided by unilateral light and temperature gradients. Bonner and co-workers (1988, 1989) recently proposed that ammonia may also mediate the phototactic and thermotactic response. The general effect of ammonia in all these responses is considered to be enhanced cell locomotion at sites of relatively high ammonia production.

Slug and fruiting body size are determined by the ability of tips to inhibit the formation of secondary tips in their vicinity. Tips were shown to secrete a diffusible autoinhibitor (Durston, 1976; MacWilliams, 1982; Kopachik, 1982), which is most likely the cAMP degradation product adenosine (Schaap and Wang, 1986). Adenosine inhibits the initiation of autonomous oscillations (Newell and Ross, 1982) and is formed from cAMP by cAMP-PDE and 5'-nucleotidase, two cell surface enzymes which are preferentially active at the tip region (Armant and Rutherford, 1979; Armant et al., 1980; Tsang and Bradbury, 1981; Schaap and Spek, 1984; Otte et al., 1986; Mee et al., 1986).

To conclude, the present data indicate that different cellular slime mold species secrete distinct chemoattractants during the aggregation process, but probably all utilize oscillatory cAMP signalling to control multicellular morphogenetic movement. Adenosine levels may control the size of the organism, and gradients of ammonia or STF or both may determine the orientation of the structure during migration and culmination.

GENE REGULATION DURING DEVELOPMENT

Spore Germination and Growth

The expression of different classes of genes during *D. discoideum* development is schematically represented in Fig. 1. The initiation of development is arbitrarily fixed at the germination of spores. Spores in the fruiting body are constitutively dormant due to the presence of an autoinhibitor of spore germination, which has been identified as an adenine derivative called discadenine (Obata et al., 1973; Abe et al., 1976). Spore germination is, under natural conditions, most likely induced by heat shock, bacterial secretion products, and autoactivators produced during the germination process (Cotter and Raper, 1966; Hashimoto et al., 1976; Dahlberg and Cotter, 1978). Both protein synthesis and RNA synthesis are required for spore germination (Giri and Ennis, 1977). A number of gene products specifically accumulate during germination, while others increase from low levels present in the spores. Germination-specific transcripts disappear rapidly during vegetative growth (Giri and Ennis, 1978; Kelly et al., 1983).

The newly emerged amoebae enter a period of growth and cell division and express a large number of growth-specific genes. About one-third of the genes transcribed in the vegetative stage turn on after spore germination and turn off during early development (Kopachik et al., 1985a). Transcripts which decrease after cessation of feeding appear to fall into at least two classes. The major class turns off in the absence of protein synthesis, and a minor class requires additional protein synthesis (Singleton et al., 1988b). Some form of intercellular interaction is required to turn off these genes, since most genes continue to be transcribed

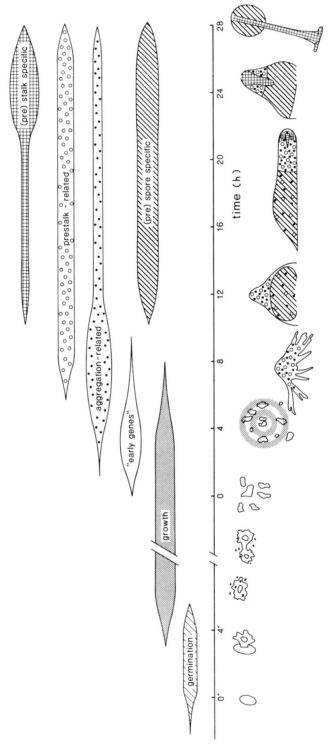

Figure 1. Gene regulation during development. The expression of classes of coregulated genes during development is roughly represented by the thickness of the shaded bars. In the multicellular structures, the spatial patterns of expression of specific classes of genes are indicated by corresponding shading patterns.

when intercellular communication is counteracted (Singleton et al., 1987). For at least two vegetative genes, this interaction is the chemotactic signal of the aggregative stage, i.e., nanomolar cAMP pulses (Kimmel and Carlisle, 1986; Hassanain and Kopachik, 1989).

Preaggregative Gene Expression

Starvation, or more specifically amino acid deprivation (Marin, 1976; Darmon and Klein, 1978), initiates multicellular development. Immediately following starvation, a large number of new genes are being expressed, including the lectin discoidin I and the so-called *I* genes (Ma and Firtel, 1978; Cardelli et al., 1985; Singleton et al., 1988a). Besides starvation, the expression of these genes seems to be dependent on attainment of a certain critical cell density, which is signalled by one or more cellular secretion products of unknown identity (Grabel and Loomis, 1978; Margolskee et al., 1980; Clarke et al., 1987; Mann and Firtel, 1989). The expression of many of the early genes is transient, but repression may be controlled by different mechanisms. Some genes, such as *K5*, are repressed by cAMP pulses (Mann et al., 1988), gene *18* and the discoidin I gene are repressed by high cAMP concentrations (Singleton et al., 1988a; Bozzone and Berger, 1987), and two other *I* genes are repressed by both pulses and high concentrations (Singleton et al., 1988a).

After a few hours of starvation, a number of genes associated with the aggregation process are being transcribed, such as those encoding cAMP receptors, cAMP-PDE, adhesive contact sites A, and a G-protein α-subunit (Gα2). Expression of these genes is accelerated by nanomolar cAMP pulses (Darmon et al., 1975; Gerisch et al., 1975; Noegel et al., 1986; Lacombe et al., 1988; Kumagai et al., 1989) and decreases when aggregation is completed. During aggregation, a second set of transcripts appear, which are later preferentially associated with the anterior prestalk cells. Synthesis of these prestalk-related transcripts can be induced both by nanomolar cAMP pulses and by micromolar cAMP concentrations (Barklis and Lodish, 1983; Mehdy et al., 1983; Mehdy and Firtel, 1985; Chisholm et al., 1984; Peters et al., 1989).

Postaggregative Gene Expression

The attainment of multicellularity coincides with another burst of gene expression (Alton and Lodish, 1977; Cardelli et al., 1985). A major subclass of the postaggregative genes is transcribed in cells which will later become spores (Barklis and Lodish, 1983; Mehdy et al., 1983; Borth and Ratner, 1983; Morrissey et al., 1984; Cardelli et al., 1985). Cells containing prespore-specific gene products are first visible at the basal/central region of the late aggregate and are absent from the tip and from cells which have recently entered the aggregate (Krefft et al., 1984; Williams et al., 1989; Haberstroh and Firtel, 1990).

During slug formation, the prespore phenotype becomes very distinct in the posterior 60 to 80% and can easily be identified by condensed cytoplasm and the presence of prespore vesicles, which contain highly antigenic spore coat proteins and enzymes required for polysaccharide synthesis (Takeuchi, 1963; Hayashi and

Takeuchi, 1976; Schaap et al., 1982; Devine et al., 1983). Intermixed with the posterior prespore cells is a class of cells which strongly resemble the anterior cells but show somewhat reduced expression of prestalk-related genes (Sternfeld and David, 1981, 1982; Devine and Loomis, 1985).

Cells at the anterior 20 to 40% of the slug do not show major changes in phenotype and typically retain the functions and gene products of aggregating cells such as chemotactic responsiveness, surface cAMP-PDE, and high-affinity cAMP binding sites (Tsang and Bradbury, 1981; Schaap and Spek, 1984; Otte et al., 1986; Mee et al., 1986) as well as the cAMP-induced prestalk-related gene transcripts (Barklis and Lodish, 1983; Mehdy et al., 1983). Anterior cells pass through major changes in gene transcription and phenotype at the early culmination stage. About 10 to 20 stalk-specific proteins are being synthesized (Kopachik et al., 1985b; Morrissey et al., 1984; Cardelli et al., 1985), and starting from the tip the anterior cells vacuolate and synthesize the characteristic cellulose cell wall of the stalk cells.

Two stalk-specific genes, pDd63 and pDd56, which code for stalk and sheath matrix proteins, are unlike the other stalk genes expressed as early as the late aggregation stage (Jermyn et al., 1987; McRobbie et al., 1988). Cells expressing pDd63 are first evident at the center of the late aggregate and move subsequently to the tip region of the slug. Cells expressing pDd56 first appear at the base of the aggregate, but are lost in the slime trail during migration. pDd56 expression reappears as a short central funnel at the tip region, reminiscent of the stalk tube formed during early culmination (Jermyn et al., 1989; Williams et al., 1989).

Spore maturation is the final phase of development and does not involve significant changes in gene transcription (Morrissey et al., 1984; Cardelli et al., 1985). The prespore vesicles fuse with the plasma membrane and initiate synthesis of the multilayered spore coat (Hohl and Hamamoto, 1969), which completes the differentiation of the dormant spore.

Signals Involved in Postaggregative Gene Expression

The ubiquitous signal molecule cAMP is also essential for the induction and maintenance of prespore-specific gene expression (Kay, 1982; Barklis and Lodish, 1983; Mehdy et al., 1983; Schaap and Van Driel, 1985; Oyama and Blumberg, 1986). Besides cAMP, at least two other cellular secretion products, PIF (prespore inducing factor) and SPIF (spore inducing factor), appear to be required for spore differentiation (Sternfeld and David, 1979; Weeks, 1984; Wilkinson et al., 1985; Kumagai and Okamoto, 1986). SPIF is either methionine or a close analog of this compound (Gibson and Hames, 1988). PIF is distinct from SPIF and has not yet been identified (Kumagai and Okamoto, 1986). Prespore gene expression is inhibited by the stalk inducing factor DIF (Kay and Jermyn, 1983; Wang et al., 1986; Early and Williams, 1988) and by the cAMP hydrolysis product adenosine (Weijer and Durston, 1985; Schaap and Wang, 1986; Spek et al., 1988).

Stalk cell differentiation can be induced in vitro by DIF (Town et al., 1976; Kay and Jermyn, 1983), which has been identified as a chlorinated hexanone (Morris et al., 1987). DIF also induces expression of the early stalk genes pDd56 and *pDd63* (Jermyn et al., 1987; Williams et al., 1987). cAMP and ammonia antagonize DIF-induced stalk cell differentiation in vitro (Gross et al., 1983; Berks and Kay, 1988), but only ammonia, and not cAMP, was found to counteract stalk cell differentiation in vivo (Wang and Schaap, 1989).

Role of Direct Cell-Cell Interactions during *Dictyostelium* Development

During development, *Dictyostelium* cells show several types of direct intercellular interactions. Vegetative cells express a glucose-recognizing receptor and a nonspecific receptor which mediate phagocytosis of bacteria (Vogel et al., 1980). Cells at this stage show some weak cohesion which is EDTA sensitive and is mediated by the so-called contact sites B (Beug et al., 1973). These sites were identified as glycoproteins of 126 kDa (Chadwick and Garrod, 1983) and are possibly identical to the nonspecific phagocytosis receptors (Vogel et al., 1980; Chadwick et al., 1984).

Upon starvation, cells become increasingly adhesive. Early postvegetative cell adhesion is also disrupted by EDTA and is mediated by a 24-kDa cell surface glycoprotein (gp24). This glycoprotein is the product of two tandemly repeated genes with a 90% conserved nucleotide sequence (Loomis and Fuller, 1990). The gene is maximally expressed after 8 h of development, remains high during the early slug stage, and thereafter decreases (Knecht et al., 1987; Loomis and Fuller, 1990). Antibodies raised against purified gp24 completely inhibit the aggregation process and further development (Knecht et al., 1987).

Somewhat later during development, the cells display an EDTA-insensitive adhesion system, the so-called contact sites A (Beug et al., 1973), which were identified as 80-kDa cell surface glycoproteins (Müller and Gerisch, 1978; Gerisch, 1980). These glycoproteins (gp80) link cells together by means of a homophilic interaction (Siu et al., 1987), which was recently mapped to the octapeptide sequence YKLNVDNS (Kamboj et al., 1989). gp80 transcripts and glycoprotein peak at about 12 h of development and decrease during slug formation (Noegel et al., 1986; Siu et al., 1988). Mutants lacking gp80, and cells treated with monoclonal antibodies raised against gp80, do not acquire EDTA-resistant adhesiveness during aggregation, but show nevertheless virtually normal postaggregative development (Murray et al., 1984; Siu et al., 1985), indicating that this adhesion system is not strictly essential for development.

During the slug stage of development, a third type of cell adhesion is evident which is EDTA resistant but cannot be blocked by antibodies against gp80 and gp24 (Knecht et al., 1987). Antibodies and Fab fragments were raised that specifically block slug stage cell adhesion. Plasma membrane glycoproteins of around 95 kDa were reported to neutralize the inhibitory effects of the antisera (Steinemann and Parish, 1980; Wilcox and Sussman, 1982; Saxe and Sussman, 1982). Recent studies suggest that several proteins may be involved in slug stage cell adhesion (Loomis, 1988).

The different adhesive systems are essential to maintain the integrity of the multicellular structure at different stages of development, but it is as yet not clear whether the formation of cell-cell contacts can directly induce cellular responses such as gene expression. Establishment of cell-cell contacts was recently reported to induce and modulate cAMP signalling in starving cells (Fontana and Price, 1989), which might have further effects on other cellular responses.

Several studies showed that the formation of close cell-cell associations is essential for the induction of postaggregative gene expression (Alton and Lodish, 1977; Chung et al., 1981; Blumberg et al., 1982; Kaleko and Rothman, 1982; Mehdy et al., 1983). It was, however, not clear whether gene regulation was a direct response to cell contact formation or whether the establishment of close proximity was required for efficient accumulation of diffusible signal molecules. Later studies showed that the requirement for close cell interactions could in most cases be mimicked by cAMP and/or other cellular secretion products (Kay, 1982; Mehdy and Firtel, 1985; Wilkinson et al., 1985; Kumagai and Okamoto, 1986; Schaap et al., 1986).

The current consensus in the *Dictyostelium* field is that the major aspects of *Dictyostelium* gene regulation are controlled by diffusible signals rather than by direct cell interactions.

Gene Regulation and Pattern Formation in Other Cellular Slime Mold Species

D. discoideum is the only species in which the regulation of gene expression during development is known in such detail. Still, comparison with the current knowledge of pattern formation in other species is useful to discriminate between important general strategies and less important species-specific characteristics.

All investigated species, i.e., *Polysphondylium, D. minutum, D. lacteum, D. vinaceo-fuscum, D. mucoroides,* and *D. purpureum,* form the typical prespore vesicles after aggregation (Takeuchi, 1963; Hohl et al., 1977; Schaap et al., 1985). However, while the latter two species show an anteroposterior pattern similar to that of *D. discoideum,* although smaller, the first four show no pattern and contain prespore vesicles up to the tip region. During culmination, and in many species also during slug migration, cells which have reached the tip differentiate into stalk cells. In species such as *Polysphondylium, D. minutum,* and *D. lacteum* this process is accompanied by continuous dedifferentiation of prespore cells into stalk cells (Hohl et al., 1977; Schaap et al., 1985). In *D. mucoroides* and *D. purpureum,* dedifferentiation during slug migration occurs at the boundary between the prespore and prestalk region (Gregg and Davis, 1982; Schaap et al., 1985).

Very little is known with regard to signals controlling gene expression in other cellular slime mold species. High concentrations of cAMP derange morphogenesis in *Polysphondylium* and *Dictyostelium* species and result in random differentiation of stalk cells (Hohl et al., 1977; George, 1977). *Polysphondylium* cells secrete a number of dialyzable molecules, called D-factors, which induce competence for the aggregation process (Hanna and Cox, 1978; Hanna et al., 1983). These factors also induce synthesis of cAMP by preaggregative *Polysphondylium* cells (Hanna et al., 1984). At this stage cAMP does not act as a chemoattractant, so its function is not clear. Some evidence is presented that cAMP alters patterns of protein synthesis in *Polysphondylium* (Francis, 1975). Glorin, the *Polysphondylium* chemoattractant, also induces synthesis of specific proteins (Kopachik, 1990).

To my knowledge nothing is known about the involvement of DIF, ammonia, or adenosine in pattern formation in other cellular slime mold species.

GENERATION OF MORPHOGENETIC SIGNALLING IN *D. DISCOIDEUM*

The formation of patterns of differentiated cells can formalistically result from two alternative mechanisms: (i) the cells are determined for a specific fate early in development and sort out to form a pattern, or (ii) cells are totipotent and differentiate according to specific positional cues. Abundant support for involvement of either of the two mechanisms during *Dictyostelium* development is available, which suggests that multiple or hybrid mechanisms may be utilized. Predetermination is suggested by observations that cell cycle phase at the onset of development is correlated with cell fate. Cells starved in early cell cycle phase form slugs with a relatively large percentage of prestalk cells and sort to the prestalk region when mixed with nonsynchronized cells (McDonald and Durston, 1984; Weijer et al., 1984). When supplied with cAMP and conditioned medium, early cell cycle cells (E-cells) preferentially synthesize prestalk antigens and late cell cycle cells (L-cells) synthesize prespore antigens (Gomer and Firtel, 1987). During growth, cells express a surface glycoprotein at higher levels during early than during late cell cycle phase. This glycoprotein is, in the slug, preferentially present on prestalk cells (Krefft and Weijer, 1989).

Arguing against predetermination are observations that prestalk and prespore isolates regulate to almost normal proportions (Bonner et al., 1955; Sakai, 1973; Oyama et al., 1983). Some mutants show normal anteroposterior patterns in the slug stage, but form fruiting structures consisting entirely of stalk cells or of spore cells, which indicates that cell fate can be altered until very late in development (Morrissey et al., 1981). Regulation by predetermination would be limited to species that form a prepattern since, as described above, in many species all cells first differentiate into prespore cells and then redifferentiate by positional cues into stalk cells.

It is therefore more likely that cell cycle phase at the onset of starvation imposes a relative preference for a specific cell fate rather than an absolute predetermination. Early (E) and late (L) cell cycle cells show pronounced differences in aggregation-related properties, which may account for their sorting behavior and more importantly provide the basis for generation of morphogenetic gradients (see Fig. 2). Following starvation, E-cells initiate aggregation centers much more rapidly than L-cells (McDonald, 1986) and furthermore exhibit cAMP receptors and surface cAMP-PDE several hours earlier and to higher levels than L-cells (Wang et al., 1988a). The E-cells initiate the aggregation process and most likely accumulate at the aggregation center and tip because they are most responsive to the chemotactic signal. This is probably the reason why anterior cells in slugs show much higher levels of cAMP binding activity and cAMP-PDE than posterior cells (Tsang and Bradbury, 1981; Schaap and Spek, 1984; Otte et al., 1986; Mee et al., 1986). Sorting of E-cells with high PDE activity to the tip region sets up a gradient of cAMP degrading activity. Since the ability to relay cAMP pulses is similar in prestalk and prespore cells (Otte et al., 1986), this PDE gradient results

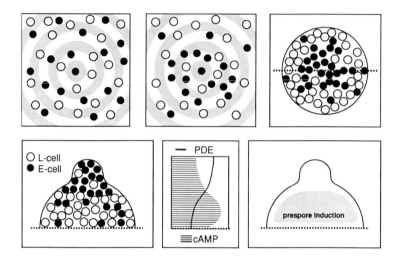

FIGURE 2. Model for initial establishment of pattern by cell sorting. Cells differing in cell cycle phase vary in expression of aggregation competence. Early cell cycle cells initiate cAMP oscillations and exhibit relatively high levels of cAMP receptors and cAMP-PDE. These cells preferentially accumulate at the aggregation center and tip region, resulting in a shallow apical/basal gradient of cAMP degrading activity. Continued cAMP signalling, combined with low levels of cAMP-PDE at the basal region, results in local induction of prespore gene expression.

in combination with continued cAMP signalling in a cAMP gradient which is low at the tip and sufficiently high at the basal part of the aggregate to induce the local expression of prespore-specific genes in this region, as was previously described (Krefft et al., 1984; Williams et al., 1989; Haberstroh and Firtel, 1990).

The anteroposterior pattern becomes more pronounced during slug formation. This is probably due to the establishment of opposite gradients of the prespore inducer cAMP and the prespore inhibitor adenosine (Fig. 3). Besides cAMP-PDE, surface 5'-nucleotidase is also specifically active at the anterior region (Armant and Rutherford, 1979; Armant et al., 1980), which results in the effective degradation of cAMP into adenosine. High cAMP levels combined with low adenosine levels permit prespore differentiation at the posterior region. At the anterior, low cAMP levels combined with high adenosine levels are not conducive for prespore differentiation. The validity of this hypothesis was demonstrated by experiments in which extracellular cAMP or adenosine levels are enzymatically depleted (Schaap and Wang, 1986; Wang et al., 1988b). cAMP depletion results in the complete loss of prespore antigens, while adenosine depletion results in synthesis of prespore antigens in the anterior prestalk region (Fig. 4). This model does not account for the presence of anteriorlike cells at the prespore region. Prespore cells were recently shown to secrete an autoinhibitor which is not adenosine but could be the stalk-inducing factor DIF (Inouye, 1989). DIF synthesis increases during aggregation to reach maximal levels at the slug stage (Brookman et al., 1982). DIF is present at somewhat higher levels in the prespore than in the prestalk region (Brookman et al., 1987), and DIF synthesis is induced by high

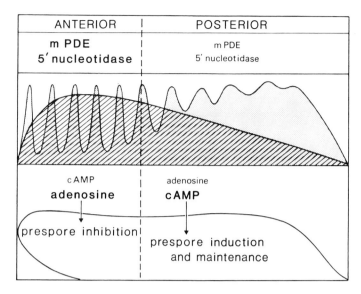

FIGURE 3. Model for maintenance of pattern by opposite cAMP and adenosine gradients. High levels of cell surface cAMP-PDE and 5′-nucleotidase at the anterior region result in degradation of cAMP (shaded area) to the prespore antagonist adenosine (hatched area). Low levels of cAMP, combined with high levels of adenosine at the anterior, inhibit differentiation of prespore cells (from Wang et al., 1988b).

cAMP concentrations (Brookman et al., 1982; Kwong and Weeks, 1989). The regulation of prespore (PSP) and anteriorlike (AL) cells could be due to a mechanism in which cAMP promotes conversion of anteriorlike cells into prespore cells, while DIF produced by prespore cells promotes the opposite conversion (MacWilliams et al., 1985; Schaap, 1986):

$$\text{PSP} \underset{\underset{\text{DIF}}{\oplus}}{\overset{\overset{\text{cAMP}}{\oplus}}{\rightleftarrows}} \text{AL}$$

The culmination process adds further complexity to the pattern. Starting at the tip, cells enter a prefabricated cellulose tube and differentiate into stalk cells. The signal for position-dependent stalk cell differentiation is most likely not DIF, since DIF levels reach their maximum at a much earlier stage and are higher at the posterior region than at the tip (Brookman et al., 1987). Stimulation of intact slugs with DIF does not induce stalk cell differentiation (Inouye, 1988; Wang and Schaap, 1989). Rapid stalk cell differentiation can be induced when the endogenous levels of the DIF antagonist ammonia are depleted simultaneously with DIF stimulation (Wang and Schaap, 1989). This suggests the following model for stalk induction (Fig. 5). During development, cells produce large amounts of ammonia

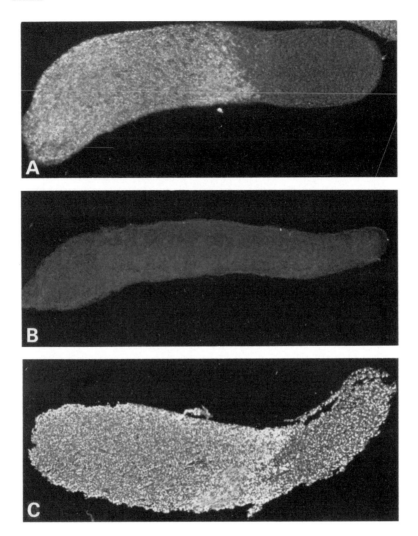

FIGURE 4. Effects of cAMP and adenosine depletion on pattern formation in slugs. Slugs were submerged either in phosphate buffer (A) or in buffer containing cAMP phosphodiesterase to inactivate endogenous cAMP (B), or in buffer containing adenosine deaminase to inactivate endogenous adenosine (C). After 4 h of incubation, slugs were sectioned and stained with prespore-specific antiserum. cAMP depletion results in complete loss of prespore antigens, while adenosine depletion results in synthesis of prespore antigens in the former prestalk region.

by degradation of protein. Migrating slugs move close to the substratum, which may contain fairly high ammonia levels. DIF levels in slugs are optimal, but DIF-induced stalk cell differentiation is inhibited by ammonia. During early culmination, the slug lifts its tip in the air and now loses ammonia by evaporation. Ammonia levels at the tip decrease, and stalk cell differentiation can proceed.

This model is probably too simplistic; ammonia depletion may act as a trigger

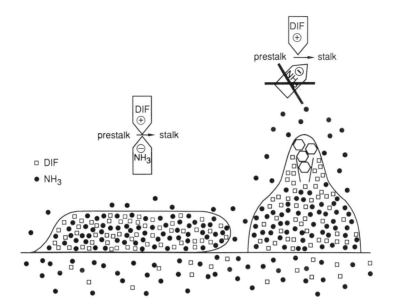

FIGURE 5. Model for induction of stalk cell differentiation by ammonia depletion. Migrating slugs contain high levels of the stalk-inducing factor DIF and of the stalk antagonist ammonia. During culmination, the cell mass erects itself and lifts its tip into the air. Ammonia evaporates most effectively from the tip region, which has a relatively high surface-to-volume ratio. The block on stalk cell differentiation is released, and stalk cells differentiate.

for stalk cell differentiation, but actual stalk formation most likely requires more complex regulation. Stalk formation is initiated by the formation of a central cellulose tube before actual stalk cell differentiation starts (Raper and Fennell, 1952). Evidently this requires coordinate synthetic activities of a group of cells at the core of the tip. The early stalk gene *pDd56,* which codes for stalk matrix proteins, is characteristically expressed at this position long before stalk cell differentiation occurs (Jermyn et al., 1989; Williams et al., 1989). Apparently, the expression of this gene is controlled by a specific local signal which is neither DIF nor ammonia. The premature expression of the other early stalk gene, *pDd63,* at the tip region of migrating slugs could result from less stringent inhibition by ammonia than the other stalk genes. More information with regard to regulation of these genes is required to account for their specific localization.

The above proposed models, which are at best still rough approximations of the actual mechanisms controlling *Dictyostelium* development, nevertheless demonstrate how increasing complexity can be generated from fairly small variations between individual cells, such as differences in cell cycle phase. The current data furthermore show that the major aspects of gene regulation in *D. discoideum* are controlled by hormonelike extracellular signal molecules. When comparing the regulation of gene expression with the control of morphogenetic movement during development, it is also evident that many signal molecules serve multiple functions. Most evident is cAMP, which acts as a chemoattractant and regulates gene expression in almost all stages of development. But also adenosine, which acts as both a tip inhibitor and an inhibitor of prespore differentiation, and ammonia, which is involved in slug orientation and stalk cell differentiation, are signals with multiple functions. An immediate question following this observation is how all these signals are processed by the cells to yield responses as diverse as gene expression and directed cell movement.

TRANSDUCTION OF EXTRACELLULAR SIGNALS

cAMP Signal Transduction

The discovery that the well-known intracellular messenger cAMP acts as an extracellular signal for chemotaxis in the cellular slime molds initiated efforts in many laboratories to elucidate how this signal was translated into chemotactic cell movement. Interest in cAMP signal transduction became even more intense after the observation that extracellular cAMP was also involved in gene regulation.

Initial studies were aimed at identifying the first target of cAMP and led to the identification of a highly specific cell surface cAMP receptor (Henderson, 1975; Green and Newell, 1975). This receptor has been cloned, and its amino acid sequence predicts a protein with seven putative transmembrane-spanning domains (Klein et al., 1988), which is characteristic for surface receptors interacting with G proteins, such as the β-adrenergic receptor, the muscarinic acetylcholine receptor, rhodopsin, and the serotonin receptor (see Dohlman et al., 1987; Hahn, 1989). Interaction of cAMP with cell surface receptors evokes a number of intracellular responses (Fig. 6), such as activation of adenylate cyclase (Roos et al., 1975; Shaffer, 1975), guanylate cyclase (Mato et al., 1977; Würster et al., 1977), and phospholipase C (Europe-Finner and Newell, 1987; Van Haastert et al., 1989), resulting in a rapid increase in the intracellular concentrations of, respectively, cAMP, cGMP, and inositol 1,4,5-trisphosphate (IP_3). IP_3 subsequently induces the release of Ca^{2+} from nonmitochondrial stores, resulting in an increase in cytosolic Ca^{2+} levels (Europe-Finner and Newell, 1986a).

The activation of adenylate cyclase and phospholipase C is mediated by different G proteins (Theibert and Devreotes, 1986; Van Haastert et al., 1987; Van Haastert et al., 1989). The α-subunit of the G protein which links the cAMP receptor to phospholipase C may be the recently cloned Gα2, which is inactive in *fgdA* mutants (Kumagai et al., 1989; Pupillo et al., 1989; Kesbeke et al., 1988; Snaar-Jagalska et al., 1988; Van Haastert, 1990). Besides a G protein, adenylate cyclase activation also requires a cytosolic factor, which is absent in mutant *synag7* (Theibert and Devreotes, 1986; Snaar-Jagalska and Van Haastert, 1988). Two putative intracellular targets for cAMP have been identified: a cAMP-dependent protein kinase (De Gunzburg and Veron, 1982; Rutherford et al., 1982; Mutzel et al., 1987) and a cAMP binding protein of unknown function (Tsang and Tasaka, 1986). Both proteins are present at low levels in vegetative cells. Levels increase during aggregation to reach a maximum at the slug stage (Leichtling et al., 1984; Kay et al., 1987).

Intracellular targets for Ca^{2+} are calmodulin (Clarke et al., 1980) and a re-

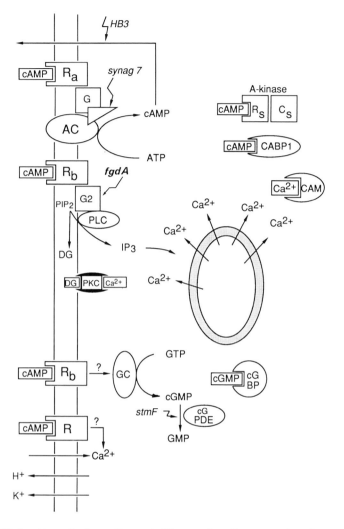

FIGURE 6. cAMP signal transduction pathways in *Dictyostelium*. Representation of currently identified components of cAMP transduction pathways. R, R_a, R_b, different subpopulations of cAMP receptors; AC, adenylate cyclase; GC, guanylate cyclase; PLC, phospholipase C; G, guanine nucleotide regulatory protein; CABP, cAMP binding protein; A-kinase, cAMP-dependent protein kinase; CAM, calmodulin; PKC, Ca^{2+}-dependent protein kinase; DS, diacylglycerol; PIP_2, phosphatidylinositol bisphosphate; R_s, regulatory subunit; C_s, catalytic subunit.

cently identified protein kinase C, which is stimulated by either Ca^{2+} or phospholipids (Ludérus et al., 1989). A phopholipid-stimulated kinase activity which is not stimulated by Ca^{2+} has also been demonstrated (Jimenez et al., 1989).

cGMP is detected by an intracellular cGMP binding protein (Mato et al., 1978; Van Haastert et al., 1982b; Parissenti and Coukell, 1989) and degraded by substrate-stimulated cGMP-PDE (Dicou and Brachet, 1980; Bulgakov and Van Haastert, 1982). Ca^{2+} and cGMP may both be involved in transduction of cAMP to the chemotactic response, because they were reported to induce actin and myosin polymerization, respectively (Europe-Finner and Newell, 1986b; Liu and Newell, 1988). cAMP-induced actin nucleation and polymerization are deficient in *fgdA* mutants, which suggests that cAMP-induced activation of the IP_3/Ca^{2+} pathway is responsible for these responses. Involvement of cGMP in chemotaxis is also evident from studies of *stmF* mutants, which lack cGMP-PDE. The prolonged presence of cGMP after stimulation with cAMP results in an extended period of chemotactic cell movement (Ross and Newell, 1981).

cAMP-induced responses involved in chemotaxis, such as cAMP, cGMP, and IP_3 accumulation, are transient and desensitize during persistent stimulation. Desensitization occurs via different mechanisms, which involve both receptor phosphorylation (Klein et al., 1985; Klein et al., 1987) and receptor down-regulation and internalization (Klein and Juliani, 1977; Van Haastert, 1987; Wang et al., 1988c).

Besides the above-mentioned responses, cAMP induces an influx of Ca^{2+} (Wick et al., 1978; Abe et al., 1988) and an efflux of K^+ ions and H^+ ions (Malchow et al., 1978; Aeckerle et al., 1985). The cAMP-induced Ca^{2+} influx does not adapt during persistent stimulation (Bumann et al., 1986), while the cAMP-induced H^+ efflux shows both a rapid transient and a more slow sustained component (Aerts et al., 1987; Van Lookeren Campagne et al., 1989).

The second final target for extracellular cAMP signals is the control of gene expression in various stages of development. As described in the previous chapters, cAMP pulses inhibit transcription of early genes and induce transcription of aggregation-associated and prestalk-related genes. Steady-state cAMP levels in the micromolar range induce expression of both prespore- and prestalk-related genes. Although the levels of some prespore transcripts are regulated by mRNA stabilization (Mangiarotti et al., 1985; Mangiarotti et al., 1989), most cAMP-regulated genes are controlled at the transcription level. A number of coregulated genes typically contain one or more 5'-flanking G/C-rich control elements, which are required for optimal expression (Datta and Firtel, 1987, 1988; Pears and Williams, 1987, 1988; Pavlovic et al., 1989). Recently a *trans*-acting factor was identified which recognizes dissimilar G/C-rich elements and confers regulated expression in correlation with its relative affinity for the respective elements (Hjorth et al., 1989; Hjorth et al., 1990). The expression of the *trans*-acting factor is itself cAMP regulated, which suggests that expression of this and related factors may control expression of classes of coregulated genes (Hjorth et al., 1989).

Studies using cAMP analogs have shown that both prespore- and prestalk-related gene expression show nucleotide specificity similar to that of the surface cAMP receptor, suggesting that a surface cAMP receptor mediates cAMP-induced gene expression (Van Haastert and Kien, 1983; Schaap and Van Driel, 1985; Oyama and Blumberg, 1986; Haribabu and Dottin, 1986). It is not yet known which intracellular messengers control the different types of gene expression. The induction of aggregation-related and prespore genes by cAMP pulses or persistent cAMP stimulation, respectively, is normal in mutants blocked in adenylate cyclase activation (Schaap et al., 1986; Bozzaro et al., 1987; Mann et al., 1988), suggesting that intracellular cAMP does not mediate expression of these genes. Recent studies suggest negative regulation of aggregation-related genes by cAMP-dependent protein kinase; cells transformed with regulatory subunits lacking cAMP binding sites, and therefore incapable of activating catalytic subunits, were found to overexpress cAMP-PDE and the pulse-induced gene *D2* (Firtel and Chapman, 1990). Intracellular cAMP may mediate the repression of vegetative genes by cAMP pulses, because this response is blocked by caffeine, which inhibits adenylate cyclase activation (Kimmel, 1987). Intracellular cAMP may also play a role in the maturation of spore cells; high concentrations of agonists of cAMP protein kinase, which supposedly pass the plasma membrane, were reported to promote the differentiation of mature spore cells (Kay, 1989; Riley et al., 1989).

Induction of prespore and aggregation-related gene expression by cAMP is not altered in *stmF* mutants with strongly altered cGMP metabolism, suggesting that intracellular cGMP does not control expression of these genes (Schaap et al., 1986; Mann et al., 1988).

Prestalk-related and prespore gene expression can both be induced by constant cAMP stimulation, which is suggestive of similar signal processing. However, several experiments suggest divergent transduction pathways. Prespore induction is strongly inhibited by Ca^{2+} antagonists (Schaap et al., 1986), while prestalk-related gene expression is much less sensitive (Blumberg et al., 1989). LiCl, an inhibitor of cAMP-induced IP_3 accumulation (Peters et al., 1989), inhibits cAMP-induced prespore gene expression and promotes prestalk-related gene expression (Van Lookeren Campagne et al., 1989). The cAMP antagonist adenosine, which inhibits cAMP binding to chemotactic receptors (Newell, 1982), inhibits cAMP induction of prespore gene expression, but slightly promotes cAMP induction of prestalk-related gene expression (Spek et al., 1988).

The sensitivity of cAMP-induced prespore gene expression to Ca^{2+} antagonists and LiCl is suggestive of involvement of intracellular IP_3/Ca^{2+} in regulation of these genes. Ginsburg and Kimmel (1989) recently showed that IP_3 combined with diacylglycerol can induce prespore gene expression. However, stimulation with these agents was required prior to the stage where cAMP-mediated induction occurs, which makes it doubtful whether the two agents are really direct intermediates for cAMP-induced transcription. It has also not yet been shown that persistent stimulation with cAMP, which is required for prespore gene expression, induces persistent IP_3 and diacylglycerol accumulation.

Besides the transient IP_3/Ca^{2+} response, cAMP also induces a sustained influx of Ca^{2+}, which may mediate prespore induction (Bumann et al., 1986). However, the recent observation that cytosolic Ca^{2+} levels are higher in prestalk than in prespore cells makes this somewhat unlikely (Abe and Maeda, 1989).

A second sustained response is the cAMP-induced increase in intracellular pH. The cAMP-induced pH_i increase can be bypassed by weak bases and inhibited by weak acids, but neither weak bases nor weak acids promote nor inhibit prespore- or prestalk-related gene expression (Van Lookeren Campagne et al., 1989; Peters et al., 1991). This indicates that the cAMP-induced increase in pH_i does not mediate induction of prespore- or prestalk-related gene expression.

The *fgdA* mutants, which are defective in the α subunit of the G protein which supposedly links the cAMP receptor to IP_3/diacylglycerol production (Coukell et al., 1983; Kesbeke et al., 1988; Kumagai et al., 1989; Van Haastert, 1990), do not express aggregation-related genes in response to cAMP pulses (Mann et al., 1988). This implies that cAMP-induced IP_3/diacylglycerol production mediates pulse-induced gene expression. However, Li^+ ions, which inhibit cAMP-induced IP_3 accumulation, promote pulse-induced gene expression (Peters et al., 1989; Peters et al., 1991), which is again in conflict with this supposition.

In short, no definite conclusions with regard to involvement of specific intracellular messengers in gene regulation can yet be drawn, which is most likely due to the fact that we do not yet understand the entire complexity of cAMP signal processing. The present data show that the regulation of different classes of genes by extracellular cAMP utilizes divergent signal transduction mechanisms, which probably all start at a surface cAMP receptor. cAMP-induced changes in intracellular cAMP, cGMP, or pH do not appear to play a major role in regulation of the different genes, but inositolphospholipid signalling may be implicated. Knowledge of cAMP signal transduction is rapidly expanding. Currently used molecular genetic and biochemical approaches to identify and disrupt the genes coding for signal transduction components will greatly facilitate analysis of involvement of specific signalling cascades in gene regulation.

DIF and Ammonia Signal Transduction

The stalk-inducing factor DIF is a small, nonpolar molecule (Morris et al., 1987) which most likely passes through the plasma membrane and interacts with an intracellular target protein. Early data showed that cytoplasmic acidification can bypass the effects of DIF on stalk cell differentiation, which suggested that the effects of DIF were mediated by a decrease in intracellular pH (Gross et al., 1983; Town, 1984; Dominov and Town, 1986). However, DIF only induces a very minor and transient decrease in intracellular pH (Kay et al., 1986; Inouye, 1985). Recent data show that cytoplasmic acidification cannot directly induce the expression of stalk-specific genes. However, a low intracellular pH is required to enable cells to respond to DIF (Wang et al., 1990). The weak base ammonia most likely exerts its inhibitory effect on DIF-induced stalk cell differentiation by increasing intracellular pH and thereby blocking DIF signal transduction.

Some evidence suggests that the effects of ammonia on slug and fruiting body orientation are also due to cytoplasmic alkalinization. The effects of ammonia on orientation are considered to be caused by enhanced cell locomotion. A small increase in the rate of cell movement in response to ammonia has been observed (Bonner et al., 1986; Bonner et al., 1989), while CO_2, which decreases intracellular pH (Inouye, 1988), reduces cell locomotion (Bonner et al., 1989). Ammonia increases the amplitude of optical density oscillations in the aggregative field, which may reflect enhanced cell locomotion (Siegert and Weijer, 1989). Further experiments are required to establish a causal relationship between ammonia, intracellular alkalinization, and cell locomotion.

CONCLUSIONS

Intercellular communication is of vital importance for almost all aspects of cellular slime mold development. *D. discoideum* has been the principal species of interest, and the identity and function of the major signals which control its development have largely been elucidated. Autonomous cAMP oscillations and cAMP signal relay act to coordinate the movement of individual cells. cAMP also controls gene expression during several stages of development. Two other important signals are DIF and ammonia, which coordinately control the differentiation of stalk cells. Ammonia also acts as a signal for orientation of slugs and fruiting bodies. The cAMP hydrolysis product adenosine is involved in size regulation and in maintenance of the prestalk/prespore pattern in slugs. Using cell cycle-related variations in cAMP signalling competence as a starting point, it is possible to incorporate current knowledge of the spatiotemporal aspects of morphogenetic signalling into a model which can account for the major events during *D. discoideum* development.

At present very little is known about a general role of the morphogenetic signals recognized in *D. discoideum* during regulation of development in other cellular slime molds. Except for oscillatory cAMP signalling, which most likely controls morphogenetic movement in all species, the mechanisms which regulate cell differentiation in other cellular slime molds are still obscure. To realize the general importance of currently recognized signalling systems involved during regulation of *D. discoideum* development, it is essential to determine to what extent these systems are utilized by other cellular slime molds and maybe by other organisms.

Strong conservation is present between *Dictyostelium* and vertebrate systems at the level of intracellular signalling. The mechanism of G protein-mediated processing of extracellular cAMP signals to intracellular effector enzymes is highly homologous to the processing of hydrophilic signal molecules such as epinephrine, acetylcholine, and serotonin in mammalian systems.

ACKNOWLEDGMENTS. I am very grateful to many colleagues in the *Dictyostelium* field who sent me preprints of their latest studies. I further thank Peter van Haastert, Theo Konijn, Dorien Peters, and Wang Mei for critical reading of the manuscript.

REFERENCES

Abe, H., and Y. Maeda. 1986. Induction of macrocyst germination in the cellular slime mould *Dictyostelium mucoroides*. *J. Gen. Microbiol.* **132**:2787–2791.

Abe, H., M. Uchiyama, Y. Tanaka, and H. Saito. 1976. Structure of discadenine, a spore germination inhibitor from the cellular slime mould *Dictyostelium discoideum*. *Tetrahedron Lett.* **42**:3807–3810.

Abe, T., and Y. Maeda. 1989. The prestalk/prespore differentiation and polarized cell movement in *Dictyostelium discoideum* slugs. A possible involvement of the intracellular Ca^{2+}-concentration. *Protoplasma* **151**:175–178.

Abe, T., Y. Maeda, and T. Lijima. 1988. Transient increase of the intracellular Cd^{2+} concentration during chemotactic signal transduction in *Dictyostelium discoideum* cells. *Differentiation* **39**:90–96.

Aeckerle, S., B. Wurster, and D. Malchow. 1985. Oscillations and cyclic AMP induced changes of the K^+ concentration in *Dictyostelium discoideum*. *EMBO J.* **4**:39–43.

Aerts, R. J., R. J. W. De Wit, and M. M. Van Lookeren Campagne. 1987. Cyclic AMP induces a transient alkalinization in *Dictyostelium*. *FEBS Lett.* **220**:366–370.

Alton, T. H., and H. F. Lodish. 1977. Synthesis of developmentally regulated proteins in *Dictyostelium discoideum* which are dependent on continued cell-cell interaction. *Dev. Biol.* **60**:207–216.

Amagai, A. 1984. Induction by ethylene of macrocyst formation in the cellular slime mould *Dictyostelium mucoroides*. *J. Gen. Microbiol.* **130**:2961–2965.

Amagai, A. 1987. Regulation of the developmental modes in *Dictyostelium mucoroides* by cAMP and ethylene. *Differentiation* **36**:111–115.

Amagai, A., and M. F. Filosa. 1984. The possible involvement of cyclic AMP and volatile substance(s) in the development of a macrocyst forming strain of *Dictyostelium mucoroides*. *Dev. Growth Differ.* **26**:583–589.

Armant, D. R., and C. L. Rutherford. 1979. 5′-AMP nucleotidase is localized in the area of the cell-cell contact of prespore and prestalk regions during culmination of *Dictyostelium discoideum*. *Mech. Ageing Dev.* **10**:199–217.

Armant, D. R., D. A. Stetler, and C. L. Rutherford. 1980. Cell surface localization of 5′AMP nucleotidase in prestalk cells of *Dictyostelium discoideum*. *J. Cell Sci.* **45**:119–129.

Barklis, E., and H. F. Lodish. 1983. Regulation of *Dictyostelium discoideum* mRNAs specific for prespore or prestalk cells. *Cell* **32**:1139–1148.

Berks, M., and R. R. Kay. 1988. Cyclic AMP is an inhibitor of stalk cell differentiation in *Dictyostelium discoideum*. *Dev. Biol.* **126**:108–114.

Beug, H., F. E. Katz, and G. Gerisch. 1973. Dynamics of antigenic membrane sites relating to cell aggregation in *Dictyostelium discoideum*. *J. Cell Biol.* **56**:647–658.

Blaskovics, J. C., and K. B. Raper. 1957. Encystment stages in *Dictyostelium*. *Biol. Bull.* **113**:58–88.

Blumberg, D. D., J. F. Comer, and E. M. Walton. 1989. Ca^{++} antagonists distinguish different requirements for cAMP-mediated gene expression in the cellular slime mold, *Dictyostelium discoideum*. *Differentiation* **41**:14–21.

Blumberg, D. D., J. P. Margolskee, E. Barklis, S. N. Chung, N. S. Cohen, and H. F. Lodish. 1982. Specific cell-cell contacts are essential for induction of gene expression during differentiation of *Dictyostelium discoideum*. *Proc. Natl. Acad. Sci. USA* **79**:127–131.

Bonner, J. T., A. Chiang, J. Lee, and H. B. Suthers. 1988. The possible role of ammonia in phototaxis of migrating slugs of *Dictyostelium discoideum*. *Proc. Natl. Acad. Sci. USA* **85**:3885–3887.

Bonner, J. T., A. D. Chiquoine, and M. Q. Kolderie. 1955. A histochemical study of differentiation in the cellular slime molds. *J. Exp. Zool.* **130**:133–158.

Bonner, J. T., W. W. Clarke, C. L. Neely, and M. K. Slifkin. 1950. The orientation to light and the extremely sensitive orientation to temperature gradients in the slime mold *Dictyostelium discoideum*. *J. Cell. Comp. Physiol.* **36**:149–158.

Bonner, J. T., and M.R. Dodd. 1962. Evidence for gas-induced orientation in the cellular slime molds. *Dev. Biol.* **5**:344–361.

Bonner, J. T., D. Har, and H. B. Suthers. 1989. Ammonia and thermotaxis: further evidence for a central role of ammonia in the directed cell mass movements of *Dictyostelium discoideum*. *Proc. Natl. Acad. Sci. USA* **86**:2733–2736.

Bonner, J. T., H. B. Suthers, and G. M. Odell. 1986. Ammonia orients cell masses and speeds up aggregating cells of slime moulds. *Nature* (London) **323**:630–632.

Borth, W., and D. Ratner. 1983. Different synthetic profiles and developmental fates of prespore versus prestalk proteins of *Dictyostelium*. *Differentiation* **24**:213–219.

Bozzaro, S., J. Hagmann, A. Noegel, M. Westphal, E. Calautti, and E. Bogliolo. 1987. Cell differentiation in the absence of intracellular and extracellular cyclic AMP pulses in *Dictyostelium discoideum*. *Dev. Biol.* **123**:540–548.

Bozzone, D. M., and E. A. Berger. 1987. Distinct developmental regulation and properties of the responsiveness of different genes to cyclic AMP in *Dictyostelium discoideum*. *Differentiation* **33**:197–206.

Brookman, J. J., K. A. Jermyn, and R. R. Kay. 1987. Nature and distribution of the morphogen DIF in the *Dictyostelium* slug. *Development* **100**:119–124.

Brookman, J. J., C. D. Town, K. A. Jermyn, and R. R. Kay. 1982. Developmental regulation of stalk cell differentiation-inducing factor in *Dictyostelium discoideum*. *Dev. Biol.* **91**:191–196.

Bulgakov, R., and P. J. M. Van Haastert. 1982. Isolation and partial characterization of a cGMP-dependent cGMP-specific phosphodiesterase of *Dictyostelium discoideum*. *Biochim. Biophys. Acta* **756**:56–66.

Bumann, J., D. Malchow, and B. Wurster. 1986. Oscillations of Ca^{++} concentration during cell differentiation of *Dictyostelium discoideum*. *Differentiation* **31**:85–91.

Cardelli, J. A., D. A. Knecht, R. Wunderlich, and R. L. Dimond. 1985. Major changes in gene expression occur during at least four stages of development of *Dictyostelium discoideum*. *Dev. Biol.* **110:**147–156.

Chadwick, C. M., J. E. Ellison, and D. R. Garrod. 1984. Dual role for *Dictyostelium* contact sites B in phagocytosis and developmental size regulation. *Nature* (London) **307:**646–647.

Chadwick, C. M., and D. R. Garrod. 1983. Identification of the cohesion molecule, contact sites B, of *Dictyostelium discoideum*. *J. Cell Sci.* **60:**251–266.

Chisholm, R. L., E. Barklis, and H. F. Lodish. 1984. Mechanism of sequential induction of cell type specific mRNAs in *Dictyostelium* differentiation. *Nature* (London) **310:**67–69.

Choi, A. H., and D. H. O'Day. 1982. Ammonia and the induction of microcyst differentiation in wild type and mutant strains of the cellular slime mold Polysphondylium pallidum. *Dev. Biol.* **92:**356–364.

Chung, S., S. M. Landfear, D. D. Blumberg, N. S. Cohen, and H. F. Lodish. 1981. Synthesis and stability of developmentally regulated *Dictyostelium* mRNAs are affected by cell-cell contact and cAMP. *Cell* **24:**785–797.

Clarke, M., W. I. Bazari, and S. C. Kayman. 1980. Isolation and properties of calmodulin from *Dictyostelium discoideum*. *J. Bacteriol.* **141:**397–400.

Clarke, M., S. C. Kayman, and K. Riley. 1987. Density-dependent induction of discoidin-I synthesis in exponentially growing cells of *Dictyostelium discoideum*. *Differentiation* **34:**79–87.

Cotter, D. A., and K. B. Raper. 1966. Spore germination in *Dictyostelium discoideum*. *Proc. Natl. Acad. Sci. USA* **56:**880–887.

Coukell, M. B., S. Lappano, and A. M. Cameron. 1983. Isolation and characterization of cAMP unresponsive (frigid) aggregation-deficient mutants of *Dictyostelium discoideum*. *Dev. Genet.* **3:**283–297.

Dahlberg, K. R., and D. A. Cotter. 1978. Autoactivation of spore germination in mutant and wild-type strains of *Dictyostelium discoideum*. *Microbios* **23:**153–166.

Darmon, M., P. Brachet, and L. H. Pereira da Silva. 1975. Chemotactic signals induce cell differentiation in *Dictyostelium discoideum*. *Proc. Natl. Acad. Sci. USA* **72:**3163–3166.

Darmon, M., and C. Klein. 1978. Effects of amino acids and glucose on adenylate cyclase and cell differentiation of *Dictyostelium discoideum*. *Dev. Biol.* **63:**377–389.

Datta, S., and R. A. Firtel. 1987. Identification of the sequences controlling cyclic AMP regulation and cell-type-specific expression of a prestalk-specific gene in *Dictyostelium discoideum*. *Mol. Cell. Biol.* **7:**149–159.

Datta, S., and R. A. Firtel. 1988. An 80-bp cis-acting regulatory region controls cAMP and development regulation of a prestalk gene in *Dictyostelium*. *Genes Dev.* **2:**294–304.

De Gunzburg, J., and M. Véron. 1982. A cAMP-dependent protein kinase is present in differentiating *Dictyostelium discoideum* cells. *EMBO J.* **1:**1063–1068.

Devine, K. M., J. E. Bergmann, and W. F. Loomis. 1983. Spore coat proteins of *Dictyostelium discoideum* are packaged in prespore vesicles. *Dev. Biol.* **99:**437–446.

Devine, K. M., and W. F. Loomis. 1985. Molecular characterization of anterior-like cells in *Dictyostelium discoideum*. *Dev. Biol.* **107:**364–372.

De Wit, R. J. W., and T. M. Konijn. 1983. Identification of the acrasin of *Dictyostelium minutum* as a derivative of folic acid. *Cell Differ.* **12:**205–210.

De Wit, R. J. W., M. X. P. Van Bemmelen, L. C. Penning, J. E. Pinas, T. D. Calandra, and J. T. Bonner. 1988. Studies of cell-surface glorin receptors, glorin degradation, and glorin-induced cellular responses during development of *Polyspondylium violaceum*. *Exp. Cell Res.* **179:**332–343.

Dicou, E. L., and P. Brachet. 1980. A separate phosphodiesterase for the hydrolysis of cyclic guanosine 3'5' monophosphate in growing *Dictyostelium discoideum* amoebae. *Eur. J. Biochem.* **109:**507–514.

Dohlman, H. H., M. G. Caron, and R. J. Lefkowitz. 1987. A family of receptors coupled to guanine nucleotide regulatory proteins. *Biochemistry* **26:**2657–2664.

Dominov, J. A., and C. D. Town. 1986. Regulation of stalk and spore antigen expression in monolayer cultures of *Dictyostelium discoideum* by pH. *J. Embryol. Exp. Morphol.* **96:**131–150.

Durston, A. J. 1976. Tip formation is regulated by an inhibitory gradient in the *Dictyostelium discoideum* slug. *Nature* (London) **263:**126–129.

Durston, A. J., and F. Vork. 1979. A cinematographical study of the development of vitally stained *Dictyostelium discoideum*. *J. Cell Sci.* **36:**261–279.

Early, V. E., and J. G. Williams. 1988. A *Dictyostelium* prespore-specific gene is transcriptionally repressed by DIF *in vitro*. *Development* **103**:519–524.

Erdos, G. W., A. W. Nickerson, and K. B. Raper. 1972. The fine structure of macrocysts in *Polysphondylium violaceum*. *Cytobiology* **6**:351–366.

Europe-Finner, G. N., and P. C. Newell. 1986a. Inositol 1,4,5 trisphosphate induces calcium release from a non-mitochondrial pool in amoebae of *Dictyostelium*. *Biochim. Biophys. Acta* **887**:335–340.

Europe-Finner, G. N., and P. C. Newell. 1986b. Inositol 1,4,5 trisphosphate and calcium stimulate actin polymerization in *Dictyostelium discoideum*. *J. Cell Sci.* **82**:41–51.

Europe-Finner, G. N., and P. C. Newell. 1987. Cyclic AMP stimulates accumulation of inositol trisphosphate in *Dictyostelium*. *J. Cell Sci.* **87**:221–229.

Feit, I. N., and R. L. Sollitto. 1987. Ammonia is the gas used for the spacing of fruiting bodies in the cellular slime mold *Dictyostelium discoideum*. *Differentiation* **33**:193–196.

Filosa, M. F., and R. E. Dengler. 1972. Ultrastructure of macrocyst formation in the cellular slime mold *Dictyostelium mucuroides:* extensive phagocytosis of amoebae by a specialized cell. *Dev. Biol.* **29**:1–16.

Firtel, R. A., and J. Bonner. 1972. Characterization of the genome of the cellular slime mold *Dictyostelium discoideum*. *J. Mol. Biol.* **66**:49–55.

Firtel, R. A., and A. L. Chapman. 1990. A role for cAMP-dependent protein kinase A in early *Dictyostelium* development. *Genes Dev.* **4**:18–28.

Fisher, F. R., E. Smith, and K. L. Williams. 1981. An extracellular chemical signal controlling phototactic behavior by *D. discoideum* slugs. *Cell* **23**:799–807.

Fontana, D. R., and P. L. Price. 1989. Cell-cell contact elicits cAMP secretion and alters cAMP signaling in *Dictyostelium discoideum*. *Differentiation* **41**:184–192.

Francis, D. 1975. Cyclic AMP-induced changes in protein synthesis in a cellular slime mould, *Polysphondylium pallidum*. *Nature* (London) **258**:763–765.

George, R. P. 1977. Disruption of multicellular organization in the cellular slime molds by cyclic AMP. *Cell Differ.* **5**:293–300.

Gerisch, G. 1964. Die Bildung des Zellverbandes bei *Dictyostelium minutum*. I. Übersicht über die Aggregation und den Funktionswechsel der Zellen. *Roux' Arch. Entw. Mech.* **155**:342–357.

Gerisch, G. 1980. Univalent antibody fragments as tools for the analysis of cell interactions in *Dictyostelium*. *Curr. Top. Dev. Biol.* **14**:243–270.

Gerisch, G., H. Fromm, A. Huesgen, and U. Wick. 1975. Control of cell-contact sites by cyclic AMP pulses in differentiating *Dictyostelium* cells. *Nature* (London) **255**:547–549.

Gibson, F. P., and D. Hames. 1988. Characterization of a spore protein inducing factor from *Dictyostelium discoideum*. *J. Cell Sci.* **89**:387–395.

Ginsburg, G., and A. R. Kimmel. 1989. Inositol trisphosphate and diacylglycerol can differentially modulate gene expression in *Dictyostelium*. *Proc. Natl. Acad. Sci. USA* **86**:9332–9336.

Giri, J. G., and H. L. Ennis. 1977. Protein and RNA synthesis during spore germination in the cellular slime mold *Dictyostelium discoideum*. *Biochem. Biophys. Res. Commun.* **77**:282–289.

Giri, J. G., and H. L. Ennis. 1978. Developmentally regulated changes in RNA and protein synthesis during germination of *Dictyostelium discoideum* spores. *Dev. Biol.* **67**:189–201.

Gomer, R. H., and R. A. Firtel. 1987. Cell-autonomous determination of cell-type choice in *Dictyostelium* development by cell-cycle phase. *Science* **237**:758–762.

Grabel, L., and W. F. Loomis. 1978. Effector controlling accumulation of N-acetylglucosaminidase during development of *Dictyostelium discoideum*. *Dev. Biol.* **64**:203–209.

Green, A. A., and P. C. Newell. 1975. Evidence for the existence of two types of cAMP binding sites in aggregating cells of *Dictyostelium discoideum*. *Cell* **6**:129–136.

Gregg, J. H., and R. W. Davis. 1982. Dynamics of cell redifferentiation in *Dictyostelium mucoroides*. *Differentiation* **21**:200–205.

Gross, J. D., J. Bradbury, R. R. Kay, and M. J. Peacey. 1983. Intracellular pH and the control of cell differentiation in *Dictyostelium discoideum*. *Nature* (London) **303**:244–245.

Haberstroh, L., and R. A. Firtel. 1990. A spatial gradient of expression of a cAMP-regulated prespore cell-type-specific gene in *Dictyostelium*. *Genes Dev.* **4**:596–612.

Hahn, M. G. 1989. Animal receptors—examples of cellular signal perception molecules. *In* B. J. J.

Lugtenberg (ed.), *Signal Molecules in Plants and Plant-Microbe Interactions.* NATO ASI series vol. H36. Springer Verlag, Berlin.

Hall, A. L., V. Warren, and J. Condeelis. 1989. Transduction of the chemotactic signal to the actin cytoskeleton of *Dictyostelium discoideum. Dev. Biol.* **136:**517–525.

Hanna, M. H., and E. C. Cox. 1978. The regulation of cellular slime mold development: a factor causing development of *Polysphondylium violaceum* aggregation defective mutants. *Dev. Biol.* **62:**206–214.

Hanna, M. H., J. L. Gardner, and J. Nowicki. 1983. The timing of phenotypic suppression of an aggregation defect by an aggregation-stimulating factor from *Polysphondylium violaceum. Differentiation* **25:**88–92.

Hanna, M. H., J. J. Nowicki, and M. A. Fatone. 1984. Extracellular cyclic AMP during development of the cellular slime mold *Polysphondylium violaceum:* comparison of accumulation in the wild type and an aggregation-defective mutant. *J. Bacteriol.* **157:**345–349.

Haribabu, B., and R. P. Dottin. 1986. Pharmacological characterization of cyclic AMP receptors mediating gene regulation in *Dictyostelium discoideum. Mol. Cell. Biol.* **6:**2402–2408.

Hashimoto, Y., Y. Tanaka, and T. Yamada. 1976. Spore germination promotor of *Dictyostelium discoideum* excreted by *Aerobacter aerogenes. J. Cell Sci.* **21:**261–271.

Hassanain, H. H., and W. Kopachik. 1989. Regulatory signals affecting a selective loss of mRNA in *Dictyostelium discoideum. J. Cell Sci.* **94:**501–509.

Hayashi, M., and I. Takeuchi. 1976. Quantitative studies on cell differentiation during morphogenesis of the cellular slime mold *Dictyostelium discoideum. Dev. Biol.* **50:**302–309.

Henderson, E. J. 1975. The cyclic adenosine 3′,5′-monophosphate receptor of *Dictyostelium discoideum. J. Biol. Chem.* **250:**4730–4736.

Hjorth, A. L., N. C. Khanna, and R. A. Firtel. 1989. A trans-acting factor required for cAMP-induced gene expression in *Dictyostelium* is regulated developmentally and induced by cAMP. *Genes Dev.* **3:**747–759.

Hjorth, A. L., C. Pears, J. G. Williams, and R. A. Firtel. 1990. A developmentally regulated transacting factor(s) recognizes dissimilar G/C-rich elements controlling a class of cAMP-inducible *Dictyostelium* genes. *Genes Dev.* **4:**419–432.

Hohl, H. R., and S. T. Hamamoto. 1969. Ultrastructure of spore differentiation in *Dictyostelium:* the prespore vacuole. *J. Ultrastruct. Res.* **26:**442–453.

Hohl, H. R., R. Honegger, F. Traub, and M. Markwalder. 1977. Influence of cAMP on cell differentiation and morphogenesis in *Polysphondylium,* p. 149–172. In P. Cappuccinelli and S. M. Ashworth (ed.), *Development and Differentiation in the Cellular Slime Moulds.* Elsevier/North-Holland Biomedical Press, Amsterdam.

Hohl, H. R., L. Y. Miura-Santo, and D. A. Cotter. 1970. Ultrastructural changes during formation and germination of microcysts in *Polysphondylium pallidum,* a cellular slime mould. *J. Cell Sci.* **7:**285–306.

Inouye, K. 1985. Measurements of intracellular pH and its relevance to cell differentiation in *Dictyostelium discoideum. J. Cell Sci.* **76:**235–245.

Inouye, K. 1988. Induction by acid load of the maturation of prestalk cells in *Dictyostelium discoideum. Development* **104:**669–681.

Inouye, K. 1989. Control of cell type proportions by a secreted factor in *Dictyostelium discoideum. Development* **107:**605–609.

Jermyn, K. A., M. Berks, R. R. Kay, and J. G. Williams. 1987. Two distinct classes of prestalk-enriched mRNA sequences in *Dictyostelium discoideum. Development* **100:**745–755.

Jermyn, K. A., K. T. Duffy, and J. G. Williams. 1989. A new anatomy of the prestalk zone in *Dictyostelium. Nature* (London) **340:**144–146.

Jimenez, B., A. Pestaña, and M. Fernandez-Renart. 1989. A phospholipid-stimulated protein kinase from *Dictyostelium discoideum. Biochem. J.* **260:**557–561.

Kaleko, M., and F. G. Rothman. 1982. Membrane sites regulating developmental gene expression in *Dictyostelium discoideum. Cell* **2:**801–811.

Kamboj, R. K., J. Gariepy, and C.-H. Siu. 1989. Identification of an octapeptide involved in homophilic interaction of the cell adhesion molecule gp80 of *Dictyostelium discoideum. Cell* **59:**615–625.

Kay, C. A., T. Noce, and A. S. Tsang. 1987. Translocation of an unusual cAMP receptor to the nucleus during development of *Dictyostelium discoideum*. *Proc. Natl. Acad. Sci. USA* **84**:2322–2326.

Kay, R. R. 1982. cAMP and spore differentiation in *Dictyostelium discoideum*. *Proc. Natl. Acad. Sci. USA* **79**:3228–3231.

Kay, R. R. 1989. Evidence that elevated intracellular cyclic AMP triggers spore maturation in *Dictyostelium*. *Development* **105**:753–759.

Kay, R. R., D. G. Gadian, and S. R. Williams. 1986. Intracellular pH in *Dictyostelium*: a ^{31}P nuclear magnetic resonance study of its regulation and possible role in controlling cell differentiation. *J. Cell Sci.* **83**:165–179.

Kay, R. R., and K. A. Jermyn. 1983. A possible morphogen controlling differentiation in *Dictyostelium*. *Nature* (London) **303**:242–244.

Kelly, L. J., R. Kelly, and H. L. Ennis. 1983. Characterization of cDNA clones specific for sequences developmentally regulated during *Dictyostelium discoideum* spore germination. *Mol. Cell. Biol.* **3**:1943–1948.

Kesbeke, F., B. E. Snaar-Jagalska, and P. J. M. Van Haastert. 1988. Signal transduction in *Dictyostelium* fgd A mutants with a defective interaction between surface cAMP receptors and a GTP-binding regulatory protein. *J. Cell Biol.* **107**:521–528.

Kesbeke, F., P. J. M. Van Haastert, and P. Schaap. 1986. Cyclic AMP relay and cyclic AMP-induced cyclic GMP accumulation during development of *Dictyostelium discoideum*. *FEMS Lett.* **34**:85–90.

Kimmel, A. R. 1987. Different molecular mechanisms for cAMP regulation of gene expression during *Dictyostelium* development. *Dev. Biol.* **122**:163–171.

Kimmel, A. R., and B. Carlisle. 1986. A gene expressed in undifferentiated vegetative *Dictyostelium* is repressed by developmental pulses of cAMP and reinduced during dedifferentiation. *Proc. Natl. Acad. Sci. USA* **83**:2506–2510.

Kitami, M. 1984. Chemotactic response of *Dictyostelium discoideum* cells to c-AMP at the culmination stage. *Cytology* **49**:257–264.

Klein, C., and M. H. Juliani. 1977. cAMP-induced changes in cAMP-binding sites on *D. discoideum* amoebae. *Cell* **10**:329–335.

Klein, P., A. Theibert, D. Fontana, and P. N. Devreotes. 1985. Identification and cyclic-AMP induced modification of the cyclic AMP receptor in *Dictyostelium discoideum*. *J. Biol. Chem.* **260**:1757–1764.

Klein, P., R. Vaughan, J. Borleis, and P. N. Devreotes. 1987. The surface cyclic AMP receptor in *Dictyostelium*. Levels of ligand-induced phosphorylation, solubilization, identification of primary transcript, and developmental regulation of expression. *J. Biol. Chem.* **262**:358–364.

Klein, P. S., T. J. Sun, C. L. Saxe III, A. R. Kimmel, R. L. Johnson, and P. N. Devreotes. 1988. A chemoattractant receptor controls development in *Dictyostelium discoideum*. *Science* **241**:1467–1472.

Knecht, D. A., D. Fuller, and W. F. Loomis. 1987. Surface glycoprotein, gp24, involved in early adhesion of *Dictyostelium discoideum*. *Dev. Biol.* **121**:277–283.

Konijn, T. M., J. G. C. Van De Meene, J. T. Bonner, and D. S. Barkley. 1967. The acrasin activity of adenosine 3′,5′-cyclic phosphate. *Proc. Natl. Acad. Sci. USA* **58**:1152–1154.

Kopachik, W. 1990. Glorin-regulated protein synthesis in *Polyspondylium violaceum*. *Exp. Cell Res.* **186**:394–397.

Kopachik, W., L. G. Bergen, and S. L. Barclay. 1985a. Genes selectively expressed in proliferating *Dictyostelium* amoebae. *Proc. Natl. Acad. Sci. USA* **82**:8540–8544.

Kopachik, W. J. 1982. Size regulation in *Dictyostelium*. *J. Embryol. Exp. Morphol.* **68**:23–35.

Kopachik, W. J., B. Dhokia, and R. R. Kay. 1985b. Selective induction of stalk-cell-specific proteins. *Differentiation* **28**:209–216.

Krefft, M., L. Voet, J. H. Gregg, H. Mairhofer, and K. L. Williams. 1984. Evidence that positional information is used to establish the prestalk-prespore pattern in *Dictyostelium discoideum* aggregates. *EMBO J.* **3**:201–206.

Krefft, M., and C. J. Weijer. 1989. Expression of a cell surface antigen in *Dictyostelium discoideum* in relation to the cell cycle. *J. Cell Sci.* **93**:199–204.

Kumagai, A., and K. Okamoto. 1986. Prespore-inducing factors in *Dictyostelium discoideum*. Developmental regulation and partial purification. *Development* **31**:79–84.

Kumagai, A., M. Pupillo, R. Gundersen, R. Miake-Lye, P. N. Devreotes, and R. A. Firtel. 1989. Regulation and function of Gα protein subunits in *Dictyostelium*. *Cell* **57**:265–275.

Kwong, L., and G. Weeks. 1989. Studies on the accumulation of the differentiation-inducing factor (DIF) in high-cell-density monolayers of *Dictyostelium discoideum*. *Dev. Biol.* **132**:554–558.

Lacombe, M. L., G. J. Podgorski, J. Franke, and R. H. Kessin. 1988. Molecular cloning and developmental expression of the cyclic nucleotide phosphodiesterase gene of *Dictyostelium discoideum*. *J. Biol. Chem.* **261**:16811–16817.

Leichtling, B. H., I. H. Majerfeld, E. Spitz, K. L. Schaller, C. Woffendin, S. Kakinuma, and H. V. Rickenberg. 1984. A cytosolic cyclic AMP-dependent protein kinase in *Dictyostelium discoideum*. II. Developmental regulation. *J. Biol. Chem.* **259**:662–668.

Liu, G., and P. C. Newell. 1988. Evidence that cyclic GMP regulates myosin interaction with the cytoskeleton during chemotaxis of *Dictyostelium*. *J. Cell Sci.* **90**:123–129.

Loomis, W. F. 1988. Cell-cell adhesion in *Dictyostelium discoideum*. *Dev. Genet.* **9**:549–559.

Loomis, W. F., and D. L. Fuller. 1990. A pair of tandemly repeated genes code for gp24, a putative adhesion protein of *Dictyostelium discoideum*. *Proc. Natl. Acad. Sci. USA* **87**:886–890.

Ludérus, M. E. E., R. G. Van der Most, A. P. Otte, and R. Van Driel. 1989. A protein kinase C-related enzyme activity in *Dictyostelium discoideum*. *FEBS Lett.* **253**:71–75.

Ma, G. C. L., and R. A. Firtel. 1978. Regulation of the synthesis of two carbohydrate-binding proteins in *Dictyostelium discoideum*. *J. Biol. Chem.* **253**:3924–3932.

MacInnes, M. A., and D. Francis. 1974. Meiosis in *Dictyostelium mucoroides*. *Nature* (London) **251**:321–323.

MacWilliams, H., A. Blaschke, and I. Prause. 1985. Two feedback loops may regulate cell-type proportions in *Dictyostelium*. *Cold Spring Harbor Symp. Quant. Biol.* **50**:779–785.

MacWilliams, H. K. 1982. Transplantation experiments and pattern mutants in cellular slime mold slugs, p. 463–483. *In* S. Subtelny (ed.), *Developmental Order: Its Origin and Regulation*. Alan R. Liss, Inc., New York.

Malchow, D., B. Nägele, H. Schwartz, and G. Gerisch. 1972. Membrane-bound cyclic AMP phosphodiesterase in chemotactically responding cells of *Dictyostelium discoideum*. *Eur. J. Biochem.* **28**:136–142.

Malchow, D., V. Nanjundiah, B. Wurster, F. Eckstein, and G. Gerisch. 1978. Cyclic AMP-induced pH changes in *Dictyostelium discoideum* and their control by calcium. *Biochim. Biophys. Acta* **538**:473–480.

Mangiarotti, G., S. Bulfone, R. Giorda, P. Morandini, A. Ceccarelli, and B. D. Hames. 1989. Analysis of specific mRNA destabilization during *Dictyostelium* development. *Development* **106**:473–481.

Mangiarotti, G., R. Giorda, A. Ceccarelli, and C. Perlo. 1985. mRNA stabilization controls the expression of a class of developmentally regulated genes in *Dictyostelium discoideum*. *Proc. Natl. Acad. Sci. USA* **82**:5786–5790.

Mann, S. K. O., and R. A. Firtel. 1989. Two-phase regulatory pathway controls cAMP receptor-mediated expression of early genes in *Dictyostelium*. *Proc. Natl. Acad. Sci. USA* **86**:1924–1928.

Mann, S. K. O., C. Pinko, and R. A. Firtel. 1988. cAMP regulation of early gene expression in signal transduction mutants of *Dictyostelium*. *Dev. Biol.* **130**:294–303.

Margolskee, J. P., S. Froshauer, R. Skrinska, and H. F. Lodish. 1980. The effects of cell density and starvation on early developmental events in *Dictyostelium discoideum*. *Dev. Biol.* **74**:409–421.

Marin, F. T. 1976. Regulation of development in *Dictyostelium discoideum*. I. Initiation of the growth to developmental transition by amino acid starvation. *Dev. Biol.* **48**:110–117.

Mato, J. M., F. A. Krens, P. J. M. Van Haastert, and T. M. Konijn. 1977. 3':5'-Cyclic AMP-dependent 3':5'-cyclic GMP accumulation in *Dictyostelium discoideum*. *Proc. Natl. Acad. Sci. USA* **74**:2348–2351.

Mato, J. M., H. Woelders, P. J. M. Van Haastert, and T. M. Konijn. 1978. Cyclic GMP binding activity in *Dictyostelium discoideum*. *FEBS Lett.* **90**:261–264.

Matsukuma, S., and A. J. Durston. 1979. Chemotactic cell sorting in *Dictyostelium discoideum*. *J. Embryol. Exp. Morphol.* **50**:243–251.

McDonald, S. A. 1986. Cell-cycle regulation of center initiation in *Dictyostelium discoideum*. *Dev. Biol.* **117**:546–549.

McDonald, S. A., and A. J. Durston. 1984. The cell cycle and sorting behaviour in *Dictyostelium discoideum*. *J. Cell Sci.* **66**:195–204.

McRobbie, S. J., K. A. Jermyn, K. Duffy, K. Blight, and J. G. Williams. 1988. Two DIF-inducible, prestalk-specific mRNAs of *Dictyostelium* encode extracellular matrix proteins of the slug. *Development* **104**:275–284.

Mee, J. D., D. M. Tortolo, and M. B. Coukell. 1986. Chemotaxis-associated properties of separated prestalk and prespore cells of *Dictyostelium discoideum*. *Biochem. Cell. Biol.* **64**:722–732.

Mehdy, M. C., and R. A. Firtel. 1985. A secreted factor and cyclic AMP jointly regulate cell-type-specific gene expression in *Dictyostelium discoideum*. *Mol. Cell. Biol.* **5**:705–713.

Mehdy, M. C., D. Ratner, and R. A. Firtel. 1983. Induction and modulation of cell-type-specific gene expression in *Dictyostelium*. *Cell* **32**:763–771.

Morris, H. R., G. W. Taylor, M. S. Masento, K. A. Jermyn, and R. R. Kay. 1987. Chemical structure of the morphogen differentiation inducing factor from *Dictyostelium discoideum*. *Nature* (London) **328**:811–814.

Morrissey, J. H., K. M. Devine, and W. F. Loomis. 1984. The timing of cell-type-specific differentiation in *Dictyostelium discoideum*. *Dev. Biol.* **103**:414–424.

Morrissey, J. H., P. A. Farnsworth, and W. F. Loomis. 1981. Pattern formation in *Dictyostelium discoideum*: an analysis of mutants altered in cell proportioning. *Dev. Biol.* **83**:1–8.

Müller, K., and G. Gerisch. 1978. A specific glycoprotein as the target site of adhesion blocking Fab in aggregating *Dictyostelium* cells. *Nature* (London) **274**:445–449.

Murray, B. A., S. Wheeler, T. Jongens, and W. F. Loomis. 1984. Mutations affecting a surface glycoprotein, gp80, of *Dictyostelium discoideum*. *Mol. Cell. Biol.* **4**:514–519.

Mutzel, R., M. L. Lacombe, M. N. Simon, J. De Gunzburg, and M. Véron. 1987. Cloning and cDNA sequence of the regulatory subunit of cAMP-dependent protein kinase from *Dictyostelium discoideum*. *Proc. Natl. Acad. Sci. USA* **84**:6–10.

Newell, P. C. 1982. Cell surface binding of adenosine to *Dictyostelium* and inhibition of pulsatile signalling. *FEMS Microbiol. Lett.* **13**:417–421.

Newell, P. C., and F. M. Ross. 1982. Inhibition by adenosine of aggregation centre initiation and cyclic AMP binding in *Dictyostelium*. *J. Gen. Microbiol.* **128**:2715–2724.

Noegel, A., G. Gerisch, J. Stadler, and M. Westphal. 1986. Complete sequence and transcript regulation of a cell adhesion protein from aggregating *Dictyostelium* cells. *EMBO J.* **5**:1473–1476.

Obata, Y., H. Abe, Y. Tanaka, K. Yanagisawa, and M. Uchiyama. 1973. Isolation of a spore germination inhibitor from a cellular slime mold, *Dictyostelium discoideum*. *Agric. Biol. Chem.* **37**:1989–1990.

Odell, G. M., and J. T. Bonner. 1986. How the *Dictyostelium discoideum* grex crawls. *Phil. Trans. R. Soc. London B* **312**:487–525.

Otte, A. P., M. J. E. Plomp, J. C. Arents, P. M. W. Janssens, and R. Van Driel. 1986. Production and turnover of cAMP signals by prestalk and prespore cells in *Dictyostelium discoideum* cell aggregates. *Differentiation* **32**:185–191.

Oyama, M., and D. D. Blumberg. 1986. Interaction of cAMP with the cell-surface receptor induces cell-type specific mRNA accumulation in *Dictyostelium discoideum*. *Proc. Natl. Acad. Sci. USA* **83**:4819–4823.

Oyama, M., K. Okamoto, and I. Takeuchi. 1983. Proportion regulation without pattern formation in *Dictyostelium discoideum*. *J. Embryol. Exp. Morphol.* **75**:293–301.

Pan, P., E. M. Hall, and J. T. Bonner. 1972. Folic acid as second chemotactic substance in the cellular slime moulds. *Nature* (London) *New Biol.* **237**:181–182.

Pan, P., E. M. Hall, and J. T. Bonner. 1975. Determination of the active portion of the folic acid molecule in cellular slime mold chemotaxis. *J. Bacteriol.* **122**:185–191.

Pan, P., and B. Wurster. 1978. Inactivation of the chemoattractant folic acid by cellular slime molds and identification of the reaction product. *J. Bacteriol.* **136**:955–959.

Parissenti, A. M., and M. B. Coukell. 1989. Identification of a nucleic acid-regulated cyclic GMP-binding activity in *Dictyostelium discoideum*. *J. Cell Sci.* **92**:291–301.

Pavlovic, J., B. Haribabu, and R. P. Dottin. 1989. Identification of a signal transduction response sequence element necessary for induction of a *Dictyostelium discoideum* gene by extracellular cyclic AMP. *Mol. Cell. Biol.* **9**:4660–4669.

Pears, C. J., and J. G. Williams. 1987. Identification of a DNA sequence element required for efficient expression of a developmentally regulated and cAMP-inducible gene of *Dictyostelium discoideum.* *EMBO J.* **6:**195–200.

Pears, C. J., and J. G. Williams. 1988. Multiple copies of a G-rich element upstream of a cAMP-inducible *Dictyostelium* gene are necessary but not sufficient for efficient gene expression. *Nucleic Acids Res.* **16:**8467–8486.

Peters, D. J. M., M. Cammans, S. Smit, W. Spek, M. M. Van Lookeren Campagne, and P. Schaap. 1991. Control of cAMP-induced gene expression by divergent signal transduction pathways. *Dev. Genet.* **12:**25–34.

Peters, D. J. M., M. M. Van Lookeren-Campagne, P. J. M. Van Haastert, W. Spek, and P. Schaap. 1989. Lithium ions induce prestalk-associated gene expression and inhibit prespore gene expression in *Dictyostelium discoideum.* *J. Cell Sci.* **93:**205–210.

Pupillo, M., A. Kumagai, G. S. Pitt, R. A. Firtel, and P. N. Devreotes. 1989. Multiple α subunits of guanine nucleotide-binding proteins in *Dictyostelium.* *Proc. Natl. Acad. Sci. USA* **86:**4892–4896.

Raper, K. B. 1940. Pseudoplasmodium formation and organization in *Dictyostelium discoideum.* *J. Elisha Mitchell Sci. Soc.* **56:**241–282.

Raper, K. B. 1984. *The Dictyostelids.* Princeton University Press, Princeton, N.J.

Raper, K. B., and D. I. Fennell. 1952. Stalk formation in *Dictyostelium.* *Bull. Torrey Bot. Club* **79:**25–51.

Riley, B. B., B. R. Jensen, and S. L. Barclay. 1989. Conditions that elevate intracellular cyclic AMP levels promote spore formation in *Dictyostelium.* *Differentiation* **41:**5–13.

Roos, W., V. Nanjundiah, D. Malchow, and G. Gerisch. 1975. Amplification of cyclic-AMP signals in aggregating cells of *Dictyostelium discoideum.* *FEBS Lett.* **53:**139–142.

Ross, F. M., and P. C. Newell. 1981. Streamers: chemotactic mutants of *Dictyostelium discoideum* with altered cyclic GMP metabolism. *J. Gen. Microbiol.* **127:**339–350.

Rubin, J., and A. Robertson. 1975. The tip of the *Dictyostelium discoideum* pseudoplasmodium as an organizer. *J. Embryol. Exp. Morphol.* **33:**227–241.

Rutherford, C. L., R. D. Taylor, L. T. Frame, and R. L. Auck. 1982. A cyclic AMP dependent protein kinase in *Dictyostelium discoideum.* *Biochem. Biophys. Res. Commun.* **108:**1210–1220.

Saga, Y., H. Okada, and K. Yanagisawa. 1983. Macrocyst development in *Dictyostelium discoideum.* II. Mating-type-specific cell fusion and acquisition of fusion competence. *J. Cell Sci.* **60:**157–168.

Sakai, Y. 1973. Cell type conversion in isolated prestalk and prespore fragments of the cellular slime mold *Dictyostelium discoideum.* *Dev. Growth Differ.* **15:**11–19.

Saxe, C. L., III, and M. Sussman. 1982. Induction of stage-specific cell cohesion in *Dictyostelium discoideum* by a plasma membrane associated moiety reactive with wheat germ agglutinin. *Cell* **29:**755–759.

Schaap, P. 1986. Regulation of size and pattern in the cellular slime molds. *Differentiation* **33:**1–16.

Schaap, P., T. M. Konijn, and P. J. M. Van Haastert. 1984. cAMP pulses coordinate morphogenetic movement during fruiting body formation of *Dictyostelium minutum.* *Proc. Natl. Acad. Sci. USA* **81:**2122–2126.

Schaap, P., J. E. Pinas, and M. Wang. 1985. Patterns of cell differentiation in several cellular slime mold species. *Dev. Biol.* **111:**51–61.

Schaap, P., and W. Spek. 1984. Cyclic-AMP binding to the cell surface during development of *Dictyostelium discoideum.* *Differentiation* **27:**83–87.

Schaap, P., L. Van der Molen, and T. M. Konijn. 1982. Early recognition of prespore differentiation in *Dictyostelium discoideum* and its significance for models on pattern formation. *Differentiation* **22:**1–5.

Schaap, P., L. Van der Molen, and T. M. Konijn. 1983. The organisation of fruiting body formation in *Dictyostelium minutum.* *Cell Differ.* **12:**287–297.

Schaap, P., and R. Van Driel. 1985. Induction of post-aggregative differentiation in *Dictyostelium discoideum* by cAMP. *Exp. Cell Res.* **159:**388–398.

Schaap, P., M. M. Van Lookeren Campagne, R. Van Driel, W. Spek, P. J. M. Van Haastert, and J. Pinas. 1986. Postaggregative differentiation induction by cyclic AMP in *Dictyostelium:* intracellular transduction pathway and requirement for additional stimuli. *Dev. Biol.* **118:**52–63.

Schaap, P., and M. Wang. 1984. The possible involvement of oscillatory cAMP signaling in multicellular morphogenesis of the cellular slime molds. *Dev. Biol.* **105:**470–478.

Schaap, P., and M. Wang. 1986. Interactions between adenosine and oscillatory cAMP signaling regulate size and pattern in *Dictyostelium*. *Cell* **45**:137–144.

Schindler, J., and M. Sussman. 1977. Ammonia determines the choice of morphogenetic pathways in *Dictyostelium discoideum*. *J. Mol. Biol.* **116**:161–169.

Shaffer, B. M. 1975. Secretion of cyclic AMP induced by cyclic AMP in the cellular slime mould *Dictyostelium discoideum*. *Nature* (London) **255**:549–552.

Shimamura, O., H. L. B. Suthers, and J. T. Bonner. 1982. Chemical identity of the acrasin of the cellular slime mold, *Polysphondylium violaceum*. *Proc. Natl. Acad. Sci. USA* **79**:7376–7379.

Siegert, F., and C. Weijer. 1989. Digital image processing of optical density wave propagation in *Dictyostelium discoideum* and analysis of the effects of caffeine and ammonia. *J. Cell Sci.* **93**:325–335.

Singleton, C. K., R. L. Delude, and C. E. McPherson. 1987. Characterization of genes which are deactivated upon the onset of development in *Dictyostelium discoideum*. *Dev. Biol.* **119**:433–441.

Singleton, C. K., P. A. Gregoli, S. S. Manning, and S. J. Northington. 1988a. Characterization of genes which are transiently expressed during the preaggregative phase of development of *Dictyostelium discoideum*. *Dev. Biol.* **129**:140–146.

Singleton, C. K., S. S. Manning, and Y. Feng. 1988b. Effect of protein synthesis inhibition on gene expression during early development of *Dictyostelium discoideum*. *Mol. Cell. Biol.* **8**:10–16.

Siu, C.-H., A. Cho, and H. C. Choi. 1987. The contact site A glycoprotein mediates cell-cell adhesion by homophilic binding in *Dictyostelium discoideum*. *J. Cell Biol.* **105**:2523–2533.

Siu, C.-H., T. Y. Lam, and A. H. C. Choi. 1985. Inhibition of cell-cell binding at the aggregation stage of *Dictyostelium discoideum* development by monoclonal antibodies directed against an 80,000-dalton surface glycoprotein. *J. Biol. Chem.* **260**:16030–16036.

Siu, C.-H., T. Y. Lam, and L. M. Wong. 1988. Expression of the contact site A glycoprotein in *Dictyostelium discoideum*: quantitation and developmental regulation. *Biochim. Biophys. Acta* **968**:283–290.

Snaar-Jagalska, B. E., F. Kesbeke, M. Pupillo, and P. J. M. Van Haastert. 1988. Immunological detection of G-protein α-subunits in *Dictyostelium discoideum*. *Biochem. Biophys. Res. Commun.* **156**:757–761.

Snaar-Jagalska, B. E., and P. J. M. Van Haastert. 1988. *Dictyostelium discoideum* mutant *synag* 7 with altered G-protein-adenylate cyclase interaction. *J. Cell Sci.* **91**:287–294.

Spek, W., K. Van Drunen, R. Van Eijk, and P. Schaap. 1988. Opposite effects of adenosine on two types of cAMP-induced gene expression in *Dictyostelium* indicate the involvement of at least two different intracellular pathways for the transduction of cAMP signals. *FEBS Lett.* **228**:231–234.

Spemann, H. 1938. *Embryonic Development and Induction*. Yale University Press, New Haven, Conn.

Steinemann, C., and R. W. Parish. 1980. Evidence that a developmentally regulated glycoprotein is target of adhesion-blocking Fab in reaggregating *Dictyostelium*. *Nature* (London) **286**:621–623.

Sternfeld, J., and C. N. David. 1979. Ammonia plus another factor are necessary for differentiation in submerged clumps of *Dictyostelium*. *J. Cell Sci.* **38**:181–191.

Sternfeld, J., and C. N. David. 1981. Cell sorting during pattern formation in *Dictyostelium*. *Differentiation* **20**:10–21.

Sternfeld, J., and C. N. David. 1982. Fate and regulation of anterior-like cells in *Dictyostelium* slugs. *Dev. Biol.* **93**:111–118.

Sussman, R., and E. P. Rayner. 1971. Physical characterization of deoxyribonucleic acids in *Dictyostelium discoideum*. *Arch. Biochem. Biophys.* **144**:127–137.

Szabo, S. P., D. H. O'Day, and A. H. Chagla. 1982. Cell fusion, nuclear fusion, and zygote differentiation during sexual development of *Dictyostelium discoideum*. *Dev. Biol.* **90**:375–382.

Takeuchi, I. 1963. Immunochemical and immunohistochemical studies on the development of the cellular slime mold *Dictyostelium mucoroides*. *Dev. Biol.* **8**:1–26.

Theibert, A., and P. N. Devreotes. 1986. Surface receptor-mediated activation of adenylate cyclase in wild-type cells and aggregation deficient mutants. *J. Biol. Chem.* **261**:15121–15125.

Tomchik, K. J., and P. N. Devreotes. 1981. Adenosine 3',5'-monophosphate waves in *Dictyostelium discoideum*: a demonstration by isotope dilution-fluorography. *Science* **212**:443–446.

Town, C. D. 1984. Differentiation of *Dictyostelium discoideum* in monolayer cultures and its modification by ionic conditions. *Differentiation* **27**:29–35.

Town, C. D., J. D. Gross, and R. R. Kay. 1976. Cell differentiation without morphogenesis in *Dictyostelium discoideum*. *Nature* (London) 262:717–719.

Tsang, A., and J. M. Bradbury. 1981. Separation and properties of prestalk and prespore cells of *Dictyostelium discoideum*. *Exp. Cell Res.* 132:433–441.

Tsang, A. S., and M. Tasaka. 1986. Identification of multiple cyclic AMP binding proteins in developing *Dictyostelium discoideum* cells. *J. Biol. Chem.* 261:10753–10759.

Van Haastert, P. J. M. 1987. Down-regulation of cell surface cyclic AMP receptors and desensitization of cyclic AMP-stimulated adenylate cyclase by cyclic AMP in *Dictyostelium discoideum*. *J. Biol. Chem.* 262:7700–7704.

Van Haastert, P. J. M. 1990. Signal transduction and the control of development in *Dictyostelium*. *Semin. Dev. Biol.* 1:159–167.

Van Haastert, P. J. M., M. J. De Vries, L. C. Penning, E. Roovers, J. Van der Kaay, C. Erneux, and M. M. Van Lookeren Campagne. 1989. Chemoattractant and guanosine 5'-[τ-thio]triphosphate induce the accumulation of inositol 1,4,5-trisphosphate in *Dictyostelium* cells that are labelled with [^{3}H]-inositol by electroporation. *Biochem. J.* 258:577–586.

Van Haastert, P. J. M., R. J. W. De Wit, Y. Grijpma, and T. M. Konijn. 1982a. Identification of a pterin as the acrasin of the cellular slime mold *Dictyostelium lacteum*. *Proc. Natl. Acad. Sci. USA* 79:6270–6274.

Van Haastert, P. J. M., and E. Kien. 1983. Binding of cAMP derivatives to *Dictyostelium discoideum* cells. Activation mechanism of the cell surface cAMP receptor. *J. Biol. Chem.* 258:9636–9642.

Van Haastert, P. J. M., B. E. Snaar-Jagalska, and P. M. W. Janssens. 1987. The regulation of adenylate cyclase by guanine nucleotides in *Dictyostelium discoideum* membranes. *Eur. J. Biochem.* 162:251–258.

Van Haastert, P. J. M., H. Van Walsum, and F. Pasveer. 1982b. Nonequilibrium kinetics of a cyclic GMP binding protein in *Dictyostelium discoideum*. *J. Cell Biol.* 94:271–278.

Van Lookeren, M. M., M. Wang, W. Spek, D. Peters, and P. Schaap. 1988. Lithium respecifies cyclic AMP-induced cell-type specific gene expression in *Dictyostelium*. *Dev. Genet.* 9:589–596.

Van Lookeren Campagne, M. M., R. J. Aerts, W. Spek, R. A. Firtel, and P. Schaap. 1989. Cyclic-AMP-induced elevation of intracellular pH precedes, but does not mediate, the induction of prespore differentiation in *Dictyostelium discoideum*. *Development* 105:401–406.

Vardy, P. H., L. R. Fisher, E. Smith, and K. L. Williams. 1986. Traction proteins in the extracellular matrix of *Dictyostelium discoideum* slugs. *Nature* (London) 320:526–529.

Vogel, G., L. Thilo, H. Schwarz, and R. Steinhart. 1980. Mechanism of phagocytosis in *Dictyostelium discoideum*: phagocytosis is mediated by different recognition sites as disclosed by mutants with altered phagocytotic properties. *J. Cell Biol.* 86:456–465.

Wang, M., R. J. Aerts, W. Spek, and P. Schaap. 1988a. Cell cycle phase in *Dictyostelium discoideum* is correlated with the expression of cyclic AMP production, detection, and degradation. *Dev. Biol.* 125:410–416.

Wang, M., J. H. Roelfsema, J. G. Williams, and P. Schaap. 1990. Cytoplasmic acidification facilitates but does not mediate DIF-induced prestalk gene expression in *Dictyostelium discoideum*. *Dev. Biol.* 140:182–188.

Wang, M., and P. Schaap. 1989. Ammonia depletion and DIF trigger stalk cell differentiation in intact *Dictyostelium discoideum* slugs. *Development* 105:569–574.

Wang, M., R. Van Driel, and P. Schaap. 1988b. Cyclic AMP-phosphodiesterase induces dedifferentiation of prespore cells in *Dictyostelium discoideum* slugs: evidence that cyclic AMP is the morphogenetic signal for prespore differentiation. *Development* 103:611–618.

Wang, M., P. J. M. Van Haastert, P. N. Devreotes, and P. Schaap. 1988c. Localization of chemoattractant receptors on *Dictyostelium discoideum* cells during aggregation and down-regulation. *Dev. Biol.* 128:72–77.

Wang, M., P. J. M. Van Haastert, and P. Schaap. 1986. Multiple effects of differentiation-inducing factor on prespore differentiation and cyclic-AMP signal transduction in *Dictyostelium*. *Differentiation* 33:24–28.

Weeks, G. 1984. The spore cell induction activity of conditioned media and subcellular fractions of *Dictyostelium discoideum*. *Exp. Cell Res.* 153:81–90.

Weijer, C. J., and A. J. Durston. 1985. Influence of cAMP and hydrolysis products on cell type regulation in *Dictyostelium discoideum*. *J. Embryol. Exp. Morphol.* **86**:19–37.

Weijer, C. J., G. Duschl, and C. N. David. 1984. Dependence of cell-type proportioning and sorting on cell cycle phase in *Dictyostelium discoideum*. *J. Cell Sci.* **70**:133–145.

Wick, U., D. Malchow, and G. Gerisch. 1978. Cyclic-AMP stimulated calcium influx into aggregating cells of *Dictyostelium discoideum*. *Cell Biol. Int. Rep.* **2**:71–79.

Wilcox, D. K., and M. Sussman. 1982. Serologically distinguishable alterations in the molecular specificity of cell cohesion during morphogenesis in *Dictyostelium discoideum*. *Proc. Natl. Acad. Sci. USA* **78**:358–362.

Wilkinson, D. G., J. Wilson, and B. D. Hames. 1985. Spore coat protein synthesis during development of *Dictyostelium discoideum* requires a low-molecular-weight inducer and continued multicellularity. *Dev. Biol.* **107**:38–46.

Williams, J. G., A. Ceccarelli, S. J. McRobbie, H. Mahbubani, M. M. Berks, R. R. Kay, and K. A. Jermyn. 1987. Direct induction of *Dictyostelium* prestalk induction by DIF provides evidence that DIF is a morphogen. *Cell* **49**:185–192.

Williams, J. G., K. T. Duffy, D. P. Lane, S. J. McRobbie, A. J. Harwood, D. Traynor, R. R. Kay, and K. A. Jermyn. 1989. Origins of the prestalk-prespore pattern in *Dictyostelium* development. *Cell* **59**:1157–1163.

Würster, B., K. Schubiger, U. Wick, and G. Gerisch. 1977. Cyclic GMP in *Dictyostelium discoideum*: oscillations and pulses in response to folic acid and cyclic AMP signals. *FEBS Lett.* **76**:141–144.

Microbial Cell-Cell Interactions
Edited by Martin Dworkin
© 1991 American Society for Microbiology, Washington, DC 20005

Chapter 7

Cell-Cell Interactions in Myxobacteria

Martin Dworkin

INTRODUCTION

Bacteria are customarily thought of as unicellular organisms, and in general, that is the case. As such, they lack the higher-order organization intrinsic to multicellular organisms. The myxobacteria, however, manifest cell-cell interactions as part of their complex life cycle (Rosenberg, 1984), and these occur throughout the organism's cycles of growth and development (Dworkin and Kaiser, 1985). These cell-cell interactions play a role in the organism's growth and motility, and during the developmental phase of the life cycle they participate in the transition to a rudimentary multicellular state.

In general, the behavior of the myxobacteria is strikingly similar to that of the cellular slime molds (chapter 6), even though, as prokaryotes, they are taxonomically and phylogenetically unrelated to these eukaryotic organisms. Kaiser has compared *Dictyostelium* and *Myxococcus* and has pointed out that the similarities between these two groups of organisms reflect the fact that they share a common ecological niche and have separately evolved similar solutions to the problems of dwelling in the soil and feeding on other microbes (Kaiser, 1986).

Over the past decade or so, work on the developmental biology of *Myxococcus xanthus* has increasingly focused on characterizing the nature and mechanisms of cell-cell interactions (Rosenberg, 1984). There are a number of reasons for this. For one thing, it is this aspect of their development and behavior that is unique among the bacteria. Other fundamental aspects of prokaryotic development are being intensively examined in other organisms. For example, understanding of the relationship between development and gene expression has been brought to a fairly sophisticated stage by those developmental biologists who study endospore formation in *Bacillus* (Losick et al., 1989), and the regulation of the placement of cellular organelles and the relationship between the growth cycle and developmental events in *Caulobacter* is being intensively examined (Shapiro, 1985; Newton, 1989). Second, a number of the myxobacteria in general, and *M. xanthus* in particular, have by now been thoroughly domesticated; it is possible to handle many of them experimentally with an ease that is generally characteristic of the bacteria. This includes the fact that it is now possible to subject *M. xanthus* to the sort of genetic analysis that is indispensable for reaching deeply into the understanding of an organism's regulatory mechanisms (Kaiser, 1989).

The goals of this chapter are to describe those cell-cell interactions in the myxobacteria that already are, or could be, model experimental cell-cell interaction systems, and to describe the experimental strategies and results of contemporary research in this area. Most of the work in this area has been done with one species, *M. xanthus*. However, a second organism, *Stigmatella aurantiaca*, manifests a number of cell-cell interactions not found in *M. xanthus* and is emerging as an additional model system. The discussion in this chapter will, therefore, be limited to these two organisms.

DESCRIPTION OF THE MYXOBACTERIA

The myxobacteria are gram-negative soil bacteria. They are distinguished from other gram-negative rods found in the soil by two properties, one of which is unusual, the other unique.

Gliding Motility

The unusual property of the myxobacteria is that they move by gliding over a solid surface. They cannot swim through an aqueous medium and there are no visible organelles of locomotion, despite numerous attempts to find such organelles (Burchard, 1984) and occasional claims to have done so (Burchard et al., 1977; Lünsdorf and Reichenbach, 1989). They share this property with a larger group of phylogenetically and taxonomically heterogeneous organisms colloquially referred to as the gliding bacteria (Reichenbach and Dworkin, 1981). While there have been a number of proposals and hypotheses attempting to describe the mechanism of gliding (Keller et al., 1983; Dworkin et al., 1983; Pate and Chang, 1979; Lapidus and Berg, 1982; Lünsdorf and Reichenbach, 1989; Burchard et al., 1977), each of these models seems to be supported only by those who have proposed it. Nevertheless, movement by gliding over a solid surface is admirably adaptive for the myxobacteria, which live in a terrestrial milieu and whose sustenance is derived from the hydrolysis of insoluble macromolecular debris rather than from the lower-molecular-weight, soluble substrates characteristically utilized by the swimming bacteria.

Social Behavior

The unique property of the myxobacteria is their social behavior, which is characterized by cell-cell interactions throughout both the growth and developmental portions of their complex life cycle. The life cycles of *M. xanthus* and *S. aurantiaca* are illustrated in Figs. 1 and 2. The complete life cycle consists of two interlocking and alternative cycles, one of growth and the other of development. When the organisms are in a nutritionally sufficient milieu they will grow and divide by binary transverse fission in a manner generally indistinguishable from the growth of other eubacteria. Under optimal conditions in the laboratory, *M. xanthus* will grow with a generation time of about 3.5 h. Most myxobacteria, when freshly isolated from nature, are unable to grow in liquid media in the dispersed fashion that microbiologists are so fond of. However, many of them can be trained to do so by repeated passage and selection. The commonly used and domesticated strains of *M. xanthus* and *S. aurantiaca* are able to do so and in general are amenable to all of the routine microbiological manipulations that make bacteria such effective experimental systems.

Induction of Development

There are three sets of conditions that must be satisfied in order to induce *M. xanthus* to shift from its growth mode to development (Wireman and Dworkin, 1975). The cells must sense a nutritional shift-down (Dworkin, 1963); specifically, they must undergo a partial starvation for any of the required amino acids, isoleucine, leucine, or valine, or for phosphate (Manoil and Kaiser, 1980a). (This is something of an oversimplification; for a more complete and accurate description of the relation between nutrition and development, see Manoil and Kaiser [1980a, 1980b]). Second, the cells must be on a solid surface. This reflects the fact that development requires movement (Kim and Kaiser, 1990c), which in turn requires a solid surface. In the laboratory, this customarily takes place on the surface of an agar medium. However, Kuner and Kaiser (1982) have developed a technique, based on an earlier observation by Fluegel (1964), that allows development to be done under submerged conditions, but still on a solid glass or plastic surface. Third, the cells must be present at a high cell density (Shimkets and Dworkin, 1981). For development to occur optimally, the cells must literally be piled atop one another. This is a very characteristic feature of the behavior of *M. xanthus* and reflects the fact that contact-mediated interactions seem to be an important aspect of cell-cell communication in these organisms. This is in interesting contrast

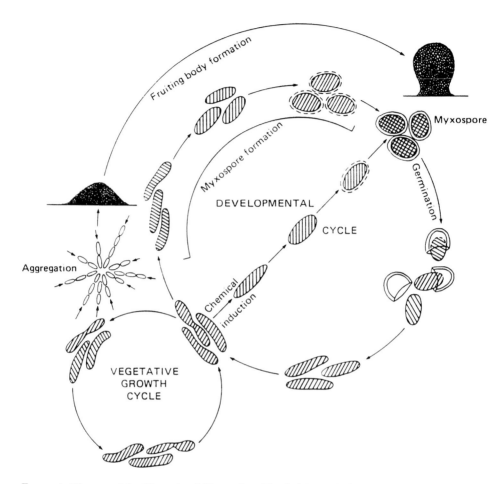

FIGURE 1. Diagram of the life cycle of *M. xanthus*. The fruiting body is not drawn to scale but is a few hundredths of a millimeter in diameter, in contrast to the vegetative cells, which are about 5 to 7 by 0.7 μm (from Dworkin, 1985).

to the behavior of *Dictyostelium*, whose amoebae can initiate development even though the individual cells are quite distant from each other.

If these three conditions are satisfied, the cells begin the process of aggregation that leads eventually to the formation of fruiting bodies and myxospores.

Tactic Behavior

The early events in aggregation have not been defined morphologically or behaviorally. Nevertheless, it seemed intuitively reasonable that aggregation was initiated by a tactic signal of some sort and that it would be a chemotactic response. In fact, on this basis, an excellent chapter on motility and fruiting body formation in *M. xanthus* (Clarke, 1981) was included in the book *Biology of the Chemotactic*

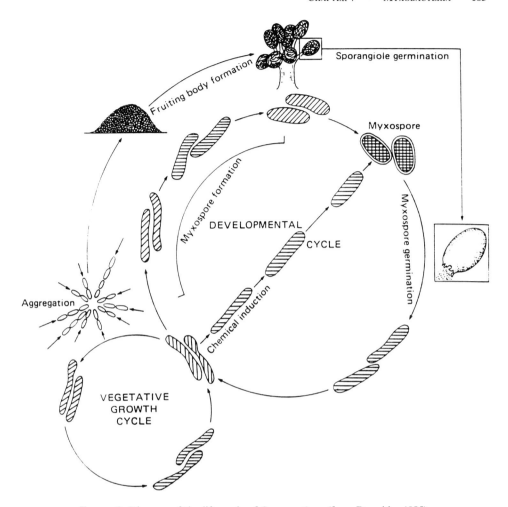

FIGURE 2. Diagram of the life cycle of *S. aurantiaca* (from Dworkin, 1985).

Response (Lackie and Wilkinson, 1981). However, it has not been possible to demonstrate that *M. xanthus* is capable of a chemotactic response (Dworkin and Eide, 1983). Not only have extensive attempts to demonstrate chemotaxis been unsuccessful, but a number of post facto theoretical arguments tend to rationalize those negative experimental observations. These are as follows. One may assume that the reasoning that led Macnab and Koshland (1972) to exclude a spatial sensing model for chemotaxis in *Escherichia coli* also applies to *M. xanthus*. One must then apply the temporal sensing model to a gliding bacterium, whose motility differs from that of *E. coli* in one critical aspect: *M. xanthus* moves at a speed approximately 1/3,000 that of *E. coli* (approximately 1 μm/min as compared to 50 μm/s). The result of this is that even though *M. xanthus* is moving down a concentration gradient, the molecules of the putative attractant will be diffusing past it so that it will sense an increasing concentration with time. Second, from a developmental point of view, the need for long-distance signalling is not as obvious for *M. xanthus* as it is for *Dictyostelium*. In the latter case, myxamoebae are able to aggregate when they occupy as little as 3.5% of the surface area (Sussman and Noel, 1952). Long-distance signalling and the ability to perceive a gradient of attractant are obvious necessities. On the other hand, in the case of *M. xanthus,* optimal development occurs when the cells are at a high density; in fact, the cells must be close enough to be touching one another (Shimkets and Dworkin, 1981). Thus, the need in *M. xanthus* for a diffusible, extracellular signal and thus a chemotactic mechanism with which to guide cells into an aggregation center is not apparent. Nevertheless, the cells do find their way into aggregation centers, and the question of the mechanism by which this occurs is still unanswered.

Blackhart and Zusman (1985a) have made the interesting and potentially important observation that a set of genes termed *frz* (for "frizzy") control the directional movement of the cells. The *frz* mutants are unable to form proper fruiting bodies and, instead, form aggregates that were described as frizzy (Zusman, 1982). Instead of aggregating normally, frizzy mutants form long streams of cells, giving the colonies their characteristic frizzy appearance. While wild-type cells were shown to reverse the direction of their gliding movement approximately every 6.8 min, the frizzy mutants only reversed their direction every 2 h. The authors concluded (Blackhart and Zusman, 1985a) that this was the most likely explanation of the inability of the mutants to form normal aggregates. The frizzy genes were subsequently cloned (Blackhart and Zusman, 1985b) and were shown to share some sequence similarity with the genes of enteric bacteria that control directional movement (McBride et al., 1989). This should not be interpreted to indicate that chemotaxis does or does not occur in *M. xanthus;* there is still no direct evidence to indicate that chemotaxis occurs in *M. xanthus*. In fact, as indicated above, the experimental evidence and the theoretical arguments favor an alternative mechanism of tactic behavior. Nevertheless, the results of Zusman's work lead to the important suggestion that there may indeed be gene products and processes in *M. xanthus* that are closely related to the processes of sensory transduction, as have been demonstrated in the enteric bacteria.

The morphological sequence of events leading to aggregation and fruiting body formation in *M. xanthus* and *S. aurantiaca* is illustrated in Figs. 3 and 4.

There is now ample evidence, based both on subjective observations of the behavior of the organisms (e.g., Reichenbach, 1965) and on experimental demonstration of signal exchange (Dworkin and Kaiser, 1985; Shimkets, 1990), that cell-cell communications play a key role in myxobacterial development and behavior.

BIOLOGICAL RATIONALE FOR THE LIFE CYCLE AND CELL-CELL INTERACTIONS IN THE MYXOBACTERIA

It seems important to establish an understanding of the biological functions of the myxobacterial life cycle as a backdrop for attempts to understand the roles and mechanisms of cell-cell interactions. This can be divided into three questions:

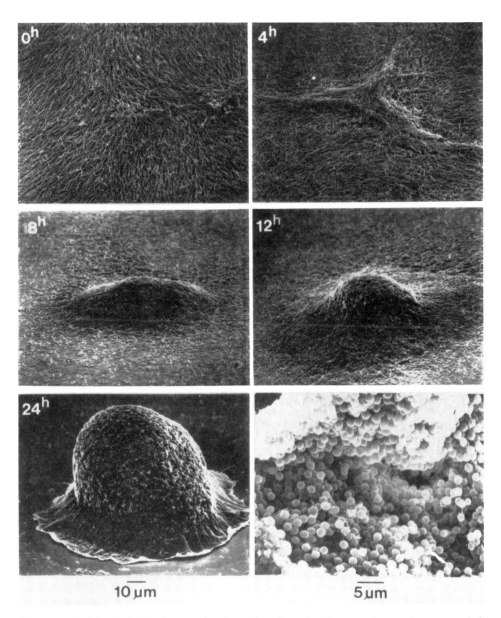

FIGURE 3. Fruiting body development in *M. xanthus*. Scanning electron micrographs were made in submerged culture by Kuner and Kaiser (1982). The time marked in the upper left-hand corner of each frame shows the interval measured from the beginning of induction. The 10-μm marker applies to these photographs. The lower right-hand frame shows a mature fruiting body that has cracked open, revealing its myxospores (from Kaiser et al., 1986).

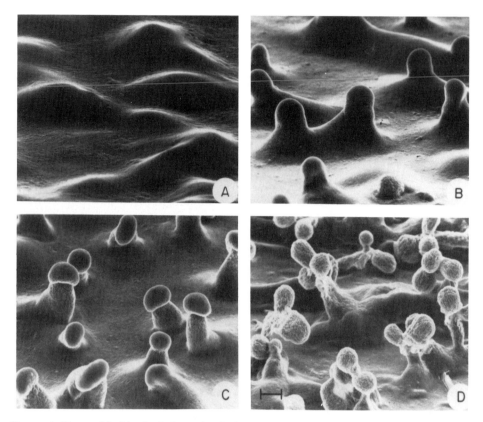

FIGURE 4. Stages of fruiting body formation in *S. aurantiaca*. (A) Early aggregates (9 h); (B) early stalks (12 h); (C) late stalks (15 h); (D) mature fruiting bodies (24 h). Bar, 20 μm (from Qualls et al., 1978).

(i) what is the function of the myxospores?; (ii) what is the function of the primitive myxobacterial multicellularity?; and (iii) what is the function of the fruiting body structure?

Myxospore Function

The first question is the easiest to answer; myxospores of *M. xanthus* have been shown to be resistant to a variety of environmental extremes. These include elevated temperatures, desiccation, sonic oscillation, and UV light irradiation (Bauer, 1905; Reichenbach and Dworkin, 1969; Sudo and Dworkin, 1969). In addition, the myxospores of *M. xanthus* have been shown to be metabolically quiescent (Dworkin and Niederpruem, 1964) even though their adenylate energy charge is essentially the same as that of the vegetatively growing cells (Smith and Dworkin, 1980). It is not unreasonable to conclude that the function of the myxospores of *M. xanthus* is that of a resistant, resting cell. Furthermore, since the myxospores of all the myxobacteria are contained in their fruiting bodies (Reichenbach, 1984), as is the case with the myxospores of *M. xanthus,* it is not a matter of excessive faith to suggest that the functions of resistance and metabolic quiescence attributable to the myxospore of *M. xanthus* may be generalized to all of the myxobacteria. However, this has not been demonstrated experimentally.

Myxobacterial Multicellularity

With regard to the function of the myxobacterial multicellularity, it is believed that in nature the myxobacteria feed on the low-molecular-weight products of hydrolysis of macromolecular debris, produced by the action of their excreted hydrolytic enzymes (Reichenbach and Dworkin, 1981). It has been demonstrated that cells of *M. xanthus,* when using casein as a growth substrate, will not grow at low cell densities ($<10^4$ cells per ml), and that at cell densities greater than 10^4/ml, growth is cooperative or synergistic; that is, the growth rate of cells on casein is strongly dependent on the cell density of the population. Growth on low-molecular-weight substrates manifested no such cell density dependence (Rosenberg et al., 1977). This provided evidence for the earlier suggestion that the primary function of the myxobacterial life cycle was to maintain the high cell densities which would optimize cooperative feeding (Dworkin, 1972), subsequently referred to as the myxobacterial "wolf-pack effect" (Dworkin, 1973).

Function(s) of the Fruiting Bodies

"The most important question involving life history strategies is, Why does one have fruiting bodies at all. . . ?" (Bonner, 1982a). The function(s) of the myxobacterial fruiting body is considerably more difficult to rationalize than that of the myxospores or of multicellularity in general. Bonner has suggested that the major function of the fruiting bodies of the *Dictyosteliaceae* is for spore dispersal and that, furthermore, a branched fruiting body may be more effective at spore dispersal than a simple, unbranched structure (Bonner, 1982b). This may apply to the fruiting bodies of the myxobacteria, which vary from the simple mounds of *Myxococcus fulvus, M. virescens,* and *M. xanthus* (Fig. 5A) to the elevated mounds of *M. stipitatus* (Fig. 5B) and *Mellitanguim* (Fig. 5C), to the clusters of sporangioles of *Cystobacter* (Fig. 5D) and *Sorangium* (Fig. 5E), and finally the elaborately sculptured fruiting bodies of *Chondromyces* (Fig. 5F) and *Stigmatella* (Fig. 5G).

Another function of the myxobacterial fruiting body may be to ensure that when spores germinate, they do so at a high cell density, thus guaranteeing an instant swarm for maximizing the cooperative feeding referred to above. In fact, the differentiation of the fruiting body of organisms such as *Chondromyces* and *Stigmatella* into sporangioles or packets containing a subdivision of the total spore population of the fruiting body may serve the function of generating an optimal size for the germinated swarm.

THE VARIOUS CELL-CELL INTERACTIONS IN MYXOBACTERIA

There have been four general approaches or strategies that have been used to identify and define the various cell-cell interactions in the myxobacteria, i.e.,

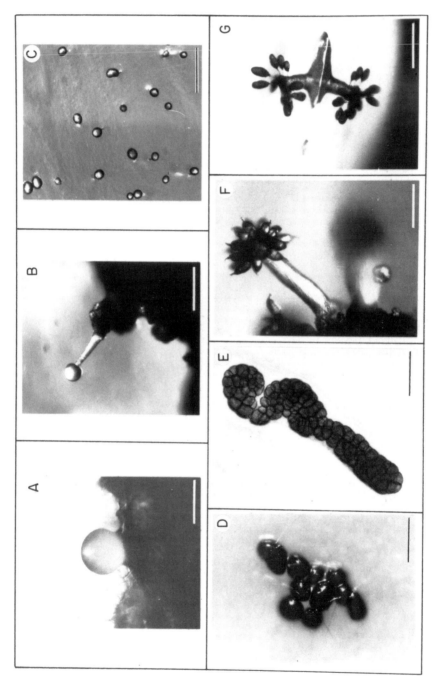

FIGURE 5. Myxobacterial fruiting bodies. (A) *Myxococcus fulvus* (bar, 35 μm); (B) *Myxococcus stipitatus* (bar, 125 μm); (C) *Mellitangium lichenicola* (bar, 140 μm); (D) *Cystobacter fuscus* (bar, 135 μm); (E) *Sorangium cellulosum* (bar, 115 μm); (F) *Chondromyces apiculatus* (bar, 105 μm); (G) *S. aurantiaca* (bar, 120 μm) (from Reichenbach, 1984).

(i) extracellular complementation of sporulation, (ii) circumventing density-dependent development, (iii) the use of monoclonal antibodies directed against cell surface antigens, and (iv) subjective, microscopic observation of motility and development. Although none of these has yet culminated in the complete definition of a particular cell-cell communication event, each has opened up an experimental path to that goal. Furthermore, it is now reasonably clear that there are both contact-mediated events as well as events mediated by soluble, extracellular signals. The remainder of this chapter will focus on the various cell-cell interactions that occur in the myxobacteria and will be divided into those interactions that are being subjected to experimental analysis and those that are recognized to occur but are not being examined.

Extracellular Complementation of Sporulation in *M. xanthus*

There have been numerous hints in the literature that the myxobacteria engage in some sort of cell interactions involving signal exchange. For example, Lev (1954) showed that if fruiting bodies of *Myxococcus* were placed on one side of a cellophane membrane and vegetative cells on the other, the cells oriented themselves in juxtaposition to the apposed fruiting bodies. Fluegel (1963) showed that diffusible material produced by fruiting bodies of *Myxococcus fulvus* could direct the formation of fruiting bodies by other vegetative cells. However, the observation that led to the current experiments on extracellular complementation was developed by McVittie et al. (1962). These authors showed that neither of two classes of mutants of *M. xanthus* was able to develop unless it was mixed with members of the other nondeveloping mutant class. The interaction could not take place across a cellophane membrane and, in fact, seemed to require that the cells be in close proximity to each other. This clearly suggested that some sort of signalling exchange was taking place that either required physical contact between the cells or involved a nondiffusible signal. Subsequently, this phenomenon was more clearly defined and developed by Hagen et al. (1978) and has become a major aspect of signalling research in the myxobacteria. They confirmed that it was possible to isolate mutants of *M. xanthus* that were able to repair each other's developmental deficiencies without exchanging genetic material. In other words, if a mutant in group A, which was unable to complete its developmental cycle, was mixed with a group B mutant, which was likewise unable to form fruiting bodies and myxospores, both mutants were able to complete their development. The complementation was not a genetic one, in that the resultant fruiting body was found to be composed of the original A and B mutants. Hagen et al. (1978) were able to show that there were four such complementation groups, referred to as groups A, B, C, and D. The most plausible interpretation of these results was that at least four signals were required for sporulation and fruiting body formation to occur and that each of the complementation groups represented a mutation in the ability to synthesize or export one of the signals.

There have been two lines of evidence that suggest that the four mutant groups represent blocks at different stages of the developmental process. Table 1 (Kaiser, 1986) lists the effects of the mutations on four different developmental parameters

TABLE 1
The four mutant classes arrest development at different stages[a]

Class	Protein S (6 h)	Protein H (12 h)	Lysis (20 h)	Spores (24 h)
Wild type	+	+	+	+
Class A	Delayed	−	−	−
Class B	Reduced	Reduced	−	−
Class D	+	−	−	−
Class C	+	+	−	−

[a] From Kaiser, 1986. Development, initiated by starvation, requires a high density of cells, and their presence on a solid surface, to permit gliding.

that are expressed at different times during development. Protein S is a spore coat protein that is synthesized early in development but assembled onto the spore at a later stage (Inouye et al., 1979a, 1979b). Protein H is a lectin whose function is unknown, but is synthesized at a later stage of development than protein S (Cumsky and Zusman, 1979). Developmental autolysis occurs approximately co-incidentally with aggregation (Wireman and Dworkin, 1977) and is finally followed by spore formation.

The second line of evidence is based on the expression of a series of *lacZ* fusions. Kuner and Kaiser (1981) were able to introduce the transposon Tn5 into *M. xanthus* by specialized transduction with bacteriophage P1. Coliphage P1 carries out a suicide transduction of *M. xanthus,* resulting in the introduction of the transposon but in no subsequent phage replication or establishment of lysogeny. Thus, the kanamycin resistance gene carried by the Tn5 transposon served as a selectable marker that could be inserted randomly along the genome of *M. xanthus*. If it was determined that the transposon had inserted close to a developmental gene, then cotransduction of the two genes allowed a selection by proxy of the developmental gene.

Timing of expression of the developmental genes

This approach was further modified by fusing the *lacZ* gene to the Tn5 transposon, thus generating a reporter of the expression of exogenous *M. xanthus* promoters that was also transportable and selectable (Downard et al., 1984; Kroos and Kaiser, 1984). A total of 2,374 Tn5 *lac* insertion mutants were isolated, and of these, 36 showed a substantial increase in the expression of β-galactosidase during development. The timing of expression of the β-galactosidase genes in these developing Tn5 *lac* strains is illustrated in Fig. 6. These Tn5 *lac* strains have been used, in a series of epistasis experiments, to determine the timing of expression of the A-, B-, C-, and D-signal genes (Kroos et al., 1986; Kuspa et al., 1986; Kroos and Kaiser, 1987). The essence of the experiment is that after the timing of expression during development of the 36 Tn5 *lac* genes was determined, A-, B-, and C-signal mutations (*asg, bsg,* and *csg*) were transduced into the Tn5 *lac* mutants. Then, whether or not expression of β-galactosidase occurred during

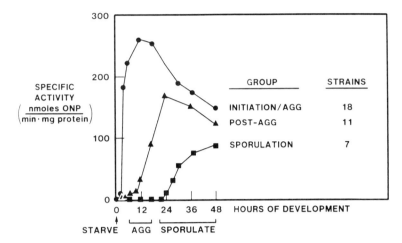

FIGURE 6. β-Galactosidase synthesis in developing strains of *M. xanthus* containing Tn*5 lac* transposons. The specific activity of β-galactosidase in sonic extracts of cells harvested at different times during development is shown for three representative strains. AGG, aggregate; ONP, *ortho*-nitrophenol (from Kaiser et al., 1986).

development was a reflection of whether or not it was independent of or dependent on expression of the *asg, bsg,* or *csg* genes. Since the timing of expression of the individual Tn*5 lac* genes had already been determined, this permitted the determination of the kinetics of the expression of the *asg, bsg,* and *csg* genes during development. Figure 7 presents the data for 31 Tn*5 lac* strains; the time of expression of each Tn*5 lac* fusion is presented as a function of time of development, and the dependence of each of these expressions on expression of the A, B, or C factors is represented by the position on the ordinate. The generalizations that emerged are that the sequence of expression is B signal, A signal, and then C signal; that these are dependent on each other in a linear fashion; and that there seem to be three branches off the main pathway. Recent results (Cheng and Kaiser, 1989a) have indicated that the D signal is expressed prior to aggregation and may be the earliest expressed of the four signals. It is also interesting that the *dsg* gene is required for the viability of vegetative cells and may thus play a role in vegetative growth (Cheng and Kaiser, 1989b). Likewise, the *csg* gene has recently been shown to be expressed at a low level in vegetatively growing cells (Shimkets and Rafiee, 1990).

The A signal

When the Tn*5 lac* transposon is inserted in the genome under the control of a developmentally regulated promoter whose expression is dependent on the prior production of the A signal, β-galactosidase production then becomes a parameter of the proper receipt of the A signal. If the *asg* gene is mutated, then the expression of β-galactosidase is prevented and can be rescued by the addition of the *asg* gene product, i.e., the A signal. Thus, this system has been used as a convenient assay

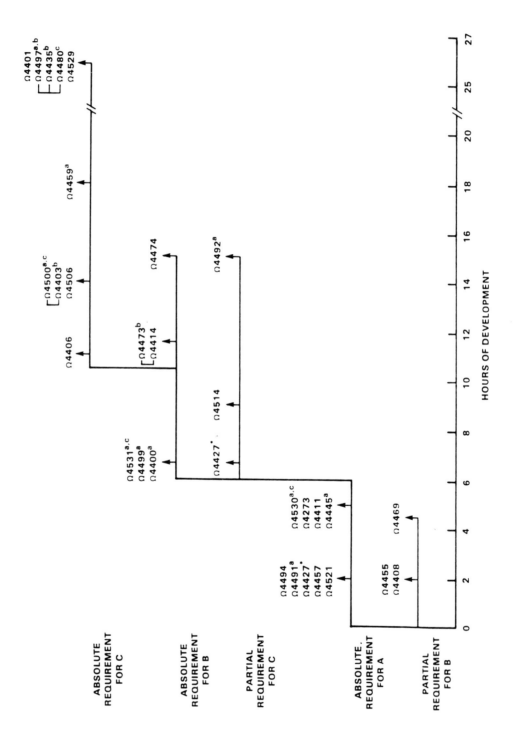

system for the isolation of the A signal (Kuspa et al., 1986). With this assay, it has been shown that the wild-type cells release A factor at the very beginning of development (1.5 to 2 h after initiation) and that A factor is a heat-labile, non-dialyzable molecule (Kuspa et al., 1986).

The B signal

Gill's laboratory has cloned a gene involved in producing the B signal (*bsg*), has done in vitro and transposon mutagenesis of the gene, and has shown that a *bsg* null mutation transduced into wild-type cells reproduces the phenotype of the original *bsg* mutants (Gill et al., 1988). The mutants have an interesting property that was not noticed originally. When vegetatively growing mutant cells were allowed to swarm into juxtaposition with wild-type cells, the advancing edge of the mutant swarm refused to merge with the wild-type cells and instead formed a ridge at the interface of the two swarms (Fig. 8). The fact that this behavior was not manifested by two mutant swarms indicates that the behavior is not simply the result of the absence in the mutant of some sort of recognition molecule.

Gill and Bornemann (1988) have partially characterized the *bsg* gene product by using a monoclonal antibody (MAb) approach. They constructed a *bsgA lacZ* gene fusion which they then expressed in *E. coli*. The resulting fusion protein was used to generate a MAb specific for the BsgA gene product. This was then used to determine that the BsgA protein was present in growing as well as developing cells and that its concentration seemed not to change substantially during growth or development. This suggests the interesting possibility that some product of the B signal may act as an inhibitor of development that is present in the growing vegetative cells and that the *bsg* mutant contains an inhibitory excess of the B signal (Gill, personal communication). The B signal was not excreted into the medium but was located in the cytosol, with only 5 to 10% of the B protein present in the membrane fraction of the cells. This is somewhat reminiscent of the finding by Janssen and Dworkin (1985) that a factor (Dsf) that rescued the development of a group C mutant of *M. xanthus* was likewise cell-bound and present in both vegetatively growing and developing cells.

The C signal

Among the 51 developmental mutants that can be complemented extracellularly by wild-type cells, only 3 were found to belong to class C (Hagen et al., 1978), and all of these were located in a single genetic complementation group (Shimkets et al., 1983). Like the other classes of extracellularly complementable mutants, the *csg* mutants were defective in sporulation and fruiting body for-

---

FIGURE 7. Dependence of gene expression in *M. xanthus* on A, B, and C signals. Each *lac* fusion (Ω followed by a four-digit number) is shown above an arrow to indicate the time at which it is expressed. Positions on the *y* axis indicate the factor dependence of expression for each *lac* fusion. Requirements accumulate upward. Superscript a, b, or c indicates that the Tn*5 lac* insertion has not been tested in the A, B, or C mutant, respectively (from Kroos and Kaiser, 1987).

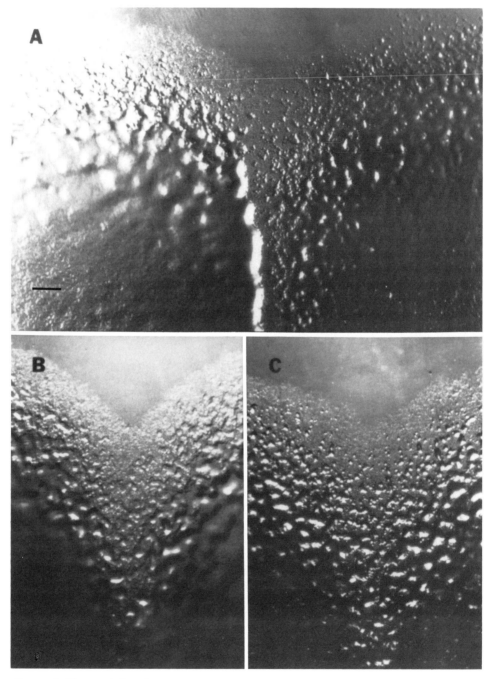

FIGURE 8. Characteristics of converging colonies of mutant and wild-type cells of *M. xanthus*. (A) Convergence of mutant (left) and wild-type (right) cells. At the leading edge the swarming mutant cells appear to form an abrupt ridge where the two cell types meet. This pattern is not seen when two mutant swarms (B) or two wild-type swarms (C) converge. Bar, 1 mm (from Gill et al., 1988).

mation, but in addition, they were unable to carry out the unusual behavior of rippling (Shimkets and Kaiser, 1982a). This is the phenomenon discovered by Reichenbach and originally described as "rhythmic oscillations" (Reichenbach, 1965). It is illustrated in Fig. 9 (Shimkets and Kaiser, 1982a).

There is an interesting relationship between the C signal on the one hand and gliding motility and development on the other. Kroos et al. found that gliding motility per se was necessary for the expression of a number of developmental genes (Kroos et al., 1988). They showed that *mgl* mutants, which are defective in the motility machinery itself, and A⁻ S⁻ mutants, which are double mutants defective in both the A (adventurous) and the S (social) motility, prevented the expression of β-galactosidase by a group of Tn*5 lac* mutants whose expression normally showed developmental kinetics. When *csg* mutants were used instead of the motility mutants, the pattern of effects on expression of the Tn*5 lac* fusions was identical. Kroos et al. drew the interesting tentative conclusion that motility is thus required for the proper functioning of the *csg*-mediated signal exchange. This implies that it is necessary for the cells to move into actual contact with each other in order for the signal exchange to be effective and is consistent with

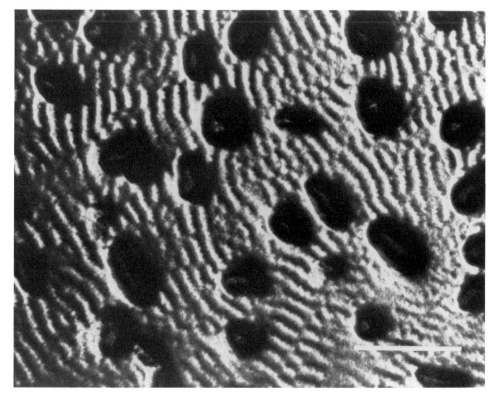

FIGURE 9. Ripples generated during fruiting body formation by *M. xanthus* DK1622. The dark masses are fruiting bodies. Bar, 400 μm (from Shimkets and Kaiser, 1982a).

the earlier observation by Janssen and Dworkin (1985) that physical contact between wild-type and mutant cells was necessary for complementation of the group C mutants to occur. These results also suggested to Kroos et al. (1986) that C signalling and movement comprised the components of a positive feedback loop whose oscillations could result in the rippling behavior.

Kim and Kaiser (1990c) have now clarified the nature of the functional relationship between motility and C-signal exchange. They have suggested that the role of motility in C-signal exchange is to align the individual cells so that they are closely apposed to each other, thus allowing exchange of a cell surface-bound signal. They showed that this was indeed the case by the following ingenious experiment. An agar surface was lightly scored with abrasive paper so that grooves 5 to 10 μm wide were created. The test organism was a nonmotile (*mgl*) mutant containing a Tn*5 lac* transposon positioned so that it was acting as a *csg*-dependent gene. The nonmotile mutant was ordinarily unable to express its *csg* gene.

The nonmotile cells, which ordinarily did not respond to the C signal, were then deposited in these grooves and became longitudinally oriented and tightly packed upon each other. Under these conditions, the cells expressed the C signal-dependent Tn*5 lac* gene. The fact that the C signal has now been shown to be a protein associated with surface material (Kim and Kaiser, 1990a, 1990b) is consistent with the need for the cells to be in close proximity to each other for the interaction to take place.

Kim and Kaiser (1991) have recently shown that C protein has two distinct developmental functions. At lower concentrations it exerts a positional effect, controlling aggregation, while at higher concentrations it controls cell differentiation into myxospores. They showed that isolated C signal, at subnanomolar levels, regulates the expression of early C factor-dependent genes and aggregation, while at higher concentrations it controls the expression of both early and late C-dependent genes as well as sporulation (Kim and Kaiser, 1991).

The D signal

The group of D mutants contains 4 members of the original 51 complementable mutants and contains at least two genetic loci. Like the members of the A, B, and C groups, the D mutants behave as if they are unable to produce a signal normally produced by wild-type cells and necessary for normal development.

When the original *dsg* mutant was backcrossed into the wild-type strain, the development of the resultant isogenic mutant was somewhat less drastically impaired than that of the original mutant; sporulation and fruiting body formation were delayed but eventually reached the same level as those of the wild-type cells (Cheng and Kaiser, 1989a).

The *dsg* gene has been cloned and characterized by Cheng and Kaiser (1989a) by the analysis of point mutants, deletions, and Tn*5* insertions. Epistasis experiments with the Tn*5 lac* transposon and the *dsg* mutants have indicated that the *dsg* gene was expressed during the stage of preaggregation or early aggregation (Cheng and Kaiser, 1989a). These authors also made the interesting observation that, while point mutations in the *dsg* gene only seemed to interfere with the progress of developmental events, Tn5 insertions in the *dsg* gene were lethal. This conclusion was based on their inability to isolate mutants in which the Tn5-containing *dsg* had simply replaced the normal *dsg* gene even though tandem duplications and Tn5 transpositions could readily be obtained (Cheng and Kaiser, 1989b). It is somewhat puzzling that point mutations in the *dsg* gene only affected developmental functions, while a null mutation affected both development and the viability of vegetative cells. It is possible that the *dsg* gene product is a polypeptide with multiple functions, some of which are retained with a missense point mutation but are completely abolished by a transposon-induced null mutation.

Varon et al. (1984) have shown that cells of *M. xanthus* produce a series of substances, called autocides, that induce autolysis in the producing cultures. It has been suggested that these autocides play a role in development (Varon et al., 1986) and are, in fact, responsible for the massive autolysis that invariably accompanies myxobacterial development (Wireman and Dworkin, 1975, 1977). The autocide AM1 has been identified as a mixture of saturated and unsaturated fatty acids (Varon et al., 1986) that are released from phosphatidylethanolamine as a result of the activity of phospholipases (Gelvan et al., 1987). Rosenbluh and Rosenberg (1989) have shown that low concentrations of AM1 are able specifically to rescue aggregation and sporulation of the D mutants. These authors have suggested that the role played by the AM1 fatty acids may be to increase the permeability of the cells, facilitating the release of the D signal (Rosenbluh and Rosenberg, personal communication).

The opportunities represented by the A, B, C, and D signalling systems for understanding cell-cell communication in *M. xanthus* are considerable. Most, if not all, of the signals will eventually be isolated and characterized; the assay systems have been worked out and are relatively convenient, some of the genes controlling the interactions have already been cloned and their gene products expressed, and the system for the genetic analysis is well in place. Characterizing the signals will allow characterization of the interaction itself, the regulation of the process, and eventually an understanding of their role in the developmental process.

Density-Dependent Behavior in the Myxobacteria

Density-dependent aggregation in *M. xanthus*

Almost everyone who has worked with fruiting body formation by the myxobacteria has informally noticed that, in order for successful aggregation to occur, the cells have to be present at a high cell density. Wireman and Dworkin (1975) demonstrated that this was, in fact, the case. The obvious question is, How do the bacteria know what their cell density is?

The hint as to the answer was provided by an early study of the germination of glycerol-induced myxospores. Ramsey and Dworkin (1968), working with glycerol-induced myxospores, found that these spores were able to germinate in distilled water if they were present at a cell density greater than 10^9/ml. The spores could be induced to germinate at lower cell densities if they were in the presence of 10^{-2} M phosphate or of the supernatant fraction of the distilled-water medium in which myxospores had been allowed to germinate. This supernatant fraction was found to contain phosphate ions, and the activity of this fraction was correlated with the phosphate concentration. Ramsey and Dworkin (1968) suggested that the cells used the excreted phosphate as a parameter of their cell density; this would be a simple but effective strategy whereby cells could monitor their cell density. Unfortunately it was not possible to show that myxospores derived from fruiting bodies manifested the same cunning strategy. Such cells failed to germinate in distilled water, regardless of the cell density or the presence or absence of germination factors other than nutrient medium (Dworkin, unpublished). Nevertheless, a strategy whereby myxobacterial cells could monitor their cell density had been revealed, and it did lead to the discovery that a similar strategy was used by the cells to monitor their cell density during aggregation (Shimkets and Dworkin, 1981). If fewer than 10^6 cells per cm^2 were placed on the surface of a medium that normally supported the formation of aggregates and fruiting bodies, the cells failed to enter that part of their developmental cycle. It had previously been shown that *M. xanthus* excreted large quantities of nucleic acid derivatives during the formation and germination of glycerol-induced myxospores (Hanson and Dworkin, 1974). Shimkets and Dworkin (1981) showed that during normal development also, *M. xanthus* excreted large amounts of nucleic acid derivatives and that among these, adenosine could rescue development of *M. xanthus* when the cells were plated at a low cell density. The authors concluded that adenosine was a major intercellular signal involved in fruiting body formation and that the cells measured their cell density by titering the extracellular concentration of excreted adenosine. There is no information that bears on the question of the nature of the mechanism whereby adenosine is perceived or how that perception is transduced into a developmental response. In addition, it is not clear what are the nature and role of the turnover events that occur during development and that lead to the accumulation and excretion of such substantial quantities of nucleic acid derivatives.

Rosenbluh et al. (1989) devised an elegant technique for isolating cells of *M. xanthus* in agarose microbeads and then examining the effect of various cell densities on myxosporulation and developmental lysis. They found that both myxosporulation and autolysis were strongly dependent on cell density within the beads as well as on the cell density of the culture as a whole. On the basis of these results they postulated two separate cell density signals, one that is readily diffusible through the agarose beads and another that is not.

The pheromone of *S. aurantiaca*

Qualls et al. (1978) reported that the myxobacterium *S. aurantiaca,* like *M. xanthus*, showed a marked dependence on cell density for fruiting body formation. They noticed, however, that if spots of cells that were at a cell density below the threshold necessary for normal development were within 1.5 cm of a normally fruiting swarm, the low-density population developed normally. While it was obvious that the cells developing at the high cell density were producing a diffusible factor that permitted low-density development, the matter was complicated by the observation that visible light in the region of 410 to 470 nm, or guanosine or guanine nucleotides, would also permit the formation of morphologically normal fruiting bodies at low cell densities (Stephens and White, 1980; White et al., 1980). Stephens et al. (1982) were subsequently able to isolate the diffusible signal and showed that it was a nonvolatile, low-molecular-weight lipid. They postulated that light sensitized the cells to lower concentrations of the pheromone, thus allowing development at the low cell densities. It is a bit surprising that a diffusible signal should turn out to be a lipid; one might expect that diffusibility of a hydrophobic compound might be somewhat limited. However, it may exist as a complex with a more hydrophilic compound.

The ability of guanine nucleotides and guanosine to replace the pheromone is somewhat puzzling, but there have been a number of reports that guanine-containing compounds affect development in myxobacteria (McCurdy et al., 1978; Manoil and Kaiser, 1980a, 1980b). It is possible that the pheromone functions as part of a chain of events involving guanine nucleotides and a signal transduction pathway. The discovery of GTP-binding proteins in general, and the uncovering of the role of G proteins in eukaryotic signal transduction (Freissmuth et al., 1989), suggest that they may also play a role in intercellular signalling in the myxobacteria.

Contact-Mediated Cell-Cell Interactions

The myxobacteria characteristically travel as a swarm; they are designed to feed as a group of cells, and the success of their developmental program depends on the continued physical contact between cells. In fact, it is becoming increasingly evident that cell-cell interactions in *M. xanthus* are mediated via contact signalling rather than by diffusible, extracellular signals (Janssen and Dworkin, 1985; Gill and Bornemann, 1988; Kim and Kaiser, 1990a, 1990b, 1990c; Shimkets and Rafiee, 1990). It was, therefore, a reasonable presumption that these cell-cell interactions would involve contact between molecules that were a fixed part of their cell surface structure.

Gerisch's laboratory (Gerisch, 1986) had pioneered the approach of using antibodies directed against cell surface molecules of *Dictyostelium discoideum* to define those surface antigens that played a role in contact-mediated interactions. While this approach has been extremely successful in some aspects, it has generated conflicting results in others (see chapter 6). Nevertheless, a similar approach was developed, using MAbs directed against the cell surface molecules of *M. xanthus*. The strategy was simply to use intact cells of *M. xanthus* as the immunogen for raising a series of MAbs directed at random against the various cell surface antigens of *M. xanthus*. These would, of course, include both protein and lipopolysaccharide antigens.

As the first part of the approach, the MAbs were used to try to identify which of the cell surface antigens were developmentally relevant or seemed to play some role in the cell-cell interactions. The two parameters that were used were: (i) does the MAb block or interfere with some aspect of the cell's development or interactional behavior? and (ii) do the kinetics of the antigen's synthesis or appearance on the cell relate in some interesting way to the development? For those antigens that seemed interesting or required for development, the MAbs could then be used as immunoaffinity reagents to assist in their isolation and characterization.

A second feature of the use of the MAbs was to use them to facilitate the isolation of antigen-deficient mutants, to get some better sense of the function of the antigens. Third, the MAbs could be used, in conjunction with some sort of immunolabeling process such as immunogold, to do a topographic mapping of the locations of the various relevant or interesting antigens. Finally, a long-range goal was to use the MAbs or their Fab fragments to do epitope mapping of the cell surface antigens so as to relate the functions of the molecule to specific sequences.

Some of these goals have been accomplished. However, a number of puzzling observations and contradictory results have made it difficult to interpret all of the data. A number of the MAbs were found either to disrupt various aspects of development or to block development completely, and attention was focused on one of these, MAb 1604, which, when added to developing cells, completely blocked development (Gill et al., 1987). MAb 1604 was used to purify the corresponding cell surface antigen (CSA) 1604, which when purified was itself able, at nanogram concentrations, to block development. This was interpreted as reflecting the ability of the purified CSA 1604 to bind to its cellular receptor and interfere with normal development. As such, this experiment was viewed as an important companion to the antibody-blocking experiment and a confirmation of the interpretation that the blocking was a result of the interference of the antibody with some function of CSA 1604. A set of second-generation MAbs were then raised against the purified CSA 1604, and competitive binding studies indicated that these MAbs reacted with a different set of epitopes on CSA 1604 than did the original MAb 1604. These second-generation MAbs blocked development in a somewhat different fashion than the first-generation MAb 1604 (Jarvis and Dworkin, 1989a, 1989b). Subsequently, Fink et al. (1989) isolated a mutant of *M. xanthus* that was nonreactive with MAb 1604. Surprisingly, this mutant was still able to aggregate and sporulate normally. However, when this mutant was examined with one of the second-generation MAbs against CSA 1604, MAb 4054, it was shown to react with this MAb (Jarvis and Dworkin, 1989b). Our interpretation is that the mutant, while lacking the epitope reactive with MAb 1604, still contains a portion of the peptide, sufficient to retain its developmental function. (However, see below for an alternative explanation.)

At this point it might be useful to discuss the use of mutants in assigning cause-and-effect relationships between two properties. There are a number of developmental systems in which a property has been clearly shown to play a role in development, yet mutations in that property have had no phenotypic effect on development. Kroos et al. (1986) have isolated 36 Tn*5 lac* mutants of *M. xanthus* that specifically increase expression of β-galactosidase during development. Yet only seven of these mutants could be shown to manifest any detectable effect on development. The authors emphasized the distinction between genes that may be required for development and those that are regulated during development. A further caution is that development is usually measured by some rather artificial behavior of the organism under a set of cultural conditions. It is not clear that all the subtle aspects of development that may be manifested under natural conditions can be detected without a much more careful and detailed examination of the organism's behavior in the laboratory. Obviously, the same caveats apply to experiments in which attempts to block a particular antigen with a MAb are used to infer causal relationships between that antigen and development.

Losick et al. (1989) have pointed out that mutations in genes (i.e., *cot* and *ssp*) encoding major structural proteins of the *Bacillus subtilis* endospore had no apparent effects on the resultant spores. They emphasized that a great deal of redundancy is built into the systems that control spore morphogenesis and that in many cases a mutation in any particular developmental gene will not have any detectable effect on development.

Kiff et al. (1988) have studied the *unc-22* gene of the nematode *Cenorhabditis elegans*. Loss-of-function alleles of *unc-22* typically lead to severe impairments of movement and behavior in the animal. Nevertheless, a number of *unc-22* mutants with substantial sequence changes and deletions showed normal behavior. The authors concluded that in large proteins, substantial deletions or changes can be consistent with essentially normal functioning of the protein.

A similar conclusion was reached as a result of the startling findings by Gerisch's group (André et al., 1989) that mutants of *D. discoideum* that lacked severin, an F-actin fragmenting protein, showed normal growth, development, and motility. Bray and Vasiliev (1989) have suggested that in the case of *D. discoideum,* there is extensive overlapping redundancy between the detection of environmental cues by receptors on the cell surface and the final cellular response. Such redundancies not only render the cell less susceptible to damage but may also reflect parallel pathways that confer a highly flexible potential to the cell's performance.

This notion of alternative developmental pathways is supported by the recent evidence (O'Connor and Zusman, 1990) that aggregation in *M. xanthus* is regulated by two alternative pathways, one of which is required at 34°C while the other is sufficient for aggregation at 28°C. The authors have suggested that cell density and nutrition may also influence the determination of which developmental pathway is operative.

It has been axiomatic, and a matter of simple experimental logic, that the ultimate experimental test of the relevance of a particular gene product to a process is whether or not a mutant lacking the normal gene product can still undergo the process. In one sense, the logic of this genetic Koch's postulate is unassailable; yet, like Koch's postulates and their role in understanding causal effects in infectious disease, it is somewhat of an oversimplification. Koch's postulates work where there are single causes and single effects; on the other hand, where there are alternative pathways or multiple components interacting to generate a multifactorial result, the threads that bind cause and effect become tangled.

Social Motility

One of the most interesting and dramatic aspects of myxobacterial cell-cell interactions is their social motility (Burchard, 1984). Burchard (1970) was the first to show that one could mutationally separate the movement of single cells of *M.*

xanthus from that of groups of cells. He isolated a motility mutant, designated SM (semimotile), that was able to glide only when groups of cells were in apposition to and in contact with each other (Fig. 10), as well as a completely nonmotile mutant, designated NM. Hodgkin and Kaiser (1979a, 1979b) subsequently clarified the genetics and, to some extent, the physiology of gliding motility when they showed that motility in *M. xanthus* consisted of two regulatory systems, each comprising multiple genetic loci. The social (S) system controls the movement of groups of cells, while the adventurous (A) system controls the movement of individual cells. Mutants in A motility (A$^-$) are able to move only as groups, while S mutants (S$^-$) can move only as individuals. Hodgkin and Kaiser were able to generate completely nonmotile mutants either by obtaining A$^-$ S$^-$ double mutants or by generating a third class of mutants, *mgl,* that were apparently mutated in some aspect of the motility machinery itself. The finding that the movement of cells in groups was not simply the random milling around of a crowd of cells, but rather an organized and apparently regulated social process, was extremely interesting. An important insight into the nature of the social motility was provided by the finding that a small percentage of the A and S mutations

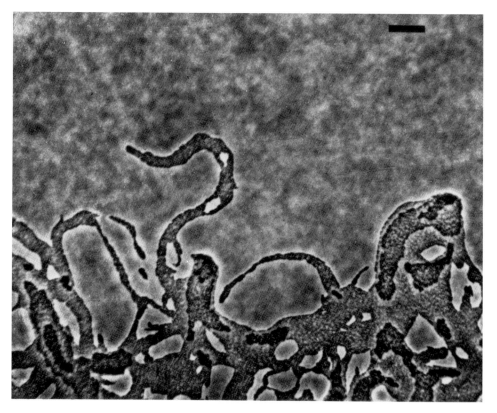

FIGURE 10. Pattern of cell movement of the semimotile mutant of *M. xanthus* gliding on a solid surface. Note the absence of isolated single cells. Bar, 25 μm (from Burchard, 1984).

were conditional; that is, the cells could transiently regain their motility if they were allowed to come into physical contact either with wild-type cells or with cells of a different mutant type (Hodgkin and Kaiser, 1977). If the cells were separated on the agar surface or if contact was prevented by a membrane filter, the complementation did not take place. A similar phenomenon was observed to take place with normal S motility. Using an $A^- S^+$ mutant that manifested only pure social motility, Kaiser and Crosby (1983) showed that if the cells were separated by more than a single cell's length social motility was prevented. These two observations were linked by the suggestion that pili played a role in the social motility. Kaiser (1979) had previously shown that there was a correlation between social motility and piliation, i.e., all S mutants lacked piliation. When the conditional S^- mutants (*tgl* locus) were stimulated, they became not only transiently motile, but also transiently piliated. Subsequent growth of the cells in the absence of stimulation resulted in the expected exponential loss of piliation. It is interesting that one group of S^- mutants can be stimulated by another group of S^- mutants. Since both groups lack pili, it is clear that the pili are not required for stimulation, even though they are necessary for the resultant social motility. Kaiser has suggested a number of possibilities that might explain the role of the pili in social motility. The pili may physically bind the cells together, or alternatively, they might serve as sensory hairs that, upon contact with a neighboring cell, may initiate the cell's movement.

What are the respective functions of A and S motility? Based on the data of Hodgkin and Kaiser (1979a), Shimkets (1987) tabulated the ability of 115 motility mutants to undergo development. Of these, none of 40 $A^- S^-$ mutants were able to undergo normal development, 35 of 39 $A^- S^+$ mutants were able to complete development, and 12 of 36 $A^+ S^-$ mutants were able to complete development.

The relatively safe conclusions are that some motility is necessary for development (which is not a surprise), that A motility is relatively unimportant to development, and that S motility seems to play some kind of role. These conclusions are necessarily tentative, as the test of development (formation of fruiting bodies) was a relatively crude one. Were there any effects of the lack of social or adventurous motility on the size, shape, or distribution of the fruiting bodies, on the efficiency of myxospore formation, or on the timing of the process?

While the precise role of social motility in the biology of the myxobacteria remains unclear, it is an excellent system for examining the role of a contact-mediated interaction on cell behavior.

Developmental Autolysis

During fruiting body formation by *M. xanthus,* approximately 80 to 90% of the vegetative cells lyse; the remaining cells complete the developmental process and end up in the fruiting body as myxospores (Wireman and Dworkin, 1977). A similar phenomenon occurs during fruiting body formation in *S. aurantiaca;* 30 to 40% of the cells undergo lysis prior to aggregation and fruiting body formation (Qualls, 1977). Developmental death is not peculiar to the myxobacteria; it is a common and normal aspect of neuronal cell development in vertebrates and invertebrates (Oppenheim, 1981). It is also associated with the normal development of *Dictyostelium*. As many as 25% of the cells of *D. discoideum* lyse during the process of assembling the fruiting body, and 90% of the cells of *D. mucoroides* or *D. purpureum* are sacrificed while they construct their fruiting bodies (Bonner, 1982a). While the function of developmental lysis in *M. xanthus* is not obvious, there are a number of formal possibilities.

O'Connor and Zusman (1988) have proposed that during development, the cells enter a stage when they are particularly fragile, and that the observed autolysis is a result of the manipulations involved in handling the cells. Rosenbluh et al. (1989) have directly tested this by means of the agarose microbead technique referred to earlier. When cells were enclosed in the microbeads, placed under developmental conditions, and observed directly with no additional manipulations, the authors found that up to 90% of the cells underwent developmental autolysis with 50% of the surviving cells forming resistant myxospores. This was consistent with other reports of developmental autolysis under different conditions (Wireman and Dworkin, 1977). While it is quite feasible that some fraction of the cells do indeed enter a fragile state during development, it does not seem that the autolysis is an artifactual consequence of the experimental procedure.

Another alternative is that autolysis plays a role in the morphogenesis of the myxospore, in that it may be part of the process for the release of the intracellular protein S that becomes the outermost coat of the mature myxospore. Protein S is a small protein found on the surface of mature myxospores of *M. xanthus*. It is synthesized by vegetative cells early in development and at that time occupies as much as 15% of the total biosynthetic capacity of the cell (Inouye et al., 1979b). In the presence of Ca^{2+} the protein eventually self-assembles onto the outer surface of the myxospore. It is interesting that protein S has been shown to share a considerable structural homology with calmodulin (Inouye et al., 1983). Another unusual feature of protein S is that it is not produced from a secretory precursor with a signal peptide (Inouye et al., 1983), despite the fact that it is produced within the vegetative cell early in development and does not appear on the outer surface of the spore until late in development (Inouye et al., 1979b).

Thus, the question is: how does protein S get out of the cell? Teintze et al. (1985) have suggested that the function of developmental autolysis is to release protein S from one subset of cells, so that that it could be self-assembled on the surface of the surviving cells. If this was indeed the function of the developmental autolysis, one would expect that inhibition of autolysis by MAb G357 would prevent release of protein S and that the myxospores formed under these conditions would thus lack protein S. When such myxospores were examined, using anti-protein S antibody (kindly supplied by David Zusman), the myxospores were found to contain a normal complement of protein S on their surface (Gill and Dworkin, unpublished). A third possibility is that autolysis may be required to release either a set of signals necessary for the proper coordination of development, or cellular nutrients necessary for the completion of development.

These possibilities have been excluded by the fact that Gill and Dworkin (unpublished) have shown that when developmental autolysis was completely inhibited by a particular MAb (G357) the cells were still able to complete the process of fruiting body and myxospore formation. These spores have not been tested extensively to determine whether or not they have the same resistance, dormancy, and germination properties as the normal myxospores. In addition, the process of aggregation and fruiting body formation was not a normal one; whereas when autolysis occurred normally, the background area between the fruiting bodies was essentially clear of remaining vegetative cells, when autolysis was inhibited the background was littered with vegetative cells and the fruiting bodies were larger and fewer in number. This leads to the possibility that the function of autolysis may be more subtle than simply controlling whether or not fruiting bodies or spores are formed, and may instead regulate the shape, size, and distribution of the fruiting bodies or the properties of the resultant myxo-spores.

An interesting analogy has been made between developmental autolysis in myxobacteria and the autolysis of the mother cell in *Bacillus* endosporulation (Dworkin, 1985). *Bacillus* sporulation is the result of an asymmetric cell division, resulting in the partition of the cell into a forespore, destined to become the mature endospore, and the mother cell, which eventually lyses during the sporulation process. Each of these cells retains its genome, and the process of spore formation involves the contribution of genetic information from both genomes; gene products from the two genomes eventually end up in the mature spore (de Lencastre and Piggot, 1979). There is substantial unidirectional protein turnover, with the mother cell contributing over 10% of its protein to the forespore (Eaton and Ellar, 1974) and eventually lysing. It has been suggested that a similar sort of cross-feeding may occur as a result of the massive autolysis that occurs during myxo-bacterial development. In other words, instead of a portion of the cell lysing as part of the sporulation process, a portion of the population lyses so that the remainder of the population can complete its development.

Whether or not the phenomenon of developmental autolysis in myxobacteria falls in the category of apoptosis (programmed cell death), a phenomenon well known among metazoan cells, still remains to be determined. Alternatively, it may be a stochastic process, determined by the position of the cells or some other essentially random factor.

Cell-Cell Cohesion

It is a common observation that freshly isolated strains of myxobacteria tend to grow in liquid media as clumps of cells. This is a reflection of the inherent tendency of the myxobacteria to cohere with each other. It has been shown by White's laboratory (Gilmore and White, 1985) that cell-to-cell cohesion in *S. aurantiaca* is an active process and is related to the developmental behavior of the organism. Likewise, Shimkets (1986a, 1986b) has shown that cell-to-cell cohesion in *M. xanthus* is associated with social motility and fruiting body formation. Thus, it has become reasonable to examine cell-to-cell cohesion as a reflection of the role of cell-cell interactions in myxobacterial development and behavior.

S. aurantiaca has been shown to posess two cohesion systems. One of these, termed class A, is constitutive, present on cells growing vegetatively or during development. It is easily reversible by mild shear forces and is indifferent to energy poisons, temperature, or the presence or absence of divalent cations. The other system, termed class B, is stable to shear forces, requires the presence of Ca^{2+} ions, and is energy dependent and its induction requires protein and RNA synthesis (Gilmore and White, 1985). In an earlier study, which did not distinguish between the class A and B systems, Qualls and White (1982) showed that the cohesion of *S. aurantiaca* cells was specific; radiolabeled cells of *M. xanthus* were not included in the induced clumps of *S. aurantiaca*. While it has not been tested directly, this would suggest that mixtures of different myxobacteria would not form mixed fruiting bodies but that the cells would recognize each other as different and sort themselves out.

Only one cohesion system has been demonstrated in *M. xanthus*. It is constitutive: the process of cohesion requires energy but not macromolecular biosynthesis (Shimkets, 1986a) and seems to require the presence on the cell surface of a network of 50-nm fibrils (Fig. 11); a *dsp* (dispersed) mutant lacking the fibrils fails to undergo cohesion, social motility, or development (Arnold and Shimkets, 1988a). It was interesting that the properties of the *dsp* mutant could be simulated by treating wild-type cells with Congo red, which is a diazo dye that binds most strongly to polysaccharides with contiguous β(1,4)-linked glucopyranosyl units (Arnold and Shimkets, 1988b). This suggested to the authors that cohesion and the resultant processes of social motility and development required the involvement of the fibrils.

Scanning electron micrographs from our laboratory, using a new ultra-high-resolution, low-voltage field emission microscope (Hitachi S-900), have revealed the fibrils and the cell surface of *M. xanthus* in even greater detail (Fig. 12).

Myxobacterial Hemagglutinin

The first myxobacterial cell surface protein shown to have some developmental relevance was the myxobacterial hemagglutinin (MBHA) first described

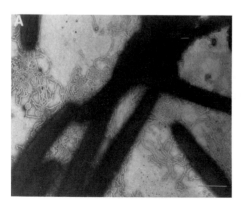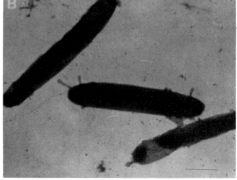

FIGURE 11. Transmission electron micrograph of negatively stained wild-type (A) and Dsp mutant (B) cells of *M. xanthus*, illustrating the presence of the extracellular fibrils in the wild-type cells and their absence in the mutant. Bar, 1 μm (from Arnold and Shimkets, 1988b).

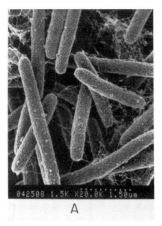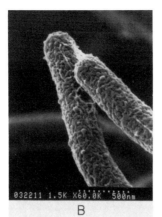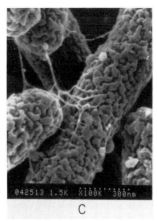

FIGURE 12. Ultra-high-resolution scanning electron micrographs of cells of *M. xanthus,* illustrating the extracellular fibrils connecting the cells. (Supplied by R. M. Behmlander and M. Dworkin.)

by Cumsky and Zusman (1979, 1981). MBHA is a lectin, located at the cell surface (Nelson et al., 1981) and synthesized during development (Cumsky and Zusman, 1979). Nelson et al. (1981), using immunofluorescence and ferritin labeling, showed that the lectin was located at one or both of the cell poles, suggesting that it may play a role in end-to-end cell-cell adhesion. The developmental phenotype of strains constructed with a null mutation in the MBHA gene was not seriously disrupted; the mutant was able to develop normally, but in the absence of Mg^{2+} its development was somewhat delayed (Romeo and Zusman, 1987). This suggests that the function of MBHA may be as a backup system that allows developmental aggregation in the absence of Mg^{2+} when normal, Mg^{2+}-mediated aggregation is not possible.

Myxobacterial Cell-Cell Interactions That Have Not Been Examined

Twenty-five years ago, Hans Reichenbach produced a remarkable series of time-lapse films of the myxobacteria that depicted the striking behavior of these social bacteria (Reichenbach et al., 1965a, 1965b, 1965c, 1965d; Reichenbach, 1966; Reichenbach et al., 1975–1976). These films attracted investigators to the field and eventually focused their attention on the cell-cell interactions of these organisms. Since that time the myxobacteria (or at least *M. xanthus* and, to some extent, *S. aurantiaca*) have been thoroughly domesticated. While there is a considerable body of information describing the fundamental properties of *M. xanthus* (e.g., Rosenberg, 1984), many of the behavioral phenomena illustrated by Reichenbach's films, especially those involving cell-cell interactions, remain a mystery. For example, there are no insights available as to the nature of the stimulus that induces developmental aggregation, nor is there any understanding of the forces that control and direct the aggregation process itself. (See Fig. 3 and 13 for an illustration of the events leading to aggregate formation.)

The films also illustrate the remarkable tendency of the cells to remain to-

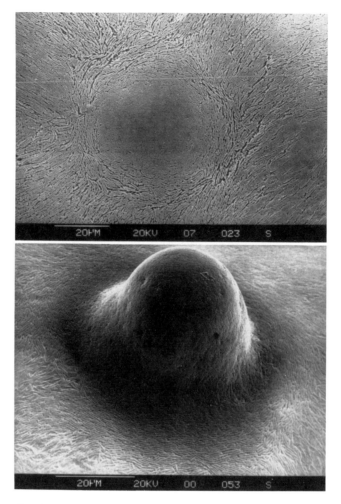

FIGURE 13. Scanning electron micrograph of cells of *S. aurantiaca* undergoing aggregation (top) and early stalk formation (bottom) (from Vasquez et al., 1985).

gether as a coherent swarm, despite the cells' ability to move as single, individual cells. While Kaiser's group has uncovered the existence of the separate adventurous (A) and social (S) systems for regulating motility (Kaiser, 1989), the nature of the forces that hold the cells together in the swarm is not understood.

It is clear from the films that the cells show directed movement, both as part of the process of developmental aggregation and as part of their ability to move toward clumps of bacterial cells during the feeding process. As indicated in the discussion on tactic behavior, while there is no evidence that the cells are using a chemotactic sensing mechanism (Dworkin and Eide, 1983), it has been demonstrated that they can perceive the presence of physical objects at a distance (Dworkin, 1983). The mechanism of this sensing ability and tactic behavior is unknown. The rhythmic oscillations illustrated in the films and described by Reichenbach (1965) have been described in some greater detail by Shimkets and Kaiser (1982a, 1982b). However, neither the function nor the mechanism of this dramatic periodic, rippling behavior is understood.

Finally, no one who has worked with the myxobacteria can fail to be intrigued by their ability to construct their characteristic multicellular fruiting bodies. While the fruiting body of *Myxococcus* is a relatively simple structure, organisms such as *Stigmatella* and *Chondromyces* construct extraordinarily complex fruiting bodies (Fig. 10 and 11). It is not clear whether the information for these structures is hard-wired into the genome or is a consequence of cell movement and a sequence of programmed but positionally dependent cell-cell interactions. If these cell-cell interactions are coordinated with the properly timed synthesis and excretion of a mixture of self-assembling polysaccharides (Rees, 1972), the resultant formation of a fruiting body begins at least to fall within the framework of a set of processes that individually can be understood.

EVOLUTION OF MULTICELLULARITY AND THE MYXOBACTERIA

Shimkets has concluded that the myxobacteria separated from their nearest relatives about 1 billion years ago (Shimkets, 1990). His argument is based on the following facts, assumptions, and conclusions. There is evidence that indicates that whereas oxygenic photosynthesis appeared on the earth about 2.8 billion years ago (Schopf et al., 1983), the concentration of atmospheric oxygen possibly sufficient to support aerobic growth did not appear until about 1.7 billion years ago (Chapman and Schopf, 1983). These data correspond to conclusions based on the phylogenetic analyses of the myxobacteria derived from the sequences of their 16S RNA. The values that compare the two major myxobacterial groups indicate that the two branches of the myxobacteria diverged about 900 million years ago (Ochman and Wilson, 1987). A similar comparison between the myxobacteria and their nearest evolutionary relatives, the sulfate-reducing bacteria and *Bdellovibrio,* reveals that they diverged about 1 billion years ago (Ludwig et al., 1983). On the basis of similar kinds of data, Kaiser (1986) has concluded that the myxobacteria diverged from their nearest bacterial relatives about 2 billion years ago. Regardless of which of the two figures is correct, since the earliest known fossils of multicellular animals have been judged to be 700 million years old (Glaessner, 1976), Kaiser (1986) has pointed out that the myxobacteria may represent the earliest evolutionary attempt at multicellularity.

It is thus especially interesting that the myxobacteria, uniquely among the prokaryotes, share a number of properties with eukaryotic cells. Protein S, a protein with a relative molecular mass of 19,000 Da, is a structural component of mature myxospores that binds Ca^{2+}. It shares some properties with bovine brain calmodulin; two of the internal domains of protein S contain a sequence of nine amino acids that is similar to the Ca^{2+}-binding sequence of calmodulin (Inouye et al., 1983). Protein S has an even more striking similarity to the β and γ crystallins of the vertebrate eye lens. Both protein S and the lens crystallins are β-sheet proteins with internally duplicated domains, and both share a significant amount of structural similarity. Based on these similarities, it has been suggested (Wistow et al., 1985) that the β and γ crystallins may have evolved from myxococcal proteins.

One of the most surprising examples of myxobacterial and eukaryotic similarity has been the recent demonstration of reverse transcriptase in *M. xanthus* (Lampson et al., 1989a). A number of myxobacteria have been shown to contain multiple copies of an unusual complex of RNA and single-stranded DNA (msDNA) (Yee et al., 1984). The properties of msDNA led to the prediction that a reverse transcriptase was required for its biosynthesis, and this was, in fact, shown to be the case (Inouye et al., 1989); the reverse transcriptase gene has been located on the chromosome immediately downstream from the msDNA coding region. The sequence of the gene product has been shown to be similar to that for other reverse transcriptases including those of retroviruses, retrotransposons, and the *Neurospora* mitochondrial plasmid enzyme. This has suggested to the authors that there may be a common evolutionary origin of all of the reverse transcriptases (Inouye et al., 1989). The authors also concluded that, based on the analysis of codon usage pattern of the gene and the distribution of the msDNA system in a variety of other, independently isolated strains of *M. xanthus,* the reverse transcriptase system in *M. xanthus* is as ancient as the other chromosomal genes and thus was not acquired recently. Subsequently, reverse transcriptase was shown to be present in some other eubacteria, including about 5% of the clinical isolates of *E. coli* (Lampson et al., 1989b). Since the codon usage pattern for the reverse transcriptase gene in *E. coli* was substantially different from that of the other *E. coli* genes, the authors concluded that the reverse transcriptase gene may have been transferred to *E. coli* relatively recently (Inouye et al., 1989).

Kaiser (1986) has argued that, while multicellularity is said to have arisen as many as 15 times (Dobzhansky et al., 1977; Cloud, 1968), the myxobacteria may represent one of the earliest of such attempts, occurring at least 2 billion years ago.

CONCLUSIONS AND EPILOGUE

Most of the contemporary research on the developmental biology of the myxobacteria is directed toward trying to understand the nature of the various cell-cell interactions that take place during growth and development. Because they combine the primitive multicellularity of a social organism with the experimental convenience of a prokaryote, the myxobacteria have offered an excellent opportunity to ask questions of cell-cell communication in a context of biochemistry, genetics, and molecular biology.

REFERENCES

André, E., M. Brink, G. Gerisch, G. Isenberg, A. Noegel, M. Schleicher, J. E. Segall, and E. Wallraf. 1989. A *Dictyostelium* mutant deficient in severin, an F-actin fragmenting protein, shows normal motility and chemotaxis. *J. Cell Biol.* 108:985–995.

Arnold, J. W., and L. J. Shimkets. 1988a. Inhibition of cell-cell interactions in *Myxococcus xanthus* by Congo red. *J. Bacteriol.* 170:5765–5770.

Arnold, J. W., and L. J. Shimkets. 1988b. Cell surface properties correlated with cohesion in *Myxococcus xanthus. J. Bacteriol.* **170:**5771–5777.

Bauer, E. 1905. Myxobakterien-Studien. *Arch. Protistenkd.* **5:**92–121.

Blackhart, B. D., and D. R. Zusman. 1985a. "Frizzy" genes of *Myxococcus xanthus* are involved in control of the frequency of reversal of gliding motility. *Proc. Natl. Acad. Sci. USA* **82:**8767–8771.

Blackhart, B. D., and D. R. Zusman. 1985b. Cloning and complementation analysis of the "frizzy" genes of *Myxococcus xanthus. Mol. Gen. Genet.* **198:**243–254.

Bonner, J. T. 1982a. Comparative biology of cellular slime molds, p. 1–33. *In* W. F. Loomis (ed.), *The Development of Dictyostelium discoideum.* Academic Press, Inc., New York.

Bonner, J. T. 1982b. Evolutionary strategies and developmental constraints in the cellular slime molds. *Am. Nat.* **119:**530–552.

Bray, D., and J. Vasiliev. 1989. Networks from mutants. *Nature* (London) **338:**203–204.

Burchard, A. C., R. P. Burchard, and J. A. Kloetzel. 1977. Intracellular, periodic structures in the gliding bacterium, *Myxococcus xanthus. J. Bacteriol.* **132:**666–672.

Burchard, R. P. 1970. Gliding motility mutants of *Myxococcus xanthus. J. Bacteriol.* **104:**940–947.

Burchard, R. P. 1984. Gliding motility and taxes, p. 139–161. *In* E. Rosenberg (ed.), *Myxobacteria: Development and Cell Interactions.* Springer-Verlag, New York.

Chapman, D. J., and W. J. Schopf. 1983. Biological and biochemical effects of the development of an aerobic environment. *In* J. W. Schopf (ed.) *Earth's Earliest Biosphere: Its Origin and Evolution.* Princeton University Press, Princeton, N.J.

Cheng, Y., and D. Kaiser. 1989a. *dsg,* a gene required for cell-cell interaction early in *Myxococcus* development. *J. Bacteriol.* **171:**3719–3726.

Cheng, Y., and D. Kaiser. 1989b. *dsg,* a gene required for *Myxococcus* development, is necessary for cell viability. *J. Bacteriol.* **171:**3727–3731.

Clarke, C. 1981. Motility and fruiting in the bacterium *Myxococcus xanthus,* p. 155–171. *In* J. M. Lackie and P. C. Wilkinson (ed.), *Biology of the Chemotactic Response.* Cambridge University Press, Cambridge.

Cloud, P. E., Jr. 1968. Pre-metazoan evolution and the origins of the metazoa, p. 1–72. *In* E. T. Drake (ed.), *Evolution and Environment.* Yale University Press, New Haven.

Cumsky, M., and D. R. Zusman. 1979. Myxobacterial hemagglutinin, a development-specific lectin of *Myxococcus xanthus. Proc. Natl. Acad. Sci. USA* **76:**5505–5509.

Cumsky, M. G., and D. R. Zusman. 1981. Purification and characterization of myxobacterial hemagglutinin, a development-specific lectin of *Myxococcus xanthus. J. Biol. Chem.* **256:**12581–12588.

de Lencastre, H., and P. J. Piggot. 1979. Identification of different sites of expression for *spo* loci by transformation of *Bacillus subtilis. J. Gen. Microbiol.* **114:**377–389.

Dobzhansky, T., F. J. Ayala, G. L. Stebbins, and J. W. Valentine. 1977. *Evolution.* W. H. Freeman & Co., San Francisco.

Downard, J. S., D. Kupfer, and D. R. Zusman. 1984. Gene expression during development of *Myxococcus xanthus:* analysis of the genes for protein S. *J. Mol. Biol.* **175:**469–492.

Dworkin, M. 1963. Nutritional regulation of morphogenesis in *Myxococcus xanthus. J. Bacteriol.* **86:**67–72.

Dworkin, M. 1972. The myxobacteria: new directions in studies of procaryotic development. *Crit. Rev. Microbiol.* **1:**435–452.

Dworkin, M. 1973. Cell-cell interactions in the myxobacteria. *Symp. Soc. Gen. Microbiol.* **23:**125–142.

Dworkin, M. 1985. *Developmental Biology of the Bacteria.* Benjamin/Cummings Publishing Co., Menlo Park, Calif.

Dworkin, M. Unpublished data.

Dworkin, M., and D. Eide. 1983. *Myxococcus xanthus* does not respond chemotactically to moderate concentration gradients. *J. Bacteriol.* **154:**437–442.

Dworkin, M., and D. Kaiser. 1985. Cell interactions in myxobacterial growth and development. *Science* **230:**18–24.

Dworkin, M., K. H. Keller, and D. Weisberg. 1983. Experimental observations consistent with a surface tension model of gliding motility of *Myxococcus xanthus. J. Bacteriol.* **155:**1367–1371.

Dworkin, M., and D. J. Niederpruem. 1964. Electron transport system in vegetative cells and microcysts of *Myxococcus xanthus*. *J. Bacteriol.* **87**:316–322.

Eaton, M. W., and D. J. Ellar. 1974. Protein synthesis and breakdown in the mother cell and forespore compartments during spore morphogenesis in *Bacillus megaterium*. *Biochem. J.* **144**:327–337.

Fink, J. M., M. Kalos, and J. F. Zissler. 1989. Isolation of cell surface antigen mutants of *Myxococcus xanthus* by use of monoclonal antibodies. *J. Bacteriol.* **171**:2033–2041.

Fluegel, W. 1963. Fruiting chemotaxis in *Myxococcus fulvus* (Myxobacteria). *Proc. Minn. Acad. Sci.* **30**:120–123.

Fluegel, W. 1964. Induced fruiting in myxobacteria. *Proc. Minn. Acad. Sci.* **31**:114–115.

Freissmuth, M., P. J. Casey, and A. G. Gilman. 1989. G proteins control diverse pathways of transmembrane signaling. *FASEB J.* **3**:2125–2131.

Gelvan, I., M. Varon, and E. Rosenberg. 1987. Cell-density-dependent killing of *Myxococcus xanthus* by autocide AMV. *J. Bacteriol.* **169**:844–848.

Gerisch, G. 1986. Inter-relation of cell adhesion and differentiation in *Dictyostelium discoideum*. *J. Cell Sci.* **4**(Suppl.):201–219.

Gill, J., and M. Dworkin. Unpublished data.

Gill, J. S., B. W. Jarvis, and M. Dworkin. 1987. Inhibition of development in *Myxococcus xanthus* by monoclonal antibody 1604. *Proc. Natl. Acad. Sci. USA* **84**:4505–4508.

Gill, R. E. 1991. Personal communication.

Gill, R. E., and C. Bornemann. 1988. Identification and characterization of the *Myxococcus xanthus* *bsgA* gene product. *J. Bacteriol.* **170**:5289–5297.

Gill, R. E., M. G. Cull, and S. Fly. 1988. Genetic identification and cloning of a gene required for developmental cell interactions in *Myxococcus xanthus*. *J. Bacteriol.* **170**:5279–5288.

Gilmore, D. F., and D. White. 1985. Energy-dependent cell cohesion in myxobacteria. *J. Bacteriol.* **161**:113–117.

Glaessner, M. F. 1976. Early phanerozoic worms and their geological and biological significance. *J. Geol. Soc. London* **132**:259–275.

Hagen, D. C., A. P. Bretscher, and D. Kaiser. 1978. Synergism between morphogenetic mutants of *Myxococcus xanthus*. *Dev. Biol.* **64**:284–296.

Hanson, C. W., and M. Dworkin. 1974. Intracellular and extracellular nucleotides and related compounds during the development of *Myxococcus xanthus*. *J. Bacteriol.* **118**:486–496.

Hodgkin, J., and D. Kaiser. 1977. Cell-to-cell stimulation of movement in nonmotile mutants of *Myxococcus*. *Proc. Natl. Acad. Sci. USA* **74**:2938–2942.

Hodgkin, J., and D. Kaiser. 1979a. Genetics of gliding motility in *Myxococcus xanthus* (Myxobacterales): two gene systems control movement. *Mol. Gen. Genet.* **171**:177–191.

Hodgkin, J., and D. Kaiser. 1979b. Genetics of gliding motility in *Myxococcus xanthus* (Myxobacterales): genes controlling movements of single cells. *Mol. Gen. Genet.* **171**:167–176.

Inouye, M., S. Inouye, and D. R. Zusman. 1979a. Gene expression during development of *Myxococcus xanthus*: pattern of protein synthesis. *Dev. Biol.* **68**:579–591.

Inouye, M., S. Inouye, and D. Zusman. 1979b. Biosynthesis and self-assembly of protein S, a development-specific protein from *Myxococcus xanthus*. *Proc. Natl. Acad. Sci. USA* **76**:209–213.

Inouye, S., T. Franceschini, and M. Inouye. 1983. Structural similarities between the development-specific protein S from a Gram-negative bacterium, *Myxococcus xanthus*, and calmodulin. *Proc. Natl. Acad. Sci. USA* **80**:6829–6833.

Inouye, S., M. Hsu, S. Eagle, and M. Inouye. 1989. Reverse transcriptase associated with the biosynthesis of the branched RNA-linked msDNA in *Myxococcus xanthus*. *Cell* **56**:709–717.

Janssen, G. R., and M. Dworkin. 1985. Cell-cell interactions in developmental lysis of *Myxococcus xanthus*. *Dev. Biol.* **112**:194–202.

Jarvis, B. W., and M. Dworkin. 1989a. Purification and properties of *Myxococcus xanthus* cell surface antigen 1604. *J. Bacteriol.* **171**:4655–4666.

Jarvis, B. W., and M. Dworkin. 1989b. Role of *Myxococcus xanthus* cell surface antigen 1604 in development. *J. Bacteriol.* **171**:4667–4673.

Kaiser, D. 1979. Social motility is correlated with the presence of pili in *Myxococcus xanthus*. *Proc. Natl. Acad. Sci. USA* **76**:5952–5956.

Kaiser, D. 1986. Control of multicellular development: *Dictyostelium* and *Myxococcus. Annu. Rev. Genet.* **20:**539–566.

Kaiser, D. 1989. Multicellular development in myxobacteria, p. 243–263. *In* D. A. Hopwood and K. F. Chater (ed.), *Genetics of Bacterial Diversity.* Academic Press, Inc. (London), Ltd., London.

Kaiser, D., and C. Crosby. 1983. Cell movement and its coordination in swarms of *Myxococcus xanthus. Cell Motil.* **3:**227–245.

Kaiser, D., L. Kroos, and A. Kuspa. 1986. Cell interactions govern the temporal pattern of *Myxococcus* development. *Cold Spring Harbor Symp. Quant. Biol.* **50:**823–830.

Keller, K. H., M. Grady, and M. Dworkin. 1983. Surface tension gradients: feasible model for gliding motility of *Myxococcus xanthus. J. Bacteriol.* **155:**1358–1366.

Kiff, J. E., D. G. Moerman, L. A. Schriefer, and R. H. Waterston. 1988. Transposon-induced deletions in unc-22 of *C. elegans* associated with almost-normal gene activity. *Nature* (London) **331:**631–633.

Kim, S. K., and D. Kaiser. 1990a. C-factor: a cell-cell signaling protein required for fruiting body morphogenesis of *M. xanthus. Cell* **61:**19–26.

Kim, S. K., and D. Kaiser. 1990b. Purification and properties of *Myxococcus xanthus* C-factor, an intercellular signaling protein. *Proc. Natl. Acad. Sci. USA* **87:**3635–3639.

Kim, S. K., and D. Kaiser. 1990c. Cell alignment required in differentiation of *Myxococcus xanthus. Science* **249:**926–928.

Kim, S. K., and D. Kaiser. 1991. C-factor has distinct aggregation and sporulation thresholds during *Myxococcus* development. *J. Bacteriol.* **173:**1722–1728.

Kroos, L., P. Hartzell, K. Stephens, and D. Kaiser. 1988. A link between cell movement and gene expression argues that motility is required for cell-cell signalling during fruiting body development. *Genes Dev.* **12A:**1677–1685.

Kroos, L., and D. Kaiser. 1984. Construction of Tn5 *lac*, a transposon that fuses *lacZ* expression to exogenous promoters, and its introduction into *Myxococcus xanthus. Proc. Natl. Acad. Sci. USA* **81:**5816–5820.

Kroos, L., and D. Kaiser. 1987. Expression of many developmentally regulated genes depends on a sequence of cell interactions. *Genes Dev.* **1:**840–854.

Kroos, L., A. Kuspa, and D. Kaiser. 1986. A global analysis of developmentally regulated genes in *Myxococcus xanthus. Dev. Biol.* **117:**252–266.

Kuner, J. M., and D. Kaiser. 1981. Introduction of transposon Tn5 into *Myxococcus* for analysis of developmental and other non-selectable mutants. *Proc. Natl. Acad. Sci. USA* **78:**425–429.

Kuner, J. M., and D. Kaiser. 1982. Fruiting body morphogenesis in submerged cultures of *Myxococcus xanthus. J. Bacteriol.* **151:**458–461.

Kuspa, A., L. Kroos, and D. Kaiser. 1986. Intercellular signalling is required for developmental gene expression in *Myxococcus xanthus. Dev. Biol.* **117:**267–276.

Lackie, J. M., and P. C. Wilkinson (ed.). 1981. *Biology of the Chemotactic Response.* Cambridge University Press, Cambridge.

Lampson, B. C., M. Inouye, and S. Inouye. 1989a. Reverse transcriptase with concomitant ribonuclease H activity in the cell-free synthesis of branched RNA-linked msDNA of *Myxococcus xanthus. Cell* **56:**701–707.

Lampson, B. C., J. Sun, M.-Y. Hsu, J. Vallejo-Ramirez, S. Inouye, and M. Inouye. 1989b. Reverse transcriptase in a clinical strain of *Escherichia coli:* production of branched RNA-linked msDNA. *Science* **243:**1033–1038.

Lapidus, I. R., and H. C. Berg. 1982. Gliding motility of *Cytophaga* sp. strain U67. *J. Bacteriol.* **151:**384–398.

Lev, M. 1954. Demonstration of a diffusible fruiting factor in myxobacteria. *Nature* (London) **173:**501.

Losick, R., L. Kroos, J. Errington, and P. Youngman. 1989. Pathways of developmentally regulated gene expression in the spore-forming bacterium *Bacillus subtilis*, p. 221–242. *In* D. A. Hopwood and K. F. Chater (ed.), *Genetics of Bacterial Diversity.* Academic Press, Inc. (London), Ltd., London.

Ludwig, W. K., H. Schleifer, H. Reichenbach, and E. Stackebrandt. 1983. A phylogenetic analysis of the myxobacteria *Myxococcus fulvus, Stigmatella aurantiaca, Cystobacter fuscus, Sorangium cellulosum* and *Nannocystis exedens. Arch. Microbiol.* **135:**58–62.

Lünsdorf, H., and H. Reichenbach. 1989. Ultrastructural details of the apparatus of gliding motility of *Myxococcus fulvus* (Myxobacterales). *J. Gen. Microbiol.* **135:**1633–1641.

Macnab, R. M., and D. E. Koshland, Jr. 1972. The gradient sensing mechanism in bacterial chemotaxis. *Proc. Natl. Acad. Sci. USA* **69:**2509–2512.

Manoil, C., and D. Kaiser. 1980a. Accumulation of guanosine tetraphosphate and guanosine pentaphosphate in *Myxococcus xanthus* during starvation and myxospore development. *J. Bacteriol.* **141:**297–304.

Manoil, C., and D. Kaiser. 1980b. Guanosine pentaphosphate and guanosine tetraphosphate accumulation and induction of *Myxococcus xanthus* fruiting body development. *J. Bacteriol.* **141:**305–315.

McBride, M. J., R. A. Weinberg, and D. R. Zusman. 1989. "Frizzy" aggregation genes of the gliding bacterium *Myxococcus xanthus* show sequence similarities to the chemotaxis genes of enteric bacteria. *Proc. Natl. Acad. Sci. USA* **86:**4224–4228.

McCurdy, H. D., J. Ho, and W. J. Dobson. 1978. Cyclic nucleotides, cyclic nucleotide phosphodiesterase and development in *Myxococcus xanthus*. *Can. J. Microbiol.* **24:**1475–1481.

McVittie, A., F. Messik, and S. A. Zahler. 1962. Developmental biology of *Myxococcus*. *J. Bacteriol.* **84:**546–551.

Nelson, D. R., M. G. Cumsky, and D. R. Zusman. 1981. Localization of myxobacterial hemagglutinin in the periplasmic space and on the cell surface of *Myxococcus xanthus* during developmental aggregation. *J. Biol. Chem.* **256:**12589–12595.

Newton, A. 1989. Differentiation in *Caulobacter:* flagellum development, motility and chemotaxis, p. 199–220. *In* D. A. Hopwood and K. F. Chater (ed.), *Genetics of Bacterial Diversity*. Academic Press, Inc. (London), Ltd., London.

Ochman, H., and A. C. Wilson. 1987. Evolution in bacteria: evidence for a universal substitution rate in cellular genomes. *J. Mol. Evol.* **26:**74–86.

O'Connor, K. A., and D. Zusman. 1988. Reexamination of the role of autolysis in the development of *Myxococcus xanthus*. *J. Bacteriol.* **170:**4103–4112.

O'Connor, K. A., and D. Zusman. 1990. Genetic analysis of *tag* mutants of *Myxococcus xanthus* provides evidence for two developmental aggregation systems. *J. Bacteriol.* **172:**3868–3878.

Oppenheim, R. W. 1981. Neuronal cell death and some related regressive phenomena during neurogenesis: a selective historical review and progress report, p. 74–133. *In* W. M. Cowan (ed.), *Studies in Developmental Neurobiology: Essays in Honor of Viktor Hamburger*. Oxford University Press, Oxford.

Pate, J. L., and L. E. Chang. 1979. Evidence that gliding motility in prokaryotic cells is driven by rotary assemblies in the cell envelopes. *Curr. Microbiol.* **2:**59–64.

Qualls, G. 1977. M. A. thesis. Indiana University, Bloomington.

Qualls, G. T., K. Stephens, and D. White. 1978. Morphogenetic movements and multicellular development in the fruiting myxobacterium, *Stigmatella aurantiaca*. *Dev. Biol.* **66:**270–274.

Qualls, G. T., and D. White. 1982. Developmental cell cohesion in *Stigmatella aurantiaca*. *Arch. Microbiol.* **131:**334–337.

Ramsey, W. S., and M. Dworkin. 1968. Microcyst germination in *Myxococcus xanthus*. *J. Bacteriol.* **95:**2249–2257.

Rees, D. A. 1972. Shapely polysaccharides. *Biochem. J.* **26:**257–273.

Reichenbach, H. 1965. Rhythmische Vorgange bei der Schwarmentfaultung von Myxobakterien. *Ber. Dtsch. Bot. Ges.* **78:**102–105.

Reichenbach, H. 1966. *Myxococcus* spp. (Myxobacteriales [sic]) Schwarmentwicklung und Bildung von Protocysten. Encyclop. Cinematogr. Film E778/1965. Inst. Wiss. Film, Göttingen, Federal Republic of Germany.

Reichenbach, H. 1984. Myxobacteria: a most peculiar group of social prokaryotes, p. 1–50. *In* E. Rosenberg (ed.), *Myxobacteria: Development and Cell Interactions*. Springer-Verlag, New York.

Reichenbach, H., and M. Dworkin. 1969. Studies on *Stigmatella aurantiaca* (Myxobacterales). *J. Gen. Microbiol.* **58:**3–14.

Reichenbach, H., and M. Dworkin. 1981. The order Myxobacterales, p. 328–355. *In* M. P. Starr et al. (ed.), *The Prokaryotes,* Springer-Verlag, Berlin.

Reichenbach, H., H. K. Galle, and H. H. Heunert. 1975–1976. *Stigmatella aurantiaca* spp. (Myxobacteriales [sic]) Schwarmentwicklung und Fruchtkorperbildung. Encyclop. Cinematogr. Film E2421. Inst. Wiss. Film, Göttingen, Federal Republic of Germany.

Reichenbach, H., H. H. Heunert, and H. Kuczka. 1965a. Schwarmentwicklung and Morphogenese bei Myxobakterien—*Archangium, Myxococcus, Chondrococcus* und *Chondromyces.* Encyclop. Cinematogr. Film C893. Inst. Wiss. Film, Göttingen, Federal Republic of Germany.

Reichenbach, H., H. H. Heunert, and H. Kuczka. 1965b. *Chondromyces apiculatus* (Myxobacterales)— Schwarmentwicklung und Morphogenese. Encyclop. Cinematogr. Film E779. Inst. Wiss. Film, Göttingen, Federal Republic of Germany.

Reichenbach, H., H. H. Heunert, and H. Kuczka. 1965c. *Myxococcus* spp. (Myxobacterales)—Schwarmentwicklung und Bildung von Protocysten. Encyclop. Cinematogr. Film E778. Inst. Wiss. Film, Göttingen, Federal Republic of Germany.

Reichenbach, H., H. H. Heunert, and H. Kuczka. 1965d. *Archangium violaceum* (Myxobacterales)— Schwarmentwicklung und Bildung von Protocysten. Encyclop. Cinematogr. Film E778. Inst. Wiss. Film, Göttingen, Federal Republic of Germany.

Romeo, J. M., and D. R. Zusman. 1987. Cloning of the gene for myxobacterial hemagglutinin and isolation and analysis of structural gene mutations. *J. Bacteriol.* **169:**3801–3808.

Rosenberg, E. (ed.). 1984. *Myxobacteria: Development and Cell Interactions.* Springer-Verlag, New York.

Rosenberg, E., K. Keller, and M. Dworkin. 1977. Cell density-dependent growth of *Myxococcus xanthus* on casein. *J. Bacteriol.* **129:**770–777.

Rosenbluh, A., R. Nir, E. Sahar, and E. Rosenberg. 1989. Cell-density dependent lysis and sporulation of *Myxococcus xanthus* in agarose microbeads. *J. Bacteriol.* **171:**4923–4929.

Rosenbluh, A., and E. Rosenberg. 1989. Autocide AM1 rescues development in *dsg* mutants of *Myxococcus xanthus. J. Bacteriol.* **171:**1513–1518.

Rosenbluh, A., and E. Rosenberg. 1991. Personal communication.

Schopf, J. W., J. M. Hayes, and M. R. Walter. 1983. Evolution of earth's earliest ecosystems: recent progress and unsolved problems. *In* J. W. Schopf (ed.), *Earth's Earliest Biosphere: Its Origin and Evolution.* Princeton University Press, Princeton, N.J.

Shapiro, L. 1985. Generation of polarity during *Caulobacter* differentiation. *Annu. Rev. Cell Biol.* **1:**173–207.

Shimkets, L. J. 1986a. Correlation of energy-dependent cell cohesion with social motility in *Myxococcus xanthus. J. Bacteriol.* **166:**837–841.

Shimkets, L. J. 1986b. Role of cell cohesion in *Myxococcus xanthus* fruiting body formation. *J. Bacteriol.* **166:**842–848.

Shimkets, L. J. 1987. Control of morphogenesis in myxobacteria. *Crit. Rev. Microbiol.* **14:**195–227.

Shimkets, L. J. 1990. Social and developmental biology of the myxobacteria. *Microbiol. Rev.* **54:**473–501.

Shimkets, L. J., and M. Dworkin. 1981. Excreted adenosine is a cell density signal for the initiation of fruiting body formation in *Myxococcus xanthus. Dev. Biol.* **84:**51–60.

Shimkets, L. J., R. E. Gill, and D. Kaiser. 1983. Developmental cell interactions in *Myxococcus xanthus* and the *spoC* locus. *Proc. Natl. Acad. Sci. USA* **80:**1406–1410.

Shimkets, L. J., and D. Kaiser. 1982a. Induction of coordinated cell movement in *Myxococcus xanthus. J. Bacteriol.* **152:**451–461.

Shimkets, L. J., and D. Kaiser. 1982b. Murein components rescue developmental sporulation of *Myxococcus xanthus. J. Bacteriol.* **152:**462–470.

Shimkets, L. J., and H. Rafiee. 1990. CsgA, an extracellular protein essential for *Myxococcus xanthus* development. *J. Bacteriol.* **142:**5299–5306.

Smith, B. A., and M. Dworkin. 1980. Adenylate energy charge during fruiting body formation by *Myxococcus xanthus. J. Bacteriol.* **142:**1007–1009.

Stephens, K., G. D. Hegeman, and D. White. 1982. Pheromone produced by the myxobacterium *Stigmatella aurantiaca. J. Bacteriol.* **149:**739–747.

Stephens, K., and D. White. 1980. Morphogenetic effects of light and guanine derivatives on the fruiting myxobacterium *Stigmatella aurantiaca. J. Bacteriol.* **144:**322–326.

Sudo, S. Z., and M. Dworkin. 1969. Resistance of vegetative cells and microcysts of *Myxococcus xanthus. J. Bacteriol.* **98:**883–887.

Sussman, M., and E. Noel. 1952. An analysis of the aggregation stage in development of the slime molds Dictyosteliaceae. I. The populational distribution of the ability to initiate aggregation. *Biol. Bull.* **103**:259–268.

Teintze, M., T. Feruichi, R. Thomas, M. Inouye, and S. Inouye. 1985. Differential expression of two homologous genes coding for spore-specific proteins in *Myxococcus xanthus*, p. 253–260. *In* J. A. Hoch and P. Setlow (ed.), *Molecular Biology of Microbial Differentiation.* American Society for Microbiology, Washington, D.C.

Varon, M., S. Cohen, and E. Rosenberg. 1984. Autocides produced by *Myxococcus xanthus. J. Bacteriol.* **160**:1146–1160.

Varon, M., A. Teitz, and E. Rosenberg. 1986. *Myxococcus xanthus* autocide AM1. *J. Bacteriol.* **167**:356–361.

Vasquez, G. M., F. Qualls, and D. White. 1985. Morphogenesis of *Stigmatella aurantiaca* fruiting bodies. *J. Bacteriol.* **163**:515–521.

White, D., W. Shropshire, Jr., and K. Stephens. 1980. Photocontrol of development in *Stigmatella aurantiaca. J. Bacteriol.* **142**:1023–1024.

Wireman, J. W., and M. Dworkin. 1975. Morphogenesis and developmental interactions in myxobacteria. *Science* **189**:516–523.

Wireman, J. W., and M. Dworkin. 1977. Developmentally induced autolysis during fruiting body formation by *Myxococcus xanthus. J. Bacteriol.* **129**:796–802.

Wistow, G., L. Summers, and T. Blundell. 1985. *Myxococcus xanthus* spore coat protein S may have a similar structure to vertebrate lens β,γ-crystallins. *Nature* (London) **315**:771–773.

Yee, T., T. Furuichi, S. Inouye, and M. Inouye. 1984. Multicopy single-stranded DNA isolated from a gram-negative bacterium, *Myxococcus xanthus. Cell* **38**:203–209.

Zusman, D. R. 1982. "Frizzy" mutants: a new class of aggregation-defective developmental mutants of *Myxococcus xanthus. J. Bacteriol.* **150**:1430–1437.

Ecological/Colonization Interactions

Microbial Cell-Cell Interactions
Edited by Martin Dworkin
© 1991 American Society for Microbiology, Washington, DC 20005

Chapter 8

Role of Intercellular Chemical Communication in the *Vibrio fischeri*-Monocentrid Fish Symbiosis

Paul V. Dunlap and E. P. Greenberg

INTRODUCTION AND SCOPE

A variety of marine fishes and cephalopods form bioluminescent symbioses with certain species of luminous bacteria. The animal host harbors a pure culture of its bacterial symbiont in specialized glandlike light organs and provides the bacteria with nutrients for growth. In turn, the animal host utilizes the light produced by the bacteria in a wide array of behaviors (Herring and Morin, 1978; Hastings and Nealson, 1981; Dunlap and McFall-Ngai, 1987). These associations have been recognized as having the potential to provide insight into cellular interactions between the bacterial symbiont and its animal host. This potential is based on the observations that the associations are specific (a single bacterial species is found as the symbiont of members of a given family of fishes or cephalopods), the symbiotic bacteria are present as a pure culture, they occur in very high numbers in the light organs, and in many cases the bacteria can be grown in laboratory culture. Furthermore, much has been learned about the biochemistry, physiology, and, more recently, the molecular genetics of bacterial luminescence. Bacterial light production, the defining characteristic of the light organ symbioses, is highly responsive to environmental conditions and therefore provides a window on the symbiosis and a guide in attempts to address more deeply the biochemical, physiological, and genetic interactions between host and symbiont.

Of the three known species of light organ-symbiotic luminous bacteria (*Vibrio fischeri, Photobacterium leiognathi,* and *Photobacterium phosphoreum*), *V. fischeri* is at present the most thoroughly understood with regard to its luminescence system. From this base of knowledge, we address in this chapter what is known about cellular interactions among *V. fischeri* and between *V. fischeri* and its fish host, fishes of the family Monocentridae, especially from the perspective of the control of luminescence. For the most part, the nature of the cellular interactions in this association, including those that influence bacterial light production, remains obscure. However, luminescence in *V. fischeri* is subject to a cell density-dependent autoregulation, a specific and sensitive mechanism that allows *V. fischeri* to sense its local population density and activate transcription of the luminescence genes at high cell density. Since the fish maintains an exceptionally high *V. fischeri* cell density in the light organ, this form of intercellular chemical communication in *V. fischeri* also permits the bacterium to sense the fish host as an environment and respond to it with a high level of the appropriate symbiotic product, bioluminescence. Besides cell density, other physiological factors that could be key features of the light organ environment (e.g., the availability of iron and oxygen) regulate the expression of luminescence in *V. fischeri*. Consequently, a knowledge of the mechanisms by which cell density and these other physiological factors regulate luminescence provides a starting point for understanding the nature of host-symbiont cellular interactions in light organ symbioses. Furthermore, in describing the present state of knowledge of the *V. fischeri*-monocentrid fish symbiosis and emphasizing the mechanism for cell density-dependent autoregulation of luminescence, this chapter has as a goal the stimulation of deeper interest in questions regarding how the association is initiated, how a pure culture is maintained in the fish light organ, and ultimately which *V. fischeri* genes other than those of the luminescence system are involved in the symbiosis. Our view is that this association, because of the depth of knowledge of the luminescence system, can contribute significantly to a general understanding of differential expression of symbiont and host genes upon the interaction of a bacterial symbiont with its animal host.

Consistent with the scope and goal of this chapter, no attempt is made here to review in depth the biochemistry of bacterial luminescence, the general ecology and systematics of luminous bacteria, bioluminescent symbiosis in general, or the burgeoning area of applications of luminescence and the luminescence genes. These subjects have been addressed in recent contributions (Hastings and Nealson, 1981; Hastings et al., 1985; Hastings et al., 1987; Dunlap and McFall-Ngai, 1987; Meighen, 1988), and updates of certain areas have been prepared (Meighen, 1991; Nealson and Hastings, in press). However, comparative information on other bioluminescent symbioses and the luminescence systems of other bacteria is presented here to provide a current and balanced perspective on *V. fischeri*. A brief review of the bacterial bioluminescent reaction follows below. The intent here is to provide a basic background so that the molecular details of the interaction between *V. fischeri* cells important for cell density-dependent autoregulation of luminescence can be fully appreciated as pertaining to the light organ symbiosis.

OVERVIEW OF THE BACTERIAL LUMINESCENCE REACTION

Light emission in *V. fischeri* is catalyzed by bacterial luciferase, a mixed-function oxidase that uses O_2, a long-chain fatty aldehyde, and reduced flavin mononucleotide ($FMNH_2$) as substrates. In the reaction, the O_2 is reduced to H_2O, the $FMNH_2$ and aldehyde are oxidized to FMN and the corresponding long-chain fatty acid, and photons, generally with a wavelength maximum of approximately 490 nm (blue-green), are emitted (Nealson and Hastings, 1979; Hastings et al., 1985). Luciferase synthesis is tightly regulated in *V. fischeri* by a cell density-dependent mechanism termed autoinduction. This mechanism activates the luminescence system when *V. fischeri* cell density is high and keeps cellular luminescence very low when cell density is low (described below).

A possible function for this regulation is to conserve energy when the *V. fischeri* cell density is low, since luminescence appears to be energetically expensive. Apparently the adaptive significance of luminescence relates to conditions in which the cell density of *V. fischeri* is high. Light emission represents an energy expenditure of approximately 6 ATP molecules per photon, assuming an efficiency for the reaction of 100% (Hastings and Nealson, 1977). Estimates of the quantum yield for luciferase range from 1 to 0.1 photon per cycle of the enzymatic reaction (Karl and Nealson, 1980), so depending on the quantum yield in vivo, luminescence can account for 6 to 60 molecules of ATP per photon of light emitted. Because luminescence in *V. fischeri* ranges from 0.01 photon $\cdot s^{-1} \cdot cell^{-1}$ in uninduced cultures to 1,000 photons $\cdot s^{-1} \cdot cell^{-1}$ in fully induced cultures, emission of light can represent a large number of ATP molecules in

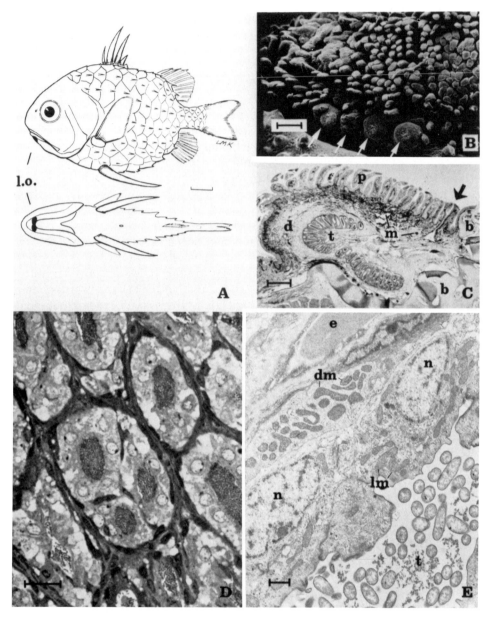

FIGURE 1. The light organ system of the Japanese pinecone fish *M. japonicus*. (A) Line drawing of fish and ventral view of lower jaw, showing location of light organ (l.o.). Bar, 1.0 cm. (B) Scanning electron micrograph of the dorsal surface of the light organ. Numerous dermal papillae can be seen. The emissary ducts from the light organ emerge at the tips of the four large dermal papillae (arrows). Bar, 0.2 mm. (C) Light micrograph of a sagittal section of the lower jaw. m, melanocytes; t, tubules with bacteria; b, mandibular bone; d, dermal layer; p, dermal papillae. Arrow points to emissary duct. Bar, 50 μm. (D) Light micrograph showing the light organ tubules filled with bacteria. Tubules are lined with a single layer of cuboidal epithelial cells that display loose nuclear chromatin and prominent nucleoli supported by connective tissue cells. Blood capillaries are sparse and not readily visible. Bar,

induced cultures. Furthermore, for induced cultures, luciferase accounts for a few to several percent of the soluble cellular protein (Hastings et al., 1965; Henry and Michelson, 1970). Since other proteins are coinduced with luciferase (Michaliszyn and Meighen, 1976; Ne'eman et al., 1977; Boylan et al., 1985), a substantial amount of cellular energy is consumed in the synthesis of proteins. Moreover, up to 20% of the oxygen taken up by fully induced cells may be consumed in the luminescence reaction (Eymers and van Schouwenberg, 1936; Watanabe et al., 1975; Makemson, 1986). Thus, it is reasonable that to conserve energy, expression of the luminescence system in *V. fischeri* would be tightly regulated, with luminescence expressed only under those environmental conditions where it has functional significance. With regard to the significance of the luminescence system and its regulation, the light-emitting enzyme luciferase may function as an alternative electron carrier to the cytochrome system and in so doing promote better survival and growth of *V. fischeri* at low ambient oxygen tensions (Ulitzur et al., 1981).

BIOLOGICAL ASPECTS OF THE MONOCENTRID FISH-*V. FISCHERI* SYMBIOSIS

Systematics, Distribution, and Ecology of Monocentrid Fish

Fishes of the beryciform family Monocentridae (commonly referred to as pinecone fish, knight fish, pineapple fish, and matsukasa-uwo) are placed in two genera, *Monocentris* and *Cleidopus*, with four species, *M. japonicus*, *M. neozelanicus*, *M. reedi*, and *C. gloriamaris*, presently recognized (Kotlyar, 1985). The geographical range for the family extends from northern and central Japan (*M. japonicus*) south to China, the Philippine Islands, and Indonesia (*M. japonicus*), to the east and west coasts of Australia (*C. gloriamaris*) and New Zealand (*M. neozelanicus*), westward to Ceylon, eastern Africa, and the Red Sea (*M. japonicus*), and eastward to the coast of Chile (*M. reedi*) (Herring and Morin, 1978; Kotlyar, 1985, 1988). These nocturnally active fish occur in shallow benthic habitats (generally 5 to 200 m in depth), often in large schools, where they feed primarily on benthic and planktonic crustaceans (Okada, 1955; Schultz, 1956; Graham et al., 1972; Paxton, 1973; Kropach, 1975). Adult monocentrids typically range in standard length from 90 to 150 mm (e.g., Okada, 1955; Schultz, 1956) (Fig. 1), with some specimens attaining a standard length of 230 mm (Stead, 1906). *M. japonicus* specimens in Japan typically are found in rocky reef areas at 20 to 30 m depth, with reproductively mature individuals occurring deeper (75 to 100

---

15 μm. (E) Electron micrograph showing the major features of tubule epithelium. Epithelial cells that make up the lining of the tubules have light-staining mitochondria (lm) with fine cristae. Epithelial cells that are further away from the tubule lumen next to the blood capillaries have dark-staining mitochondria (dm) with thick cristae. t, tubule containing luminous bacteria; e, erythrocyte visible in capillary; n, nucleus of tubule epithelium cells. Bar, 1 μm. Reprinted with permission from Hastings and Nealson, 1981.

m depth) and juveniles inhabiting shallow seagrass beds; gravid females of *M. japonicus* are found in October to November (Haygood, 1984).

Aspects of the Biology of *V. fischeri*

V. fischeri is a gram-negative, facultatively anaerobic marine bacterium that is prototrophic and grows readily and produces luminescence in a variety of culture media containing seawater levels of salts (Nealson and Hastings, 1979; Baumann and Baumann, 1981). This species is commonly encountered in temperate and subtropical coastal waters (e.g., Ruby and Nealson, 1978; Baumann and Baumann, 1981; Hastings and Nealson, 1981), and isolates have been characterized from the light organs of monocentrid fish (Yasaki, 1928; Haneda, 1966; Yoshiba and Haneda, 1967; Yoshiba, 1970; Graham et al., 1972; Ruby and Nealson, 1976; Fitzgerald, 1977) and from the light organs of certain sepiolid squids (Leisman et al., 1980; Hastings et al., 1987; Wei and Young, 1989; Boettcher and Ruby, 1990), where they apparently occur as a pure culture.

Developmental Biology of the *V. fischeri*-Monocentrid Fish Symbiosis

Little information is presently available on the timing of initiation of the symbiotic infection by *V. fischeri*. Because monocentrid (and other fish) light organs are open to the external environment (Fig. 1) and because *V. fischeri* (and other species of luminous bacteria) are present in the water column, it is reasonable to assume that establishment of the symbiosis occurs with each new host generation by infection from the environment. Apparently, eggs of monocentrids do not contain *V. fischeri* (Yamada et al., 1979). Similarly, for leiognathid fish (ponyfish) which harbor *P. leiognathi* as the light organ symbiont, eggs of gravid females apparently do not contain *P. leiognathi* (Dunlap, 1991).

With regard to development of the light organ, larval specimens of *M. japonicus* 11.5 mm in length resemble adult fish in general appearance and bear light organs that emit light (Uchida, 1932; Okada, 1955). However, *M. japonicus* reared from artificially fertilized eggs reached a length of 6 mm in 21 days but lacked light organs and luminescence (Yamada et al., 1979). Thus, development of the light organ and its infection by *V. fischeri* might occur in specimens between these sizes or possibly before the 6-mm stage if a particular habitat or an association with adult fish enriches the presence of *V. fischeri*. In the apogonid fish *Siphamia versicolor,* which bears a light organ, colonized by *P. leiognathi,* that is derived from the mid-gut, larvae 2.8 mm in length contained light organs, but the presence of (unidentified) bacterial cells was first noted in specimens 3.5 mm in length (Leis and Bullock, 1986).

At present, no published information is available for monocentrid fish on whether the presence of *V. fischeri* is necessary for development of the light organ or for survival and maturation of the fish. It is reasonable to assume that, since the bacterial light is used in the fish's feeding behavior (see below), fish that lack the symbiosis would be selected against. Anecdotal evidence from the leiognathid fish symbiosis suggests indirectly that this may be the case. Of a few hundred adult leiognathids examined, only one specimen lacked a robust, brightly luminescent light organ. The light organ of that individual was atrophied (or never developed) and luminous bacteria (*P. leiognathi*) were not present (Dunlap, personal observation). Furthermore, leiognathid fish held in a laboratory aquarium without food retained a normal-appearing light organ with a dense culture of brightly luminescent bacteria up to the point of death by starvation and cannibalism (3 weeks) (Dunlap, 1984a, 1984b). These observations suggest that the light organ symbiosis is an integral and probably indispensable feature of the biology of the fish.

Functions of the Association for the Fish and for *V. fischeri*

As suggested by Stead (1906), monocentrid fish use the luminescence generated from their light organs for feeding. For example, specimens of *M. japonicus* held in a darkened aquarium (observed with dim red light) use the light from their light organs to orient to and ingest live brine shrimp introduced into the aquarium as food (Dunlap, personal observation). It is not known if monocentrids use the light for other functions (e.g., signalling, antipredation), as has been described for certain other bacterially bioluminescent fishes, the anomalopids and leiognathids (Morin et al., 1975; McFall-Ngai and Dunlap, 1983; Dunlap and McFall-Ngai, 1987). The observations of Okada (1926) indicate that monocentrids are able to emit light, day or night, after a period of time in darkness, and that the fish, besides using the light in feeding, can emit light in response to stress. Apparently, there are no published observations of monocentrid fish under natural conditions using luminescence.

The bacteria colonizing the light organs also benefit from the symbiotic association; they are provided with a habitat that supports their growth while excluding other species of bacteria, and through their growth and release from the light organ they are disseminated into other habitats (e.g., seawater, gut tract of fish). There is no evidence that *V. fischeri* cells are digested within the monocentrid fish light organ. However, in the leiognathid fish symbiosis, bacterial "ghost" cells have been observed in a few cases, which suggests that, at least in the *P. leiognathi*-leiognathid fish association, some digestion of the bacteria in the light organ does occur (Bassot, 1975; Dunlap, 1984a).

Light Organ Location, Morphology, and Ultrastructure

The light organs of monocentrid fish have been described by various workers (Yoshizawa, 1916; Okada, 1926; Yasaki, 1928; Haneda, 1966; Herring and Morin, 1978; Tebo et al., 1979). The information given here summarizes these studies. The light organs of *M. japonicus* are a pair of small (3 to 4 mm by 2 to 3 mm) bulbous protuberances at the anterior end of the lower jaw (mandible), surrounded by dark, chromatophore-rich tissue (Fig. 1A). In *C. gloriamaris* the light organs are similar but are located more laterally, are larger, are covered by a concavity in the upper jaw below the eye, and bear a reddish orange filter of skin. Expansion and contraction of the chromatophores apparently controls light emission in monocentrid fish, and *C. gloriamaris* also can block the light with the upper jaw by closing its mouth. The light emitted from these organs shines forward, downward, and laterally, thereby illuminating an area in front of and below the fish. *C. gloriamaris* (and possibly also *M. japonicus* [Yasaki, 1928]) bear a third luminous area in the floor of the anterior portion of the mouth (Graham et al., 1972; Paxton, 1973).

The light organs of monocentrid fish are open to the external environment through emissary ducts (Fig. 1B). Through these ducts, the bacteria are released from the light organ to the external environment (Haygood et al., 1984). Internally, the light organs exhibit an excretory gland-like morphology (Fig. 1C) in that they are composed of numerous tubules formed by epithelial cells rich in mitochondria and Golgi apparatus (Okada, 1926; Tebo et al., 1979) (Fig. 1D and E). The nature of the material released by these cells, if any, is unknown. The numerous smaller tubules (approximately 15 μm in diameter) that compose much of the bulk of the light organ coalesce into larger tubules which join to form reservoirs that empty via the emissary ducts into the seawater (Okada, 1926; Yasaki, 1928; Tebo et al., 1979) (Fig. 1C and D).

The individual tubules are composed of a single layer of cuboidal epithelial cells (Fig. 1D), except that at the basal end the epithelial cells form two, three, or more cell layers. Okada (1926) noted that the epithelial cells lyse, especially those at the basal end of the tubules, and are continually replaced by growth of underlying cells. It is possible that this cell lysate provides the source of nutrition for the bacteria within the tubules. In this regard, the apparently novel and highly specific ability of *V. fischeri* to utilize 3',5'-cyclic AMP (cAMP) as a growth substrate (Dunlap et al., unpublished) is intriguing and may have implications for *V. fischeri*-host cell chemical communication, host provision of growth substrates for *V. fischeri* in the light organ, and the bacterial species specificity of the association.

Alternatively, Nealson (1979) proposed a model in which nutrients (e.g., glucose) from the circulatory system cross through the epithelial cells to the bacteria in the lumina of the tubules. In the Nealson (1979) model, the bacteria metabolize the glucose to pyruvate, which they excrete. The mitochondria of the tubule epithelial cells consume oxygen to metabolize this pyruvate, thereby maintaining the oxygen tension in the tubules at low levels. This model is based on the observations that in culture *V. fischeri* cells growing on glucose release high levels of pyruvate into the medium (Ruby and Nealson, 1977), that under low oxygen levels *V. fischeri* cells grow more slowly and produce more light than under high oxygen (Nealson and Hastings, 1977), and that the epithelial cells of the monocentrid light organ contain numerous mitochondria (Tebo et al., 1979).

Symbiotic State of *V. fischeri* within Monocentrid Fish Light Organs

Within the tubules, the bacteria occur extracellularly (with some exceptions [Tebo et al., 1979]) at high cell density (at least 10^9 to 10^{10} cells·ml^{-1} of light organ fluid) and in close apposition to the host cells (Okada, 1926; Ruby and Nealson, 1976; Tebo et al., 1979) (Fig. 1D and E). It is not known if *V. fischeri* cells within monocentrid light organ tubules are packaged within the light organ saccules observed to constrain cells of *P. leiognathi* in leiognathid light organs

(Dunlap, 1984a) and *P. phosphoreum* in light organs of other fishes (Kishitani, 1930; Yasaki and Haneda, 1935; Haneda, 1938). The bacteria in monocentrid light organ tubules apparently are immotile and lack flagella (Tebo et al., 1979), whereas in culture, *V. fischeri* cells are motile by means of a tuft of polar flagella (Baumann and Baumann, 1981). However, Yasaki (1928) indicated that the bacteria in suspensions released from the organ were motile. The difference in these observations might relate to the length of time the bacterial suspensions sat before being observed by Yasaki (1928). *P. leiognathi* cells in leiognathid light organs also lack flagella, but when released from the light organ they develop flagella and motility within a few hours (Dunlap, 1984a). Cells of *V. fischeri* and *P. leiognathi* in colonies on agar plates are flagellated and actively motile, so it is not yet clear if the absence of flagellation in fish light organs is a cell density effect or if other factors of the light organ might be involved in repressing the synthesis of flagella.

Within the light organs, the bacteria grow slowly compared to growth rates of *V. fischeri* in culture. Tebo et al. (1979) found that the percentage of *V. fischeri* cells in division in the light organ was markedly lower than in culture (2% compared to 10 to 15%, respectively), which is consistent with a slower growth rate in the light organ. The rate of release of bacteria from light organs of *M. japonicus* was found to be 10^6 to 10^7 bacteria·h^{-1} from a total population of 1.5×10^8 cells, giving a doubling time for the population in the light organ ranging from 7.5 to 135 h, markedly longer than for *V. fischeri* in culture (approximately 0.75 h) (Haygood et al., 1984). Similar release rates (1.6×10^6 to 2×10^7 cells·h^{-1}) were obtained for *C. gloriamaris* (Nealson et al., 1984). It is clear from these studies that *V. fischeri* cells in the monocentrid light organ grow more slowly than in culture and that through release of bacteria from the light organ, the fish makes a substantial contribution to the density and dissemination of *V. fischeri* in the environment (Nealson et al., 1984).

Based on these observations, the symbiotic state of *V. fischeri* within monocentrid light organs can be described as one of very high cell density and slow growth. Furthermore, the luminescence system is probably fully induced. This conclusion is based on the observation that the monocentrid light organs emit a high level of light and on the assumption that the concentration of the inducer for luminescence (autoinducer, see below) is high due to the high density of *V. fischeri* cells within the light organ tubules. Moreover, by inference from studies of growth rate and light production of *V. fischeri* in culture, it is possible that the light organ environment also is low in oxygen and iron (Nealson, 1979; Haygood and Nealson, 1985a). Low oxygen and iron restriction both have been shown to restrict the growth rate and promote luminescence of *V. fischeri* in culture (Nealson, 1979; Haygood and Nealson, 1985a). To date, no measurements of the oxygen tension, iron content, or biological components of the light organ fluid have been made. Also, no information yet is available concerning the level of luminescence of *V. fischeri* cells in the light organ. Consequently, it is difficult to speculate beyond these observations as to the nature of physiological conditions within the light organ tubules.

However, information from the leiognathid light organ symbiosis supports the general concept of the symbiotic state of the bacteria in fish light organs being one of restricted growth rate and maximal light production. The bacterial population in leiognathid light organs has been estimated to double in approximately 23 h, 20 to 30 times slower than in culture (Dunlap, 1984a). *P. leiognathi* cells recovered without subculture from leiognathid light organs produced 10- to 100-fold more light and had a markedly lower respiration rate than *P. leiognathi* cells grown in culture (Dunlap, 1984a). In this species, growth and respiration rates in culture were restricted by low osmolarity, which stimulated luminescence to a level essentially identical to that of bacteria taken without subculture from the light organ (Dunlap, 1985). However, osmolarity has little or no differential effect on the growth and luminescence of *V. fischeri* from monocentrid light organs (Dunlap, 1984b). Furthermore, growth under low oxygen does not stimulate luminescence in *P. leiognathi* (Nealson and Hastings, 1977), and the leiognathid fish has been shown to deliver a high concentration of oxygen from its gas bladder to the bacteria in the light organ (McFall-Ngai, 1983; Dunlap and McFall-Ngai, 1987). Consequently, each group of fishes apparently has its own physiological mechanisms for restricting bacterial growth rate while maximizing bacterial light production.

MOLECULAR-GENETIC ANALYSIS OF THE *V. FISCHERI* LUMINESCENCE SYSTEM

Autoinduction of the *V. fischeri* Luminescence System

In *V. fischeri,* luciferase synthesis and its physiological consequence, cellular luminescence, are controlled by a genetic regulatory mechanism called autoinduction (Fig. 2). Control is mediated by the concentration of a metabolite produced by *V. fischeri* during growth, termed autoinducer [*N*-(3-oxo-hexanoyl)homoserine lactone] (Fig. 3), which can diffuse across biological membranes and accumulates gradually in the medium and in cells during growth. When autoinducer reaches a threshold concentration, which generally occurs above a cell density of 10^7 cells·ml^{-1}, it triggers the synthesis of luciferase and other enzymes involved in luminescence. *V. fischeri* autoinducer, which can be thought of as a bacterial pheromone, is species specific; other species of luminous bacteria neither respond to the pure molecule nor produce a compound that stimulates the luminescence of *V. fischeri* (Eberhard, 1972; Nealson, 1977; Rosson and Nealson, 1981; Eberhard et al., 1981; Hastings and Nealson, 1981; Kaplan and Greenberg, 1985; Boettcher and Ruby, 1990).

The autoinduction phenomenon can be observed readily during growth of *V. fischeri* in batch culture (Fig. 2); upon transfer of luminescent cells to fresh medium, luciferase levels per milliliter of culture remain constant for 2 to 4 h (i.e., luciferase synthesis stops), and during this time luminescence levels decrease markedly (10- to 100-fold). After sufficient growth of *V. fischeri* has occurred (i.e., after accumulation of sufficient autoinducer), the lag in luciferase synthesis is followed by a rapid increase, resulting in a 100- to 1,000-fold-higher level of luciferase and a 1,000- to 10,000-fold increase in luminescence over preinduction levels (Coffey, 1967; Kempner and Hanson, 1968; Eberhard, 1972). The lag in

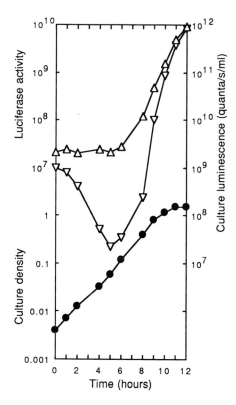

FIGURE 2. Autoinduction of luminescence and luciferase synthesis in *V. fischeri* MJ-1. The *V. fischeri* culture was grown in a seawater complete medium (Dunlap, 1989) at 22°C. Culture density (●) was measured as absorbance at 660 nm, luciferase activity (△) is in relative units per milliliter, and culture luminescence (▽) is reported as quanta per second per milliliter.

luciferase synthesis has been attributed to the requirement for metabolism of an inhibitor of luciferase synthesis present in rich media and for accumulation of sufficient autoinducer to activate transcription of genes encoding luciferase. Both of these activities occur during growth of *V. fischeri* (Kempner and Hanson, 1968; Nealson et al., 1970; Eberhard, 1972). Middleton and Smith (1976) have indicated that the decrease in luminescence before induction of luciferase synthesis relates to an insufficiency of $FMNH_2$ and aldehyde, substrates for luciferase, during the lag period.

Because of the sensitivity of the luminescence system to autoinducer, and because of the permeability of autoinducer and its concentration-dependent effect, autoinducer can be considered as the signal molecule of a system highly responsive to the cell density of *V. fischeri*. In habitats where *V. fischeri* cell density is high, such as in the monocentrid fish light organ ($\sim 10^{10}$ cells·ml^{-1} of light organ fluid

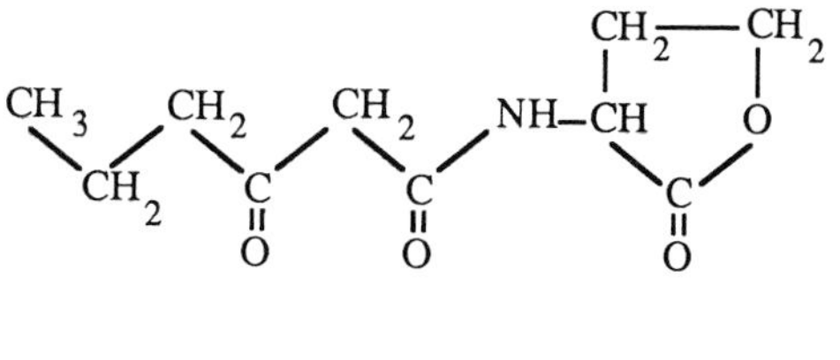

FIGURE 3. Chemical structure of *N*-(3-oxo-hexanoyl)homoserine lactone, the autoinducer of the *V. fischeri* luminescence system.

[Ruby and Nealson, 1976]), autoinducer is thought to accumulate above its threshold concentration and induce the luminescence system. In seawater, where *V. fischeri* occurs at much lower cell densities (i.e., $<10^2$ cells·ml^{-1} [Ruby and Nealson, 1978; Ruby et al., 1980]), autoinducer would not accumulate to a sufficiently high concentration, and induction of luminescence would not be expected. Thus, expression of luminescence is coupled to the symbiotic state (although not exclusively so) in that the monocentrid fish maintains a very high concentration of *V. fischeri* cells within its light organs, a concentration that apparently ensures a high level of induction of the luminescence system.

V. fischeri is known also to enter into light organ symbioses with certain sepiolid squids found in coastal waters of the Hawaiian Islands and Japan (Leisman et al., 1980; Hastings et al., 1987; Wei and Young, 1989; Boettcher and Ruby, 1990). However, differences in the expression of luminescence have been recognized in *V. fischeri* strains from different squids and fishes, and these differences might relate to symbiosis-specific chemical communication between the animal and the bacteria. Isolates, tentatively identified as *V. fischeri,* from the light organs of the Japanese squid *Euprymna morsei* produce a high level of luminescence in culture (Dunlap and Fukasawa, unpublished), whereas strains from the light organs of the Hawaiian squid *Euprymna scolopes* produce 1,000-fold-less light in laboratory culture than in the squid, due to a lower level of luciferase synthesis (Boettcher and Ruby, 1990). The *E. scolopes* strains also produce little or no autoinducer in culture but respond to exogenous autoinducer, which suggests either that autoinducer (or a precursor) is supplied to the bacteria by the squid or that in culture the level of autoinducer synthase (product of the *luxI* gene, see below) is low in these strains (Boettcher and Ruby, 1990). The genes responsible for luminescence (*lux* genes, see below) from the *E. scolopes* strain ES114 have been cloned, and *Escherichia coli* containing these genes on recombinant plasmids expresses a high level of luminescence (Gray and Greenberg, 1991). Thus, a functional autoinducer synthase gene apparently is present in the cloned DNA. This suggests that the host squid might provide a precursor produced by *E. coli* but not produced by the *V. fischeri* strain isolated from the squid or that the squid might in some way influence the synthesis or activity of the *luxI* gene product (Boettcher and Ruby, 1990). The differences between light production by symbionts of *E. scolopes* and *E. morsei* are intriguing, and comparative analyses of these strains at the genetic level could provide further insight into this aspect of possible host-symbiont chemical communication.

In this regard, at present little is known about how autoinducer is made. Concerning whether autoinducer synthesis is constitutive (Nealson, 1977) or inducible, physiological studies with *V. fischeri* led to the suggestion that autoinducer synthesis was under the same transcriptional control as luciferase synthesis (Friedrich and Greenberg, 1983), and this was borne out by molecular genetic studies (Engebrecht et al., 1983). Furthermore, biochemical studies with *V. fischeri* grown in the presence of radioactively labeled methionine, a precursor of autoinducer, indicate that autoinducer is synthesized autocatalytically (Eberhard et al., 1991). Since only one *V. fischeri* gene, the *luxI* gene (the autoinducer synthase gene, see below), is necessary for *E. coli* to synthesize autoinducer

(Engebrecht et al., 1983), the substrates used by autoinducer synthase must be present in *E. coli*. Results of studies with crude enzyme preparations from *V. fischeri* indicate that [α-^{3}H]methionine is incorporated into autoinducer, whereas acetate is poorly incorporated, and that substrates for autoinducer synthase may be *S*-adenosylmethionine and either 3-oxo-hexanoyl coenzyme A or 3-oxo-hexanoyl acyl carrier protein (Eberhard et al., 1991). The connection between these compounds and amino acid and lipid biosynthesis may provide insight into the pathway of autoinducer synthesis. Analogs of the *V. fischeri* autoinducer exhibit a wide range of activities; some are strong agonists of autoinducer and others are antagonists (Eberhard et al., 1986).

Organization of the Luminescence Genes

Rapid advances in molecular and genetic analyses of the *V. fischeri* luminescence system were made possible by the isolation of an 8.8-kb fragment of *V. fischeri* DNA that encodes luminescence enzymes and contains regulatory elements necessary for their expression in *E. coli* (Engebrecht et al., 1983). The strain from which the original DNA clone was obtained, *V. fischeri* MJ-1, was isolated by Ruby and Nealson (1976) from the light organ of a specimen of *M. japonicus*, and, in contrast to strains isolated from the *E. scolopes* light organs (Boettcher and Ruby, 1990), it produces a high level of luminescence in culture. The luminescence genes (*lux* genes) comprise two transcriptional units, *luxR* and the *lux* operon (*luxICDABE*) (operon L and operon R, respectively, in the terminology of Engebrecht et al. [1983]), which are divergently transcribed from an intermediate regulatory region (Fig. 4). In the *lux* operon, *luxI* encodes autoinducer synthase; *luxC, luxD,* and *luxE* specify components of a fatty acid reductase

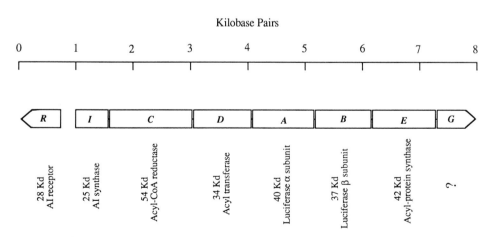

FIGURE 4. The *V. fischeri lux* genes and their products. The products of *luxC, -D,* and *-E* are required for the synthesis of the aldehyde substrate for luciferase. The sequence of *luxG* encodes a polypeptide with a predicted molecular weight of 26,000; however, this product has not been demonstrated to occur in *V. fischeri* or *E. coli* carrying the *lux* genes. In *E. coli, luxG* is not required for luminescence or autoinduction of luminescence. Thus, a function has not been assigned for this gene.

system for synthesis of the aldehyde substrate of luciferase (a reductase, an acyl-transferase, and an acyl protein synthase, respectively); and *luxA* and *luxB* encode the α and β subunits of luciferase (Boylan et al., 1985; Boylan et al., 1989; Engebrecht et al., 1983; Engebrecht and Silverman, 1984). The *luxR* gene encodes a protein (LuxR protein) which, along with autoinducer, is necessary for activation of *lux* operon transcription (Engebrecht et al., 1983). An additional *lux* gene, *luxG*, of undefined function, recently has been identified downstream of *luxE* (Swartzman et al., 1990a; Swartzman et al., 1990b). This gene is not required for expression of the luminescence phenotype in *E. coli*. Downstream of *luxG* in the cloned *V. fischeri* DNA, a bidirectional terminator similar to that defined for the *lux* operon of *V. harveyi* (see below) has been identified (Swartzman et al., 1990a).

The organization of the *lux* genes in the two most thoroughly examined strains of *V. fischeri* (MJ-1 [from the fish *M. japonicus*] and ATCC 7744 [isolated from seawater]) exhibits only minor differences in restriction endonuclease cleavage sites (Engebrecht and Silverman, 1987; Devine et al., 1988). However, the organization of the *lux* genes of other *V. fischeri* strains might differ substantially. As described above, strain ES114 from the squid *E. scolopes* produces a low level of light in culture (Boettcher and Ruby, 1990), yet *E. coli* containing the cloned *lux* genes of this strain is brightly luminescent and produces autoinducer (Gray and Greenberg, 1991). Therefore, a *luxI* gene must be present. The restriction pattern of the cloned ES114 *lux* DNA is quite different from the similar pattern exhibited by the MJ-1 and ATCC 7744 *lux* DNA (Gray and Greenberg, 1991). Consequently, organization of the *lux* genes in *V. fischeri* ES114 may differ from that in *V. fischeri* MJ-1.

Molecular Mechanism of the Cell Density-Dependent Autoinduction of Luminescence in *V. fischeri*

At the molecular level, autoinducer, the signal molecule for cell density-dependent autoinduction of luminescence, is thought to interact with the product of the *luxR* gene (LuxR protein), binding to it and forming a complex that then binds to the *lux* regulatory region (see Fig. 6). The autoinducer-LuxR protein complex is thought then to activate transcription from the *lux* operon promoter (Engebrecht et al., 1983; Engebrecht and Silverman, 1984). Tritiated autoinducer of high specific activity has been synthesized (Kaplan et al., 1985), but formation of an autoinducer-LuxR protein complex has not yet been demonstrated (Kaplan and Greenberg, 1987). Studies with the radiolabeled autoinducer (Kaplan et al., 1985), however, revealed that 1 to 2 molecules of autoinducer per cell give some induction of luminescence in *V. fischeri* and that a maximal rate of induction requires only 40 molecules of autoinducer per cell (Kaplan and Greenberg, 1985). Apparently, the rapid diffusion of autoinducer out of cells (Kaplan and Greenberg, 1985) prevents the basal rate of autoinducer synthesis from leading to induction of *lux* operon transcription at a low cell density.

Point mutational analyses support the hypothesis that the LuxR protein binds both autoinducer and *lux* DNA (Shadel et al., 1990b; Slock et al., 1990). Two clusters of point mutations have been described, thus leading to the suggestion

FIGURE 5. Schematic diagram of the 250-amino-acid-residue LuxR protein. Numbers on bottom indicate amino acid residue number. The four proposed regions, the N-terminal arm, the autoinducer (AI) binding region, the linker, and the DNA binding region, of the polypeptide are labeled.

that there are two domains critical for the activity of the LuxR protein (Fig. 5). One domain, which spans residues 79 to 127 of this 250-amino-acid-residue protein, is the proposed autoinducer binding domain. This proposal was based on the finding that the addition of high levels of autoinducer in vivo resulted in activity of some but not all of the mutant LuxR proteins with amino acid substitutions in this region. The reversal of the mutant phenotype by high concentrations of autoinducer led to the suggestion that these mutations resulted in a decreased affinity for autoinducer (Slock et al., 1990). The second domain spanned residues 184 to 230. Because the amino acid sequence of this region showed significant similarity to the DNA binding domains of a number of other transcriptional regulators, the region was assigned as the DNA binding or DNA recognition domain (Henikoff et al., 1990; Shadel et al., 1990b; Slock et al., 1990). Interestingly, many of the transcriptional regulatory proteins that show sequence similarity with LuxR in the DNA binding regions are a subclass of the so-called two-component (sensor kinase/transcriptional regulator) environmental sensing systems (Miller et al., 1989; Deretic et al., 1989; Henikoff et al., 1990). However, it should be pointed out that the transcriptional regulators in this family of proteins all show similarity in their N-terminal regions, but LuxR does not show this similarity (Henikoff et al., 1990; Slock et al., 1990). This N-terminal region is subject to phosphorylation by the sensor kinase component of the two-component systems. LuxR does not appear to be phosphorylated (Slock et al., 1990). Rather than activation by a phosphate group, it is activated by autoinducer. In the *V. fischeri* luminescence system, then, the receiver and the transmitter of the signal are a single molecule, LuxR. Because the *V. fischeri* autoinducer is produced only by *V. fischeri* and because it rapidly diffuses across cell membranes, the response of the system is specific and sensitive, allowing *V. fischeri* to sense its own local cell density.

Point mutations that map to the N-terminal 79 amino acid residues of LuxR were not identified, nor were point mutations that mapped to the area between the two critical domains (Shadel et al., 1990b; Slock et al., 1990). What role do these regions of the protein play in autoinduction? One can easily imagine that the region between residues 127 and 184 serves as a linker that maintains the two domains in a single polypeptide. The N-terminal arm (Fig. 5) could perhaps serve to block DNA binding or transcriptional activation in the absence of autoinducer; the function of autoinducer binding could be to alter the folding of the N-terminal arm such that it does not hinder the interaction of the DNA binding domain with the putative LuxR protein binding site (Choi and Greenberg, 1991). A prediction of this hypothesis is that the N-terminal arm of the protein is dispensable, and in its absence, autoinducer should not be required for *lux* transcriptional activation.

Although this has not yet been tested completely, the results of a *luxR* 5'-deletion analysis have demonstrated that LuxR proteins without the N-terminal arm show good activity, and in fact LuxR proteins without the N-terminal arm, the proposed autoinducer binding domain, and most of the linker region are capable of *lux* transcriptional activation (Choi and Greenberg, 1991).

DNA sequence analysis, S1 nuclease mapping, and deletion mapping of the cloned *lux* DNA from *V. fischeri* MJ-1 and ATCC 7744 have defined transcriptional start sites and open reading frames for *luxR* and *luxI* flanking a 219-bp regulatory region (Fig. 6). The regulatory region contains at least two promoters. The *luxR* promoter exhibits -10 and -35 regions very similar to consensus se-

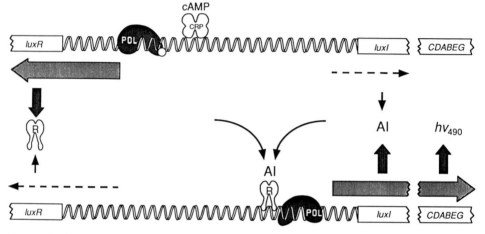

FIGURE 6. The proposed cAMP-CRP/autoinducer-LuxR protein regulatory circuit for control of *V. fischeri lux* gene transcription. There are 219 bp between the *luxR* and *luxI* structural genes (drawn approximately to scale). The start of *luxR* transcription (leftward-pointing horizontal arrows) has been mapped to a site 42 bp upstream of the *luxR* structural gene, and the start of *luxICDABEG* transcription (rightward-pointing horizontal arrows) has been located 21 bp upstream of the *luxI* structural gene. A sequence matching the consensus sequence for the CRP binding site (indicated by cAMP-CRP bound to DNA in upper diagram) has been located at position -59 from the transcriptional start of *luxR* (Engebrecht and Silverman, 1987; Devine et al., 1988), and a mutational analysis indicated that the *luxICDABEG* operator (putative LuxR protein binding site, indicated by AI/R bound to DNA in lower diagram) is a 20-bp palindrome centered around position -40 from the *luxI* transcriptional start site (Devine et al., 1989). As shown in the upper diagram, cAMP-CRP activates transcription of *luxR* (indicated by the thick leftward-pointing arrow) and decreases transcription of *luxICDABEG* to a very low basal level (indicated by the dashed rightward-pointing arrow). RNA polymerase (POL) with sigma factor (white ball) is shown bound at the -10 and -35 regions defining the *luxR* promoter. Activation of *luxR* transcription increases the concentration of the LuxR protein (R) to a sufficient level for interaction with the low concentration of autoinducer (AI) that accumulates due to basal expression of *luxI*. As shown in the lower diagram, the LuxR protein and autoinducer (AI/R) then bind at the *luxICDABEG* operator and activate transcription of *luxICDABEG* (indicated by the thick rightward-pointing arrow). Note that no sigma factor is associated with POL bound at the *luxI* promoter, which contains a -10 region but no obvious -35 region. Transcriptional activation of *luxICDABEG* leads to increased levels of AI and induced levels of luminescence (hv_{490}). LuxR and AI also repress transcription of *luxR* (indicated by the dashed leftward-pointing arrow). This should effect a decrease in cellular levels of the LuxR protein. However, due to the regulatory circuit, this repression will be relieved as the concentration of LuxR protein begins to drop, thereby maintaining LuxR protein at a concentration appropriate for high-level expression of luminescence (Dunlap and Greenberg, 1988).

quences of *E. coli* "housekeeping" promoters, whereas the *lux* operon promoter contains a −10 region, but no −35 region that would be similar to consensus sequences seen in *E. coli* (Fig. 6). Instead, a 20-bp inverted repeat, the putative LuxR protein binding site (termed the *lux* regulon operator [Baldwin et al., 1989; Devine et al., 1989; Shadel et al., 1990a]), is present at this position. Deletion analysis of the region upstream of the *lux* operator provided evidence for a *cis*-acting element involved in repressing *lux* operon transcription in the absence of LuxR and autoinducer (Devine et al., 1989). In addition, a sequence identical to the consensus *E. coli* cAMP receptor protein (CRP) binding site occurs approximately midway between the two promoters (Engebrecht and Silverman, 1987; Devine et al., 1988; Baldwin et al., 1989; Devine et al., 1989) (Fig. 6).

Because the gene for autoinducer synthase (*luxI*) is part of the *lux* operon, autoinducer controls its own synthesis through a positive feedback circuit (Engebrecht et al., 1983; Friedrich and Greenberg, 1983). Furthermore, control of the luminescence system by autoinducer is complex. Expression of *luxR* is negatively autoregulated by autoinducer and the LuxR protein at posttranscriptional (Engebrecht and Silverman, 1986) and transcriptional (Dunlap and Greenberg, 1985, 1988; Dunlap and Ray, 1989) levels. Furthermore, the presence of the *luxI* gene supresses the synthesis of LuxR (Engebrecht and Silverman, 1984; Kaplan and Greenberg, 1987). Regardless of the complexity of the system, the *luxR* and *luxI* genes, as the genetic elements controlling the expression of luminescence, provide the genetic basis for the cell-cell chemical communication system exhibited by *V. fischeri*.

Control of Autoinduction by cAMP

Besides control by autoinducer and LuxR protein, autoinduction of the *V. fischeri* luminescence system is subject to at least one additional form of control at the molecular level, that of a requirement for cAMP and CRP. This form of control apparently provides an interface between the intercellular cell density-sensing mechanism, mediated by autoinducer, and the physiological state of *V. fischeri* cells, mediated by carbon and energy availability. Early studies on cAMP control in *V. fischeri* demonstrated that autoinduction of luminescence exhibits catabolite repression by glucose, but in an apparently atypical fashion. Glucose represses luminescence temporarily in batch culture; however, this repression is not reversed by addition of cAMP, and prior growth of *V. fischeri* on glucose eliminates the repression (Ruby and Nealson, 1976). In contrast, in phosphate-limited chemostat culture, glucose repression is permanent and addition of cAMP (or autoinducer) reverses that repression (Friedrich and Greenberg, 1983).

With the cloning and expression of the *V. fischeri lux* genes in *E. coli*, it became possible to examine more directly the question of an involvement of cAMP and CRP in *lux* gene regulation, through the use of *E. coli* mutants (Dunlap and Greenberg, 1985, 1988). Studies with the cloned *V. fischeri lux* genes in *E. coli* adenylate cyclase (*cya*) and CRP (*crp*) mutants demonstrated that autoinduction of the luminescence system requires cAMP and CRP. The parent (*cya*⁺ *crp*⁺) strain exhibited characteristic autoinduction of the luminescence system and produced levels of luciferase and luminescence similar to those of *V. fischeri*. Compared with the parent strain, the mutants produced very low levels of luciferase and luminescence. Exogenous addition of cAMP stimulated luciferase and luminescence in the *cya* mutant to levels approaching those in the parent, but had no effect in the *crp* mutant. However, as with *V. fischeri*, the temporary glucose repression of luminescence observed in the *E. coli* parent strain was not reversed by exogenous cAMP (Dunlap and Greenberg, 1985, 1988).

Mu dI(*lacZ*) fusion technology was utilized to determine how cAMP and CRP function in this system (Dunlap and Greenberg, 1985, 1988). These studies revealed that cAMP and CRP activate transcription from the *luxR* promoter by 5- to 10-fold (Fig. 6). This activation presumably increases the level of the LuxR protein in cells, potentiating the system for autoinduction (Dunlap and Greenberg, 1985, 1988). Activation of transcription from the *luxR* promoter by cAMP-CRP, however, decreases transcription from the *lux* operon promoter by two- to fivefold (Dunlap and Greenberg, 1985, 1988; Dunlap and Ray, 1989). Conversely, during activation of the *lux* operon transcription, the LuxR protein, along with autoinducer, counters the effect of cAMP-CRP by repressing transcription from the *luxR* promoter, thereby possibly modulating the level of LuxR protein in the cell (Dunlap and Greenberg, 1985, 1988; Dunlap and Ray, 1989) (Fig. 6). The level of LuxR protein in the cell also is modulated by a posttranscriptional *luxR* negative autoregulation (Engebrecht and Silverman, 1986).

A consensus CRP binding site is present in the *lux* regulatory region (Devine et al., 1988; Engebrecht and Silverman, 1987) (Fig. 6), and a cAMP-dependent binding of the *E. coli* CRP to this site has been demonstrated by in vitro footprinting (Shadel et al., 1990a). Furthermore, studies with *cya*-like and *crp*-like mutants of *V. fischeri* have confirmed a physiological requirement for cAMP and CRP in autoinduction of luminescence. Like the *E. coli* mutants, the *V. fischeri* *cya*-like and *crp*-like mutants produced a very low level of luciferase and luminescence, and exogenous addition of cAMP stimulated luciferase synthesis and luminescence in the *cya*-like mutant to parental levels but had no effect in the *crp*-like mutant (Dunlap, 1989). Thus, expression of luminescence in *V. fischeri* is subject to two interacting forms of regulatory control at the molecular level, autoinduction (a cell density sensing mechanism) and cAMP-CRP control (a cellular global regulatory system for sensing of carbon and energy availability). The physiological significance of control by cAMP-CRP is not yet clear.

Much of the information described above on the regulation of *V. fischeri lux* gene expression was obtained from studies of the cloned *lux* genes on multicopy recombinant plasmids in *E. coli*. Recently, methods were developed to mobilize plasmid DNA into *V. fischeri*, and these methods have been combined with gene replacement procedures to construct *V. fischeri lux::lacZ* fusion mutants in which mapped Mu dI(*lacZ*) insertions in the *lux* genes have replaced the chromosomal *lux* genes (Dunlap, 1991). Studies with these strains confirm in part results with the cloned *lux* genes in *E. coli*. *V. fischeri luxC::lacZ* and *luxD::lacZ* fusion mutants exhibited autoinduction of β-galactosidase synthesis analogous to autoinduction of luciferase synthesis in the wild-type strain, and a *luxI::lacZ* fusion mutant required addition of exogenous autoinducer to produce high levels of β-galactosidase. These results demonstrate that the *lacZ* fusions reliably report *lux* gene expression. With regard to cAMP control of expression from the *luxR* and *lux* operon promoters, in a *V. fischeri luxR::lacZ* fusion mutant, β-galactosidase synthesis (transcription from the *luxR* promoter) was decreased in the presence of glucose but was stimulated four- to sixfold by the addition of cAMP, which simultaneously decreased luciferase synthesis (transcription from the *lux* operon promoter) two- to threefold (Dunlap, 1991). These results are similar to those obtained with *E. coli* strains containing plasmids with *lacZ* fusions in the cloned *lux* genes (Dunlap and Greenberg, 1985, 1988; Dunlap and Ray, 1989). The construction of the *V. fischeri* mutants will facilitate the identification of other cellular and environmental factors that control *lux* gene expression in the native organism. Furthermore, through application of the plasmid mobilization and transposon mutagenesis procedures it should be possible to identify other *V. fischeri* genes whose expression is controlled directly by autoinducer (and the LuxR protein) and by factors other than autoinducer, such as host fish (or squid) chemicals and growth conditions of the light organ environment (Dunlap, unpublished). These studies should provide additional insight into aspects of chemical communication between the host fish and *V. fischeri*.

On the basis of results from studies with *E. coli* and with *V. fischeri lux::lacZ* fusion mutants, the atypical nature of catabolite repression of luminescence in *V. fischeri* described above could result in part from the repressive action of cAMP-CRP on *lux* operon transcription (and thereby on autoinducer synthesis), delaying the accumulation of autoinducer. In the presence of glucose, cAMP levels would be lower, and this would delay synthesis of LuxR protein. The lower level of LuxR protein would delay induction of *lux* operon transcription even though autoinducer is accumulating at a normal rate. This notion assumes that the LuxR protein, like autoinducer, must reach a threshold concentration in cells in order for autoinduction of *lux* operon transcription to occur. Conversely, addition of cAMP would stimulate transcription from the *luxR* promoter, increasing the level of LuxR protein, but would repress the basal level of *lux* operon transcription, decreasing the rate of autoinducer synthesis. Consequently, autoinduction of *lux* operon transcription would still be delayed, giving the appearance that exogenous addition of cAMP had no effect. However, these considerations are not fully satisfying as to the basis for the atypical catabolite repression of luminescence; other effects of glucose appear to be involved (Dunlap, 1991), and the contrast in results with batch- and chemostat-grown *V. fischeri* cells is perplexing.

Furthermore, it is not yet clear how cAMP-CRP represses *lux* operon transcription. Since this effect occurs both in the presence and in the absence of the LuxR protein (Dunlap and Greenberg, 1985, 1988), and since the CRP binding site (positions 92 to 112, counting rightward from the start of *luxR* transcription; see Fig. 6) and the LuxR protein binding site (positions 148 to 168; Fig. 6) (Devine et al., 1988) are separated by 36 bp, it is possible that the mild effect of cAMP-CRP on *lux* operon transcription results simply from a preferential association of RNA polymerase with the *luxR* promoter when the cAMP-CRP complex is bound in the regulatory region, rather than from a steric interaction between cAMP-CRP and autoinducer-LuxR. Similarly, the negative autoregulation of transcription from the *luxR* promoter by autoinducer-LuxR protein could result from a preferential association of RNA polymerase with the *lux* operon promoter when the putative autoinducer-LuxR complex is bound in the regulatory region. In vitro runoff transcription studies would be valuable to help determine if this is the case or if there may in fact be a steric interaction between these two antagonistically acting regulatory proteins. Additional unresolved aspects of *lux* gene regulation are the mechanism of posttranscriptional negative autoregulation of *luxR* (Engebrecht and Silverman, 1986), the possibility of the involvement of DNA looping (Devine et al., 1989), and whether autoinducer interacts directly with the LuxR protein or instead operates through a second messenger cascade system at the cell membrane. Figure 6 is a model summarizing current understanding of *lux* gene transcriptional control and showing the positive and negative transcriptional effects of cAMP-CRP and autoinducer-LuxR.

A general similarity in regulatory motifs is seen for the *V. fischeri lux* system and certain *E. coli* operons, such as the *araC araBAD* and *malT malPQ* systems (Dunlap and Greenberg, 1985; Baldwin et al., 1989; Devine et al., 1989). More striking, however, is the similarity between the *lux* system and the *E. coli* threonine dehydratase (*tdc*) operon (Schweizer and Datta, 1989). In both *lux* and *tdc*, a gene (*luxR* and *tdcR,* respectively) encoding a *trans*-acting positive regulatory protein is transcribed divergently, via an intermediate regulatory region, from an operon that encodes the light-producing or catabolic enzymes. The *tdcR* gene promoter, like that for *luxR,* contains a recognizable −35 region (Schweizer and Datta, 1989). In both cases, the positive activator protein is thought to bind to a physiologically produced inducer (i.e., an autoinducer sensu Nealson, 1977), recognize and bind to a specific DNA sequence, and activate transcription of the operon encoding light-producing or catabolic enzymes. In both cases, the regulatory region contains a consensus CRP binding site. However, relative to the positions of the binding sites for LuxR protein and CRP in the *lux* regulatory region, the positions of the binding sites for TdcR protein and CRP in the *tdc* regulatory region are reversed, and cAMP and CRP apparently activate transcription of *tdcABC* instead of *tdcR* (Schweizer and Datta, 1989). Thus, the mechanics of regulation differ substantially in these two systems. At present the *tdc* inducer is not known, nor is it known if cells are permeable to the inducer and therefore if the inducer could mediate a cell density-dependent induction like that of the *lux* system. Compared to the *E. coli* operons mentioned, unique features of the *lux* system include its function (light production instead of catabolism of sugars or an amino acid) and the inclusion of a regulatory gene (*luxI*, autoinducer synthase gene) in the operon encoding the light-generating enzymes. In this regard, divergent promoters represent a general type of gene organization found in many procaryotic and eucaryotic organisms and their viruses (Beck and Warren, 1988). Since, in our view, both *luxR* and *luxI* encode polypeptides involved in regulation, the *V. fischeri* divergent *lux* promoter arrangement apparently is of the regulatory-regulatory type (R-R, in the terminology of Beck and Warren [1988]), a type previously not known to occur in procaryotes (Beck and Warren, 1988). Furthermore, since LuxR and autoinducer both repress transcription from the *luxR* promoter and activate transcription from the *luxICDABEG* promoter, this system can be described more specifically as $R_{I/A}$-$R_{I/A}$ (i.e., repressor/activator) in the terminology of Beck and Warren (1988).

Other Factors Affecting Autoinduction of the *V. fischeri* Luminescence System

In culture, the growth and luminescence of *V. fischeri* are differentially affected by oxygen and iron in ways that may have importance for understanding control of these activities in the symbiosis. At low oxygen concentrations, *V. fischeri* cultures grow more slowly than fully aerated cultures, but induction of luciferase synthesis occurs at a lower cell density (Nealson and Hastings, 1977). Consequently, cultures grown under low oxygen conditions contain more luciferase. The mechanism by which growth under low oxygen effects this response is not yet known. Based on these observations, Nealson (1979) proposed that oxygen limitation may be a control factor in the monocentrid symbiosis (see above).

For iron, its addition to low-iron minimal medium delays induction of luciferase synthesis, whereas in complete medium, restriction of iron availability by addition of iron chelators results in slower growth and induction of luminescence and luciferase synthesis at a lower cell density (Haygood and Nealson, 1985a). Thus, as for cultures grown under low oxygen, cultures grown under iron restriction have a higher luciferase content. The iron repression of luminescence in *V. fischeri* was suggested to involve either iron interfering with autoinducer transport or the activity of an iron-binding repressor protein that blocks *lux* operon transcription (Haygood and Nealson, 1985b). However, autoinducer can enter *V. fischeri* cells by simple diffusion (Kaplan and Greenberg, 1985), so iron is not likely to interfere with its entry into cells. Recent studies with *E. coli* iron transport (*tonB*) and iron regulatory (*fur*) mutants containing the *V. fischeri lux* genes on recombinant plasmids, however, indicate that the availability of iron affects expression of the luminescence system indirectly, possibly through an influence on DNA content, rather than through a direct control of transcription from the *luxR* or *lux* operon promoters (Dunlap, unpublished). In the symbiosis, restriction of iron availability by the host fish could be an important way of limiting the growth of *V. fischeri* while maintaining a high level of bacterial light production (Haygood and Nealson, 1985a). Thus either oxygen or iron restriction, or both, could be functioning in the *V. fischeri*-monocentrid fish symbiosis as a host-mediated regulatory factor.

Possible Involvement of *lexA* and *htpR*

Besides the regulatory factors described above, other genetic regulatory factors might be involved in controlling *lux* gene expression. For example, the putative LuxR protein binding site (Fig. 6), a 20-bp inverted repeat in the *lux* regulatory region, bears strong sequence similarity to the *E. coli* LexA protein recognition sequence (Ulitzur and Kuhn, 1988; Baldwin et al., 1989). Binding of *E. coli* LexA protein to this region has been demonstrated by DNase I protection experiments (Shadel et al., 1990a). It is possible that an interaction between LuxR protein and a *V. fischeri* LexA protein plays a role in controlling *lux* operon transcription (Shadel et al., 1990a). Alternatively, the similarity between the sequences of the putative LuxR protein binding site and the LexA protein binding site could be accidental. At present, no published information is available on the SOS response or a LexA protein in *V. fischeri*. In addition to LexA, a possible promoter site typical of σ^{32} has been identified within the *luxI* gene; it has been suggested that the HtpR protein plays a role in *lux* gene regulation (Ulitzur and Kuhn, 1988; Ulitzur, 1989), but this issue remains to be resolved.

COMPARISON OF *V. FISCHERI lux* GENE ORGANIZATION AND REGULATION WITH THAT OF OTHER SPECIES

The *V. harveyi* Luminescence System

Besides studies of *lux* gene organization and regulation in *V. fischeri,* work on the luminescence system of *V. harveyi* has advanced rapidly in the past several years. Substantial similarities and differences exist between the *V. fischeri* and *V. harveyi* luminescence systems with regard to *lux* gene organization and regulation. Because *V. harveyi* is a "free-living" marine bacterium (i.e., not known to enter into light organ symbiosis with an animal host), it provides a valuable comparison for the information from *V. fischeri,* and these similarities and differences are presented in some detail here.

V. harveyi is common in coastal waters and sediments, generally in areas somewhat warmer than those in which *V. fischeri* is found (Nealson and Hastings, 1979; Baumann and Baumann, 1981). In *V. harveyi,* as in *V. fischeri,* luminescence is controlled by a cell density-dependent autoinduction at the level of transcription (Nealson et al., 1970; Eberhard, 1972; Ulitzur and Hastings, 1979; Rosson and Nealson, 1981; Barak and Ulitzur, 1981). Luminescence in *V. harveyi* also exhibits catabolite repression by glucose, but in this species glucose repression of luminescence is permanent in batch culture and is reversible by cAMP (Nealson et al., 1972). A mutant of *V. harveyi* that requires cAMP for luminescence has been isolated (Ulitzur and Yashphe, 1975), and CRP from *V. harveyi* has been purified and characterized (Chen et al., 1985). Iron can repress luminescence in this species and is involved in some way in catabolite repression (Makemson and Hastings, 1982). During induction, other polypeptides, both cytoplasmic and membrane associated, are coinduced with luciferase (Michaliszyn and Meighen, 1976; Ne'eman et al., 1977).

With regard to the cellular location of luciferase in *V. harveyi* and other species, some controversy exists. Although bacterial luciferases in general are found in the cytoplasmic fraction of cell extracts, the apparent relationship to the electron transport system (Ulitzur et al., 1981) suggests the possibility of association with the cytoplasmic membrane. Support for a membrane association was obtained for the luciferases of *P. leiognathi* (Balakrishnan and Langerman, 1977) and *V. harveyi* and *V. fischeri* (Angell et al., 1989). However, in another study, the luciferase of *V. harveyi* was shown to be cytoplasmic, with no indication of a loose membrane association or association with a membrane-bound protein complex (Colepicolo et al., 1989b). This result is consistent with the apparent lack of a leader sequence on the α or β subunit of the *V. harveyi* luciferase (Colepicolo et al., 1989b; Cohn et al., 1985; Johnston et al., 1986). Recently, however, Tn*phoA* fusion technology was applied to the question of the cellular location of the *V. fischeri lux* gene products (Kolibachuk and Greenberg, unpublished). Tn*phoA* creates fusions between a target gene and the gene encoding the periplasmic enzyme alkaline phosphatase such that if the target gene is exported from the cytoplasm into the cytoplasmic membrane or into the periplasm, the alkaline phosphatase becomes active. *E. coli* strains containing Tn*phoA* fusions in the *V. fischeri luxA* gene (encoding the α subunit of luciferase) and in *luxR* (encoding the autoinducer receptor) exhibited alkaline phosphatase activity, which suggests that these proteins associate with the cytoplasmic membrane. This approach may ultimately prove useful in resolving the controversy about the cellular location of luciferase (Kolibachuk and Greenberg, unpublished). Resolution of this ongoing controversy may have significance for the physiological role of luciferase and possibly also for *lux* gene regulation, if the LuxR protein is confirmed as being membrane associated.

With regard to autoinducer, *V. harveyi*, like *V. fischeri*, produces during growth a factor that accumulates in the medium and stimulates expression of the luminescence system (Eberhard, 1972). In contrast to the *V. fischeri* autoinducer, the *V. harveyi* factor is heat sensitive (Eberhard, 1972) and is also produced by some but not all other marine bacteria, including nonluminous species (it is not produced by *E. coli*) (Greenberg et al., 1979). The *V. fischeri* autoinducer has no effect on *V. harveyi* (Eberhard, 1972; Greenberg et al., 1979). Recently, a molecule that stimulates luminescence in *V. harveyi* was purified from *V. harveyi*-conditioned media and was identified as *N*-(β-hydroxybutyryl)homoserine lactone (Cao and Meighen, 1989). The chemical similarity of this molecule to the *V. fischeri* autoinducer, *N*-(3-oxo-hexanoyl)homoserine lactone, is striking. Chemically synthesized *N*-(β-hydroxybutyryl)homoserine lactone produces a detectable stimulation of luminescence in *V. harveyi* at 0.1 mg·ml^{-1} (0.53 μM), and the effect is saturated (5,000-fold stimulaton of luminescence) at 100 mg·ml^{-1} (0.53 mM) (Cao and Meighen, 1989). These concentrations are approximately 100- to 250-fold higher than the minimal effective and saturating concentrations typical of the *V. fischeri* autoinducer (approximately 2 to 5 nM and 0.2 to 5 μM, respectively) (Eberhard et al., 1981; Kaplan and Greenberg, 1985; Dunlap and Ray, 1989; Eberhard, personal communication).

The arrangement of genes in *V. harveyi* that encodes the three fatty acid reductase polypeptides and the α and β subunits of luciferase (*luxCDABE*) is identical to that in *V. fischeri* (Miyamoto et al., 1988a), and substantial sequence similarity exists at the DNA and protein levels (e.g., Johnston et al., 1989). Each system contains a newly described *lux* gene, *luxG* (also present in *P. phosphoreum*), which as yet has no known function (Swartzman et al., 1990a; Swartzman et al., 1990b). The *V. harveyi lux* operon also contains another newly identified gene, *luxH,* such that the organization of genes in the *V. harveyi lux* operon is *luxCDABEGH;* neither of the new genes is necessary for luminescence or for regulation of *lux* operon expression (Swartzman et al., 1990b). A classical rho-independent terminator was identified downstream of *luxH* in *V. harveyi* and *luxG* in *V. fischeri* (Swartzman et al., 1990a; Swartzman et al., 1990b). Furthermore, a potential stem-loop structure, possibly involved in mRNA stability, occurs in the sequence between *luxB* and *luxE* in *V. harveyi* and *V. fischeri* (Sugihara and Baldwin, 1988; Baldwin et al., 1989). Removal of this region in *V. harveyi* decreases luciferase expression (Sugihara and Baldwin, 1988).

With regard to transcription initiation of the *luxCDABE(GH)* genes in *V. harveyi,* several mRNA species are produced by this DNA (Miyamoto et al., 1985); the expression of luminescence by *E. coli* containing fragments of *V. harveyi luxAB* DNA up to 18 kb long (thereby encompassing at least *luxCDABE* [and *luxGH*]) is constitutive, and light levels are low except when the cloned insert is under the control of a strong promoter and aldehyde is added (Belas et al., 1982; Cohn et al., 1983; Baldwin et al., 1984; Gupta et al., 1986), when the DNA is present in certain *E. coli* mutants (Miyamoto et al., 1987), or when expression is under the control of T7 RNA polymerase (Miyamoto et al., 1988b). The results of the latter studies suggest that low expression of luminescence by *E. coli* containing *V. harveyi* DNA relates to difficulties in transcription of *V. harveyi* DNA by *E. coli* RNA polymerase (Miyamoto et al., 1988b).

In *V. harveyi,* the expression of the fatty acid reductase enzymes is regulated by glucose and cAMP in the same way as that of luciferase (Byers et al., 1988). Consistent with these data, a promoter upstream of *luxC* was identified and was found to be regulated by cell density and by glucose (Miyamoto et al., 1990). The promoter contained a typical − 10 region but no recognizable − 35 region, and *lux* operon mRNA started 26 bases before the translation initiation codon for *luxC* (Swartzman et al., 1990b). Synthesis of mRNA encoded by this region was found to be inducible in a cell density-dependent manner (Swartzman et al., 1990b).

An additional difference between the *V. fischeri* and *V. harveyi* systems is that the region upstream of *luxC* in *V. harveyi* bears only slight resemblance to the *V. fischeri luxR*-regulatory region-*luxI* arrangement. A leftward-oriented open reading frame of approximately 410 bp occurs 640 bp upstream of *luxC* in *V. harveyi* (compared to approximately 850 bp for *luxR* in *V. fischeri*), and it is preceded by a potential CRP binding site approximately 200 bp from the start of the open reading frame (compared to approximately 100 bp in *V. fischeri*) (Miyamoto et al., 1988b; Engebrecht and Silverman, 1987; Baldwin et al., 1989). However, the region adjacent to *luxCDABE* that corresponds to the *V. fischeri luxI* gene contains numerous stop codons on both strands in each reading frame (Miyamoto et al., 1988b), so a comparably located *luxI* gene apparently does not exist in *V. harveyi* (Miyamoto et al., 1988a). The lack of a *luxI* gene upstream of *luxC* is consistent with the location of the *luxCDABEGH* promoter just upstream of *luxC* (Swartzman et al., 1990b).

Attempts to identify regions of the *V. harveyi* chromosome necessary for a high level of luminescence expression, using complementation between a 13-kb fragment of *V. harveyi* DNA encompassing the *luxCDABE* genes and shotgun-cloned genomic DNA, were unsuccessful (Miyamoto et al., 1988a). However, Martin et al. (1989), using transposon mutagenesis, identified two regions of the *V. harveyi* chromosome involved in luminescence: region I, which encodes *luxCDABE,* and region II, which lies at least 10 kb from *luxCDABE*. Expression of β-galactosidase from *lacZ* fusions in region I exhibits an induction pattern and response to conditioned medium consistent with control by cell density-dependent accumulation of autoinducer during growth (Martin et al., 1989). Region II is required for this induction and encodes a *trans*-acting regulatory function analogous to that of the *V. fischeri luxR* gene (Showalter et al., 1990). The *V. harveyi luxR* gene contains a region with an amino acid sequence similar to that of the DNA-binding domain of Cro-like proteins, but there is no DNA or amino acid sequence similarity between the *luxR* of *V. harveyi* and that of *V. fischeri* (Showalter et al., 1990). Cells containing transposon insertions in region II produce autoinducer (Martin et al., 1989), so the region II regulatory function is thought to be other than one involved in autoinducer synthesis (i.e., other than a *luxI* gene). None of the 54 Lux⁻ mutants of *V. harveyi* generated by transposon insertion mutagenesis (of approximately 30,000 mutants recovered) was defective in autoinducer synthesis or in synthesis of cAMP or CRP (Martin et al., 1989), although mutants defective in autoinducer synthesis and cAMP synthesis have been isolated using chemical mutagenesis (Ulitzur and Yashphe, 1975; Cao and Meighen, 1989; Cao et al., 1989). It is possible that the gene for autoinducer synthesis is located in a third region of the *V. harveyi* chromosome.

In summary, there are a number of similarities and differences in *lux* gene organization and regulation between *V. fischeri* and *V. harveyi*. It is not yet clear if these differences relate to ecological differences between these species.

The Luminescence Systems of Other Bacterial Species

Much less is presently known about *lux* gene organization and regulation in other luminous bacteria. For *P. phosphoreum* and *P. leiognathi,* marine bacteria found both free-living and as the species-specific bioluminescent symbionts of certain fishes and squids (Dunlap and McFall-Ngai, 1987; Hastings et al., 1987), luminescence is subject to control by iron, osmolarity of the growth medium, Na⁺, and other factors (Haygood and Nealson, 1985a, 1985b; Dunlap, 1984b, 1985; Hastings et al., 1985; Watanabe and Hastings, 1986). An involvement of the outer membrane in regulation of luminescence in *P. phosphoreum* has been described (Lummen and Winkler, 1986). The putative autoinducers of these species have not been characterized, although substantial evidence for production of an autoinducer by *P. leiognathi* strains from light organs of leiognathid fish has been obtained (Dunlap, 1984b). Luciferase and the fatty acid reductase enzymes are induced at high cell density in some strains of *P. leiognathi* and *P. phosphoreum* (Wall et al., 1984; Delong et al., 1987; Illarionov and Protopopova, 1987), but luciferase expression is constitutive in others (Katznelson and Ulitzur, 1977; Watanabe and Hastings, 1986). Using a T7 promoter system to express cloned *P. phosphoreum* DNA, Mancini and co-workers (Mancini et al., 1988; Mancini et al., 1989) demonstrated the presence of polypeptides of the fatty acid reductase and luciferase components, plus a novel 26-kDa *lux* polypeptide. The corresponding genes occurred in the sequence *luxCDABFE,* with *luxF* encoding the novel polypeptide. However, no polypeptides from DNA upstream of *luxC* were detected. It is possible that, as in *V. harveyi,* regulatory genes corresponding to the *V. fischeri luxR* and *luxI* genes may not be present in this location in *P. phosphoreum* (Mancini et al., 1988; Mancini et al., 1989). A similar arrangement of *lux* genes including an analogous novel gene, *luxN*, between *luxB* and *luxE*, was found for *P. leiognathi (luxABNE)* (Illarionov et al., 1988; Mancini et al., 1989; Baldwin et al., 1989).

The predicted amino acid sequence of the LuxF polypeptide is similar to the known amino acid sequences of the α and β subunits of luciferase, with greater identity (approximately 30%) to the β subunit and somewhat greater similarity to the β subunits from *V. fischeri* and *V. harveyi* than to the β subunit of *P. phosphoreum* (Cohn et al., 1985; Johnston et al., 1986; Soly et al., 1988). These similarities suggest that *luxF* arose by gene duplication of *luxB* (Soly et al., 1988), as has been proposed for *luxA* and *luxB* (Baldwin et al., 1979; Foran and Brown, 1988). With regard to the function of LuxF and LuxN, the N-terminal sequences of recently described nonfluorescent flavoproteins from *P. phosphoreum* and *P. leiognathi* (O'Kane et al., 1987; Kasai et al., 1987) are strikingly similar to the first 35 amino acid residues of these polypeptides, so *luxF* and *luxN* quite probably encode nonfluorescent flavoproteins (Soly et al., 1988; Mancini et al., 1989; Baldwin et al., 1989). Consistent with this, the nonfluorescent proteins are produced coordinately with luciferase (Kasai et al., 1987). The location of the *luxF* and *luxN* genes and their sequence similarity to genes encoding the α and β subunits of luciferase suggest that they could play a role in the luminescence reaction in those strains in which they are found. It is known in this regard, however, that DNA without apparent function can be maintained and even amplified in bacteria. For example, *P. phosphoreum* contains repetitive, apparently functionless pseudogenes for tRNA (Giroux and Cedergren, 1989). A detailed screening for the distribution of this new gene in different strains of *P. phosphoreum* and *P. leiognathi* and a survey for its presence in different species of luminous bacteria would be valuable as a basis for evaluating *lux* gene evolution.

For *Xenorhabdus luminescens,* a terrestrial luminous bacterium found free in soil and as a symbiont of nematodes that infect insect larvae, luminescence and luciferase exhibit a lag in expression followed by a rapid increase at high cell density, a pattern that is consistent with autoinduction (Colepicolo et al., 1989a; Schmidt et al., 1989). Growth and luminescence are not differentially controlled by oxygen or by iron in this species (Colepicolo et al., 1989a). Results of studies on *X. luminescens lux* gene organization and expression are consistent with the presence of the *luxABCDE* genes, with the genes involved in aldehyde synthesis flanking those for luciferase (Frackman et al., 1990; Johnston et al., 1990; Szittner and Meighen, 1990).

A number of other species of luminous bacteria have been described, including *Vibrio logei, Vibrio orientalis, Vibrio splendidus, Shewanella (Alteromonas) hanedai,* and luminous strains of *Vibrio cholerae* and *Vibrio vulnificus* (Baumann and Baumann, 1981; Hastings and Nealson, 1981). At present, very little is known about the ecology of these bacteria or the control of their luminescence systems. They present fertile ground for comparative analyses of *lux* gene organization and regulation, and their study may provide additional insight into the physiological function and ecological role of bacterial luminescence.

FUTURE DIRECTIONS

In this chapter, we have presented an overview of intercellular chemical communication involved in the light organ symbiosis of *V. fischeri* with monocentrid fish. Although much has been learned in the past several years about this association, especially in the area of cell density-dependent *lux* gene regulation, our understanding of the nature and dynamics of cellular interactions between *V. fischeri* cells in the fish light organ and between cells of *V. fischeri* and cells of the host fish is clearly at the beginning stage. However, with the application of molecular approaches to the biology of *V. fischeri* and with the growing interest in fish and squid light organ symbiosis as a representative type of procaryote-eucaryote interaction, we feel that in the next several years understanding of these associations will expand broadly and deeply, particularly in the following areas.

The Infection Process

Sepiolid squids (*E. scolopes*) can be hatched aseptically and reared under laboratory conditions (Wei and Young, 1989; Boettcher and Ruby, 1990). This permits infection studies to be carried out in which juvenile, aposymbiotic animals are presented with different species or strains of luminous bacteria, such that the infection process and the factors that influence it can be examined.

Species Specificity and the Pure Culture State of the Association

At present, it is not clear how the light organs of the members of a given family of fish or squid come to harbor a pure culture of luminous bacteria, and this question is related to the infection process. Besides the possible nutritionally based specificity with cAMP described above, *V. fischeri* might elaborate adhesins specific for receptors on the surfaces of fish cells. It is known in this regard that enteric bacteria synthesize several distinct types of adhesins involved in bacterial attachment to different host surfaces and that certain of these may be induced by host factors (Finlay and Falkow, 1989; Finlay et al., 1989). Less specific but potentially equally important for this question may be the physiological conditions of the light organ, activity of the host immune system, and enrichment of the environment with the specific bacterial types from adult fish or squid with an established symbiosis.

Presence and Function of Other Bacterial Genes Involved in the Symbiosis

Some evidence exists that there should be genes other than those of the luminescence system responsive to the symbiosis. For example, the bacterial cells in the light organ exhibit distinct morphological differences from the bacteria grown in culture. As described above, they lack flagella in the light organ, and they are generally larger and more coccobacilloid than in culture (e.g., Tebo et al., 1979; Dunlap, 1984a, 1985). Furthermore, it is a reasonable expectation that genes encoding bacterial adhesins, as mentioned above, are involved in the symbiosis and that their expression is regulated in some way by the symbiosis.

Role of the Bacteria in Ontogeny of the Light Organ

The presence of the bacteria could feasibly play a role, through chemical signalling and physical contact between cells, in host developmental processes involved in light organ tissue differentiation. Work in this area will depend on rearing and development of the host animal and would benefit from development of genetic methods for assessing host gene expression.

Regulation of Bacterial Gene Expression in Response to the Host

A number of recent studies indicate that gene expression in pathogenic bacteria can be regulated in response to the host animal. Studies with several bacterial pathogens have demonstrated the presence of specific mechanisms that allow the bacteria to sense the host environment and respond by differential gene expression (Miller et al., 1989; Deretic et al., 1989). In the case of *Salmonella typhimurium*, attachment and invasion of host cells requires de novo synthesis of several bacterial proteins and depends in part on the growth state of the bacteria (Finlay et al., 1989; Lee and Falkow, 1990). Furthermore, the availability of iron and oxygen, which affect virulence in various pathogens (Brown and Williams, 1985; Lee and Falkow, 1990), influences the luminescence of *V. fischeri;* iron and oxygen may control other *V. fischeri* genes involved in the symbiosis as well. Clearly, the bacteria that establish specific, nonpathogenic, bioluminescent symbioses with marine fish and squids respond in a fashion generally similar to that of bacteria pathogenic in animals by differentially controlling the expression of genes in response to the host animal. Our understanding of the molecular basis of this differential gene expression in luminous bacteria, coupled with the specificity of the interactions between luminous bacteria and the marine animal host, establishes these systems as useful models for understanding regulation of bacterial gene expression in response to the host.

Clearly, work on questions dealing with the infection process, species specificity, bacterial involvement in light organ development, and the presence and regulation of bacterial genes involved in the symbiosis is at an initial stage, with many areas as yet unaddressed. The next several years, however, will undoubtedly provide much additional insight into cell-cell interactions and chemical signalling between the luminous bacteria and their animal hosts.

Acknowledgments. This chapter is dedicated to Professor Yata Haneda in recognition of his fundamental contributions in the areas of bioluminescence and bacterial symbiosis.

We acknowledge the support of the New York Sea Grant and the Office of Naval Research. We thank K. H. Nealson and Springer-Verlag for permission to use Fig. 1, E. A. Meighen for information on *luxG*, and A. Eberhard for information on autoinducer synthesis prior to publication.

This paper is contribution no. 7524 from the Woods Hole Oceanographic Institution.

REFERENCES

Angell, P., D. Langley, and A. H. L. Chamberlain. 1989. Localisation of luciferase in luminous marine bacteria by gold immunocytochemical labelling. *FEMS Microbiol. Lett.* **65:**177–182.

Balakrishnan, C. V., and N. Langerman. 1977. A glycoprotein with luciferase activity isolated from *Photobacterium leiognathi*. Arch. Biochem. Biophys. **181:**680–682.

Baldwin, T. O., T. Berends, T. A. Bunch, T. F. Holzman, S. K. Rausch, L. Shamansky, M. L. Treat, and M. M. Ziegler. 1984. Cloning of the luciferase structural genes from *Vibrio harveyi* and expression of bioluminescence in *Escherichia coli. Biochemistry* 23:3663–3667.

Baldwin, T. O., J. H. Devine, R. C. Heckel, J.-W. Lin, and G. S. Shadel. 1989. The complete nucleotide sequence of the *lux* regulon of *Vibrio fischeri* and the *luxABN* region of *Photobacterium leiognathi* and the mechanism of control of bacterial bioluminescence. *J. Biolum. Chemilum.* 4:326–341.

Baldwin, T. O., M. M. Ziegler, and D. A. Powers. 1979. Covalent structure of subunits of bacterial luciferase: NH$_2$-terminal sequence demonstrates subunit homology. *Proc. Natl. Acad. Sci. USA* 76:4887–4889.

Barak, M., and S. Ulitzur. 1981. The induction of bacterial bioluminescence system on solid medium. *Curr. Microbiol.* 5:299–301.

Bassot, J.-M. 1975. Les organes lumineux à bactéries symbiotiques de quelques Teleosteens leiognathides. *Arch. Zool. Exp. Gen.* 116:359–373.

Baumann, P., and L. Baumann. 1981. The marine Gram-negative eubacteria: genera *Photobacterium, Beneckea, Alteromonas,* and *Alcaligenes,* p. 1302–1331. *In* M. P. Starr, H. Stolp, H. G. Truper, A. Balows, and H. G. Schlegel (ed.), *The Prokaryotes: a Handbook on Habitats, Isolation and Identification of Bacteria.* Springer-Verlag, New York.

Beck, C. F., and R. A. J. Warren. 1988. Divergent promoters, a common form of gene organization. *Microbiol. Rev.* 52:318–326.

Belas, R., A. Mileham, D. Cohn, M. Hilmen, M. Simon, and M. Silverman. 1982. Bacterial bioluminescence: isolation and expression of the luciferase genes from *Vibrio harveyi. Science* 218:791–793.

Boettcher, K. J., and E. G. Ruby. 1990. Depressed light emission by symbiotic *Vibrio fischeri* of the sepiolid squid *Euprymna scolopes. J. Bacteriol.* 172:3701–3706.

Boylan, M., A. F. Graham, and E. A. Meighen. 1985. Functional identification of the fatty acid reductase components encoded in the luminescence operon of *Vibrio fischeri. J. Bacteriol.* 163:1186–1190.

Boylan, M., C. Miyamoto, L. Wall, A. Graham, and E. Meighen. 1989. Lux C, D and E genes of the *Vibrio fischeri* luminescence operon code for the reductase, transferase, and synthetase enzymes involved in aldehyde biosynthesis. *Photochem. Photobiol.* 49:681–688.

Brown, M. R. W., and P. Williams. 1985. The influence of environment on envelope properties affecting survival of bacteria in infections. *Annu. Rev. Microbiol.* 39:527–556.

Byers, D. M., A. Bognar, and E. A. Meighen. 1988. Differential regulation of enzyme activities involved in aldehyde metabolism in the luminescent bacterium *Vibrio harveyi. J. Bacteriol.* 170:967–971.

Cao, J.-G., and E. A. Meighen. 1989. Purification and structural identification of an autoinducer for the luminescence system of *Vibrio harveyi. J. Biol. Chem.* 264:21670–21676.

Cao, J. G., E. Swartzman, C. Miyamoto, and E. Meighen. 1989. Regulatory mutants of the *Vibrio harveyi Lux* system. *J. Biolum. Chemilum.* 3:207–212.

Chen, P.-F., S.-C. Tu, N. Hagag, F. Y.-H. Wu, and C.-W. Wu. 1985. Isolation and characterization of a cyclic AMP receptor protein from luminous *Vibrio harveyi* cells. *Arch. Biochem. Biophys.* 241:425–431.

Choi, S.-H., and E. P. Greenberg. 1991. Activation of the *Vibrio fischeri* luminescence genes by truncated LuxR proteins is autoinducer independent, abstr. H-86, p. 169. Abstr. 91st Gen. Meet. Am. Soc. Microbiol.

Coffey, J. J. 1967. Inducible synthesis of bacterial luciferase: specificity and kinetics of induction. *J. Bacteriol.* 94:1638–1647.

Cohn, D. H., A. J. Mileham, M. I. Simon, K. H. Nealson, S. K. Rausch, D. Bonam, and T. O. Baldwin. 1985. Nucleotide sequence of the *luxA* gene of *Vibrio harveyi* and the complete amino acid sequence of the α subunit of bacterial luciferase. *J. Biol. Chem.* 260:6139–6146.

Cohn, D. H., R. C. Ogden, J. N. Abelson, T. O. Baldwin, K. H. Nealson, M. I. Simon, and A. J. Mileham. 1983. Cloning of the *Vibrio harveyi* luciferase genes: use of a synthetic oligonucleotide probe. *Proc. Natl. Acad. Sci. USA* 80:120–123.

Colepicolo, P., K.-W. Cho, G. O. Poinar, and J. W. Hastings. 1989a. Growth and luminescence of the bacterium *Xenorhabdus luminescens* from a human wound. *Appl. Environ. Microbiol.* 55:2601–2606.

Colepicolo, P., M.-T. Nicolas, J.-M. Bassot, and J. W. Hastings. 1989b. Expression and localization of bacterial luciferase determined by immunogold labeling. *Arch. Microbiol.* **152**:72–76.

Delong, E. F., D. Steinhauer, A. Israel, and K. H. Nealson. 1987. Isolation of the *lux* genes from *Photobacterium leiognathi* and expression in *Escherichia coli. Gene* **54**:203–210.

Deretic, V., R. Dikshit, W. M. Konyescsni, A. M. Chakrabarty, and T. K. Misra. 1989. The *algR* gene, which regulates mucoidy in *Pseudomonas aeruginosa*, belongs to a class of environmentally responsive genes. *J. Bacteriol.* **171**:1278–1283.

Devine, J. H., C. Countryman, and T. O. Baldwin. 1988. Nucleotide sequence of the *luxR* and *luxI* genes and the structure of the primary regulatory region of the *lux* regulon of *Vibrio fischeri* ATCC 7744. *Biochemistry* **27**:837–842.

Devine, J. H., G. S. Shadel, and T. O. Baldwin. 1989. Identification of the operator of the *lux* regulon from the *Vibrio fischeri* strain ATCC7744. *Proc. Natl. Acad. Sci. USA* **86**:5688–5692.

Dunlap, P. V. 1984a. Physiological and morphological state of the symbiotic bacteria from light organs of ponyfish. *Biol. Bull.* **167**:410–425.

Dunlap, P. V. 1984b. Ph.D. dissertation. University of California, Los Angeles.

Dunlap, P. V. 1985. Osmotic control of luminescence and growth in *Photobacterium leiognathi* from ponyfish light organs. *Arch. Microbiol.* **141**:44–50.

Dunlap, P. V. 1989. Regulation of luminescence by cyclic AMP in *cya*-like and *crp*-like mutants of *Vibrio fischeri. J. Bacteriol.* **171**:1199–1202.

Dunlap, P. V. 1991. Control of luminescence gene expression in *lux::lacZ* fusion mutants of *Vibrio fischeri*, abstr. H-87, p. 169. Abstr. 91st Gen. Meet. Am. Soc. Microbiol.

Dunlap, P. V. Personal observation.

Dunlap, P. V., and S. Fukasawa. Unpublished data.

Dunlap, P. V., and E. P. Greenberg. 1985. Control of *Vibrio fischeri* luminescence gene expression in *Escherichia coli* by cyclic AMP and cyclic AMP receptor protein. *J. Bacteriol.* **164**:45–50.

Dunlap, P. V., and E. P. Greenberg. 1988. Analysis of the mechanism of *Vibrio fischeri* luminescence gene regulation by cyclic AMP and cyclic AMP receptor protein in *Escherichia coli. J. Bacteriol.* **170**:4040–4046.

Dunlap, P. V., and M. J. McFall-Ngai. 1987. Initiation and control of the bioluminescent symbiosis between *Photobacterium leiognathi* and leiognathid fish. *Ann. N.Y. Acad. Sci.* **503**:269–283.

Dunlap, P. V., U. Mueller, K. Lundberg, and T. A. Lisa. Unpublished data.

Dunlap, P. V., and J. M. Ray. 1989. Requirement for autoinducer in transcriptional negative autoregulation of the *Vibrio fischeri luxR* gene in *Escherichia coli. J. Bacteriol.* **171**:3549–3552.

Eberhard, A. 1972. Inhibition and activation of bacterial luciferase synthesis. *J. Bacteriol.* **109**:1101–1105.

Eberhard, A., T. Longin, C. A. Widrig, and S. J. Stranick. 1991. Synthesis of the *lux* gene autoinducer in *Vibrio fischeri* is positively autoregulated. *Arch. Microbiol.* **155**:294–297.

Eberhard, A. 1990. Personal communication.

Eberhard, A., A. L. Burlingame, C. Eberhard, G. L. Kenyon, K. H. Nealson, and N. J. Oppenheimer. 1981. Structural identification of autoinducer of *Photobacterium fischeri. Biochemistry* **20**:2444–2449.

Eberhard, A., C. A. Widrig, P. McBath, and J. B. Schineller. 1986. Analogs of the autoinducer of bioluminescence in *Vibrio fischeri. Arch. Microbiol.* **146**:35–40.

Engebrecht, J., K. Nealson, and M. Silverman. 1983. Bacterial bioluminescence: isolation and genetic analysis of functions from *Vibrio fischeri. Cell* **32**:773–781.

Engebrecht, J., and M. Silverman. 1984. Identification of genes and gene products necessary for bacterial bioluminescence. *Proc. Natl. Acad. Sci. USA* **81**:4154–4158.

Engebrecht, J., and M. Silverman. 1986. Regulation of expression of bacterial genes for bioluminescence, p. 31–44. *In* J. K. Setlow and A. Hollaender (ed.), *Genetic Engineering*, vol. 8. Plenum Press, New York.

Engebrecht, J., and M. Silverman. 1987. Nucleotide sequence of the regulatory locus controlling expression of bacterial genes for bioluminescence. *Nucleic Acids Res.* **15**:10455–10467.

Eymers, J. G., and K. L. van Schouwenburg. 1936. On the luminescence of bacteria. II. Determination of the oxygen consumed in the light emitting process of *Photobacterium phosphoreum. Enzymologia* **1**:328–340.

Finlay, B. B., F. Heffron, and S. Falkow. 1989. Epithelial cell surfaces induce *Salmonella* proteins required for bacterial adherence and invasion. *Science* 243:940–943.

Finlay, B. B., and S. Falkow. 1989. Common themes in microbial pathogenicity. *Microbiol. Rev.* 53:210–230.

Fitzgerald, J. M. 1977. Classification of luminous bacteria from the light organ of the Australian pine-cone fish, *Cleidopus gloriamaris*. *Arch. Microbiol.* 112:153–156.

Foran, D. R., and W. M. Brown. 1988. Nucleotide sequence of the LuxA and LuxB genes of the bioluminescent marine bacterium *Vibrio fischeri*. *Nucleic Acids Res.* 16:777.

Frackman, S., M. Anhalt, and K. H. Nealson. 1990. Cloning, organization, and expression of the bioluminescence genes of *Xenorhabdus luminescens*. *J. Bacteriol.* 172:5767–5773.

Friedrich, W. F., and E. P. Greenberg. 1983. Glucose repression of luminescence and luciferase in *Vibrio fischeri*. *Arch. Microbiol.* 134:87–91.

Giroux, S., and R. Cedergren. 1989. Evolution of a tRNA operon in gamma purple bacteria. *J. Bacteriol.* 171:6446–6454.

Graham, P. H., J. R. Paxton, and K. Y. Cho. 1972. Characterization of luminescent bacteria from the light organs of the Australian pine cone fish (*Cleidopus gloriamaris*). *Arch. Mikrobiol.* 81:305–308.

Gray, K. M., and E. P. Greenberg. 1991. Cloning and characterization of the luminescence gene cluster from *Vibrio fischeri* ES114, the light organ symbiont of the sepiolid squid *Euprymna scolopes*, abstr. I-23, p. 194. Abstr. 91st Gen. Meet. Am. Soc. Microbiol.

Greenberg, E. P., J. W. Hastings, and S. Ulitzur. 1979. Induction of luciferase synthesis in *Beneckea harveyi* by other marine bacteria. *Arch. Microbiol.* 120:87–91.

Gupta, S. C., C. P. Reese, and J. W. Hastings. 1986. Mobilization of cloned luciferase genes into *Vibrio harveyi* luminescence mutants. *Arch. Microbiol.* 143:325–329.

Haneda, Y. 1938. Uber den Leuchtfisch, *Malacocephalus laevis* (Lowe). *Jpn. J. Med. Sci.* 5:355–366.

Haneda, Y. 1966. On a luminous organ of the Australian pine-cone fish, *Cleidopus gloriamaris* De Vis, p. 547–555. *In* F. H. Johnson and Y. Haneda (ed.), *Bioluminescence in Progress*. Princeton University Press, Princeton, N.J.

Hastings, J. W., J. Makemson, and P. V. Dunlap. 1987. How are growth and luminescence regulated independently in light organ symbionts? *Symbiosis* 4:3–24.

Hastings, J. W., and K. H. Nealson. 1977. Bacterial bioluminescence. *Annu. Rev. Microbiol.* 31:549–595.

Hastings, J. W., and K. H. Nealson. 1981. The symbiotic luminous bacteria, p. 1332–1345. *In* M. P. Starr, H. Stolp, H. G. Truper, A. Balows, and H. G. Schlegel (ed.), *The Prokaryotes: a Handbook on Habitats, Isolation and Identification of Bacteria*. Springer-Verlag, New York.

Hastings, J. W., C. J. Potrikus, S. C. Gupta, M. Kurfurst, and J. C. Makemson. 1985. Biochemistry and physiology of bioluminescent bacteria. *Adv. Microb. Physiol.* 26:236–291.

Hastings, J. W., W. H. Riley, and J. Massa. 1965. The purification, properties and chemiluminescent quantum yield of bacterial luciferase. *J. Biol. Chem.* 240:1473–1481.

Haygood, M. G. 1984. Ph.D. dissertation. University of California, San Diego.

Haygood, M. G., and K. H. Nealson. 1985a. The effect of iron on the growth and luminescence of the symbiotic bacterium *Vibrio fischeri*. *Symbiosis* 1:39–51.

Haygood, M. G., and K. H. Nealson. 1985b. Mechanisms of iron regulation of luminescence in *Vibrio fischeri*. *J. Bacteriol.* 162:209–216.

Haygood, M. G., B. M. Tebo, and K. H. Nealson. 1984. Luminous bacteria of a monocentrid fish (*Monocentris japonicus*) and two anomalopid fishes (*Photoblepharon palpebratus* and *Kryptophaneron alfredi*): population sizes and growth within the light organs, and rates of release into the seawater. *Mar. Biol.* 78:249–254.

Henikoff, S., J. C. Wallace, and J. P. Brown. 1990. Finding protein similarities with nucleotide sequence databases. *Methods Enzymol.* 183:111–132.

Henry, J. P., and A. M. Michelson. 1970. Etudes de bioluminescence. Regulation de la bioluminescence bactérienne. *C. R. Acad. Sci.* 270:1947–1949.

Herring, P. J., and J. G. Morin. 1978. Bioluminescence in fishes, p. 273–329. *In* P. J. Herring (ed.), *Bioluminescence in Action*. Academic Press, Inc. (London), Ltd., London.

Illarionov, B. A., and M. V. Protopopova. 1987. Cloning and expression of genes of the luminescent system of *Photobacterium leiognathi*. *Mol. Gen. Mikrobiol. Virusol.* 8:41–46.

Illarionov, B. A., M. V. Protopopova, V. A. Karginov, M. P. Mertvetsov, and I. I. Gitelson. 1988. Nucleotide sequence of genes of the luciferase α and β subunits from *Photobacterium leiognathi*. *Bioorgan. Khim.* **14**:412–415.

Johnston, T. C., K. S. Hruska, and L. F. Adams. 1989. The nucleotide sequence of the *luxE* gene of *Vibrio harveyi* and a comparison of the amino acid sequences of the acyl-protein synthetases from *V. harveyi* and *V. fischeri. Biochem. Biophys. Res. Commun.* **163**:93–101.

Johnston, T. C., E. B. Rucker, L. Cochrum, K. S. Hruska, and V. Vandegrift. 1990. The nucleotide sequence of the *luxA* and *luxB* genes of *Xenorhabdus luminescens* HM and a comparison of the amino acid sequences of luciferases from four species of bioluminescent bacteria. *Biochim. Biophys. Acta* **170**:407–415.

Johnston, T. C., R. B. Thompson, and T. O. Baldwin. 1986. Nucleotide sequence of the *luxB* gene of *Vibrio harveyi* and the complete amino acid sequence of the β subunit of bacterial luciferase. *J. Biol. Chem.* **261**:4805–4811.

Kaplan, H. B., A. Eberhard, C. Widrig, and E. P. Greenberg. 1985. Synthesis of *N*-[3-oxo-(4,5-^{3}H$_2$)-hexanoyl] homoserine lactone: biologically active tritium-labelled *Vibrio fischeri* autoinducer. *J. Labelled Comp. Radiopharm.* **22**:387–395.

Kaplan, H. B., and E. P. Greenberg. 1985. Diffusion of autoinducer is involved in regulation of the *Vibrio fischeri* luminescence system. *J. Bacteriol.* **163**:1210–1214.

Kaplan, H. B., and E. P. Greenberg. 1987. Overproduction and purification of the *luxR* gene product: the transcriptional activator of the *Vibrio fischeri* luminescence system. *Proc. Natl. Acad. Sci. USA* **84**:6639–6643.

Karl, D. M., and K. H. Nealson. 1980. Regulation of cellular metabolism during synthesis and expression of the luminous system in *Beneckea* and *Photobacterium. J. Gen. Microbiol.* **117**:357–368.

Kasai, S., K. Matsui, and T. Nakamura. 1987. Purification and some properties of FP$_{390}$ from *P. phosphoreum*, p. 647–650. *In* D. E. Edmondson and D. B. McCormick (ed.), *Flavin and Flavoproteins*. Walter de Gruyter, Berlin.

Katznelson, R., and S. Ulitzur. 1977. Control of luciferase synthesis in a newly isolated strain of *Photobacterium leiognathi. Arch. Microbiol.* **115**:347–351.

Kempner, E. S., and F. E. Hanson. 1968. Aspects of light production by *Photobacterium fischeri. J. Bacteriol.* **95**:975–979.

Kishitani, T. 1930. Studien über die Leuchtsymbiose in *Physiculus japonicus* (Hilgendorf), mit der Beilage der zwei neuen Arten der Leuchtbakterien. *Sci. Rep. Tohoku Univ. Sect. 4* **5**:801–823.

Kolibachuk, D., and E. P. Greenberg. Unpublished data.

Kotlyar, A. N. 1985. Taxonomy and distribution of the fishes of the family Monocentridae (Beryciformes). *Vopr. Ikhtiol.* **25**:531–545.

Kotlyar, A. N. 1988. Data on the systematics and biology of *Monocentris reedi* and *Polymixia yuri* from the underwater mountain range Naska. *Vopr. Ikhtiol.* **28**:853–856.

Kropach, C. 1975. The pinecone fish *Monocentris japonicus* (Houttuyn), a first live record from the Red Sea. *Isr. J. Zool.* **24**:194–196.

Lee, C. A., and S. Falkow. 1990. The ability of *Salmonella* to enter mammalian cells is affected by bacterial growth state. *Proc. Natl. Acad. Sci. USA* **87**:4304–4308.

Leis, J. M., and S. Bullock. 1986. The luminous cardinalfish *Siphamia* (Pisces, Apogonidae): development of larvae and the luminous organ, p. 703–714. *In* T. Uyeno, R. Arai, T. Taniuchi, and K. Matsuura (ed.), *Indo-Pacific Fish Biology: Proceedings of the Second International Conference on Indo-Pacific Fishes*. Ichthyological Society of Japan, Tokyo.

Leisman, G., D. H. Cohn, and K. H. Nealson. 1980. Bacterial origin of luminescence in marine animals. *Science* **208**:1271–1273.

Lummen, P., and U. K. Winkler. 1986. Bioluminescence of outer membrane defective mutants of *Photobacterium phosphoreum. FEMS Microbiol. Lett.* **37**:293–298.

Makemson, J. C. 1986. Luciferase-dependent oxygen consumption by bioluminescent vibrios. *J. Bacteriol.* **165**:461–466.

Makemson, J. C., and J. W. Hastings. 1982. Iron represses bioluminescence and affects catabolite repression in *Vibrio harveyi. Curr. Microbiol.* **7**:181–186.

Mancini, J., M. Boylan, R. Soly, S. Ferri, R. Szittner, and E. Meighen. 1989. Organization of the *Lux* genes of *Photobacterium phosphoreum. J. Biolum. Chemilum.* **3**:201–205.

Mancini, J. A., M. Boylan, R. R. Soly, A. F. Graham, and E. A. Meighen. 1988. Cloning and expression of the *Photobacterium phosphoreum* luminescence system demonstrates a unique *lux* gene organization. *J. Biol. Chem.* **263**:14308–14314.

Martin, M., R. Showalter, and M. Silverman. 1989. Identification of a locus controlling expression of the luminescence genes in *Vibrio harveyi*. *J. Bacteriol.* **171**:2406–2414.

McFall-Ngai, M. J. 1983. The gas bladder as a central component of the leiognathid bacterial light organ symbiosis. *Am. Zool.* **23**:907.

McFall-Ngai, M. J., and P. V. Dunlap. 1983. Three new modes of luminescence in the leiognathid fish *Gazza minuta* (Perciformes: Leiognathidae): discrete projected luminescence, ventral body flash and buccal luminescence. *Mar. Biol.* **73**:227–237.

Meighen, E. A. 1988. Enzymes and genes from the *lux* operons of bioluminescent bacteria. *Annu. Rev. Microbiol.* **42**:151–176.

Meighen, E. A. 1991. Molecular biology of bacterial bioluminescence. *Microbiol. Rev.* **55**:123–142.

Michaliszyn, G. A., and E. A. Meighen. 1976. Induced polypeptide synthesis during the development of bacterial bioluminescence. *J. Biol. Chem.* **251**:2541–2549.

Middleton, A. J., and E. B. Smith. 1976. General anaesthetics and bacterial luminescence. I. The effect of diethyl ether on the *in vivo* light emission of *Vibrio fischeri*. *Proc. R. Soc. London Ser. B* **193**:159–171.

Miller, J. F., J. J. Mekalanos, and S. Falkow. 1989. Coordinate regulation and sensory transduction in the control of bacterial virulence. *Science* **243**:916–922.

Miyamoto, C. M., M. Boylan, A. F. Graham, and E. A. Meighen. 1988a. Organization of the *lux* structural genes of *Vibrio harveyi*. *J. Biol. Chem.* **263**:13393–13399.

Miyamoto, C. M., D. Byers, A. F. Graham, and E. A. Meighen. 1987. Expression of bioluminescence by *Escherichia coli* containing recombinant *Vibrio harveyi* DNA. *J. Bacteriol.* **169**:247–253.

Miyamoto, C. M., A. D. Graham, M. Boylan, J. F. Evans, K. W. Hasel, E. A. Meighen, and A. F. Graham. 1985. Polycistronic mRNAs code for polypeptides of the *Vibrio harveyi* luminescence system. *J. Bacteriol.* **161**:995–1001.

Miyamoto, C. M., A. F. Graham, and E. A. Meighen. 1988b. Nucleotide sequence of the *luxC* gene and the upstream DNA from the bioluminescent system of *Vibrio harveyi*. *Nucleic Acids Res.* **16**:1551–1562.

Miyamoto, C. M., E. A. Meighen, and A. F. Graham. 1990. Transcriptional regulation of *lux* genes transferred into *Vibrio harveyi*. *J. Bacteriol.* **172**:2046–2054.

Morin, J. G., A. Harrington, K. Nealson, N. Krieger, T. O. Baldwin, and J. W. Hastings. 1975. Light for all reasons: versatility in the behavioral repertoire of the flashlight fish. *Science* **190**:74–76.

Nealson, K. H. 1977. Autoinduction of bacterial luciferase. Occurrence, mechanism and significance. *Arch. Microbiol.* **112**:73–79.

Nealson, K. H. 1979. Alternative strategies of symbiosis of marine luminous fishes harboring light-emitting bacteria. *Trends Biochem. Sci.* **4**:105–110.

Nealson, K. H., A. Eberhard, and J. W. Hastings. 1972. Catabolite repression of bacterial bioluminescence: functional implications. *Proc. Natl. Acad. Sci. USA* **69**:1073–1076.

Nealson, K. H., and J. W. Hastings. 1977. Low oxygen is optimal for luciferase synthesis in some bacteria. Ecological implications. *Arch. Microbiol.* **112**:9–16.

Nealson, K. H., and J. W. Hastings. 1979. Bacterial bioluminescence: its control and ecological significance. *Microbiol. Rev.* **43**:496–518.

Nealson, K. H., and J. W. Hastings. 1991. The luminous bacteria. *In* A. Balows, H. G. Trüper, M. Dworkin, W. Harder, and K.-H. Schleifer (ed.), *The Prokaryotes, a Handbook on the Biology of Bacteria: Ecophysiology, Isolation, Identification, Applications*, 2nd ed. Springer-Verlag, New York.

Nealson, K. H., M. G. Haygood, B. M. Tebo, M. Roman, E. Miller, and J. E. McCosker. 1984. Contribution by symbiotically luminous fishes to the occurrence and bioluminescence of luminous bacteria in seawater. *Microb. Ecol.* **10**:69–77.

Nealson, K. H., T. Platt, and J. W. Hastings. 1970. Cellular control of the synthesis and activity of the bacterial luminescent system. *J. Bacteriol.* **104**:313–322.

Ne'eman, Z., S. Ulitzur, D. Branton, and J. W. Hastings. 1977. Membrane polypeptides co-induced with the bacterial bioluminescent system. *J. Biol. Chem.* **252**:5150–5154.

Okada, Y. 1955. *Fishes of Japan*. Maruzen Co., Tokyo.

Okada, Y. K. 1926. On the photogenic organ of the knight-fish *Monocentris japonicus* (Houttuyn). *Biol. Bull.* **50:**365–373.

O'Kane, D. J., J. Vervoort, F. Muller, and J. Lee. 1987. Purification and characterization of an unusual non-fluorescent flavoprotein from *Photobacterium leiognathi*, p. 641–645. *In* D. E. Edmondson and D. B. McCormick (ed.), *Flavin and Flavoproteins*. Walter de Gruyter, Berlin.

Paxton, J. R. 1973. Bioluminescence in the Australian monocentrid fish, *Cleidopus gloriamaris*, p. 521. *In* R. Fraser (comp.), *Oceanography of the South Pacific, 1972*. New Zealand National Commission for UNESCO, Wellington.

Rosson, R. A., and K. H. Nealson. 1981. Autoinduction of bacterial bioluminescence in a carbon-limited chemostat. *Arch. Microbiol.* **129:**299–304.

Ruby, E. G., E. P. Greenberg, and J. W. Hastings. 1980. Planktonic marine luminous bacteria: species distribution in the water column. *Appl. Environ. Microbiol.* **39:**302–306.

Ruby, E. G., and K. H. Nealson. 1976. Symbiotic association of *Photobacterium fischeri* with the marine luminous fish *Monocentris japonica:* a model of symbiosis based on bacterial studies. *Biol. Bull.* **151:**574–586.

Ruby, E. G., and K. H. Nealson. 1977. Pyruvate production and excretion by the luminous marine bacteria. *Appl. Environ. Microbiol.* **34:**164–169.

Ruby, E. G., and K. H. Nealson. 1978. Seasonal changes in the species composition of luminous bacteria in nearshore seawater. *Limnol. Oceanogr.* **23:**530–533.

Schmidt, T. M., K. Kopecky, and K. H. Nealson. 1989. Bioluminescence in the insect pathogen *Xenorhabdus luminescens. Appl. Environ. Microbiol.* **55:**2607–2612.

Schultz, L. P. 1956. A new pinecone fish, *Monocentris reedi*, from Chile, and new family record for the eastern Pacific. *Proc. U. S. Natl. Mus.* **106:**237–239.

Schweizer, H. P., and P. Datta. 1989. Identification and DNA sequence of *tdcR*, a positive regulatory gene of the *tdc* operon of *Escherichia coli. Mol. Gen. Genet.* **218:**516–522.

Shadel, G. S., J. H. Devine, and T. O. Baldwin. 1990a. Control of the *lux* regulon of *Vibrio fischeri. J. Biolum. Chemilum.* **5:**99–106.

Shadel, G. S., R. Young, and T. O. Baldwin. 1990b. Use of regulated cell lysis in a lethal genetic selection in *Escherichia coli:* identification of the autoinducer-binding region of the LuxR protein from *Vibrio fischeri* ATCC 7744. *J. Bacteriol.* **172:**3980–3987.

Showalter, R. E., M. O. Martin, and M. R. Silverman. 1990. Cloning and nucleotide sequence of *luxR*, a regulatory gene controlling bioluminescence in *Vibrio harveyi. J. Bacteriol.* **172:**2946–2954.

Slock, J., D. VanRiet, D. Kolibachuk, and E. P. Greenberg. 1990. Critical regions of the *Vibrio fischeri* LuxR protein defined by mutational analysis. *J. Bacteriol.* **172:**3974–3979.

Soly, R. R., J. A. Mancini, S. R. Ferri, M. Boylan, and E. A. Meighen. 1988. A new *lux* gene in bioluminescent bacteria codes for a protein homologous to the bacterial luciferase subunits. *Biochem. Biophys. Res. Commun.* **155:**351–358.

Stead, D. G. 1906. *Fishes of Australia*. William Brooks & Co., Sydney.

Sugihara, J., and T. O. Baldwin. 1988. Effects of 3' end deletions from the *Vibrio harveyi luxB* gene on luciferase subunit folding and enzyme assembly. Generation of temperature-sensitive polypeptide folding mutants. *Biochemistry* **27:**2872–2880.

Swartzman, E., S. Kapoor, A. F. Graham, and E. A. Meighen. 1990a. A new *Vibrio fischeri lux* gene precedes a bidirectional termination site for the *lux* operon. *J. Bacteriol.* **172:**6797–6802.

Swartzman, E., C. Miyamoto, A. Graham, and E. Meighen. 1990b. Delineation of the transcriptional boundaries of the *lux* operon of *Vibrio harveyi* demonstrates the presence of two new *lux* genes. *J. Biol. Chem.* **265:**3513–3517.

Szittner, R., and E. Meighen. 1990. Nucleotide sequence, expression, and properties of luciferase coded by *lux* genes from a terrestrial bacterium. *J. Biol. Chem.* **265:**16581–16587.

Tebo, B. M., D. S. Linthicum, and K. H. Nealson. 1979. Luminous bacteria and light emitting fish: ultrastructure of the symbiosis. *BioSystems* **11:**269–280.

Uchida, K. 1932. Juvenile stages of *Monocentris japonicus. Dobutsugaku Zasshi* **44:**366–367.

Ulitzur, S. 1989. The regulatory control of the bacterial luminescence system—a new view. *J. Biolum. Chemilum.* **4:**317–325.

Ulitzur, S., and J. W. Hastings. 1979. Autoinduction in a luminous bacterium: a confirmation of the hypothesis. *Curr. Microbiol.* **2:**345–348.

Ulitzur, S., and J. Kuhn. 1988. The transcription of bacterial luminescence is regulated by sigma 32. *J. Biolum. Chemilum.* **2:**81–93.

Ulitzur, S., A. Reinhertz, and J. W. Hastings. 1981. Factors affecting the cellular expression of bacterial luciferase. *Arch. Microbiol.* **129:**67–71.

Ulitzur, S., and J. Yashphe. 1975. An adenosine 3′,5′-monophosphate-requiring mutant of the luminous bacteria *Beneckea harveyi. Biochim. Biophys. Acta* **404:**321–328.

Wall, L. A., D. M. Byers, and E. A. Meighen. 1984. In vivo and in vitro acylation of polypeptides in *Vibrio harveyi:* identification of proteins involved in aldehyde production for bioluminescence. *J. Bacteriol.* **159:**720–724.

Watanabe, H., and J. W. Hastings. 1986. Expression of luminescence in *Photobacterium phosphoreum:* Na$^+$ regulation of in vivo luminescence appearance. *Arch. Microbiol.* **145:**342–346.

Watanabe, H., N. Mimura, A. Takimoto, and T. Nakamura. 1975. Luminescence and respiratory activities of *Photobacterium phosphoreum. J. Biochem.* **77:**1147–1155.

Wei, S. L., and R. E. Young. 1989. Development of symbiotic bacterial bioluminescence in a nearshore cephalopod, *Euprymna scolopes. Mar. Biol.* **103:**541–546.

Yasaki, Y. 1928. On the nature of the luminescence of the knight-fish (*Monocentris japonicus*) (Houttuyn). *J. Exp. Zool.* **50:**495–505.

Yasaki, Y., and Y. Haneda. 1935. On the luminescence of the deep-sea fishes, family Macrouridae. *Oyo Dobutsu Zasshi Tokyo* **7:**165–177.

Yamada, K., M. Haygood, and H. Kabasawa. 1979. On fertilization and early development in the pinecone fish, *Monocentris japonicus. Annu. Rep. Keikyu Aburatsubo Mar. Park Aquar.* **10:**31–38.

Yoshiba, S. 1970. Saprophytic luminescence by the symbiotic luminous bacteria of the Japanese Knight-fish, *Monocentris japonicus,* observed in an aquarium. *Sci. Rep. Yokosuka City Mus.* **16:**83–89.

Yoshiba, S., and Y. Haneda. 1967. Bacteriological study of the symbiotic luminous bacteria cultivated from the luminous organs of the apogonid fish, *Siphamia versicolor,* and the Australian pine cone fish, *Cleidopus gloria-maris. Sci. Rep. Yokosuka City Mus.* **13:**82–84.

Yoshizawa, S. 1916. Matsukasa-uwo no hakko ni tsuite (On the luminescence of the fish *Monocentris japonicus*). *Dobutsugaku Zasshi* **28:**411–412.

Microbial Cell-Cell Interactions
Edited by Martin Dworkin
© 1991 American Society for Microbiology, Washington, DC 20005

Chapter 9

Rhizobium-Legume Symbiosis

L. Evans Roth and Gary Stacey

INTRODUCTION

Rhizobium, Azorhizobium, and *Bradyrhizobium* are all genera of bacteria that possess the ability to infect and establish a nitrogen-fixing symbiosis with leguminous plants. Although sharing this common trait, these organisms are taxonomically distinct. The taxonomic relationships of these organisms have unfortunately been confused by the traditional classification of species based on their plant host range (Table 1). Such a classification scheme accentuates the fact that legume infection by rhizobia is host specific but says little about the true genetic relatedness between strains. We urge readers as they proceed with this review to be cautious in generalizing results with one species to all genera. A thorough, recent review of the phylogeny of rhizobia is given by Young (in press).

The rhizobium-plant relationship is actually a model for other endosymbiotic relationships. Numerous cases of prokaryotes as endocytobionts are known in free-living protozoa (Gortz, 1983; Jeon, 1987), parasitic protozoa, such as those causing malaria (Aikawa, 1980) and leishmaniasis (Chang, 1983), and human cells (Horwitz, 1983), while in other cases, the symbiont is a eukaryote (Cook, 1980). In all these cases, the symbiont is intracellular, usually endocytoplasmic, and enclosed in a membrane that allows it to thrive in its own special conditions. This

TABLE 1
Rhizobium, Azorhizobium, and *Bradyrhizobium* species and their hosts

Species	Example of host plant (genus)[a]
Rhizobium leguminosarum bv. viciae	Pea (*Pisum*)
R. leguminosarum bv. trifolii	Clover (*Trifolium*)
R. leguminosarum bv. phaseoli	Bean (*Phaseolus*)
R. meliloti	Alfalfa (*Medicago*)
R. loti	Lotus (*Lotus*)
R. fredii (alias *R. japonicum, Sinorhizobium sinensis*)	Soybean (*Glycine*)
R. galegae	Goat's rue (*Galega*)
Rhizobium sp. (strain NGR234)	Siratro (*Macroptilium*)
Azorhizobium caulinodans	Sesbania (*Sesbania*)
Bradyrhizobium japonicum	Soybean (*Glycine*)
Bradyrhizobium sp.	Lupine (*Lupinus*), peanut (*Arachis*), cowpea (*Vigna*), parasponia (*Parasponia*)

[a] Some rhizobial species can nodulate multiple hosts. Only the primary host is listed here.

recurrence of membrane-bound vesicles in many endosymbiotic systems suggests similarities in the cellular events of infection and symbiotic maintenance. To draw attention to these similarities and stimulate comparative research between diverse endosymbiotic systems, Roth et al. (1988) proposed the term "symbiosome" to refer to the membrane-bound compartment containing one or more symbionts and located in the cytoplasm of eukaryotic cells (Fig. 1 and 2). This terminology will be used throughout this review. Therefore, in some ways, rhizobial symbiosomes are exemplary of many other symbioses. In other ways, rhizobium-plant relationships appear more complex, are established by more steps and signals, and are more highly regulated than other symbioses are known to be. Nevertheless, we believe that our discussion does have broader-ranging importance than rhizobia and legumes. Such a statement should in no way be taken to diminish the importance of legumes; soybean alone is the second largest crop in the United States and is of equal importance in many other countries. If other legume forage crops are added, the agronomic magnitude becomes clear. The combined impor-

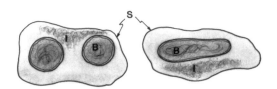

FIGURE 1. Conceptualized symbiosomes. This fundamental unit of nitrogen fixation and endosymbioses of almost all kinds is characterized by having a symbiosome membrane (S), included bacteroids (B) or other organisms, and a space containing various included substances (I).

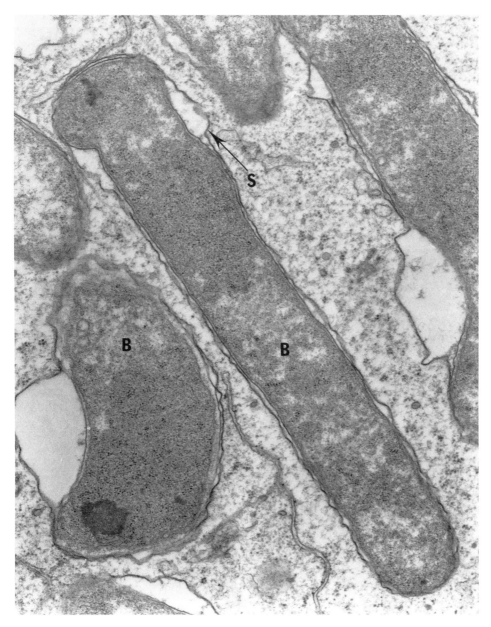

FIGURE 2. Symbiosomes in a nodule cell of alfalfa. *R. meliloti* is found as a single organism in each symbiosome, with a comparatively small space between the symbiosome membrane (S) and the bacteroids (B). ×31,100.

tance—medical, agronomic, and marine, etc.—of such endosymbioses is very great.

The process of rhizobium-legume infection is made up of successive, discrete recognition events involving interactive, complementary plant and bacterial functions. It is our contention that knowledge of each discrete step will eventually allow the complete description of the molecular mechanism of legume infection. Therefore, we will use as an outline for this review the successive stages of infection which have thus far been characterized.

A rhizobium inoculant industry has existed in the United States since the late 19th century to provide farmers with symbiotic bacteria to enhance legume growth. A considerable amount of literature describing efforts to improve legume nitrogen fixation by manipulating either the plant or the bacterial partner has been written. We will not review this literature here but will take a more basic approach, i.e., describing that information which will undoubtedly serve as the basis for any successful effort to engineer the symbiosis. Moreover, we especially want to point out the broad ramifications of research on symbiotic nitrogen fixation. Methods first developed for the investigation of rhizobia and their hosts are now finding broad applicability in the study of plant (and animal) pathogens and plant biology. This applicability is especially evident in the areas of bacterial genetics and molecular biology.

Two fields of study, rhizobium-legume interaction and pathogen-plant interaction, are drawing closer and being enriched by mutual information exchange and comparison. Eventually, such research will allow the formulation of a variety of general principles which will allow a better understanding of plant-microbe interactions. An example is the presence of bacterial genes transcriptionally regulated by secondary metabolites produced by the plant host (see below). In rhizobia, these secondary metabolites are flavonoid compounds which specifically induce the transcription of *nod* genes. Analogous gene regulatory schemes have been described for *Agrobacterium tumefaciens* (Melchers et al., 1989; Stachel et al., 1985; Stachel et al., 1986a; Stachel and Zambryski, 1986) and other plant pathogens (Osbourn et al., 1987).

Finally, in this review we hope to point out the utility of research on the rhizobium-legume interaction as a model system for the study of plant and bacterial development. An obvious example is the study of the signal pathways operable in the induction of nodules and the genetic mechanisms underlying these pathways. However, another fruitful area of research is the cell biology of the interaction, e.g., such processes as membrane biogenesis, protein transport, and cell structure. Comparative studies with other endosymbiotic systems may be important in the eventual understanding of the rhizobium-legume interaction and the underlying cell biology.

SUMMARY OF RHIZOBIUM INFECTION OF LEGUME ROOTS

The establishment of a nitrogen-fixing symbiosis between rhizobia and legumes is a complex, multistep process. The interaction shows a degree of host specificity in that only certain bacterial species will infect certain plant species.

For example, *Rhizobium leguminosarum* bv. viciae nodulates pea, lentil, and common vetch but not clover or bean, whereas *R. leguminosarum* bv. trifolii nodulates only clover. The taxonomy of rhizobia has traditionally been based on this host specificity; however, now that this is known to be an imperfect system, the taxonomy is in dire need of revision (Young, in press). Generally, rhizobia are grouped into two broad classes (Table 1), the fast-growing rhizobia (e.g., *R. leguminosarum* and *R. meliloti*) and the slow-growing rhizobia (e.g., *Bradyrhizobium japonicum*). In the fast-growing rhizobia, the genes important for establishing the symbiosis are primarily plasmid (Sym plasmid) encoded, whereas in the slow-growing rhizobia, these genes are thought to be chromosomally encoded. A more recent revelation is that a number of genes important for symbiotic development have been found in fast-growing rhizobia on other than the Sym plasmid or the chromosome (reviewed by Long, 1984; and Appelbaum, 1989).

Rhizobia in the rhizosphere of the plant are chemoattracted to compounds excreted by legume roots (Gaworzewska and Carlile, 1982; Currier and Strobel, 1976, 1977; Gittie et al., 1978; Hunter and Fahring, 1980). The bacteria attach to the root hair (Fig. 3) and induce a marked curling of the hair into a "shepherd's crook" (Fig. 4). The bacteria within the cavity formed by the curling of the root hair penetrate the hair cell and induce the formation of an infection thread. Before

FIGURE 3. Attachment of *B. japonicum* to root and root hair surfaces, the first step in infection. This scanning electron micrograph shows that attachment is at one end (arrows), indicating a polar differentiation of the organism. ×1,550.

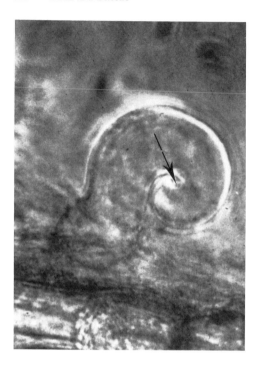

FIGURE 4. Root hair curling, the second step in infection. The root hair curls at the site of bacterial accumulation, and in the tightest part of the curl (arrow) infection begins. This light micrograph is of living material. Ca. ×300.

or during infection thread formation, rhizobial factors likely act as signals to initiate root nodule meristem formation. Two types of nodule structures are formed, depending on the plant species: indeterminate nodules, formed on such plants as alfalfa and clover, arise from ground meristem cells via a process resembling lateral root formation, and determinate nodules, formed on such plants as soybean and bean, arise from the division of subepidermal cortical cells (reviewed by Rolfe and Gresshoff, 1988). The infection thread grows in the direction of the prenodule meristem and is ramified into the plant cortex until it reaches a host cell, in which the bacteria are released. The bacteria released into the plant cell are contained in symbiosomes, the functional unit for nitrogen fixation. Subsequent to release, numerous metabolic processes are integrated to produce needed substances to maintain the symbiosis. Finally, depending on the plant species, the nodule can senesce, i.e., all of the symbiosomes and host cells are degraded. At this stage, it is likely that bacteria remaining in the infection thread utilize the decaying plant material as a food source until they are released into the soil environment.

It should be obvious from the above description that the plant undergoes several developmental changes leading to the formation of the nodule structure. Indeed, during the infection process, a number of plant gene products, termed nodulins, are specifically expressed (reviewed by Govers et al., 1987; Verma and Long, 1983; Verma et al., 1986; and Gloudemans and Bisseling, 1989). Likewise, in addition to the *nod* genes, which are required for nodule formation, the bacteria specifically express *nif* and *fix* genes, which are necessary for nitrogen fixation or maintenance of the symbiotic state.

PREINFECTION EVENTS

Root Colonization

Rhizobia are rhizosphere bacteria; i.e., they are found in much higher numbers in the rhizosphere of plants (i.e., the soil environment influenced by the plant roots) than in fallow soil. In fact, data indicate that *B. japonicum* cells grow very slowly in fallow soil, having a doubling time of 241 to 361 h (Schmidt, 1974), as opposed to a doubling time of 9 to 12 h in the rhizosphere (Bowen and Rovira, 1976). A number of investigators have suggested that growth stimulation of rhizobia in the rhizosphere is host specific; particular legume species enhance the growth of their specific symbiont (reviewed by Stacey and Brill, 1982). However, rhizobia are also found in high numbers in the rhizosphere of nonhost plants, and in recent years, very little attention has been focused on the specific stimulation of rhizobial growth in the rhizosphere. An interesting paper on this subject is that of Van Egeraat (1975), who showed that *R. leguminosarum* bv. viciae can selectively utilize homoserine as a sole carbon and nitrogen source. Homoserine is the predominant ninhydrin-positive material released from pea roots (Van Egeraat, 1975). The paper by Van Egeraat has been made more interesting by the recent demonstration that the genes for the utilization of homoserine are encoded on the Sym plasmid in *R. leguminosarum* bv. viciae (Johnston et al., 1988), thereby strengthening the proposition that the utilization of homoserine has a role in the symbiotic association. Perhaps renewed attention should be paid to the role of the rhizosphere environment in nutritionally favoring one rhizobial strain over another.

Chemotaxis

In addition to growth stimulation of rhizobia, the high numbers of cells found in the rhizosphere could be due to chemoattraction of rhizobia to roots. The existence, specificity, and requirement of chemotaxis for infection of legumes by rhizobia have been controversial for some time (e.g., Wilson, 1940). Rhizobial cells are chemoattracted to compounds released from legume roots (Gaworzewska and Carlile, 1982; Currier and Strobel, 1976, 1977; Gittie et al., 1978; Hunter and Fahring, 1980). Amino acids, sugars, organic acids, and plant secondary metabolites have all been shown to be chemoattractants for rhizobia, with crude root exudates usually being found superior to individual substances (Gittie et al., 1978; Hunter and Fahring, 1980). Bacterial movement in soil is largely determined by the water content. Hamdi (1971) showed that soil moisture was the primary variable of rhizobial movement in soil and, under normal soil conditions, very little movement occurred. Considering the steep water gradient near roots generated by transpiration (Papendick and Campbell, 1975), the lack of soil moisture should limit significant movement of bacteria in the rhizosphere. However, Soby and Bergman (1983) used *R. meliloti* mutants defective in motility or chemotaxis to demonstrate a role for both functions in movement in soil under a defined soil water content. Therefore, it would appear that motility and chemotaxis might play a role in bacterial movement in soil, although the question of their occurrence or importance in a natural environment remains an open question. It is possible that chemotaxis could be important for initial bacterial colonization of the plant after sowing. Under this condition, the rhizosphere population would not be well developed and chemoattractants could diffuse into the soil. Indeed, recent data (Caetano-Anollés et al., 1988b; Mellor et al., 1987) also support the notion that chemotaxis plays an important role in early colonization. Motility and chemotaxis were found to be determinants of competition between bacterial strains and to affect the efficiency of nodulation.

Gulash et al. (1984) reported that *R. meliloti* appeared to be chemoattracted to specific sites on a sterile root surface, suggesting localized sites of chemoattractant secretion. The compounds responsible may be flavonoids, which are known to be specific inducers of *nod* gene expression and to be specific chemoattractants for *Rhizobium* species (Caetano-Anollés et al., 1988a; Aguilar et al., 1988) and *B. japonicum* (Barbour et al., 1991). Flavonoids are released from specific sites on legume roots (Djordjevic et al., 1987). Aguilar et al. (1988) reported that *R. leguminosarum* bv. phaseoli was attracted to *nod* gene inducers (i.e., apigenin, luteolin, and naringenin), inhibitors of *nod* gene induction (i.e., umbelliferone and acetosyringone), and a variety of other organic compounds which might be found in bean root exudates. Caetano-Anollés et al. (1988a) found a slight response of *R. meliloti* to *nod* gene-inducing compounds. At 10^{-6} to 10^{-8} M, cells were attracted to luteolin at a level two- to threefold over background. However, *R. meliloti* mutants defective in the *nodD*, *nodA*, or *nodC* gene were not attracted to luteolin. Transfer of the *R. meliloti* *nod* gene region to *A. tumefaciens* also conferred on these bacteria attraction to luteolin. The specific role of NodA, NodC, or NodD in chemotaxis is not known, but the importance of the *nodA*, *nodC*, and *nodD* genes and their flavonoid inducers in the early stages of legume infection correlates well with the idea that chemotaxis could be important for the establishment of rhizobia on a germinating seedling. However, recent data indicate that isoflavones, inducers of the *nod* genes of *B. japonicum*, are poor chemoattractants for these organisms (Barbour et al., 1991). Moreover, *B. japonicum* strains mutated in *nodD*, *nodA*, or *nodC* were not affected in chemotaxis to isoflavones or soybean seed exudates.

Bacterial Attachment

Rhizobial attachment to legume root hairs is thought to be an essential prerequisite for infection. Until recently, the prevailing view for the mechanism of attachment of rhizobia to the root hair surface held that attachment was mediated by lectins which were produced by the plant and which bound specifically to the bacterial cell surface (Bohlool and Schmidt, 1974; Dazzo and Hubbell, 1975; Bauer, 1981). This model was attractive because it provided a mechanism to explain host-specific attachment: the host produced a lectin specific for its compatible symbiont. Considerable circumstantial evidence has accumulated to support this "lectin hypothesis." The data are too voluminous to review in detail here, and the reader is referred to reviews detailing earlier literature on this subject (Bauer, 1981; Halverson and Stacey, 1986b; Smit and Stacey, 1990; Dazzo, 1981).

In summary, the data clearly showed that under certain circumstances only homologous rhizobia bound to host root hairs and that this attachment could be inhibited by monosaccharide haptens of the host lectin (Dazzo et al., 1976; Stacey et al., 1980). Moreover, the lectin was found to be localized on the root hair tips, the site at which infection occurs (Dazzo et al., 1978). The bacterial lectin receptor was variously identified as capsular polysaccharide (Abe et al., 1984; Mort and Bauer, 1980) or lipopolysaccharide (Hrabak et al., 1981; Kato et al., 1980). Host-specific attachment of rhizobia to root hairs appeared to depend on the growth phase of the bacteria, with optimal plant attachment and lectin binding occurring at the late log phase of growth in batch cultures, apparently reflecting the time at which the specific receptors were most abundant (Mort and Bauer, 1980; Dazzo et al., 1979).

In contrast to the above-described research, a number of investigators found results inconsistent with the involvement of lectin in attachment, with the result that the lectin hypothesis has continued to be controversial. There have been reports in which specific lectin-mediated binding of homologous rhizobia to legume roots could not be found (Mills and Bauer, 1985; Pueppke, 1984). Moreover, in some cases, heterologous rhizobia appeared to attach to legume roots to an equal or better extent than did homologous rhizobia (Smit et al., 1986b). Indeed, on some host plants, heterologous bacteria were capable of attachment to and curling of root hairs, although no infection threads were formed (Debelle et al., 1986b; Wijffelman et al., 1985). The fact that the lectin hypothesis did not appear to answer all the questions about rhizobial attachment led some investigators to look for alternative attachment mechanisms. For example, Vesper and colleagues (Vesper and Bauer, 1985, 1986; Vesper et al., 1987; Vesper and Bhuvaneswari, 1988) provided evidence that fimbriae on *B. japonicum* were a major determinant of bacterial attachment, although this attachment did not show host specificity. Mutants defective in fimbria production were able to nodulate normally but had reduced attachment ability (Vesper et al., 1987; Vesper and Bhuvaneswari, 1988). Other research led to the discovery that rhizobia produce cellulose microfibrils which anchor the bacteria to the root surface (Napoli et al., 1975; Smit et al., 1987). Again, mutants defective in the production of these fibrils were normal for nodulation but defective in firm attachment (Smit et al., 1987). The idea was proposed that attachment was a two-step process: the first, reversible, step mediated by lectin or another adhesin and the second, irreversible, step mediated by cellulose microfibrils (Dazzo et al., 1984; Smit et al., 1986a).

One can conclude from this summary that rhizobial attachment to the root hair surface is more complex than has been generally appreciated and may involve multiple mechanisms. However, this research failed to amply explain all of the conflicting observations on specific versus nonspecific attachment, the role of the bacterial growth stage in attachment, or what role lectin may play in the infection. Recent work (Smit et al., 1986a, 1987, 1989a; Smit et al., 1986b, 1989b; Kijne et al., 1988) has finally produced a clearer picture of rhizobial attachment to the plant surface which appears to explain earlier, conflicting observations. These workers focused on the interesting phenomenon that the growth phase and growth conditions had a significant effect on the ability of *R. leguminosarum* bv. viciae to attach to pea root hairs. Carbon-limited growth resulted in cells which bound to root surfaces in a non-host-specific fashion which could not be reversed by the addition of haptens for pea lectin (Smit et al., 1986a, 1987). This condition mirrors that reported by workers who concluded that lectin is not involved in rhizobial attachment. In contrast, manganese-limited growth resulted in cells with an enhanced ability to attach to pea root hairs which was significantly delayed in the presence of 3-O-methyl-D-glucose, a hapten of pea lectin (Kijne et al., 1988). This condition mirrors that reported by workers who concluded that lectin is involved in rhizobial attachment. However, in the case of *R. leguminosarum* bv. viciae, the lectin hapten did not inhibit but merely delayed attachment, suggesting that lectin does not mediate but accelerates attachment. Manganese-limited cells were found to be more infective than carbon-limited or exponentially grown cells (Kijne et al., 1988).

Calcium-limited growth of *R. leguminosarum* bv. viciae was the only condition, of those tested, which did not result in an increase in attachment to pea root hairs (Smit et al., 1986a, 1987, 1989a; Smit et al., 1989b). Indeed, the data were interpreted to suggest that calcium was essential for attachment. Subsequently, a calcium-dependent adhesin was isolated from the surface of *R. leguminosarum* bv. viciae and purified by an assay which measured the ability of fractions to inhibit attachment of carbon- or manganese-limited cells (Smit et al., 1989b). This adhesin, designated rhicadhesin, has a molecular weight of 14,000 and the interesting property of having five binding sites for Ca^{2+} per monomer. This adhesin is released from cells grown in the absence of calcium or treated with a calcium chelator (Smit et al., 1989a; Smit et al., 1989b). Rhicadhesin was found to be common among all genera of the family *Rhizobiaceae,* including *Agrobacterium* (Smit et al., 1989b). Attachment mediated by rhicadhesin is not host specific; however, since low amounts of rhicadhesin can inhibit the binding of bacteria to roots, it is likely that specific receptors exist on the plant surface (Smit et al., 1989a). The identity of the receptor site remains to be determined but possibly represents a common plant cell wall component in view of the large range of plants to which rhizobial cells attach.

With the discovery of rhicadhesin, a novel model for rhizobial attachment to legume root surfaces can be proposed. This model appears to require that rhizobia in the rhizosphere be under some type of growth limitation. Indeed, if this were the case, the symbiosis might be viewed as a mechanism by which bacteria could avoid growth limitation. Lectin involvement in attachment could not be seen in *R. leguminosarum* bv. viciae when cells were carbon limited. However, in view of the composition of plant root exudates (Boulter et al., 1966; Lipton et al., 1987), carbon is probably not growth limiting in the rhizosphere. Therefore, rhizobia likely attach to the root hair surface through an initial mechanism involving rhicadhesin, a step in attachment that appears to be enhanced by specific lectin recognition. The greater infectivity of manganese-limited than of carbon-limited *R. leguminosarum* bv. viciae cells would suggest that lectin is involved in infection. Following this initial attachment, cellulose microfibrils appear to firmly anchor the cells to the root hair surface. Fimbriae could possibility also be involved in either of these two steps. An important experimental test of this hypothesis will be a determination of the symbiotic properties of mutants defective in rhic-adhesin synthesis.

What Is the Role of Lectin in Legume Nodulation?

Considering the above discussion, it seems unlikely that lectin plays an essential role in rhizobial attachment to the plant root surface. However, a variety of evidence correlates with an important role for plant lectin in infection. What then is the role of lectin in infection? Our working hypothesis is that lectin is an essential element in the ability of the infection to proceed past the stage of root hair curling.

Halverson and Stacey (1984, 1985, 1986a) were the first to provide data showing a direct effect of lectin on nodulation, as opposed to earlier correlative data. In these experiments, pretreatment of cells with soybean lectin was found to enhance nodulation by wild-type *B. japonicum* and to restore normal nodulating ability to a mutant, strain HS111 (Halverson and Stacey, 1985, 1986a). This mutant strain attached normally to soybean roots and could stimulate root hair curling. Therefore, the action of lectin on these cells was proposed to affect some subsequent step in nodulation. The mechanism of action of soybean lectin on *B. japonicum* cells appeared to involve the induction of specific proteins (Halverson and Stacey, 1986a; Stacey et al., 1986). The experiments of Halverson and Stacey indicated that lectin could influence nodulation but provided no evidence in support of an essential role for lectin in nodulation.

Rhizobia infect roots at a large number of sites, but only a few lead to productive infection, i.e., an infection leading to nodule formation. Diaz et al. (1986) and Diaz (1989) reported that only those infections arising above the xylem vascular tissue in pea result in nodule formation. Similarly, recent localization studies with pea (Diaz et al., 1986) have shown lectin localized in cross-sections of roots only above the xylem poles. Recently, Diaz et al. (1989) reported experiments which provide the strongest evidence to date that lectin is essential for nodulation. *R. leguminosarum* bv. viciae, the symbiont of pea, attaches to and curls root hairs on clover roots; however, no infection threads are formed, and nodulation is rare. Diaz et al. (1989) constructed transgenic clover plants which expressed the pea lectin gene and tested these for nodulation by *R. leguminosarum* bv. viciae. The transgenic plants were nodulated by both *R. leguminosarum* bv. trifolii, the normal symbiont, and *R. leguminosarum* bv. viciae.

On the basis of these data, one can speculate as to the role of lectin in nodulation. As noted above, lectin appears to influence rhizobial attachment to legume root hairs but is not essential. Lectin may play a role in increasing the affinity of bacteria for the root surface through the above-mentioned interaction with bacterial polysaccharides. Moreover, the localization of lectin in roots may mark those sites in which productive infection can occur. Lectin may play an essential role beyond the steps of attachment and root hair curling, perhaps by mediating the formation of infection threads. Rhizobial proteins specifically induced by lectin may be essential for penetration into roots or stimulation of infection thread formation. It is hoped that the new lectin hypothesis outlined here will prove to be less controversial than the previous one.

EARLY NODULATION EVENTS

Nodulation Genes of Rhizobia

Long et al. (1982) were the first to isolate genes essential for nodule formation. Since that time, progress in the identification of genes essential for nodulation has been rapid. Figure 5 displays the current genetic maps of *R. leguminosarum* bv. trifolii, *R. leguminosarum* bv. viciae, *R. meliloti*, and *B. japonicum*, showing those genes necessary for nodulation (*nod*) and nitrogen fixation (*nif* and *fix*).

For the purposes of discussion, the nodulation genes can be placed into three groups, the regulatory *nodD* gene, the so-called "common" *nod* genes (*nodABCIJ*), and host-specific nodulation genes.

Regulation of *nod* Gene Expression

Figure 6 presents a generic diagram of the regulatory scheme for *nod* gene induction in rhizobia. In this model, the *nodD* gene product is constitutively expressed and is essential for the expression of the other *nod* genes. NodD is a member of a large family of prokaryotic regulatory proteins, examples of which include *Escherichia coli* LysR, IlvY, and CysB; *Salmonella typhimurium* MetR; and *Enterobacter cloacae* AmpR (Henikoff et al., 1989). The expression of the *nod* genes only occurs in the presence of the host or exudates or extracts of host plant tissue (reviewed by Long, 1989a, 1989b; and Halverson and Stacey, 1986b). The compounds that are produced by the plant and that are responsible for this induction are from the flavonoid family of plant secondary products (examples are shown in Fig. 7). The first such inducer identified was luteolin from alfalfa seed extracts, capable of inducing a *nodC-lacZ* fusion in *R. meliloti* (Peters et al., 1986). Subsequently, other flavonoid compounds that would specifically induce the *nod* genes in other *Rhizobium* and *Bradyrhizobium* species were identified (Zaat et al., 1987; Zaat et al., 1988; Djordjevic et al., 1987; Rossen et al., 1985; Kosslak et al., 1987; Maxwell et al., 1989; Gottfert et al., 1988; Hartwig et al., 1990; Banfalvi et al., 1988).

The mechanism by which these compounds induce *nod* gene expression is not completely understood. Apparently the *nodD* gene product interacts directly with the inducing flavonoid and stimulates transcription. There is only circumstantial evidence to substantiate the direct interaction of flavonoids with NodD; for example, modification of the primary sequence of NodD can change the specificity of the inducer (Burn et al., 1989; Hong et al., 1987; Spaink et al., 1987a; Spaink et al., 1989a). NodD is also apparently associated with the cytoplasmic membrane of the cell and in a good position to interact with the hydrophobic flavonoid molecules (Schlaman et al., 1989; Recourt et al., 1989).

NodD has the ability to bind to the promoter of *nod* genes in the absence of an inducer (Kondorosi et al., 1989; Fisher et al., 1988; Hong et al., 1987; Fisher and Long, 1989). Therefore, it is likely that binding of the inducer may increase the affinity of NodD for the promoter and/or induce a conformational change in the protein which is essential for transcription initiation. The binding site of NodD 5′ to the *nod* genes has been identified by DNA footprint analysis (Fisher and

R. meliloti RCR2011

nifL J N G H I S   200 kb   nifN nodD$_1$ A B C I J   Q P G E F H   syrM nodD$_3$ nifE K D H fixA B C X nifA B nodD$_2$

R. leguminosarum bv. viciae 248

nifK D H rhiC A B R nodO   nodT N M L E F D A B C I J (X)   nifB A

R. leguminosarum bv trifolii ANU843

nodN M L R E F D A B C I J T   nifB A   fixX C B A   nifH D K E N

B. japonicum USDA110

CLUSTER I

nodV W   25 kb   nifD K E N   nifS B frxA nifH   fixB C X   70 kb   hsn

CLUSTER II

hsm nod nod nolA nodD$_2$ D$_1$ Y A B C S U I J   nodZ fixR nifA fixA P

FIGURE 5. Genetic maps of the symbiosis genes of R. meliloti, R. leguminosarum bv. viciae, R. leguminosarum bv. trifolii, and B. japonicum.

nodA
nodD
nod box
Plant flavonoid
NodD protein

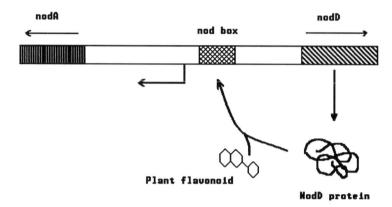

FIGURE 6. Generalized model for the induction of rhizobial nodulation genes by plant flavonoids. Constitutively expressed NodD recognizes the specific flavonoid inducer and stimulates transcription after binding to the *nod* box sequence. The arrows above the map represent the direction of transcription.

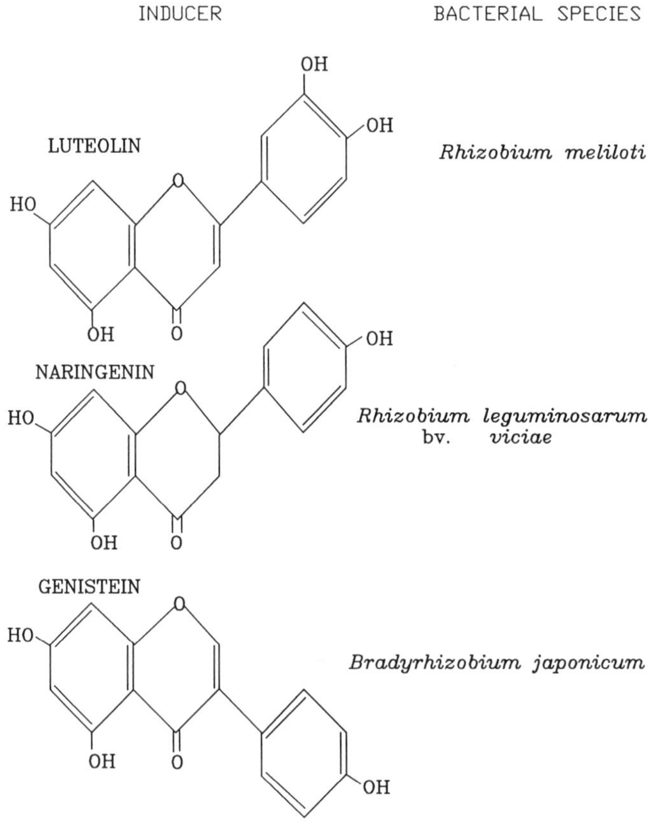

INDUCER
BACTERIAL SPECIES
LUTEOLIN
Rhizobium meliloti
NARINGENIN
Rhizobium leguminosarum bv. viciae
GENISTEIN
Bradyrhizobium japonicum
HO
OH
O

FIGURE 7. Examples of flavonoid inducers of *nod* gene expression. The common name and specific structure of each inducer are shown, as is the corresponding rhizobial species on which it acts.

Long, 1989; Kondorosi et al., 1989) and is, in general, well conserved upstream of the *nod* genes in all species of *Rhizobium* and *Bradyrhizobium* analyzed (Rostas et al., 1986; Spaink et al., 1987b). This sequence has been termed the "*nod* box." Deletion or mutation of the *nod* box sequence abolishes the ability of NodD to stimulate transcription (Hong et al., 1987; Rostas et al., 1986; Spaink et al., 1987b; Wang and Stacey, 1990a). Likewise, mutations in NodD affect the ability of the protein to bind to the *nod* box sequence (Hong et al., 1987; Burn et al., 1989; McIver et al., 1989).

NodD is an important determinant of host specificity; each rhizobial species contains a NodD which interacts optimally with the flavonoids produced by its host. Indeed, in some cases, the host range can be changed by transfer of only the *nodD* gene between *Rhizobium* species (Spaink et al., 1987a). Most *Rhizobium* and *Bradyrhizobium* species appear to possess multiple copies of *nodD;* for example, *R. meliloti* has three (Gottfert et al., 1986; Honma and Ausubel, 1987), and *R. fredii* (Appelbaum et al., 1988) and *B. japonicum* (Gottfert et al., 1989) have two. In contrast, *R. leguminosarum* bv. viciae and *R. leguminosarum* bv. trifolii possess a single *nodD* gene (Wijffelman et al., 1989). These genes are all called *nodD* because of their DNA and protein sequence conservation, although they differ with regard to specificity for the flavonoid inducers. Different *nodD* genes in a single strain may exist to interact with different inducers or inhibitors or to regulate different genes. For example, *R. meliloti* strains possessing single mutations in any of the *nodD* genes nodulate *Medicago sativa* normally; a strain possessing mutations in *nodD1* and *nodD2* nodulates with a 5- to 6-day delay (Honma and Ausubel, 1987; Gyorgypal et al., 1988). In contrast, mutants defective in NodD2 or NodD1 and NodD2 behave identically, showing a 3-day delay in nodulation of *Melilotus alba*. These data suggest that NodD2 is more critical for the nodulation of *M. alba,* presumably because of the specific flavonoid inducers produced by this plant. An *R. meliloti* strain with mutations in all three *nodD* genes is incapable of nodulation (Honma and Ausubel, 1987).

Recently, a number of additional factors have been shown to affect *nod* gene regulation, indicating that the model in Fig. 6 is not strictly applicable to every situation. For example, *nod* gene expression in *R. meliloti* (Dusha et al., 1989) and *B. japonicum* (Wang and Stacey, 1990b) is repressed by the addition of ammonia. This result suggests that a coupling of nodulation and nitrogen fixation functions may be mediated by ammonia regulation of both events. *R. meliloti* strains apparently differ in the way that *nod* gene expression is regulated. Expression of *nodABC* in strain 1021 occurs as shown in Fig. 6; however, strain AK631 regulates *nodABC* expression positively, via NodD, and negatively, via a repressor (Kondorosi et al., 1989). The repressor binds to the overlapping *nodD1* and *nodABC* promoters, at the RNA polymerase binding site, and inhibits both *nodD1* and *nodABC* expression (Kondorosi et al., 1989). Most *R. meliloti* strains and field isolates appear to possess this repressor, and the repressor-producing strains appear to be more efficient in alfalfa nodulation. The need for negative regulation of nodulation in *R. meliloti* is not clear; it may present a way to "fine-tune" *nod* gene regulation or respond to other, as-yet-unidentified, environmental or host-produced factors. The existence of the repressor binding sequence upstream of the *nodD2* gene suggests that its expression is also negatively controlled, but other genes (e.g., *nodEFGH*) lack this sequence (Kondorosi et al., 1989). Therefore, repression also may allow the differential regulation of the *nodD* genes separate from the positive regulation of other *nod* genes.

As indicated in Fig. 6, the *nodD* gene, as originally studied in *Rhizobium* species, was found to be constitutively expressed. In *R. leguminosarum* bv. viciae, the *nodD* gene is constitutively expressed, but the introduction of multiple copies of the *nodD* gene resulted in reduced expression (Rossen et al., 1985). Therefore, in this organism, NodD appears to negatively regulate its own transcription. As noted above, most strains of *R. meliloti* negatively regulate *nodD1* and *nodD2* expression by the synthesis of a repressor protein. However, in the case of *nodD3*, transcription appears to be positively regulated and dependent on the synthesis of a novel regulatory protein, SyrM (Long et al., 1989; Mulligan and Long, 1989). NodD3 is unusual in possessing the ability to induce *nod* gene expression in the absence of flavonoid inducers (Long et al., 1989c).

B. japonicum possesses two *nodD* genes arranged tandemly (Gottfert et al., 1989). The *nodD2* gene has not been extensively studied. NodD1 is required for *nod* gene expression, which is dependent on the presence of isoflavone genistein or daidzein (Kosslak et al., 1987; Banfalvi et al., 1988; Gottfert et al., 1988). However, the expression of NodD1 is itself dependent on the presence of isoflavones, and NodD1 appears to positively regulate its own expression (Banfalvi et al., 1988). Although genistein or daidzein induces both *nodD1* transcription and *nodYABC* transcription, soybean roots also excrete compounds which induce only *nodD1* expression (Smit and Stacey, submitted for publication). Therefore, the plant appears to have the potential to induce *nodD1* expression without inducing the expression of other *nod* genes. Recently, Johnston et al. (1989) reported that the *nodD1* gene of *R. leguminosarum* bv. phaseoli was induced by isoflavones, a result similar to the situation in *B. japonicum*.

Clearly, *nod* gene expression is much more complex than originally realized. The teleological reasons for this complexity and the apparent differences between *Rhizobium* and *Bradyrhizobium* species and strains are currently unknown. However, the recent revelation that *nodD* expression can be highly regulated suggests that the *nodD* gene product may be important for the regulation of other functions in the cell besides the known *nod* genes (Table 2). Indeed, Economou et al. (1989) have recently presented data showing that naringenin (a *nod* gene inducer) negatively controls the expression of the *rhiA* gene of *R. leguminosarum* bv. viciae, although this effect may be indirect.

Function of *nod* Gene Products

Table 3 lists each of the known *nod* genes and the information available concerning their possible biochemical functions. Evidence is now accumulating to suggest that many of the *nod* genes, if not all, are involved directly or indirectly in the production of a "phytohormone" substance which induces root hair curling and cortical cell division. Although numerous earlier reports had reported the isolation of material from rhizobial cultures that could affect leguminous roots,

TABLE 2
Molecular mechanisms of *nodD* regulation

Species or strain	Regulatory mechanism
R. leguminosarum bv. trifolii	Constitutive (Djordjevic et al., 1987; McIver et al., 1989)
R. leguminosarum bv. viciae	Constitutive, autoregulatory (Rossen et al., 1985)
R. meliloti	
RCR2011	
nodD1	Constitutive (Mulligan and Long, 1985)
nodD3	Controlled by SyrM (Long et al., 1989; Mulligan and Long, 1989)
AK631 *nodD1* (likely also *nodD2*)	Repressive (Banfalvi et al., 1988)
B. japonicum	Inducible, autoregulatory (Banfalvi et al., 1988)
R. phaseoli	Inducible (Johnston et al., 1989)

the first indication of the involvement of the *nod* genes in this process came from the work of van Brussel and colleagues (van Brussel et al., 1986; van Brussel et al., 1990; Zaat et al., 1987; Zaat et al., 1988). These authors reported that sterile culture supernatants from *R. leguminosarum* bv. viciae cultures elicited a thick and short root phenotype on seedlings of common vetch (*Vicia sativa* subsp. *nigra*). These studies also showed that the *nodD* and *nodABC* genes were essential for this effect and that inducers of the *nod* genes must be present. Subsequently, Schmidt et al. (1988) reported that culture supernatants from luteolin-induced *R. meliloti* cells induced cell division in soybean and alfalfa protoplasts. More recently, Banfalvi and Kondorosi (1989) reported that the expression of the *nodABC* genes of *R. meliloti* in *E. coli* resulted in the release of substances into the culture supernatants which could induce root hair curling on alfalfa and a few other plants. The compounds released by *R. meliloti* appeared to be host plant specific. Faucher et al. (1988) and Faucher et al. (1989) reported that, although culture supernatants from *R. meliloti* cells curled root hairs on a wide variety of plants, the addition of the *nodH* gene rendered the supernatants specific for alfalfa. In addition, *nodQ* gene mutants produced compounds which could curl the root hairs of vetch and alfalfa (Banfalvi and Kondorosi, 1989; Cervantes et al., 1989). At least three bioactive compounds were isolated by high-pressure liquid chromatography from culture supernatants of *R. meliloti;* when added separately, these compounds induced root hair curling, and when added together, they induced significant root cortical cell division to yield an empty, nodulelike structure (Faucher et al., 1989). The formation of infection threads was not detected. The structures of two of the compounds produced by *R. meliloti* have been determined. The first of these compounds, NodRm-1, is an *N*-acyl-tri-*N*-acetyl-β-1,4-D-glucosamine tetrasaccharide bearing a sulfate group on C-6 of the reducing sugar (Lerouge et al., 1990b). The second compound isolated, NodRm-2, is identical to NodRm-1 but lacks the sulfate substituent. It now seems likely that the NodH, NodP, and NodQ

TABLE 3
Nodulation genes[a] and their possible functions

nod gene	Approximate molecular mass (kDa)[b]	Proposed function
A	22	Essential for root hair curling and cortical cell division; likely synthesizes a phytohormonelike substance (Torok et al., 1984; Egelhoff and Long, 1985; Scott, 1986; Faucher et al., 1989; Faucher et al., 1988; Hollingsworth et al., 1989; Schmidt et al., 1988; Banfalvi and Kondorosi, 1989; Lerouge et al., 1990b)
B	24	See *nodA*
C	47	See *nodA*
D	35–39	Transcriptional regulator (Mulligan and Long, 1985; Gottfert et al., 1986; Rossen et al., 1985; Appelbaum et al., 1988)
E	43	Host range; reported sequence similarity to the *fabB* gene of *E. coli*; encodes a condensing enzyme of fatty acid synthase; encodes a membrane protein (Spaink et al., 1989b; Bibb et al., 1989; Debelle et al., 1986a; Djordjevic et al., 1986; Downie et al., 1985; Horvath et al., 1986; Djordjevic et al., 1985)
F	10	Host range; reported sequence similarity to acyl carrier protein gene (Shearman et al., 1986; Debelle et al., 1986a; Horvath et al., 1986; Djordjevic et al., 1985; Djordjevic et al., 1986; Downie et al., 1985)
G	27	Host range; reported sequence similarity to dehydrogenase genes (Horvath et al., 1986; Debelle et al., 1986a; Kannenberg and Brewin, 1989; Fisher et al., 1987)
H	29	Host range; may encode a sulfate transferase for the synthesis of NodRm-1 (Faucher et al., 1988; Faucher et al., 1989; Horvath et al., 1986; Debelle et al., 1986a; Lerouge et al., 1990b)
I	34	Reported sequence similarity to ATP binding transport protein genes (Higgins et al., 1986; Evans and Downie, 1986)
J	28	Encodes a membrane protein; may act in conjunction with *nodI* (Evans and Downie, 1986)
K	15	Reported only in *Bradyrhizobium* sp. (*Parasponia*); function unknown (Scott, 1986)
L	20	Host range; reported sequence similarity to acetyltransferase gene (Canter Cremers et al., 1989; Surin and Downie, 1988; Downie, 1989)
M	66	Host range; reported sequence similarity to amidotransferase gene (Surin and Downie, 1988)
N	18	Host range; function unknown (Surin and Downie, 1988)
O	30	Function unknown; encoded protein is exported; reported to bind calcium (Economou et al., 1990; de Maagd et al., 1989)
P	35	Host range; likely encodes a subunit of ATP sulfurylase (Schwedock and Long, 1990)
Q	71	Host range; likely encodes a subunit of ATP sulfurylase (Schwedock and Long, 1990)
R	14	Reported only in *R. leguminosarum* bv. viciae; function unknown (Rolfe et al., 1989)
S	23	Function unknown (Gottfert et al., 1990b; Lewin et al., 1990)
T		Host range; function unknown (Canter Cremers et al., 1989)
U	62	Function unknown (Gottfert et al., 1990b; Lewin et al., 1990)
V	99	Reported only in *B. japonicum*; important for host range; sequence similarity to membrane sensor family genes (Gottfert et al., 1990a)

(*Continued on next page*)

TABLE 3—*Continued*

nod gene	Approximate molecular mass (kDa)[b]	Proposed function
W	25	Reported only in *B. japonicum*; important for host range; sequence similarity to transcriptional regulatory protein family genes (Gottfert et al., 1990a)
X		Reported only in *R. leguminosarum* strains capable of nodulating Afghanistan pea (Davis et al., 1988)
Y	15	Reported only in *B. japonicum* but found in all rhizobia by hybridization; function unknown (Nieuwkoop et al., 1987; Banfalvi et al., 1988)
Z	35	Host range; function unknown (Stacey et al., 1989; Schell et al., personal communication)
nolA	27	Host range; essential for nodulation of selected soybean cultivars; reported only in *B. japonicum*; likely a transcriptional regulatory protein gene (Sadowsky et al., 1991)

[a] A variety of other genes can affect nodulation but have not been given the "*nod*" or "*nol*" designation. The reader is referred to other recent reviews which describe these related genes (e.g., Martinez et al., 1990; Appelbaum, 1989; and Long, 1989b).

[b] In most cases, molecular masses are those deduced from the DNA sequence. Molecular masses deduced by physical methods may differ.

proteins are involved in the conversion of NodRm-1 to NodRm-2 via a sulfation reaction (Lerouge et al., 1990b; Schwedock and Long, 1990). This modification is apparently important for the specificity of the former compound for *Medicago sativa*. A similar compound has very recently been identified in culture supernatants of *R. leguminosarum* bv. viciae; however, it is a pentasaccharide (Spaink et al., 1991; Spaink et al., unpublished data). Likewise, a related compound has been isolated from the supernatants of *B. japonicum* cultures induced for *nod* gene expression (Stacey et al., unpublished data).

These compounds, in either the crude or the purified state, have been shown to have a variety of biological activities. (i) They can inhibit root growth (the thick and short root phenotype) (van Brussel et al., 1986; Zaat et al., 1987). (ii) They can induce plant root hair deformation (Had[+] phenotype) (Lerouge et al., 1990a; Lerouge et al., 1990b; Banfalvi and Kondorosi, 1989). (iii) They can induce cortical cell division (Coi[+] phenotype) (Lerouge et al., 1990a; Lerouge et al., 1990b). (iv) They can induce nodule formation (Nod[+] phenotype) (Lerouge et al., 1990a; Spaink et al., 1991); of course, these nodule structures lack bacteria. (v) They can induce increased exudation of flavonoids, the inducers of *nod* gene expression (the increased *nod* gene induction phenotype) (van Brussel et al., 1990). These biological effects of the isolated *nod* factor mirror various aspects of the bacterial infection process summarized above. Therefore, it now seems likely that these plant responses to the bacteria are due to an interaction with the *nod* factor(s).

INFECTION INITIATION AND INFECTION THREAD GROWTH

Rhizobial penetration of the root hair wall appears to occur without significant degradation of the wall. Instead, the cell wall of the host invaginates at the site of infection, and new wall material is deposited to form the infection thread (Callahan and Torrey, 1981). The role of hydrolytic enzymes, of either plant or bacterial origin, in this process has been controversial for some time. The reader is referred to earlier reviews (Dazzo, 1981; Hodgson and Stacey, 1986; Bauer, 1981) for a discussion of the literature in this area. The obvious difficulty in this type of research is in determining what is occurring in a very small, discrete site. Recently, Baker and colleagues (1989) reported the isolation of a pectinase which was cell associated in *R. leguminosarum* bv. trifolii. Scanning electron microscopic examination of roots inoculated with *R. leguminosarum* bv. trifolii revealed obvious zones of cell wall degradation below the attached rhizobial cells, consistent with the activity of this enzyme.

As indicated above, the current model for legume infection suggests that the NodABC factor has phytohormone activity and is responsible for root hair curling and cortical cell division. However, rhizobia are also known to produce cytokinin and auxin, although the role of these compounds in nodulation is unclear (reviewed by Dazzo, 1981; Hodgson and Stacey, 1986; Bauer, 1981; Halverson and Stacey, 1986b; and Rolfe and Gresshoff, 1988). Recently, an auxin biosynthetic gene was cloned from *B. japonicum* (Sekine et al., 1989a, 1989b). Hybridization to genomic DNA of *B. japonicum* and other *Rhizobium* species with this cloned gene as a probe suggests that this gene may be present in several copies (Syono, personal communication). Kittell et al. (1989) recently presented similar data for *R. meliloti*. These authors identified four isoforms of aromatic aminotransferase presumptively involved in the biosynthesis of indoleacetic acid via tryptophan and an indolepyruvate intermediate. Genes encoding three of these proteins were isolated, and two enzyme mutants, AAT1 and AAT2, were constructed. The mutants continued to produce some indoleacetic acid and were fully competent for alfalfa nodulation. The presence of multiple genes for auxin synthesis might explain the difficulty found in several laboratories in trying to isolate mutants defective in this process.

Taller and Sturtevant (1991) and Sturtevant and Taller (1989) have investigated the ability of rhizobia to produce cytokinin. All of the strains examined produced significant amounts of cytokinin. Most interestingly, treatment of *B. japonicum* or *R. meliloti* with their respective *nod* gene-inducing flavonoids resulted in a shift in the profile of cytokinins produced (Taller and Sturtevant, 1991). These data suggest the possible involvement of *nod* genes in cytokinin production, perhaps regulated by NodD. However, *B. japonicum* mutants defective in *nodA*, *nodB*, *nodC*, or *nodD* did not differ significantly from the parental strain with regard to the cytokinins produced. Therefore, any possible role for cytokinin or auxin in the nodulation process has yet to be unequivocally established.

Once the bacteria penetrate the root hair, the infection thread containing the bacteria grows to the base of the root hair cell and is ramified into the cortex of the root. In general, studies of the process of infection thread initiation and growth have largely been limited to light microscopy. However, Bakhuizen (1988) recently studied this process and linked infection thread growth to the plant microtubule cytoskeleton. That work focuses on the fact that rhizobia within an infection thread trigger mitotic divisions in advance of the growing thread, perhaps in response to factors produced by the *nod* gene products or by other substances. Bakhuizen suggests that the dividing cells are arrested in the G_2 phase of the cell cycle after the nucleus migrates to the center of the cell. Microtubules radiating from the nucleus guide the infection thread via a cytoplasmic bridge, thus traversing what would otherwise be a cell wall, toward the nucleus of the next cell. This is an attractive hypothesis, since it does not require novel mechanisms; rather, it suggests that the rhizobia utilize the mitotic apparatus for directing intraroot infection thread growth and offers a possible explanation for the typical proximity of the nucleus to the growing infection thread.

Infection threads are largely composed of two structures, both thought to be contributed by the host cells. First, they are bound by a membrane that is continuous with the host cell plasma membrane, the threads thus being kept extracellular. Second, much cell wall material must be provided, since the membrane is separated from the bacteria by such a wall (Fig. 8 and 9). Although the wall is usually thought to be composed of typical cell wall material, it could be made of callose, a substance used by plants to respond quickly to wounds. Most infection threads normally span a distance of several millimeters, and the amount of material supplied by the host cells to accommodate such growth is sizeable. This is another example of the close cooperation between the two cell types; bacteria divide and signal to the host cells, which respond by providing wall material.

These host cell contributions undoubtedly result from the usual synthetic functions of the endoplasmic reticulum and Golgi bodies, both of which seem to proliferate even before bacteria are present and continue in high amounts throughout bacterium release (Fig. 8 and 9). Thus, the bacteria seem able to induce the proliferation of the synthetic apparatus, probably as a separate step beyond the earlier initiation of mitosis.

BACTERIUM RELEASE

The infection thread shows specializations for the release of bacteria. In some species, threads balloon at their ends into droplets, although this is usually not the case (see Fig. 10 for an exception) in soybean, in which single-file bacteria are usually seen. Each host cell must not only allow but also actively support this event to the extent of accommodating the needs of thousands of bacteria. By the end of the release stage, the cytoplasm of each cell is greatly expanded and almost totally filled by several thousand symbionts, one to a few per symbiosome (Fig. 10).

Bacterial release from the infection thread appears to be primarily an active process, as evidenced from studies of bacterial mutants defective in release (Regensburger et al., 1986; Fortin et al., 1987; Roth and Stacey, 1989a). However, at present, these mutants have provided little information about the bacterial signal, if one exists, that triggers release. The release process, described from a cell biology viewpoint, is a combination of both exocytic and endocytic events related closely in time and place. Electron micrographs precisely oriented enough to show the sequence are rare because of the spatial and temporal limitations

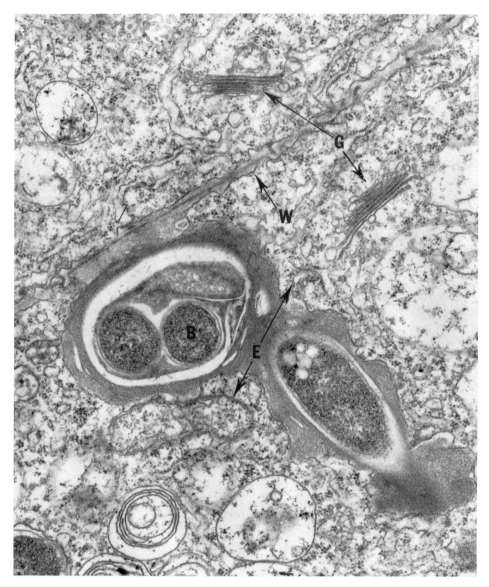

FIGURE 8. Infection thread in soybean. This transmission electron micrograph shows an infection thread containing several organisms of *B. japonicum* (B) surrounded by a cell wall, all of which is contained in a plasma membrane supplied by the host cells. Parts of two such cells are shown, along with the cell wall (W) separating them. The secretory apparatus of the cell has proliferated; hence, much endoplasmic reticulum (E) and numerous Golgi bodies (G) are present. ×31,600.

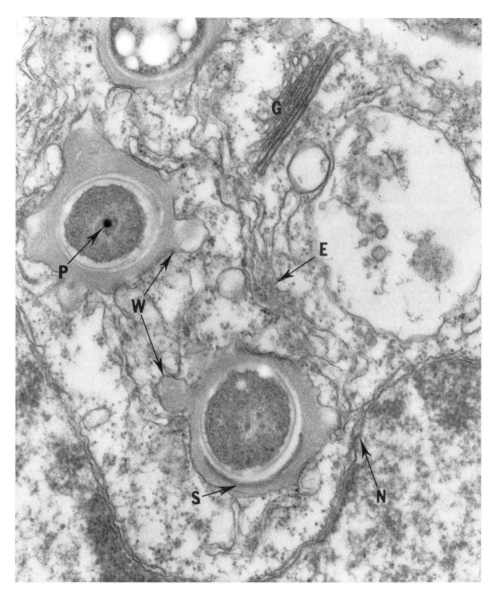

FIGURE 9. Infection threads cross-sectioned near the point of bacteroid release in soybean. The plasma membrane is convoluted, with protrusions containing portions of the cell wall (W) being endocytosed into the cytoplasm of the host cells. Deep in the infection threads, new symbiosome membrane (S) is forming for the endocytosis of the adjacent bacteroid. Endoplasmic reticulum (E) and Golgi bodies (G) are always nearby, and release of bacteroids always takes place very near the nuclear envelope (N). One bacteroid shows a small early polyphosphate body (P). ×43,000.

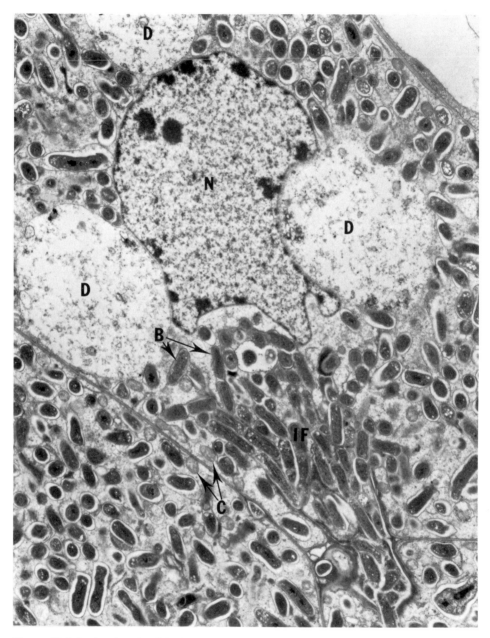

FIGURE 10. Infection thread activity in a soybean cell with digestive vacuoles. An unusually complex infection thread (IF) is releasing bacteria (B) close to the nucleus (N), which is indented by three large digestive vacuoles (D). In an adjacent cell, numerous symbiosomes can be seen packing the cytoplasm; in this case, only one bacteroid is found per symbiosome. The cytoplasm has comparatively little endoplasmic reticulum and few Golgi bodies, while other organelles, e.g., mitochondria (C), are now located just under the plasma membrane. ×6,700.

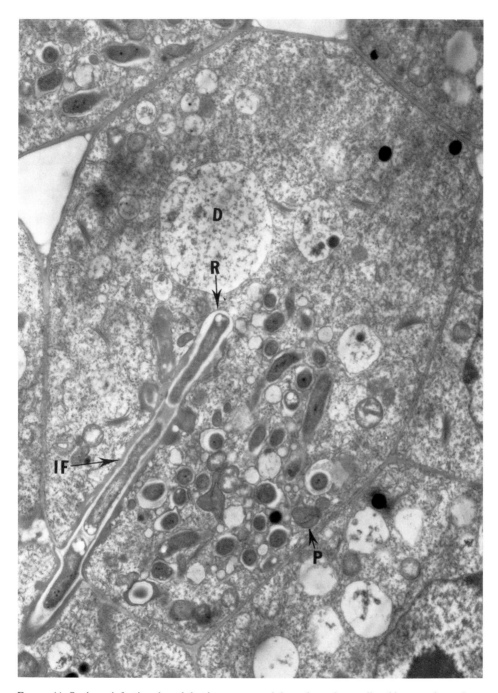

FIGURE 11. Soybean infection thread that has penetrated through one host cell and into another, where release is taking place. The single file of bacteroids has a cell wall in the infection thread (IF), except where release will take place (R). Some parts of the cell are devoid of bacteroids, while the area closer to the release area has numerous organisms in their symbiosome compartments, and the mitochondria and plastids (P) are largely displaced to the cell periphery. A prominent digestive vacuole is also present already (D). ×9,150.

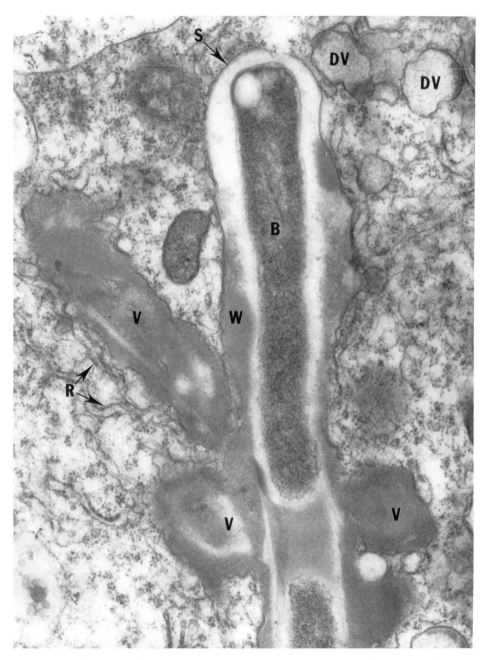

FIGURE 12. Release area of Fig. 11. The cell wall (W) is being engulfed into large endocytic vesicles (V), the membranes of which appear to be supplied by exocytosis of the endoplasmic reticulum, which has only a few ribosomes (R); cell wall digestive vacuoles (DV) are found in the adjacent cytoplasm. The bacteroid (B) has a symbiosome membrane (S) forming without cell wall material. ×38,500.

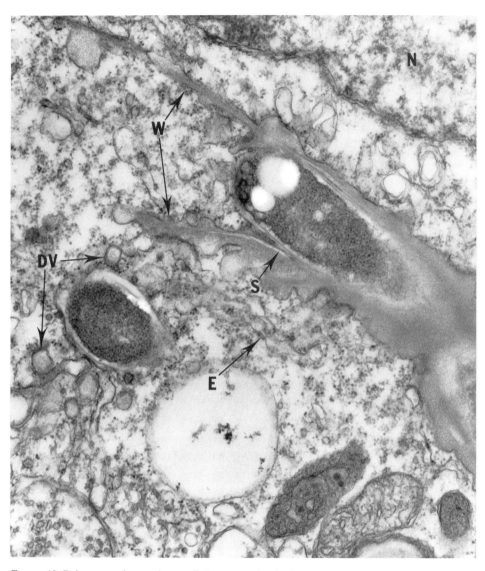

FIGURE 13. Release area in a soybean cell. Long strands of cell wall (W) protrude beyond the bacteroid being released. Many small cell wall digestive vacuoles (DV) have formed and are scattered nearby; they eventually disperse throughout the cytoplasm, where they are often seen at about the same frequency as symbiosomes. The endoplasmic reticulum (E) is plentiful, and release is taking place within a few micrometers of the nuclear envelope (N). The new symbiosome membrane (S) is seen here forming back into the infection thread. ×40,300.

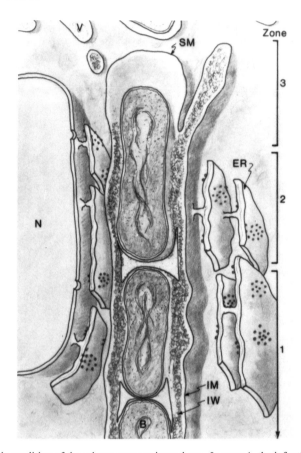

FIGURE 14. Artist's rendition of the release process in soybean. In zone 1, the infection thread is shown in its typical form. In zone 2, the endoplasmic reticulum (ER) is clustered around the infection thread; this is the exocytic zone, where membrane is being added to the thread and enzymes are being inserted into the wall to begin its degradation. In zone 3, a bacteroid (B) is being released as new symbiosome membrane (SM) is being formed. The wall is also being endocytosed into the cytoplasm as vesicles (V). Note that between the bacteroids and very deep in the infection thread, new symbiosome membrane is also being formed. N, nucleus; IM, infection thread membrane; IW, infection thread wall.

involved. Figures 11 through 13 are such sections, and Fig. 14 is an artist's conception of the process, recently described in detail (Roth and Stacey, 1989b).

The first step in release is exocytosis, when a continuous supply of new membrane is needed at an accelerated rate. Since each bacterium is contained in a membrane-bounded symbiosome, each cell must supply membrane for thousands of symbiosomes. Although estimates vary and amounts differ from one species to another, it seems plain that tens of thousands of square micrometers of new membrane must be produced by each cell within a few days. The organelle seen in profusion at the release site is the endoplasmic reticulum (Fig. 9 and 12 through 14), which apparently provides this membrane by fusion with the plasma membrane of the infection thread.

Invariably, however, these exocytic processes do not just insert new membrane; they also export synthesized proteins into the extracellular space. The contribution is probably to another event, the elimination of the infection thread wall. Such walls are totally absent in mature infections, and it is probable that the endoplasmic reticulum contributes the enzymes to at least begin this degradation.

The release event per se takes place by endocytosis, the taking of substances into the cytoplasm. This process results in bacteria being taken into or released into the host cytoplasm, a process also called phagocytosis. The bacteria seem to be pushed out of the end of the thread, which is now inhibited from further growth. The endocytic process is an active cellular function and results in each bacteroid being encased in a membrane, the symbiosome membrane, which moves away into the cytoplasm. The symbiosome membrane is therefore composed of both the plasma membrane, which was the original infection thread membrane, and the newly inserted endoplasmic reticulum (Robertson et al., 1978). Interestingly, this event always takes place very close to the nuclear envelope of the host cell, where the endoplasmic reticulum is in good supply (Fig. 9, 13, and 14); it is possible that the explanation for this positioning is that release is dependent on the endoplasmic reticulum and can be initiated only when the endoplasmic reticulum is plentiful.

It also appears that, rather deep in the infection thread, new membrane is provided de novo, a method cells use when membrane must be formed rapidly. Parts of the early symbiosome membrane are found where it is difficult to see how endoplasmic reticulum or other preformed vesicles could penetrate. Thus, a third source of new membrane is indicated (Fig. 9 and 14) (Roth and Stacey, 1989b).

What happens to the infection thread wall next? Another, parallel endocytosis takes place, this one to partition the wall into endosomes and begin the degradation of wall components. These endosomal vesicles move into the host cytoplasm in large numbers (Fig. 9 and 12 through 14), although long layers of wall may extend out beyond the release point, perhaps indicating that the process sometimes causes a lag in release (Fig. 13 and 14). This process constitutes another demand on the host cell for large amounts of membrane synthesis.

The latter situation raises the possibility that the enzymes needed at this time could be rendered defective or absent by mutations. Bacterial mutations that induce nodules dysfunctional in fixing nitrogen (Nod$^+$ Nif$^-$) are beginning to be studied productively, but results are still in the early stages (Regensburger et al., 1986; Morrison, and Verma, 1987; Roth and Stacey, 1989a; Stacey et al., 1991).

The bacteria, for a time after release, continue to divide as symbionts. In optimum growth conditions, this proliferation is vigorous enough that multiple symbionts are seen in each symbiosome (Fig. 15); in many cases, symbiosomes contain only one symbiont (Fig. 10).

Bacteria within the symbiosomes undergo important transformations and are termed bacteroids to draw attention to these changes. These transformations include an increase in the volumes of individual symbionts, especially in crops infected by the fast-growing *Rhizobium* species. With *Bradyrhizobium* species,

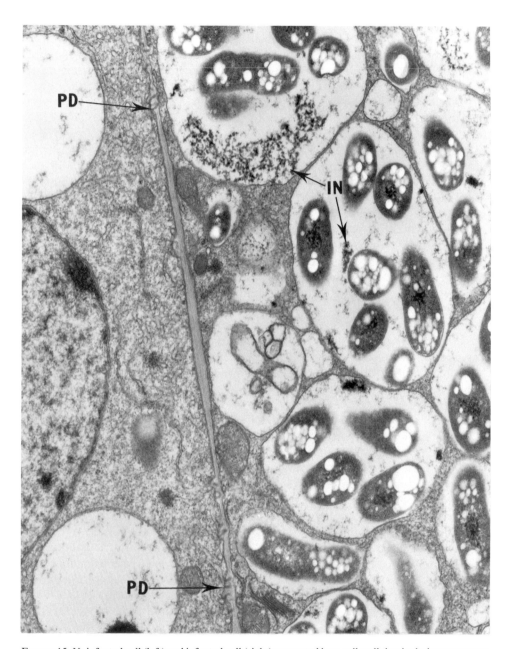

FIGURE 15. Uninfected cell (left) and infected cell (right) separated by a cell wall that includes numerous plasmodesmata (PD) in two areas of the cell wall. Symbiosomes, each with numerous bacteroids that should now be fixing nitrogen, have numerous inclusions (IN); the nitrogen fixed in the bacteroids is thought to traverse the plasmodesmata to enter the cytoplasm of the uninfected cells. ×18,500.

the volumes are only slightly increased, even though other transformations take place; e.g., polyphosphate bodies may form in each bacteroid (Fig. 10 and 15), in contrast to the bacteria in infection threads, which show no such bodies or only a few small ones (Fig. 11). (Typical polyphosphate bodies are not seen in *Rhizobium* species, even in bacteroids.) Substances of unknown composition appear in the symbiosome space (Fig. 15 and 16). The volume of this space varies from one species to another and from cell to cell; it is quite small in alfalfa (Fig. 2) but is large in soybean in comparison with bacteroid volume (Fig. 10, 15, and 16).

A series of biochemical and cellular changes also occur. Bacteroids differ from their free-living forms in the composition of their cell surface (van Brussel et al., 1977), metabolism (reviewed by Hodgson and Stacey, 1986), and chromatin structure (Wheatcroft et al., 1990). In addition, bacteroids soon stop dividing, and the number of symbionts, although increased by divisions for a time after release, soon reaches a maximum.

The host cells are best described now in terms of the organization of mature nodules at intracellular, intercellular, and whole-nodule levels. In addition to greatly increasing their volumes to accommodate many symbiosomes, plants supply nutrients to a rapidly increasing number of bacteroids. The plant cells are uniquely organized; organelles, especially mitochondria and plastids, are positioned just under the plasma membrane (Fig. 10 and 15) and are particularly concentrated under the membrane opposite intercellular spaces, where oxygen may be supplied (Fig. 17). In the cytoplasm as a whole, however, they are organized for a very low oxygen level, mediated by leghemoglobin, which is formed in high amounts (Robertson et al., 1984).

Uninfected cells are typified by having a large central vacuole with a thin layer of cytoplasm (Fig. 16). In healthy nodules, numerous plastids containing several large starch grains are seen (Fig. 16). Peroxisomes are characteristic features of uninfected cells, along with much tubular smooth endoplasmic reticulum, many small mitochondria, and numerous Golgi bodies (Fig. 17). Digestive vacuoles are also present, sometimes numbering several per cell, having quite large volumes, and occurring as early as during bacterium release (Fig. 10 and 11).

Intercellular relationships also change. In early nodule cells, plasmodesmata (Fig. 8, 15, and 17) allow cytoplasmic confluence between all cells, both infected and uninfected (Selker and Newcomb, 1985); the entire conjoined cytoplasm of numerous such cells is referred to as "symplasm." Plasmodesmata are important as routes for intercellular metabolite movement (Wolf et al., 1989; Gunning and Overall, 1983; Gharyal et al., 1989). In mature nodules, plasmodesmata are found only between uninfected cells and between uninfected and infected cells; they are conspicuously absent between infected cells (Selker, 1988). Gharyal et al. (1989) reported that *B. japonicum* lipopolysaccharides could inhibit symplastic transport in soybean tissue culture cells. Whether such bacterial products could act to control plasmodesma distribution and function within nodules remains to be tested.

The nodule diameter is now 1 to 4 mm in diameter and has the cellular organization described by Selker (1988) and Selker and Newcomb (1985). Strands

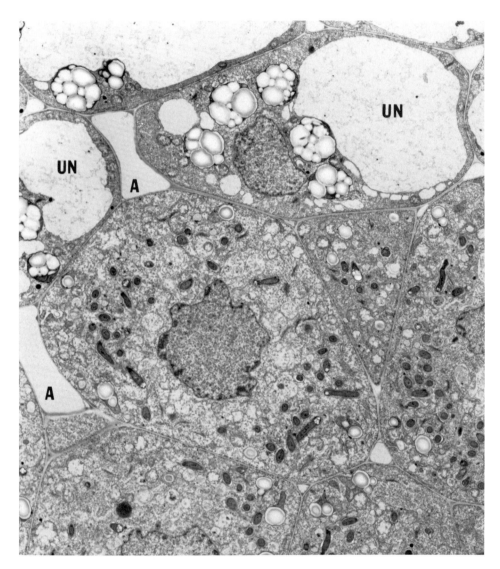

FIGURE 16. Survey of uninfected and infected cells in soybean. Note the frequency of uninfected cells (UN) that are characterized by a large vacuole and many plastids, each containing numerous starch grains. This is an early infection; later, the infected cells would be much larger and have cytoplasms engorged with symbiosomes. Air spaces (A) are found where cells do not abut; with more mature infected cells, these would still be found, and the mitochondria and plastids would tend to cluster just inside the plasma membrane of the infected cells opposite these spaces. ×3,700.

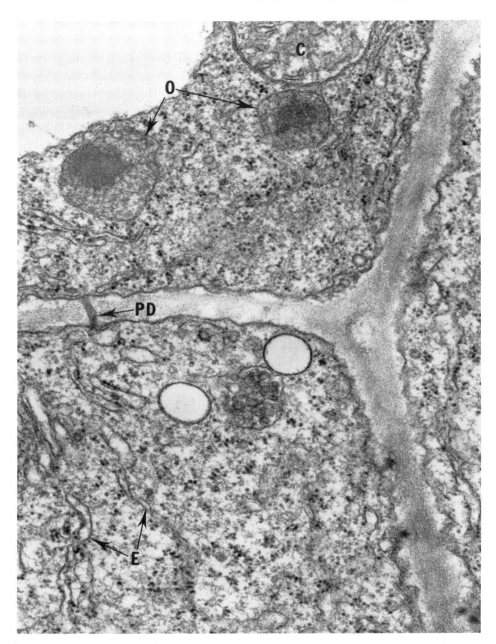

FIGURE 17. Uninfected cells of soybean. The unique organelles are the peroxisomes (O). A prominent plasmodesma (PD) connects the two cells, providing a symplasm of continuous cytoplasm between them. Through them, nitrogen fixed by the bacteroids in the symbiosomes of infected cells enters the uninfected cells, where it is synthesized into ureides in the peroxisomes, reenters the cytoplasm, and traverses numerous plasmodesmata through the symplasm of uninfected cells to the circulatory system. Uninfected cells have much endoplasmic reticulum (E), with few attached ribosomes, which are found dispersed in the cytoplasm. Mitochondria (C), with double membranes that contrast with the single membrane of the peroxisomes, are present. ×68,000.

of uninfected cells occupy about 20% of the volume (Selker and Newcomb, 1985), are arranged in a central group with roughly radial patterns, and separate larger groups of infected cells. Just under the nodule surface, several layers of uninfected cells are found along with vascular bundles that connect with the root vasculature.

Selker (1988) has hypothesized the functional significance of this organization. The fixed nitrogen produced in the symbiosomes passes across the symbiosome membrane, through the host cytoplasm, and across the plasmodesmata to the cytoplasm of adjacent uninfected cells. There the fixed nitrogen enters the peroxisomes; ureides, the form of nitrogen circulated in soybean (Blevins, 1989; Newcomb and Kowal, 1985), are synthesized and released into the cytoplasm. The ureides or asparagine (the latter being the nitrogen compound transported in plants such as alfalfa [Blevins, 1989]) can pass through successive uninfected cells in the symplasm to reach vascular bundles at the periphery of the nodule, from which they can circulate throughout the plant.

The nodule organization and transport system that must efficiently carry away fixed nitrogen is thus defined. However, the nodule organization must also facilitate the transport of carbon sources for the bacteroid, although this route is not as well understood. Gas exchange has also been studied. Oxygen is required for ATP production, but nitrogenase is inactivated by direct contact with O_2. Witty and Minchen (1990) summarized the evidence for an O_2 barrier in the outer cortical cell layers of nodules that would limit O_2 in the air spaces between nodule cells. They described a layer of cells in the nodule cortex that constitutes an aqueous diffusion barrier and suggested that it may be controlled so that O_2 availability is varied, perhaps by the regulation of the hydrophobic or hydrophilic nature of a glycoprotein.

By the mature infection stage, the host cell cytoplasm is engorged with symbiosomes, has little endoplasmic reticulum (although many free ribosomes and polyribosomes may be present), and has few Golgi bodies. The needs for new membrane and more cell wall have been met, and infection threads have disappeared. Several bacteroids and some amorphous material are contained in the symbiosomes.

Digestive vacuoles are now more common. They may contain a few identifiable structures in some cases; in others, they may contain lipid whorls, granules, and a few bacteroids (Fig. 18).

CONCEPT OF BACTEROID TURNOVER

The digestive vacuoles described above can sometimes be quite large, and several may be present in a given host cell. This may be the case in soybean even as early as when infection threads are present and bacteria are still being released (Fig. 10 and 11). In rare cases, bacteroids are seen in the digestive vacuoles, along with other material, including myelinlike whorls reminiscent of the reaggregation of membrane degradation products (Fig. 18).

Many small granules may be seen frequently, and X-ray microprobe analysis reveals that these granules are polyphosphate bodies (Roth et al., 1987), since they show the normal "signature" of such bodies: phosphate, calcium, and mag-

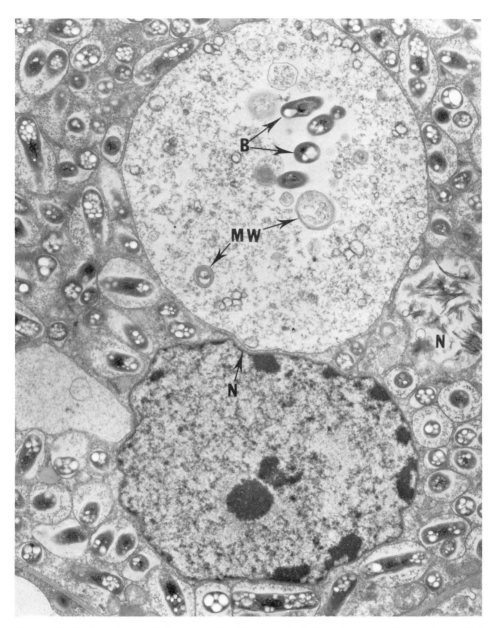

FIGURE 18. Soybean host cell showing a digestive vacuole that contains several recently engulfed bacteroids (B) and that is almost in contact with the nuclear envelope (N with arrow). Such bacteroids are rarely seen in digestive vacuoles, indicating that they are probably degraded rapidly after they are exocytosed into the vacuoles. Otherwise, the vacuoles contain myelin whorls (MW) and other debris. Other vesicles that contain needle-shaped inclusions are frequently seen in some cells (N). The spaces in the many symbiosomes in this cell contain much fine material. ×8,400.

nesium. This finding provides direct evidence for bacteroid degradation as an event that can, in at least some circumstances, take place simultaneously with bacterium release. The high number of polyphosphate bodies seen in these vacuoles indicates that hundreds of bacteroids per cell are degraded. Thus, the concept of a significant presenescence turnover of bacteroids is indicated.

Whether this turnover is a normal process or whether it is induced or greatly enhanced by stresses has not been determined. Although adverse conditions cause intracytoplasmic turnover in many cellular systems, evidence is scanty in nodules. The only direct evidence comes from a study of aluminum in soybean (Roth et al., 1987). Long known to be toxic to a variety of biological processes (Haug, 1984; Karlik et al., 1980), aluminum not only was present in the polyphosphate body remnants in digestive vacuoles but also first appeared in the polyphosphate granules of bacteroids that looked normal. Although some bacteroids were found without aluminum in younger infections, it was suggested that aluminum first appeared some time after polyphosphate bodies had formed, probably as a sequestering and inactivating place for toxic ions. Aluminum may have reacted with the bacteroid DNA to kill or alter bacteroids that were then recognized as foreign bodies by host cells, which destroyed them.

How do bacteroids normally avoid being recognized as foreign bodies? Here is another case of close interaction between the cellular processes of host and symbiont. Normally, a eukaryotic cell recognizes a bacterium and invokes its lysosomal processes to destroy the invader, but in this and many other endocytobiological cases, e.g., leishmaniasis (Rabinovitch and Alfieri, 1987), these processes are repressed. The matter has been little studied in nitrogen-fixing systems, and the reader is referred to other chapters for explanations. The turnover concept indicates, however, that conditions in which this repression becomes "leaky" or is derepressed can occur.

If such turnover occurs early in infections, while bacteroids are still dividing, the loss is only in the energy spent in useless bacteroid divisions, their subsequent degradation, and the handling of retained "garbage" from undigestible bacteroid remnants. If, however, it occurs later, when division has stopped, a depletion of bacteroid populations in host cells results, and the nitrogen-fixing capabilities of the cells are probably permanently reduced.

It is these same lysosomal processes that remove infection thread walls. Thus, concurrent with the digestion of these infection thread components, symbiosomes are usually able to withstand or repress these processes. Lysosomes have been little studied in root nodules. Mellor (1989) has suggested that lysosomal enzymes are integral components of the symbiosome space, although one could also suggest that the homology should be symbiosomes to endosomes and digestive vacuoles to lysosomes. Mellor's concept, thus modified, has merit in that normal lysosomes have pHs of about 5 and normal endosomes have pHs of about 6 (Fok et al., 1988; McCoy and Schwartz, 1988). When the carbon sources for bacteroid function are cut off, the bacteroids can no longer fix nitrogen and the symbiosome space is allowed to acidify; then, enzymatic pH optima are able to develop and lead to the degradation of the bacteroids (Brewin et al., 1990). The matter needs much more study than it has received.

SENESCENCE OF NODULES

Senescence of nodules is a well-known occurrence. It is typified by a total disintegration of nodules and is essentially an event controlled by the plant. Comparatively little cell biology study has been directed to senescence, but a brief treatment is given at the end of the review by Newcomb (1981). It is usually considered to be the only mechanism of bacteroid demise, but much earlier turnover or degradation, as discussed above, must be considered to be another mechanism, since it can affect the capacity of the nitrogen fixation system well before the total nodule destruction of senescence takes place.

SUMMARY

Intimate interrelationships between host and symbiont are found repeatedly in these systems. For example, the plant is induced to supply membrane and cell wall for infection thread growth, proliferate cells to form nodules, enlarge each cell's capacity for synthesis, facilitate bacteroid release, and provide carbon for symbiont metabolism; bacteria are induced to reduce their surface polysaccharides and transform into bacteroids.

Knowledge of the bacterial genetics of legume nodulation has advanced rapidly. Recent revelations highlight the complexity of host control of the *nod* genes and present a conundrum as to why this level of control exists. Much work remains to be done to elucidate the biochemical functions of the various genes identified. However, considerable progress has been made in this direction. New phytohormonelike compounds have been found and promise to aid our understanding of the rhizobium-legume symbiosis.

The current level of knowledge, although requiring continued basic research, is approaching a time at which applications that will improve the yields of leguminous crops are possible, both on presently used land and on land that currently provides marginal yields.

REFERENCES

Abe, M., J. E. Sherwood, R. I. Hollingsworth, and F. B. Dazzo. 1984. Stimulation of clover root hair infection by lectin-binding oligosaccharides from the capsular and extracellular polysaccharides of *Rhizobium trifolii. J. Bacteriol.* **160:**517–520.

Aguilar, J. M. M., A. M. Ashby, A. J. M. Richards, G. J. Loake, M. D. Watson, and C. H. Shaw. 1988. Chemotaxis of *Rhizobium leguminosarum* biovar phaseoli towards flavonoid inducers of the symbiotic nodulation genes. *J. Gen. Microbiol.* **134:**2741–2746.

Aikawa, M. 1980. Host cell invasion by malarial parasites, p. 31–46. *In* C. B. Cook, P. W. Pappas, and E. D. Rudolph (ed.), *Cellular Interactions in Symbiosis and Parasitism.* Ohio State University Press, Columbus.

Appelbaum, E. 1989. The Rhizobium/Bradyrhizobium-legume symbiosis, p. 131–158. *In* P. M. Gresshoff (ed.), *Molecular Biology of Nitrogen Fixation.* CRC Press, Boca Raton, Fla.

Appelbaum, E. R., D. V. Thompson, K. Idler, and N. Chartrain. 1988. *Rhizobium japonicum* USDA 191 has two *nodD* genes that differ in primary structure and function. *J. Bacteriol.* **170:**12–20.

Baker, D., M. Petersen, M. Robeles, J. Chen, F. Dazzo, and D. Hubbell. 1989. Pit erosion of root epidermal cell walls in the *Rhizobium*-white clover symbiosis, p. 96. *Proc. 12th North Am. Symbiotic Nitrogen Fixation Conf.,* 30 July to 3 August 1989, Ames, Iowa. Iowa State University, Ames.

Bakhuizen, R. 1988. Ph.D. thesis. University of Leiden, Leiden, The Netherlands.

Banfalvi, Z., and A. Kondorosi. 1989. Production of root hair deformation factors by *Rhizobium meliloti* nodulation genes in *Escherichia coli:* HsnD (NodH) is involved in the plant host-specific modification of the NodABC factor. *Plant Mol. Biol.* **13:**1–12.

Banfalvi, Z., A. J. Nieuwkoop, M. G. Schell, L. Besl, and G. Stacey. 1988. Regulation of nod gene expression in *Bradyrhizobium japonicum. Mol. Gen. Genet.* **214:**420–424.

Barbour, W. M., D. R. Hattermann, and G. Stacey. 1991. Chemotaxis of *Bradyrhizobium japonicum* to soybean exudates. *Appl. Environ. Microbiol.* **57:**2635–2639.

Bauer, W. D. 1981. Infection of legumes by rhizobia. *Annu. Rev. Plant Physiol.* **32:**407–449.

Bibb, M. J., S. Biro, H. Motamedi, J. F. Collins, and C. R. Hutchinson. 1989. Analysis of the nucleotide sequence of the *Streptomyces glaucescens*-Tcml genes provides key information about the enzymology of polyketide antibiotic biosynthesis. *EMBO J.* **8:**2727–2736.

Blevins, D. G. 1989. An overview of nitrogen metabolism in higher plants, p. 1–41. *In* J. E. Poulton, J. T. Romeo, and E. E. Conn (ed.), *Plant Nitrogen Metabolism.* Plenum Publishing Corp., New York.

Bohlool, B. B., and E. L. Schmidt. 1974. Lectins: a possible basis for specificity in the *Rhizobium*-legume root nodule symbiosis. *Science* **185:**269–271.

Boulter, D., J. J. Jeremy, and M. Wilding. 1966. Amino acids liberated into the culture medium by pea seedling roots. *Plant Soil* **24:**121–127.

Bowen, G. D., and A. D. Rovira. 1976. Microbial colonization of plant roots. *Annu. Rev. Phytopathol.* **14:**121–144.

Brewin, N. J., A. L. Rae, S. Peroto, J. P. Knox, K. Roberts, M. F. LeGal, S. S. Sindhu, E. A. Wood, and E. L. Kannenberg. 1990. Immunological dissection of the plant-microbe interface in pea nodules, p. 227–234. *In* P. M. Gresshoff, L. E. Roth, G. Stacey, and W. E. Newton (ed.), *Nitrogen Fixation: Achievements and Objectives.* Chapman and Hall, New York.

Burn, J. E., W. D. Hamilton, J. C. Wootton, and A. W. B. Johnston. 1989. Single and multiple mutations affecting properties of the regulatory gene *nodD* of *Rhizobium. Mol. Microbiol.* **3:**1567–1577.

Caetano-Anollés, G., D. K. Crist-Estes, and W. D. Bauer. 1988a. Chemotaxis of *Rhizobium meliloti* to the plant flavone luteolin requires functional nodulation genes. *J. Bacteriol.* **170:**3164–3169.

Caetano-Anollés, G., L. G. Wall, A. T. de Micheli, E. M. Macchi, W. D. Bauer, and G. Favelukes. 1988b. Role of motility and chemotaxis in efficiency of nodulation by *Rhizobium meliloti. Plant Physiol.* **86:**1228–1235.

Callahan, D. A., and J. G. Torrey. 1981. The structural basis for infection of root hairs of *Trifolium repens* by *Rhizobium. Can. J. Bot.* **59:**1647–1664.

Canter Cremers, H. C. J., H. P. Spaink, A. H. M. Wijfjes, E. Pees, C. A. Wijffelman, R. J. H. Okker, and B. J. J. Lugtenberg. 1989. Additional nodulation genes on the sym plasmid of *Rhizobium leguminosarum* biovar viciae. *Plant Mol. Biol.* **13:**163–174.

Cervantes, E., S. B. Sharma, F. Maillet, J. Vasse, G. Truchet, and C. Rosenberg. 1989. The *Rhizobium meliloti* host range *nodQ* gene encodes a protein which shares homology with translation elongation and initiation factors. *Mol. Microbiol.* **3:**745–755.

Chang, K.-P. 1983. Cellular and molecular mechanisms of intracellular symbiosis in leishmaniasis. *Int. Rev. Cytol. Suppl.* **14:**267–305.

Cook, C. B. 1980. Infection of invertebrates with algae, p. 57–73. *In* C. B. Cook, P. W. Pappas, and E. D. Rudolph (ed.), *Cellular Interactions in Symbiosis and Parasitism.* Ohio State University Press, Columbus.

Currier, W. W., and G. A. Strobel. 1976. Chemotaxis of *Rhizobium* spp. to plant root exudates. *Plant Physiol.* (Bethesda) **57:**820–823.

Currier, W. W., and G. A. Strobel. 1977. Chemotaxis of *Rhizobium* spp. to a glycoprotein produced by birdsfoot trefoil roots. *Science* **196:**434–436.

Davis, E. O., I. J. Evans, and A. W. B. Johnston. 1988. Identification of *nodX*, a gene that allows *Rhizobium leguminosarum* biovar viciae strain TOM to nodulate Afghanistan peas. *Mol. Gen. Genet.* **212:**531–535.

Dazzo, F. B. 1981. Bacterial attachment as related to cellular recognition in the Rhizobium-legume symbiosis. *J. Supramol. Struct. Cell. Biochem.* **16**:29–41.

Dazzo, F. B., and D. H. Hubbell. 1975. Cross-reactive antigens and lectin as determinants of symbiotic specificity in the *Rhizobium*-clover association. *Appl. Microbiol.* **30**:1017–1033.

Dazzo, F. B., C. A. Napoli, and D. Hubbell. 1976. Adsorption of bacteria to roots as related to host specificity in the *Rhizobium*-clover symbiosis. *Appl. Environ. Microbiol.* **32**:166–171.

Dazzo, F. B., G. L. Truchet, J. E. Sherwood, E. M. Hrabak, M. Abe, and S. H. Pankratz. 1984. Specific phases of root hair attachment in the *Rhizobium trifolii*-clover symbiosis. *Appl. Environ. Microbiol.* **48**:1140–1150.

Dazzo, F. B., M. R. Urbano, and W. J. Brill. 1979. Transient appearance of lectin receptors on *Rhizobium trifolii*. *Curr. Microbiol.* **2**:15–20.

Dazzo, F. B., W. E. Yanke, and W. J. Brill. 1978. Trifoliin: a *Rhizobium* recognition protein from white clover. *Biochim. Biophys. Acta* **539**:276–286.

Debelle, F., C. Rosenberg, J. Vasse, F. Maillet, E. Martinez, J. Denarie, and G. Truchet. 1986a. Assignment of symbiotic developmental phenotypes to common and specific nodulation (*nod*) genetic loci of *Rhizobium meliloti*. *J. Bacteriol.* **168**:1075–1086.

Debelle, F., S. B. Sharma, C. Rosenberg, J. Vasse, F. Mailet, G. Truchet, and J. Denarie. 1986b. Respective roles of common and specific *Rhizobium meliloti nod* genes in the control of lucerne infection, p. 17–28. *In* B. J. J. Lugtenberg (ed.), *Recognition in Microbe-Plant Symbiotic and Pathogenic Interactions*. Springer-Verlag, Berlin.

de Maagd, R. A., A. H. M. Wijfjes, H. P. Spaink, J. E. Ruiz-Sainz, C. A. Wijffelman, R. J. H. Okker, and B. J. J. Lugtenberg. 1989. *nodO,* a new *nod* gene of the *Rhizobium leguminosarum* biovar viciae Sym plasmid pRL1JI, encodes a secreted protein. *J. Bacteriol.* **171**:6764–6770.

Diaz, C., L. S. Melchers, P. J. J. Hooykaas, B. J. J. Lugtenberg, and J. W. Kijne. 1989. Root lectin as a determinant of host specificity in the *Rhizobium*-legume symbiosis. *Nature* (London) **338**:579–581.

Diaz, C. L. 1989. Ph.D. thesis. University of Leiden, Leiden, The Netherlands.

Diaz, C. L., P. C. Van Spronsen, R. Bakhuizen, G. J. J. Logman, B. J. J. Lugtenberg, and J. W. Kijne. 1986. Correlation between infection by *Rhizobium leguminosarum* and lectin on the surface of *Pisum sativum* L. roots. *Planta* **168**:350–359.

Djordjevic, M. A., R. W. Innes, C. A. Wijffelman, P. R. Schofield, and B. G. Rolfe. 1986. Nodulation of specific legumes is controlled by several distinct loci in *Rhizobium trifolii*. *Plant Mol. Biol.* **6**:389–394.

Djordjevic, M. A., J. W. Redmond, M. Batley, and B. G. Rolfe. 1987. Clovers secrete specific phenolic compounds which stimulate or repress *nod* gene expression in *Rhizobium trifolii*. *EMBO J.* **6**:1173–1179.

Djordjevic, M. A., P. R. Schofield, and B. G. Rolfe. 1985. Tn5 mutagenesis of *Rhizobium trifolii* host-specific genes results in mutants with altered host range ability. *Mol. Gen. Genet.* **200**:463–471.

Downie, J. A. 1989. The *nodL* gene from *Rhizobium leguminosarum* is homologous to the acetyl transferases encoded by *lacA* and *cysE*. *Mol. Microbiol.* **3**:1649–1651.

Downie, J. A., C. D. Knight, A. W. B. Johnston, and L. Rossen. 1985. Identification of genes and gene products involved in the nodulation of peas by *Rhizobium leguminosarum*. *Mol. Gen. Genet.* **198**:591–597.

Dusha, I., A. Bakos, A. Kondorosi, F. J. de Bruijn, and J. Schell. 1989. The *Rhizobium meliloti* early nodulation genes (*nodABC*) are nitrogen-regulated: isolation of a mutant strain with efficient nodulation capacity on alfalfa in the presence of ammonium. *Mol. Gen. Genet.* **219**:89–96.

Economou, A., W. D. O. Hamilton, A. W. B. Johnston, and J. A. Downie. 1990. The *Rhizobium* nodulation gene *nodO* encodes a Ca^{2+}-binding protein that is exported without N-terminal cleavage and is homologous to haemolysin and related proteins. *EMBO J.* **9**:349–354.

Economou, A., F. K. L. Hawkins, J. A. Downie, and A. W. B. Johnston. 1989. Transcription of *rhiA,* a gene on a *Rhizobium leguminosarum* bv. viciae sym plasmid, requires *rhiR* and is repressed by flavonoids that induce *nod* genes. *Mol. Gen. Genet.* **3**:87–93.

Egelhoff, T. T., and S. R. Long. 1985. *Rhizobium meliloti* nodulation genes: identification of *nodABC* gene products, purification of *nodA* protein, and expression of *nodA* in *Rhizobium meliloti*. *J. Bacteriol.* **164**:591–599.

Evans, I. J., and J. A. Downie. 1986. The *nodI* gene product of *Rhizobium leguminosarum* is closely related to ATP-binding bacterial transport proteins; nucleotide sequence analysis of the *nodI* and *nodJ* genes. *Gene* **43**:95–101.

Faucher, C., S. Camut, J. Denarie, and G. Truchet. 1989. The *nodH* and *nodQ* host range genes of *Rhizobium meliloti* behave as avirulence genes in *R. leguminosarum* bv. viciae and determine changes in the production of plant-specific extracellular signals. *Mol. Plant Microbe Interact.* **2**:291–300.

Faucher, C., F. Maillet, J. Vasse, C. Rosenberg, A. A. N. van Brussel, G. Truchet, and J. Dénarié. 1988. *Rhizobium meliloti* host range *nodH* gene determines production of an alfalfa-specific extracellular signal. *J. Bacteriol.* **170**:5489–5499.

Fisher, R. F., T. T. Egelhoff, J. T. Mulligan, and S. R. Long. 1988. Specific binding of proteins from *Rhizobium meliloti* cell-free extracts containing NodD to DNA sequences upstream of inducible nodulation genes. *Genes Dev.* **2**:282–293.

Fisher, R. F., and S. R. Long. 1989. DNA footprint analysis of the transcriptional activator proteins NodD1 and NodD3 on inducible *nod* gene promoters. *J. Bacteriol.* **171**:5492–5502.

Fisher, R. F., J. A. Swanson, J. T. Mulligan, and S. R. Long. 1987. Extended region of nodulation genes in *Rhizobium meliloti* 1021. II. Nucleotide sequence, transcription start sites and protein products. *Genetics* **117**:191–198.

Fok, A. K., M. S. Ueno, E. A. Azada, and R. D. Allen. 1988. Phagosomal acidification in *Paramecium*: effects on lysosomal fusion. *Eur. J. Cell Biol.* **43**:412–420.

Fortin, M. G., N. A. Morrison, and D. P. S. Verma. 1987. Nodulin 26, a peribacteroid membrane nodulin, is expressed independently of the development of the peribacteroid compartment. *Nucleic Acids Res.* **15**:813–824.

Gaworzewska, E. T., and M. J. Carlile. 1982. Positive chemotaxis of *Rhizobium leguminosarum* and other bacteria towards root exudates from legumes and other plants. *J. Gen. Microbiol.* **128**:1179–1188.

Gharyal, P. K., S.-C. Ho, J. L. Wang, and M. Schindler. 1989. O-antigen from *Bradyrhizobium japonicum* lipopolysaccharide inhibits intercellular (symplast) communication between soybean (*Glycine max*) cells. *J. Biol. Chem.* **264**:12119–12121.

Gittie, R. R., P. V. Rai, and R. B. Patil. 1978. Chemotaxis of *Rhizobium* sp. towards root exudate of *Cicer arietinum*. *Plant Soil* **50**:553–566.

Gloudemans, T., and T. Bisseling. 1989. Plant gene expression in early stages of *Rhizobium*-legume symbiosis. *Plant Sci.* **65**:1–14.

Gortz, H.-D. 1983. Endosymbionts of Euplotes. *Int. Rev. Cytol. Suppl.* **14**:145–176.

Gottfert, M., P. Grob, and H. Hennecke. 1990a. Proposed regulatory pathway encoded by the *nodV* and *nodW* genes, determinants of host specificity in *Bradyrhizobium japonicum*. *Proc. Natl. Acad. Sci. USA* **87**:2680–2684.

Gottfert, M., S. Hitz, and H. Hennecke. 1990b. Identification of *nodS* and *nodU*, two inducible genes inserted between the *Bradyrhizobium japonicum nodYABC* and *nodIJ* genes. *Mol. Plant Microbe Interact.* **3**:308–316.

Gottfert, M., B. Horvath, E. Kondorosi, P. Putnoky, F. Rodriquez-Quinones, and A. Kondorosi. 1986. At least two *nodD* genes are necessary for efficient nodulation of alfalfa by *Rhizobium meliloti*. *J. Mol. Biol.* **191**:411–420.

Gottfert, M., J. W. Lamb, R. Gasser, J. Semenza, and H. Hennecke. 1989. Mutational analysis of the *Bradyrhizobium japonicum* common *nod* genes and further *nod* box-linked genomic DNA regions. *Mol. Gen. Genet.* **215**:407–415.

Gottfert, M., J. Weber, and H. Hennecke. 1988. Induction of a *nodA-lacZ* fusion in *Bradyrhizobium japonicum* by an isoflavone. *J. Plant Physiol.* **132**:394–397.

Govers, G., J.-P. Nap, A. Van Kammen, and T. Bisseling. 1987. Nodulins in the developing root nodule. *Plant Physiol. Biochem.* **25**:309–322.

Gulash, M., P. Ames, R. C. Larosiliere, and K. Bergman. 1984. Rhizobia are attracted to localized sites on legume roots. *Appl. Environ. Microbiol.* **48:**149–152.

Gunning, B. E. S., and R. L. Overall. 1983. Plasmodesmata and cell-to-cell transport in plants. *BioScience* **33:**260–265.

Gyorgypal, Z., N. Iyer, and A. Kondorosi. 1988. Three regulatory *nodD* alleles of divergent flavonoid-specificity are involved in host-dependent nodulation by *Rhizobium meliloti. Mol. Gen. Genet.* **212:**85–92.

Halverson, L. J., and G. Stacey. 1984. Host recognition in the *Rhizobium*-soybean symbiosis. Detection of a protein factor present in soybean root exudate which is involved in the nodulation process. *Plant Physiol.* (Bethesda) **74:**84–89.

Halverson, L. J., and G. Stacey. 1985. Host recognition in the *Rhizobium*-soybean symbiosis. Evidence for the involvement of lectin in nodulation. *Plant Physiol.* (Bethesda) **77:**621–625.

Halverson, L. J., and G. Stacey. 1986a. Effect of lectin on nodulation by wild-type *Bradyrhizobium japonicum* and a nodulation-defective mutant. *Appl. Environ. Microbiol.* **51:**753–760.

Halverson, L. J., and G. Stacey. 1986b. Signal exchange in plant-microbe interactions. *Microbiol. Rev.* **50:**193–225.

Hamdi, Y. A. 1971. Soil-water tension and the movement of *Rhizobium. Soil Biol. Biochem.* **3:**121–126.

Hartwig, U. A., C. A. Maxwell, C. M. Joseph, and D. A. Phillips. 1990. Chrysoeriol and luteolin released from alfalfa seeds induce *nod* genes in *Rhizobium meliloti. Plant Physiol.* (Bethesda) **92:**116–122.

Haug, A. 1984. Molecular aspects of aluminum toxicity. *Crit. Rev. Plant Sci.* **1:**342–375.

Henikoff, S., G. W. Haughn, J. M. Calvo, and J. C. Wallace. 1989. A large family of bacterial activator proteins. *Proc. Natl. Acad. Sci. USA* **85:**6602–6606.

Higgins, C. F., I. D. Hiles, G. P. C. Salmond, D. R. Gill, J. A. Downie, I. J. Evans, I. B. Holland, L. Gray, S. D. Buckel, A. W. Bell, and M. A. Hermodson. 1986. A family of related ATP-binding subunits coupled to many distinct biological processes in bacteria. *Nature* (London) **323:**448–450.

Hodgson, A. L. M., and G. Stacey. 1986. Potential for *Rhizobium* improvement. *Crit. Rev. Biotechnol.* **4:**1–74.

Hollingsworth, R., A. Squartini, S. Phillip-Hollingsworth, and F. Dazzo. 1989. Root hair deforming and nodule initiating factors from *Rhizobium trifolii,* p. 387–393. *In* B. J. J. Lugtenberg (ed.), *Signal Molecules in Plants and Plant-Microbe Interactions.* Springer-Verlag, Berlin.

Hong, G.-F., J. E. Burn, and A. W. B. Johnston. 1987. Evidence that DNA involved in the expression of nodulation (*nod*) genes in *Rhizobium* binds to the product of the regulatory gene *nodD. Nucleic Acids Res.* **15:**9677–9689.

Honma, M. A., and F. M. Ausubel. 1987. *Rhizobium meliloti* has three functional copies of the *nodD* symbiotic regulatory gene. *Proc. Natl. Acad. Sci. USA* **84:**8558–8562.

Horvath, B., E. Kondorosi, M. John, J. Schmidt, I. Torok, Z. Gyorgypal, I. Barabas, U. Wieneke, J. Schell, and A. Kondorosi. 1986. Organization, structure, and symbiotic function of *Rhizobium meliloti* nodulation genes determining host specificity for alfalfa. *Cell* **46:**335–343.

Horwitz, M. A. 1983. Symbiotic interaction between *Legionella pneumoniae* and human leucocytes. *Int. Rev. Cytol. Suppl.* **14:**307–328.

Hrabak, E. M., M. R. Urbano, and F. B. Dazzo. 1981. Growth-phase-dependent immunodeterminants of *Rhizobium trifolii* lipopolysaccharide which bind trifoliin A, a white clover lectin. *J. Bacteriol.* **148:**697–711.

Hunter, W. J., and C. J. Fahring. 1980. Movement by *Rhizobium* and nodulation of legumes. *Soil Biol. Biochem.* **12:**537–542.

Jeon, K. W. 1987. Change of intracellular "pathogens" into required cell components. *Ann. N.Y. Acad. Sci.* **503:**359–371.

Johnston, A. W. B., J. E. Burn, A. Economou, E. O. Davis, F. K. L. Hawkins, and M. J. Bibb. 1988. Genetic factors affecting host-range in *Rhizobium leguminosarum,* p. 378–384. *In* R. Palacios and D. P. S. Verma (ed.), *Molecular Genetics of Plant-Microbe Interactions.* The American Phytopathological Society, St. Paul, Minn.

Johnston, A. W. B., J. W. Latchford, E. O. Davis, A. Economou, and J. A. Downie. 1989. Five genes on the sym plasmid of *Rhizobium leguminosarum* whose products are located in the bacterial membrane or periplasm, p. 311–318. *In* B. J. J. Lugtenberg (ed.), *Signal Molecules in Plants and Plant-Microbe Interactions*. Springer-Verlag, Berlin.

Kannenberg, E. L., and N. J. Brewin. 1989. Expression of a cell surface antigen from *Rhizobium leguminosarum* 3841 is regulated by oxygen and pH. *J. Bacteriol*. **171**:4543–4548.

Karlik, S. J., G. L. Eichhorn, P. N. Lewis, and D. R. Crapper. 1980. Interaction of aluminum species with deoxyribonucleic acid. *Biochemistry* **19**:5991–5998.

Kato, G., Y. Maruyama, and M. Nakamura. 1980. Role of bacterial polysaccharides in the adsorption process of the *Rhizobium*-pea symbiosis. *Agric. Biol. Chem*. **44**:2843–2855.

Kijne, J. W., G. Smit, C. L. Diaz, and B. J. J. Lugtenberg. 1988. Lectin-enhanced accumulation of manganese-limited *Rhizobium leguminosarum* cells on pea root hair tips. *J. Bacteriol*. **170**:2994–3000.

Kittell, B. L., D. R. Helinski, and G. S. Ditta. 1989. Aromatic aminotransferase activity and indoleacetic acid production in *Rhizobium meliloti*. *J. Bacteriol*. **171**:5458–5466.

Kondorosi, E., J. Gyuris, J. Schmidt, M. John, E. Duda, B. Hoffman, J. Schell, and A. Kondorosi. 1989. Positive and negative control of *nod* gene expression in *Rhizobium meliloti* is required for optimal nodulation. *EMBO J*. **8**:1331–1340.

Kosslak, R. M., R. Bookland, J. Barkei, H. E. Paaren, and E. R. Appelbaum. 1987. Induction of *Bradyrhizobium japonicum* common *nod* genes by isoflavones isolated from *Glycine max*. *Proc. Natl. Acad. Sci. USA* **84**:7428–7432.

Lerouge, P., P. Roche, C. Faucher, F. Maillet, G. Truchet, J. C. Prome, and J. Denarie. 1990a. Symbiotic host-specificity of *Rhizobium meliloti* is determined by a sulphated and acylated glucosamine oligosaccharide signal. *Nature* (London) **344**:781–784.

Lerouge, P., P. Roche, J.-C. Prome, C. Faucher, J. Vasse, F. Maillet, S. Camut, F. de Billy, D. G. Barker, J. Denarie, and G. Truchet. 1990b. *Rhizobium meliloti* nodulation genes specify the production of an alfalfa-specific sulphated lipo-oligosaccharide signal, p. 177–186. *In* P. M. Gresshoff, L. E. Roth, G. Stacey, and W. E. Newton (ed.), *Nitrogen Fixation: Achievements and Objectives*. Chapman and Hall, New York.

Lewin, A., E. Cervantes, W. Chee-Hoong, and W. J. Broughton. 1990. *nodSU*, two new *nod* genes of the broad host range *Rhizobium* strain NGR234, encode host-specific nodulation of the tropical tree *Leucaena leucocephala*. *Mol. Plant Microbe Interact*. **3**:317–326.

Lipton, D. S., R. W. Blanchar, and D. G. Blevins. 1987. Citrate, malate, and succinate concentrations in exudates from P-sufficient and P-stressed *Medicago sativa* L. seedlings. *Plant Physiol*. (Bethesda) **85**:315–317.

Long, S. R. 1984. Genetics of *Rhizobium* nodulation, p. 265–306. *In* E. Nester and T. Kosuge (ed.), *Plant-Microbe Interactions*. MacMillan Press, New York.

Long, S. R. 1989a. *Rhizobium* genetics. *Annu. Rev. Genet*. **23**:483–506.

Long, S. R. 1989b. *Rhizobium*-legume nodulation: life together in the underground. *Cell* **56**:203–214.

Long, S. R., W. Buikema, and F. M. Ausubel. 1982. Cloning of *Rhizobium meliloti* nodulation genes by direct complementation of Nod⁻ mutants. *Nature* (London) **298**:485–488.

Long, S. R., J. Schwedock, T. Egelhoff, M. Yelton, J. Mulligan, M. Barnett, B. Rushing, and R. Fisher. 1989. Nodulation genes and their regulation in *Rhizobium meliloti*, p. 145–151. *In* B. J. J. Lugtenberg (ed.), *Signal Molecules in Plants and Plant-Microbe Interactions*. Springer-Verlag, Berlin.

Martinez, E., D. Romero, and R. Palacios. 1990. The *Rhizobium* genome. *Crit. Rev. Plant Sci*. **9**:59–93.

Maxwell, C. A., U. A. Hartwig, C. M. Joseph, and D. A. Phillips. 1989. A chalcone and two related flavonoids released from alfalfa roots induce *nod* genes of *Rhizobium meliloti*. *Plant Physiol*. (Bethesda) **91**:842–847.

McCoy, K. L., and R. H. Schwartz. 1988. The role of intracellular acidification in antigen processing. *Immunol. Rev*. **106**:129–147.

McIver, J., M. A. Djordjevic, J. J. Weinman, G. L. Bender, and B. G. Rolfe. 1989. Extension of host range of *Rhizobium leguminosarum* biovar *trifolii* caused by point mutations in *nodD* which result in alterations in regulatory function and recognition of inducer molecules. *Mol. Plant Microbe Interact*. **2**:97–106.

Melchers, L. S., A. J. G. Regensburg-Tunik, R. A. Schilperoort, and P. J. J. Hooykaas. 1989. Specificity of signal molecules in the activation of *Agrobacterium* virulence gene expression. *Mol. Microbiol.* **3:**969–977.

Mellor, H. Y., A. R. Glenn, R. Arwas, and M. J. Dilworth. 1987. Symbiotic and competitive properties of motility mutants of *Rhizobium trifolii* TA1. *Arch. Microbiol.* **148:**34–39.

Mellor, R. B. 1989. Bacteroids in the *Rhizobium*-legume symbiosis inhabit a plant internal lytic compartment—implications for other microbial endosymbioses. *J. Exp. Bot.* **40:**831–839.

Mills, K. M., and W. D. Bauer. 1985. *Rhizobium* attachment to clover roots. *J. Cell Sci. Suppl.* **2:**333–345.

Morrison, N., and D. P. S. Verma. 1987. A block in the endocytosis of *Rhizobium* allows cellular differentiation in nodules but affects the expression of some peribacteroid membrane nodulins. *Plant Mol. Biol.* **9:**185–196.

Mort, A. J., and W. D. Bauer. 1980. Composition of the capsular and extracellular polysaccharides of *Rhizobium japonicum:* changes with culture age and correlation with binding of soybean seed lectin to the bacteria. *Plant Physiol.* (Bethesda) **66:**158–163.

Mulligan, J. T., and S. R. Long. 1985. Induction of *Rhizobium meliloti nodC* expression by plant exudate requires *nodD. Proc. Natl. Acad. Sci. USA* **82:**6609–6613.

Mulligan, J. T., and S. R. Long. 1989. A family of activator genes regulates expression of *Rhizobium meliloti* nodulation genes. *Genetics* **122:**7–18.

Napoli, C., F. Dazzo, and D. Hubbell. 1975. Production of cellulose microfibrils by *Rhizobium. Appl. Environ. Microbiol.* **30:**123–131.

Newcomb, E. H., and R. R. Kowal. 1985. Ultrastructural specialization for ureide production in uninfected cells of soybean nodules. *Protoplasma* **125:**1–12.

Newcomb, W. 1981. Nodule morphology and differentiation, p. 247–298. *In* K. L. Giles and A. G. Atherly (ed.), *Biology of the Rhizobiaceae.* Academic Press, New York.

Nieuwkoop, A. J., Z. Banfalvi, N. Deshmane, D. Gerhold, M. G. Schell, K. M. Sirotkin, and G. Stacey. 1987. A locus encoding host range is linked to the common nodulation genes of *Bradyrhizobium japonicum. J. Bacteriol.* **169:**2631–2638.

Osbourn, A. E., C. E. Barber, and M. J. Daniels. 1987. Identification of plant-induced genes of the bacterial pathogen *Xanthomonas campestris* pathovar campestris using a promoter-probe plasmid. *EMBO J.* **6:**23–28.

Papendick, R. I., and G. S. Campbell. 1975. Water potential in the rhizosphere and plant methods of measurement and experimental control, p. 39–49. *In* G. W. Bruehl (ed.), *Biology and Control of Soil Borne Plant Pathogens.* The American Phytopathological Society, St. Paul, Minn.

Peters, N. K., J. W. Frost, and S. R. Long. 1986. A plant flavone, luteolin, induces expression of *Rhizobium meliloti* nodulation genes. *Science* **233:**977–980.

Pueppke, S. G. 1984. Adsorption of slow- and fast-growing rhizobia to soybean and cowpea roots. *Plant Physiol.* (Bethesda) **75:**924–928.

Rabinovitch, M., and S. C. Alfieri. 1987. From lysosomes to cells, from cells to Leishmania: amino acid esters as potential chemotherapeutic agents. *Braz. J. Med. Biol. Res.* **20:**665–674.

Recourt, K., A. A. N. van Brussel, A. J. M. Driessen, and B. J. J. Lugtenberg. 1989. Accumulation of a *nod* gene inducer, the flavonoid naringenin, in the cytoplasmic membrane of *Rhizobium leguminosarum* biovar *viciae* is caused by the pH-dependent hydrophobicity of naringenin. *J. Bacteriol.* **171:**4370–4377.

Regensburger, B., L. Meyer, M. Filser, J. Weber, D. Studer, J. W. Lamb, H. M. Fischer, M. Hahn, and H. Hennecke. 1986. *Bradyrhizobium japonicum* mutants defective in root-nodule bacteroid development and nitrogen fixation. *Arch. Microbiol.* **144:**355–366.

Robertson, J. G., P. Lyttleton, S. Bullivant, and G. F. Grayston. 1978. Membranes in lupin root nodules. I. The role of Golgi bodies in the biogenesis of infection threads and peribacteroid membranes. *J. Cell Sci.* **30:**129–149.

Robertson, J. G., B. Wells, T. Bisseling, K. J. F. Farnden, and A. W. B. Johnston. 1984. Immunogold localization of leghemoglobin in cytoplasm of nitrogen-fixing root nodules of pea. *Nature* (London) **311:**254–256.

Rolfe, B. G., and P. M. Gresshoff. 1988. Genetic analysis of legume nodule initiation. *Annu. Rev. Plant Physiol. Plant Mol. Biol.* **39:**297–319.

Rolfe, B. G., C. L. Sargent, J. J. Weinman, M. A. Djordjevic, J. McIver, J. W. Redmond, M. Batley,

D. C. Yuan, and M. W. Sutherland. 1989. Signal exchange between *R. trifolii* and clovers, p. 303–310. *In* B. J. J. Lugtenberg (ed.), *Signal Molecules in Plants and Plant-Microbe Interactions.* Springer-Verlag, Berlin.

Rossen, L., C. A. Shearman, A. W. B. Johnston, and J. A. Downie. 1985. The *nodD* gene of *Rhizobium leguminosarum* is autoregulatory and in the presence of plant exudate induces the *nodABC* genes. *EMBO J.* **4**:3369–3373.

Rostas, K., E. Kondorosi, B. Horvath, A. Simoncsits, and A. Kondorosi. 1986. Conservation of extended promoter regions of nodulation genes in rhizobia. *Proc. Natl. Acad. Sci. USA* **83**:1757–1761.

Roth, L. E., J. R. Dunlap, and G. Stacey. 1987. Localizations of aluminum in soybean bacteroids and seeds. *Appl. Environ. Microbiol.* **53**:2548–2553.

Roth, L. E., K. Jeon, and G. Stacey. 1988. Homology in endosymbiotic systems: the term "symbiosome," p. 220–225. *In* R. Palacios and D. P. S. Verma (ed.), *Molecular Genetics of Plant-Microbe Interactions.* The American Phytopathological Society, St. Paul, Minn.

Roth, L. E., and G. Stacey. 1989a. Cytoplasmic membrane systems involved in bacterium release into soybean nodule cells as studied with two *Bradyrhizobium japonicum* mutant strains. *Eur. J. Cell Biol.* **49**:24–32.

Roth, L. E., and G. Stacey. 1989b. Bacterium release into host cells of nitrogen-fixing soybean nodules: the symbiosome membrane comes from three sources. *Eur. J. Cell Biol.* **49**:13–23.

Sadowsky, M. J., P. B. Cregan, M. Gottfert, A. Sharma, D. Gerhold, F. Rodriguez-Quinones, H. H. Keyser, H. Hennecke, and G. Stacey. 1991. The *Bradyrhizobium japonicum nolA* gene and its involvement in the genotype-specific nodulation of soybeans. *Proc. Natl. Acad. Sci. USA* **88**:637–641.

Schell, M., Z. Banfalvi, and G. Stacey. 1991. Personal communication.

Schlaman, H. R. M., H. P. Spaink, R. J. H. Okker, and B. J. J. Lugtenberg. 1989. Subcellular localization of the *nodD* gene product in *Rhizobium leguminosarum. J. Bacteriol.* **171**:4686–4693.

Schmidt, E. L. 1974. Quantitative autecological study of microorganisms in soil by immunofluorescence. *Soil Sci.* **118**:141–149.

Schmidt, J., R. Wingender, M. John, U. Wieneke, and J. Schell. 1988. *Rhizobium meliloti nodA* and *nodB* genes code for the synthesis of compounds which stimulate mitosis in plant cells. *Proc. Natl. Acad. Sci. USA* **85**:8578–8582.

Schwedock, J., and S. R. Long. 1990. ATP sulphurylase activity of the *nodP* and *nodQ* gene products of *Rhizobium meliloti. Nature* (London) **348**:644–646.

Scott, K. F. 1986. Conserved nodulation genes from the non-legume symbiont *Bradyrhizobium japonicum* sp. (Parasponia). *Nucleic Acids Res.* **14**:2905–2910.

Sekine, M., K. Watanabe, and K. Syono. 1989a. Nucleotide sequence of a gene for indole-3-acetamide hydrolase from *Bradyrhizobium japonicum. Nucleic Acids Res.* **17**:6400.

Sekine, M., K. Watanabe, and K. Syono. 1989b. Molecular cloning of a gene for indole-3-acetamide hydrolase from *Bradyrhizobium japonicum. J. Bacteriol.* **171**:1718–1724.

Selker, J. M. L. 1988. Three-dimensional organization of uninfected tissue in soybean root nodules and its relation to cell specialization in the central region. *Protoplasma* **147**:178–190.

Selker, J. M. L., and E. H. Newcomb. 1985. Spatial relationships between uninfected and infected cells in root nodules of soybean. *Planta* **165**:446–454.

Shearman, C. A., L. Rossen, A. W. B. Johnston, and J. A. Downie. 1986. The *Rhizobium leguminosarum* nodulation gene *nodF* encodes a polypeptide similar to acyl-carrier protein and is regulated by *nodD* plus a factor in pea root exudate. *EMBO J.* **5**:647–652.

Smit, G., J. W. Kijne, and B. J. J. Lugtenberg. 1986a. Correlation between extracellular fibrils and attachment of *Rhizobium leguminosarum* to pea root hair tips. *J. Bacteriol.* **168**:821–827.

Smit, G., J. W. Kijne, and B. J. J. Lugtenberg. 1987. Involvement of both cellulose fibrils and a Ca^{2+}-dependent adhesin in the attachment of *Rhizobium leguminosarum* to pea root hair tips. *J. Bacteriol.* **169**:4294–4301.

Smit, G., J. W. Kijne, and B. J. J. Lugtenberg. 1989a. Roles of flagella, lipopolysaccharide, and a Ca-dependent cell surface protein in attachment of *Rhizobium leguminosarum* to pea root hair tips. *J. Bacteriol.* **171**:569–572.

Smit, G., T. J. J. Logman, M. E. T. I. Boerrigter, J. W. Kijne, and B. J. J. Lugtenberg. 1989b. Purification and partial characterization of the *Rhizobium leguminosarum* biovar viciae Ca^{2+}-dependent adhesin,which mediates the first step in attachment of cells of the family *Rhizobiaceae* to plant root hair tips. *J. Bacteriol.* **171:**4054–4062.

Smit, G., and G. Stacey. 1990. Adhesion of bacteria to plant cells: role of specific interactions versus hydrophobicity, p. 179–210. *In* R. J. Doyle and M. Rosenberg (ed.), *Microbial Cell Surface Hydrophobicity.* American Society for Microbiology, Washington, D.C.

Smit, G., and G. Stacey. Submitted for publication.

Smit, G., A. A. van der Baan, J. W. Kijne, and B. J. J. Lugtenberg. 1986b. The attachment mechanism of *Rhizobium. Antonie van Leeuwenhoek* **52:**362–363.

Soby, S., and K. Bergman. 1983. Motility and chemotaxis of *Rhizobium meliloti* in soil. *Appl. Environ. Microbiol.* **46:**995–998.

Spaink, H. Unpublished data.

Spaink, H. P., O. Geiger, D. M. Sheeley, A. A. N. van Brussel, W. S. York, V. N. Reinhold, B. J. J. Lugtenberg, and E. P. Kennedy. 1991. The biochemical function of the *Rhizobium leguminosarum* proteins involved in the production of host specific signal molecules, p. 142–149. *In* H. Hennecke and D. P. S. Verma (ed.), *Advances in Molecular Genetics of Plant-Microbe Interactions,* vol. 1. Kluwer Academic Publishers, Dordrecht, The Netherlands.

Spaink, H. P., R. J. H. Okker, C. A. Wijffelman, E. Pees, and B. J. J. Lugtenberg. 1987a. Promoters in the nodulation region of the *Rhizobium leguminosarum* Sym plasmid pRL1JI. *Plant Mol. Biol.* **9:**27–39.

Spaink, H. P., J. Weinman, M. A. Djordjevic, C. A. Wijffelman, R. J. H. Okker, and B. J. J. Lugtenberg. 1989a. Genetic analysis and cellular localization of the *Rhizobium* host specificity-determining NodE protein. *EMBO J.* **8:**2811–2818.

Spaink, H. P., C. A. Wijffelman, R. J. H. Okker, and B. J. J. Lugtenberg. 1989b. Localization of functional regions of the *Rhizobium nodD* product using hybrid *nodD* genes. *Plant Mol. Biol.* **12:**59–73.

Spaink, H. P., C. A. Wijffelman, W. Pees, R. J. H. Okker, and B. J. J. Lugtenberg. 1987b. *Rhizobium* nodulation gene *nodD* as a determinant of host specificity. *Nature* (London) **328:**337–340.

Stacey, G., and W. J. Brill. 1982. Nitrogen-fixing bacteria: colonization of the rhizosphere and roots, p. 225–247. *In* M. S. Mount and G. H. Lacy (ed.), *Phytopathogenic Prokaryotes,* vol. I. Academic Press, New York.

Stacey, G., L. J. Halverson, A. J. Nieuwkoop, Z. Banfalvi, M. G. Schell, D. Gerhold, N. Deshmane, J.-S. So, and K. M. Sirotkin. 1986. Nodulation of soybean: *Bradyrhizobium japonicum* physiology and genetics, p. 87–99. *In* B. J. J. Lugtenberg (ed.), *Recognition in Microbe-Plant Symbiotic and Pathogenic Interactions.* Springer-Verlag, Berlin.

Stacey, G., A. Paau, and W. J. Brill. 1980. Host recognition in the *Rhizobium*-soybean symbiosis. *Plant Physiol.* (Bethesda) **66:**609–614.

Stacey, G., M. G. Schell, and N. Deshmane. 1989. Determinants of host specificity in the *Bradyrhizobium japonicum*-soybean symbiosis, p. 394–399. *In* B. J. J. Lugtenberg (ed.), *Signal Molecules in Plants and Plant-Microbe Interactions.* Springer-Verlag, Berlin.

Stacey, G., J.-S. So, L. E. Roth, B. S. K. Lakshmi, and R. W. Carlson. 1991. A lipopolysaccharide mutant of *Bradyrhizobium japonicum* that uncouples plant from bacterial differentiation. *Mol. Plant Microbe Interact.* **4:**332–340.

Stacey, G., H. Spaink, and R. Carlson. Unpublished data.

Stachel, S. E., E. Messens, M. Van Montagu, and P. Zambryski. 1985. Identification of the signal molecules produced by wounded plant cells that activate T-DNA transfer in *Agrobacterium tumefaciens. Nature* (London) **318:**624–629.

Stachel, S. E., E. W. Nester, and P. Zambryski. 1986. A plant cell factor induces *Agrobacterium tumefaciens vir* gene expression. *Proc. Natl. Acad. Sci. USA* **83:**379–383.

Stachel, S. E., and P. Zambryski. 1986. *virA* and *virG* control the plant-induced activation of the T-DNA transfer process of *Agrobacterium tumefaciens. Cell* **46:**325–333.

Sturtevant, D. B., and B. J. Taller. 1989. Cytokinin production by *Bradyrhizobium japonicum. Plant Physiol.* (Bethesda) **89:**1247–1252.

Surin, B. P., and J. A. Downie. 1988. Characterization of the *Rhizobium leguminosarum* genes *nodLMN* involved in efficient host-specific nodulation. *Mol. Microbiol.* **2:**173–183.

Syono, K. 1988. Personal communication.

Taller, B. J., and D. B. Sturtevant. 1991. Cytokinin production by rhizobia, p. 215–221. *In* H. Hennecke and D. P. S. Verma (ed.), *Advances in Molecular Genetics of Plant-Microbe Interactions,* vol. 1. Kluwer Academic Publishers, Dordrecht, The Netherlands.

Torok, I., E. Kondorosi, T. Stepkowski, J. Postfai, and A. Kondorosi. 1984. Nucleotide sequence of *Rhizobium meliloti* nodulation genes. *Nucleic Acids Res.* **12**:9509–9524.

van Brussel, A. A. N., K. Panque, and A. Quispel. 1977. The wall of *Rhizobium leguminosarum* in bacteroid and free-living forms. *J. Gen. Microbiol.* **101**:51–56.

van Brussel, A. A. N., K. Recourt, E. Pees, H. P. Spaink, T. Tak, C. A. Wijffelman, J. W. Kijne, and B. J. J. Lugtenberg. 1990. A biovar-specific signal of *Rhizobium leguminosarum* bv. viciae induces increased nodulation gene-inducing activity in root exudate of *Vicia sativa* subsp. *nigra. J. Bacteriol.* **172**:5394–5401.

van Brussel, A. A. N., S. A. J. Zaat, H. C. J. Canter-Cremers, C. A. Wijffelman, E. Pees, T. Tak, and B. J. J. Lugtenberg. 1986. Role of plant root exudate and Sym plasmid-localized nodulation genes in the synthesis by *Rhizobium leguminosarum* of Tsr factor, which causes thick and short roots on common vetch. *J. Bacteriol.* **165**:517–522.

Van Egeraat, A. W. 1975. The possible role of homoserine in the development of R. *leguminosarum* in the rhizosphere of pea seedlings. *Plant Soil* **42**:381–386.

Verma, D. P. S., M. G. Fortin, J. Stanley, V. P. Mauro, S. Purohit, and N. Morris. 1986. Nodulins and nodulin genes of *Glycine max. Plant Mol. Biol.* **7**:51–61.

Verma, D. P. S., and S. R. Long. 1983. The molecular biology of *Rhizobium*-legume symbiosis. *Int. Rev. Cytol. Suppl.* **14**:211–245.

Vesper, S. J., and W. D. Bauer. 1985. Characterization of *Rhizobium* attachment to soybean roots. *Symbiosis* **1**:139–162.

Vesper, S. J., and W. D. Bauer. 1986. Role of pili in *Rhizobium japonicum* attachment to soybean roots. *Appl. Environ. Microbiol.* **52**:134–141.

Vesper, S. J., and T. V. Bhuvaneswari. 1988. Nodulation of soybean roots by an isolate of *Bradyrhizobium japonicum* with reduced firm attachment ability. *Arch. Microbiol.* **150**:15–19.

Vesper, S. J., N. S. A. Malik, and W. D. Bauer. 1987. Transposon mutants of *Bradyrhizobium japonicum* altered in attachment to host roots. *Appl. Environ. Microbiol.* **53**:1959–1961.

Wang, S.-P., and G. Stacey. 1990a. A divergent *nod* box sequence is essential for *nodD*1 induction in B. japonicum, p. 600. *In* P. M. Gresshoff, L. E. Roth, G. Stacey, and W. E. Newton (ed.), *Nitrogen Fixation: Achievements and Objectives.* Chapman and Hall, New York.

Wang, S.-P., and G. Stacey. 1990b. Ammonia regulation of *nod* genes in *Bradyrhizobium japonicum. Mol. Gen. Genet.* **223**:329–331.

Wheatcroft, R., D. G. McRae, and R. W. Miller. 1990. Changes in the *Rhizobium meliloti* genome and the ability to detect supercoiled plasmids during bacteroid development. *Mol. Plant Microbe Interact.* **3**:9–17.

Wijffelman, C., H. Spaink, H. Schlaman, B. Zaat, K. Recourt, R. de Maagd, R. Okker, and B. J. J. Lugtenberg. 1989. Regulation of *nod* gene expression: the role of NodD protein, p. 137–144. *In* B. J. J. Lugtenberg (ed.), *Signal Molecules in Plants and Plant-Microbe Interactions.* Springer-Verlag, Berlin.

Wijffelman, C. A., E. Pees, A. A. N. van Brussel, R. J. H. Okker, and B. J. J. Lugtenberg. 1985. Genetic and functional analysis of the nodulation region of the *Rhizobium leguminosarum* Sym plasmid pRL1JI. *Arch. Microbiol.* **143**:225–232.

Wilson, P. W. 1940. *The Biochemistry of Symbiotic Nitrogen Fixation.* University of Wisconsin Press, Madison.

Witty, J. F., and F. R. Minchen. 1990. Oxygen diffusion in the legume root nodule, p. 285–292. *In* P. M. Gresshoff, L. E. Roth, G. Stacey, and W. E. Newton (ed.), *Nitrogen Fixation: Achievements and Objectives.* Chapman and Hall, New York.

Wolf, S., C. M. Deom, R. N. Beachy, and W. J. Lucas. 1989. Movement protein of tobacco mosaic virus modifies plasmodesmatal size exclusion limit. *Science* **246**:377–379.

Young, J. P. W. *In* G. Stacey, R. H. Burris, and H. J. Evans (ed.), *Biological Nitrogen Fixation,* in press. Chapman and Hall, New York.

Zaat, S. A. J., A. A. N. van Brussel, T. Tak, E. Pees, and B. J. J. Lugtenberg. 1987. Flavonoids induce

Rhizobium leguminosarum to produce *nodDABC* gene-related factors that cause thick, short roots and root hair responses on common vetch. *J. Bacteriol.* **169:**3388–3391.

Zaat, S. A. J., C. A. Wijffelman, I. M. Mulders, A. A. N. van Brussel, and B. J. J. Lugtenberg. 1988. Root exudates of various host plants of *Rhizobium leguminosarum* contain different sets of inducers of *Rhizobium* nodulation genes. *Plant Physiol.* (Bethesda) **86:**1298–1303.

Microbial Cell-Cell Interactions
Edited by Martin Dworkin
© 1991 American Society for Microbiology, Washington, DC 20005

Chapter 10

Coaggregation: Adherence in the Human Oral Microbial Ecosystem

Paul E. Kolenbrander

INTRODUCTION

Intergeneric coaggregation is defined as cell-to-cell recognition and adherence between bacterial pairs from different genera and is exhibited by nearly all human oral bacteria tested to date (Kolenbrander, 1988, 1989; Kolenbrander et al., 1989). More than 700 strains representing 18 genera have been examined. When a suspension (10^9 cells per ml) of one cell type is mixed with an equal volume of a suspension of a genetically distinct partner cell type, large clumps or coaggregates consisting of an interactive network of both cell types are formed, leaving a clear supernatant. Coaggregation is not random, and the mediators of coaggregation in some cases have been identified.

Bacteria occupy all of the oral surfaces. Bacterial accretion is most noticeable on teeth and occurs as dental plaque. The mechanisms of initial attachment to teeth have been investigated by studying bacterial adherence to saliva-coated spheroidal hydroxyapatite (SHA), a model substratum for the tooth surface (Clark et al., 1978). It appears that adherence occurs by a multiple-step process (Germaine and Schachtele, 1976; Morris and McBride, 1984; Staat et al., 1980), which may include ionic (Cowan et al., 1987), hydrophobic (Nesbitt et al., 1982), and lectin-receptor mechanisms (Morris and McBride, 1984). Once bacteria have occupied the SHA surface, further accretion may occur by intergeneric coaggregation (Ciardi et al., 1987; Schwarz et al., 1987). A significant and perhaps critical role for coaggregation in accreting dental plaque in vivo is supported by the observations that only a coaggregating partner cell type can colonize a gnotobiotic animal (McBride and van der Hoeven, 1981) and that, in humans, coaggregating partners occupy the same oral surfaces (Hughes et al., 1988; Kolenbrander et al., 1983).

COAGGREGATION: POTENTIAL ROLE IN CHEMOSENSING AND IN TISSUE COLONIZATION

Adherence is an important function for oral bacteria. The oral cavity is a lotic, or flowing, environment, which contains a fluid volume of about 0.5 to 1.0 ml and has a normal salivary flow of 0.5 to 1 ml/min. Adherence mechanisms effective against this natural dilution are critical. Hard and soft tissue surfaces are abundant, and bacteria successfully colonize these surfaces by a combination of adherence mechanisms and nutritional relationships with the already attached population.

Release of cells and reattachment at a different site may be coordinated with nutritional requirements of cells and chemosensing of a favorable habitat in accreting dental plaque. Some frequently isolated bacteria, such as the treponemes, selenomonads, and capnocytophagae, are motile by means of axial fibrils, flagella, and gliding, respectively. If adherence opportunities are abundant, as they may be at the periphery of accreting dental plaque, motility may not be required. While in search of a desirable habitat, bacteria may use chemosensing to coordinate cycles of cellular attachment and release from the plaque surface. The dynamics of chemosensing and adherence, without the requirement for motility, add another dimension to the already intriguing area of chemosensing among motile bacteria. Bacterial sensing of environmental signals, coupled with a responsive gene expression system, for example, osmotic pressure and synthesis of altered ratios of the porins OmpF and OmpC, operates in the absence of motility, as do certain other related signal transducing systems (Stock et al., 1990). Regulation and expression of oral bacterial adhesin genes are beginning to be investigated, and their relationship to the broader phenomenon of bacterial adherence in the ecosystem is becoming an active area of research.

Distinct populations of bacteria are recovered from hard and soft oral surfaces. A study of mixtures of different bacterial species isolated from hard and soft surfaces showed that the bacteria redistribute and adhere to hard and soft surfaces, respectively, in the same relative proportions as those in which they are found naturally on these surfaces (Gibbons and van Houte, 1971). The relationship of intergeneric coaggregation to colonization of these surfaces in the human oral cavity was shown by the observation that veillonellae isolated from the tongue coaggregate with other bacteria that also inhabit the tongue (e.g., *Streptococcus salivarius*) but not with most bacteria found in subgingival dental plaque (e.g., *S. sanguis*) (Hughes et al., 1988). Conversely, subgingival bacteria coaggregate with each other, including veillonellae isolated from subgingival plaque, but not with veillonellae isolated from the tongue. It appears, therefore, that intergeneric coaggregation is a contributing factor in the establishment of distinct populations on hard and soft oral surfaces.

COAGGREGATION: CELL SURFACE RECOGNITION

Intergeneric Coaggregation

Partner specificity

Over the course of a 10-year study, more than 400 distinct bacterial taxa have been identified in dental plaque samples taken from 176 people (Moore et al., 1988). They represent 37 genera, but the genera most frequently identified among the 30,000 isolates examined are *Actinobacillus*, *Actinomyces*, *Bifidobacterium*, *Capnocytophaga*, *Eikenella*, *Eubacterium*, *Fusobacterium*, *Haemophilus*, *Lactobacillus*, *Leptotrichia*, *Neisseria*, *Peptostreptococcus*, *Porphyromonas* (*Bacteroides*, in part), *Prevotella* (*Bacteroides*, in part), *Propionibacterium*, *Selenomonas*, *Streptococcus*, *Treponema*, *Veillonella*, and *Wolinella* (Moore et al., 1987; Moore et al., 1985). Except for isolates in the genera *Bifidobacterium* and *Lactobacillus*, which have been tested with a very limited number of possible partner strains, and *Leptotrichia* and *Neisseria*, which have not yet been studied, isolates in the remaining genera coaggregate with isolates in the other genera (Kolenbrander, 1988, 1989; Kolenbrander and Andersen, unpublished results). Such intergeneric coaggregation or partner cell surface recognition can be very selective; e.g., *Capnocytophaga gingivalis* has only a few coaggregation partners (Kolenbrander and Andersen, 1986). Other species, such as *Fusobacterium nucleatum*, coaggregate with all gram-positive and gram-negative partner cell types currently tested (Kolenbrander et al., 1989; Kolenbrander and Andersen, unpublished results).

Mixed cell suspensions of coaggregating pairs form visibly discernable aggregates consisting of both cell types. If the two cell types tested are not partners, the mixed cell suspensions remain turbid. This simple visual assay has been used to screen over 700 strains for their ability to be coaggregation partners (Kolenbrander, 1988, 1989). Each strain has its own specific partners, but often several isolates of a genus exhibit the same set of partners, a property that has been used to organize isolates into coaggregation groups. The parameters tested and used to characterize the coaggregation groups are (i) ability of a pair to recognize each other, (ii) inhibition of coaggregation by the addition of lactose, (iii) inhibition of coaggregation by heat or protease treatment of one or both partners, and (iv) simultaneous loss of the ability of a coaggregation-defective mutant to recognize all members of a partner coaggregation group.

The coaggregation properties of the oral streptococci and actinomyces have been most thoroughly examined and used to characterize six coaggregation groups of streptococci and six coaggregation groups of actinomyces (Cisar et al., 1979; Kolenbrander et al., 1983; Kolenbrander and Williams, 1981, 1983). More than 300 streptococcal and actinomyces strains have been surveyed; 84% coaggregate and, of these, 92% exhibit lactose-inhibitable coaggregation. For each coaggregating pair, three possible kinds of coaggregation can occur. First, the streptococcus is inactivated by heat or protease treatment, while the actinomyces is insensitive to these treatments. Second, the actinomyces is inactivated by heat or protease treatment, while the streptococcus is not. Third, both cell types must be heated or protease treated to prevent coaggregation. In the first two kinds, the protease-inactivated partner is thought to bear an adhesin and the insensitive cell type to express the adhesin's complementary carbohydrate receptor. The third kind is a combination of the first two, in that each cell type possesses a distinct adhesin as well as the receptor for its partner's adhesin. When the coaggregation is inhibited by a sugar, such as lactose, the adhesin appears to function as a lectin, a nonantibody protein that binds specifically to sugar moieties. Eight complementary sets of adhesin-carbohydrate interactions can account for all of the interactions observed with over 300 strains. Thus, coaggregation mechanisms among these two genera appear to be limited and highly specific.

After potential partners are screened by the visual coaggregation assay, quantitative measurements of specific coaggregation partners can be taken by a radioactivity-based assay (Kolenbrander and Andersen, 1986). One cell type is radioactively labeled and its number is kept constant, while the number of the partner cell type is progressively increased (Fig. 1). Gentle centrifugation (100 × g for 1 min) separates the coaggregates from free cells. As the ratio of the two cell types is changed, the percentage of input radioactivity of the labeled cells in coaggregates increases with increasing numbers of unlabeled partner cells. Complete transfer of labeled cells from the free state to coaggregates occurs when labeled cells are saturated with unlabeled cells. Saturation is observed at about a fivefold excess of unlabeled fusobacteria. Many pairs have been examined by this method, and most show a saturation near 100% of labeled cells with unlabeled cells in coaggregates. Situations which differ have been used to demonstrate the formation of rosettes and competition between different cell types for a partner that they have in common (Kolenbrander and Andersen, 1988).

The radioactivity-based assay is useful in determining the effects of specific coaggregation-blocking antisera or sugar inhibitors, etc., on the ability of the labeled cell type to coaggregate with its partner. Turbidometric and visual assays are useful in determining the overall contributions of both cell types to the settling coaggregates. For comparison, in turbidometric and visual assays most pairs exhibit maximal coaggregation when the ratio of cell types is between 0.5 and 2. By microscopic examination of coaggregated cell suspensions at these ratios, coaggregates are clearly composed of a mixture of both cell types (Fig. 1B and C). Free cells are few. In Fig. 1B, streptococci are in a twofold excess, whereas in Fig. 1C, fusobacteria are in a fourfold excess. The compositions of the coaggregates clearly change as the cell ratios change. At ratios of 0.1 or 10, the coaggregates are obviously composed primarily of one cell type, which surrounds its partner cell type. Free or noncoaggregated cells are numerous. In the phase-contrast micrographs shown here, the coaggregates appear as rosettes when the central cell is a streptococcus (Fig. 1D) but appear as "corncobs" when the central cell is a fusobacterium (Fig. 1A).

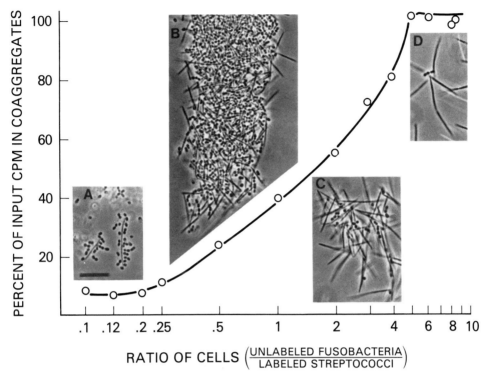

FIGURE 1. Percentage of input radioactivity in coaggregates formed between radioactively labeled *S. sanguis* C104 and unlabeled *F. nucleatum* PK1909. The number of labeled cells was constant and was mixed with increasing numbers of unlabeled cells. The logarithmic plot is used to expand the region of cell ratios between 0.5 and 4, in which the greatest changes in coaggregation occur. The phase-contrast micrographs show the typical cell-to-cell arrangements in the cell suspensions before centrifugation at $100 \times g$ for 1 min. (A) Corncob configuration at a cell ratio of 0.1 showing streptococci surrounding slender fusobacterial cells. (B) Tightly arranged networks of both cell types at a cell ratio of 0.5. (C) Loosely arranged networks of both cell types at a cell ratio of 4. (D) Rosette at a cell ratio of 10 showing streptococcal cells surrounded by fusobacteria. Note the numerous free cells in panels A and D and the lack of noncoaggregated cells in panels B and C. Bar, 10 μm.

Intergeneric coaggregation with members of most genera exhibits properties like those shown in Fig. 1. Such cell-to-cell interactions are commonly seen by microscopic examination of dental plaque samples. A juxtapositioning of fluorescein- and rhodamine-labeled cells in dental plaque has been observed by using fluorescein- and rhodamine-conjugated antibodies against streptococcus and actinomyces coaggregation partners, respectively (Cisar et al., 1985). Micromanipulation of a corncob from dental plaque was used to separate a streptococcus from *Corynebacterium (Bacterionema) matruchotii* (Mouton et al., 1980).

Coaggregation-defective mutants

For studying the mechanisms of coaggregation, coaggregation-defective (Cog) mutants have been isolated and characterized (Kolenbrander, 1982). All mutants are obtained without the use of mutagens and are selected by mixing the parent cell type with an appropriate partner cell type. Coaggregates are removed by low-speed centrifugation, and the cycle of adding a partner, mixing, and centrifuging is repeated several times. The final supernatant contains the desired Cog mutants. The coaggregation profiles of the parent and its partners versus those of the Cog mutants and the parent's partners have shown, for example, that all lactose-inhibitable coaggregations between the parent and its partners are often altered simultaneously in the phenotype of the Cog mutant. Such results suggest that the lactose-inhibitable coaggregations exhibited by the parent strain are mediated by the same lectin or adhesin. Several examples of these results have been reported for both gram-positive (Cisar et al., 1983; Kolenbrander, 1982) and gram-negative (Hughes et al., 1990; Weiss et al., 1987a; Weiss et al., 1987b; Ganeshkumar et al., 1989) cell types.

In a recent study of Cog mutants of *Veillonella* species (Hughes et al., 1990), it was shown not only that lactose-inhibitable coaggregations were lost simultaneously in one mutant, *Veillonella atypica* PK2726, but also that lactose-noninhibitable coaggregations were lost simultaneously in another mutant, *V. atypica* PK1885 (Fig. 2). The parent, *V. atypica* PK1910, coaggregated with *S. oralis* 34 and *Streptococcus* SM PK509 by a lactose-inhibitable mechanism (rectangular symbols) but coaggregated with *S. gordonii* DL1 and *S. gordonii* PK488 by a lactose-noninhibitable mechanism (triangular symbols). No coaggregation was observed with *S. oralis* H1. Coaggregation with *S. oralis* J22 was mediated by both kinds of interactions. *V. atypica* PK2726 was selected for its inability to coaggregate with *S. oralis* 34, but it still coaggregated with *S. gordonii* DL1 and PK488 as well as with *S. oralis* J22. Similarly, *V. atypica* PK1885 failed to coaggregate with *S. gordonii* DL1 but exhibited lactose-reversible coaggregation with strains 34, PK509, and J22. No coaggregation with *S. oralis* J22 or any of the streptococci was observed when Cog double mutants were selected; these mutants are obtained by selecting a mutant of *V. atypica* PK1885 that fails to coaggregate with *S. oralis* 34 or by selecting a mutant of *V. atypica* PK2726 with *S. gordonii* DL1 as the partner (Hughes, 1991). Thus, these two mechanisms of coaggregation, which are clearly distinguishable by sugar inhibition and partner selection, are also distinguishable by analysis with Cog mutants.

In a study of 28 strains of *F. nucleatum*, it was reported that as a group they coaggregated with all of the potential partner strains representing 12 genera (Kolenbrander et al., 1989). At first, it seemed that this recognition was nonspecific and differed from the highly specific partner recognition observed previously with the oral actinomyces, bacteroides, capnocytophagae, streptococci, and veillonellae. However, each *F. nucleatum* strain coaggregated with its own select cluster of strains. This inability to arrange the fusobacteria into coaggregation groups is consistent with the heterogeneous nature of the fusobacterial genome, as shown by DNA-DNA hybridization results (Dzink et al., 1990; Selin and Johnson, 1981).

Examination of one strain, *F. nucleatum* PK1594, revealed lactose-reversible coaggregations with *Actinobacillus actinomycetemcomitans* Y4, *Prevotella (Bacteroides) loescheii* PK1295, *C. sputigena* ATCC 33612, *Porphyromonas (Bacteroides) gingivalis* PK1924, *Selenomonas flueggei* PK1958, and *V. dispar* PK2503,

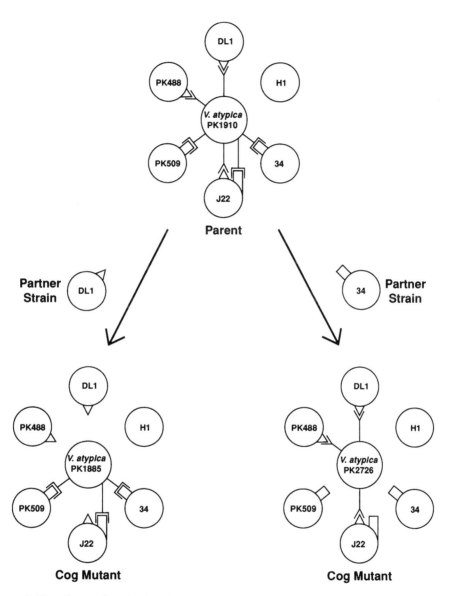

FIGURE 2. Flow diagram for selection of Cog mutants of *V. atypica* PK1910. The streptococcal strains used were *S. gordonii* DL1, *S. oralis* 34, *S. oralis* J22, *Streptococcus* SM PK509, and *S. gordonii* PK488. The Cog mutant on the bottom left, *V. atypica* PK1885, was selected as a strain that failed to coaggregate with *S. gordonii* DL1. The Cog mutant on the bottom right, *V. atypica* PK2726, was selected as a strain that failed to coaggregate with *S. oralis* 34.

all gram-negative partners. The Cog mutants of *F. nucleatum* PK1594 obtained by selections with *P. gingivalis* PK1924 were of two classes (Ganeshkumar et al., 1989). The first failed to coaggregate with all the gram-negative partners of parent strain PK1594, and the second failed to coaggregate with both gram-negative and gram-positive partners. As was observed with *V. atypica* PK1910 and its Cog mutant PK2726 (Fig. 2), a single mechanism appeared to mediate the lactose-reversible coaggregations of *F. nucleatum* PK1594 with its six gram-negative partners.

When taken in total, these results indicate that intergeneric coaggregations are not random. Each strain has a limited array of partners, and when the partners surveyed are all members of one genus, then often identical coaggregations are exhibited by several strains, which constitute a coaggregation group. Strains that belong to the same coaggregation group would, therefore, be expected to compete with each other for coaggregation with a common partner. In fact, strains belonging to different coaggregation groups but exhibiting identical coaggregations also compete with each other for coaggregation with their common partner (Kolenbrander and Andersen, 1985).

Multigeneric Coaggregation

Competition and independence of coaggregation mechanisms

Two distinct noncoaggregating cell types can compete for binding to their common partner if they coaggregate with that partner by the same mechanism. For example, *Actinomyces* serovar WVA963 strain PK1259 and *P. loescheii* PK1295 are not coaggregation partners, but they compete for coaggregation with their common partner, *S. oralis* 34, because both recognize *S. oralis* 34 by lactose-inhibitable coaggregation.

Competition is as specific as coaggregation; a Cog mutant altered in only one of the types of coaggregation exhibited by the parent cannot compete with the parent in that interaction but can compete with the parent for binding to other partners recognized by both the parent and the mutant. Results obtained with the radioactivity-based assay show that the presence of a large excess of isogenic Cog mutant cells does not affect the ability of the parent cells to coaggregate with their partner cells (Kolenbrander and Andersen, 1985). These results indicate that competition is based on the specific competitive recognition of partner cells and not on the steric hindrance possible with an excess of unreactive cells.

This independence of coaggregation is even more evident in studies of multigeneric coaggregates, in which cells coaggregating by lactose-inhibitable mechanisms are independent of those coaggregating by lactose-noninhibitable mechanisms (Kolenbrander and Andersen, 1986). The issue under consideration is the effect of multiple genera and multiple kinds of interactions on any one interaction. Do the cell-to-cell interactions that constitute the overall adherence in the consortium interfere with individual cell-to-cell adherence? The key cell type in an example of this experiment is *S. oralis* J22, which exhibits both lactose-reversible and lactose-nonreversible coaggregation with the other four members included in the multigeneric network. Two additional cell types, *S. sanguis* C104 and *Acti-*

nomyces naeslundii PK947, exhibit only lactose-reversible coaggregation. Only the latter two cell types dissociate when lactose is added to the multigeneric coaggregate. Strain J22 exhibits the independence of the lactose-reversible and lactose-nonreversible mechanisms, since it remains part of the coaggregate even when lactose-sensitive coaggregation is inhibited.

Rosettes

When the ratio of two coaggregating cell types is equal to or greater than 10:1, the coaggregates formed often appear as rosettes. The cell type in higher numbers surrounds the other cell type and competes with a third cell type for binding to the central cell (Kolenbrander and Andersen, 1988). If, on the other hand, a partner of the exterior cell type is added to a population of rosettes, large multigeneric coaggregates in which the third cell type bridges between the rosettes to form an interconnecting network of rosettes are formed. These mixed-cell-type rosettes are distinct from rosettes formed by pure cultures of bacteria with appendages or by identical cell types attached by a holdfast material (Hirsch, 1984; Poindexter, 1964). Rosettes are commonly found at the periphery of dental plaque and are probably involved in cell surface recognition and accretion of dental plaque. This relationship of intergeneric coaggregation to accretion will be discussed in detail in the last section of this chapter, in which other morphologically unusual forms of coaggregation, namely, corncob configurations, are described.

Bridging and sequential addition

Cooperation to form a multigeneric coaggregate occurs when two noncoaggregating cell types are mixed with a partner cell type of both of them and recognition of the partner cell type occurs by different mechanisms. The common partner cell type acts as a bridge between the noncoaggregating cell types (Kolenbrander and Andersen, 1984; Kolenbrander et al., 1985). For example, *P. loescheii* PK1295 bridges between *S. oralis* 34 and *A. israelii* PK14. The *P. loescheii-S. oralis* coaggregation is lactose inhibitable, and the *P. loescheii-A. israelii* coaggregation is not. Monoclonal antibodies (MAbs) that block the former coaggregation have no effect on the latter coaggregation and vice versa (Weiss et al., 1988a). Bridging occurs in the form of the sequential addition of bacteria to a growing chain of coaggregated cells in which each newly added cell type is dependent on the preceding one for accretion to the aggregate (Kolenbrander and Andersen, 1986). In the human oral cavity, sequential colonization of a surface in vivo probably occurs by multiple kinds of possible interactions among the accreting bacteria. Each new cell type might encounter by chance several different bacterial surfaces. It could employ the mechanism of adherence specific for a surface exposed in a population of bacteria in a favorable habitat.

Intrageneric Coaggregation

In a survey of isolates from 11 genera for intergeneric coaggregation (Kolenbrander, 1988, 1989), some were also screened for intrageneric coaggregation

(Kolenbrander et al., 1990). More than 100 strains, including actinobacilli, actinomyces, capnocytophagae, eubacteria, fusobacteria, porphyromonads, prevo-
tellae, rothias, selenomonads, streptococci, and veillonellae, were tested. Nine
genera showed no intrageneric coaggregation; these included 13 strains of fusobacteria which coaggregated with members of all the other genera. One actinomyces strain coaggregated with four other actinomyces strains. However, extensive coaggregation among the streptococci was observed.

Eight of the 10 strains of streptococci were coaggregation partners in several
pairings. Four of the eight streptococcal partners were either *S. oralis* or *S. sanguis,* three others were *S. gordonii,* and the remaining one was *Streptococcus*
SM PK509. These unusual interactions were highly specific in that only certain
paired strains were coaggregation partners, and all of the coaggregations were
inhibited by galactosides. Most interactions were inhibited completely by 15 mM
N-acetyl-β-D-galactosamine (GalNAc) and to a lesser degree by D-galactose (Gal),
methyl-α-D-galactopyranoside, methyl-β-D-galactopyranoside, or lactose. By
comparison, *N*-acetyl-D-glucosamine, fructose, D-glucose, D-galacturonic acid,
and many other sugars were either poor inhibitors or noninhibitory. One strain
of each pair was inactivated by heat (85°C for 30 min) or protease, while the other
was insensitive to these treatments. Such specificity only among the streptococci
and possibly actinomyces suggests an important role for intrageneric coaggregation in the oral cavity.

This role may be in the initial colonization of freshly cleaned tooth surfaces,
since the most numerous colonizers 4 h after cleaning of a human enamel surface
are streptococci, which constitute 78% of the total viable counts (Nyvad and
Kilian, 1990). Repopulation of the enamel occurs rapidly, as shown by the recovery of 0.35×10^5 to 4.7×10^5 CFU from 25 mm^2 of enamel surface in a study
of seven patients. The predominant streptococci are *S. oralis, S. mitis* biovar 1,
and *S. sanguis.* Lower proportions of *S. gordonii, S. mitis* biovar 2, and *S. salivarius* are found. Less than 2% of the early colonizing streptococci are *S. mutans,*
the etiologic agent of dental caries, even though up to 2×10^6 CFU of *S. mutans*
are found per ml of patient saliva. These data are in agreement with a large body
of literature that has shown that *S. mutans* is not a prominent member of early
microflora on human enamel surfaces (Liljemark et al., 1986; Ostrom et al., 1977;
Theilade et al., 1982; van Houte and Green, 1974). In addition, these data are
consistent with the results of a survey in which no coaggregation was observed
between 22 strains of mutans streptococci tested for coaggregation with 26 other
streptococcal strains (Crowley et al., 1987). However, a recent study in which
coaggregates were measured on a nitrocellulose membrane rather than in suspension after vigorous mixing revealed adherence between mutans streptococci
and other oral bacteria, including other streptococci (Lamont and Rosan, 1990).
While mutans streptococci do not appear able to bind directly to an enamel surface, bacteria such as *S. gordonii, S. oralis, S. mitis,* and *S. sanguis* react quickly
with an unoccupied enamel surface. Successful colonization by all streptococci,
however, may be effected by their ability to adhere to other streptococci already
attached to the acquired pellicle.

After being cleaned, the tooth surface is immediately coated with the acquired pellicle, a mixture of salivary components of host and bacterial origin. Some of the components, such as acidic proline-rich protein 1 and statherin, appear to mediate the initial attachment of bacteria to the tooth surface (Gibbons et al., 1988). Besides these important host-derived substances, saliva also contains bacterial cell wall fragments and other components present on the bacterial cell surface (for example, the adherence-mediating enzyme glucosyltransferase [McCabe and Donkersloot, 1977; Wittenberger et al., 1977]). It should be noted that living cells, dead cells (cells stored in azide-containing coaggregation buffer), and cell walls participate in coaggregation. It is possible that some initial bacterial attachment to teeth occurs by the interaction of a primary colonizing cell with cellular debris that is suspended in saliva and hence has become part of the acquired pellicle. Since streptococci constitute a major proportion of the salivary bacteria, it is likely that intrageneric recognition between dead streptococci or streptococcal cellular debris and viable streptococci may contribute to the accretion of bacteria to the cleaned tooth surface.

Mediators

The combined studies of coaggregation mediators and Cog mutants have clearly shown that bacterial cells express more than one mechanism of cell-to-cell recognition. Cells may express both adhesins and carbohydrate receptors for different adhesins found on partner cells. The adhesins may be borne on fimbriae or be part of the outer membrane of gram-negative cells. Mutants that are altered or lack one mechanism of coaggregation are still capable of coaggregating with other partners by the remaining mechanisms.

A major advance in unraveling the distinction between adhesins and fimbriae in oral bacteria was made when MAbs were selected for blocking coaggregations rather than agglutinating cells or reacting in enzyme-linked immunosorbent assays with cells as the plate-bound antigens (Weiss et al., 1988a). Studies with a coaggregation bridge organism, *P. loescheii* PK1295, revealed that a 75-kDa adhesin monomer mediates lactose-reversible coaggregation with streptococci. Its native form is a hexamer with an estimated M_r of 450,000 (London and Allen, 1990), and about 400 of these molecules are expressed on the prevotella surface (Weiss et al., 1988b). They are not detected on a Cog mutant that is unable to coaggregate with streptococci (Weiss et al., 1987a), but a second kind of adhesin, of about 45 kDa, is found on the mutant and parent cells. The 45-kDa adhesin is specific for recognizing *A. israelii,* and about 310 adhesin molecules are expressed per prevotella cell. Both adhesins are localized at the distal portion of fimbriae, and both adhesins can be found on the same cell (London et al., 1989).

The structures of three streptococcal carbohydrate receptors for coaggregation-mediating adhesins have been reported (Fig. 3) (Abeygunawardana et al., 1990; Cassels et al., 1990; McIntire et al., 1988). There is an apparent strong similarity among the structures, which are polysaccharides with repeating hexa- or heptasaccharide units. Both the hexasaccharide units from *S. oralis* 34 and the heptasaccharide units from *S. oralis* J22 are joined by phosphodiester bridges. The hexasaccharide units from *S. oralis* H1 appear not to be joined in that way,

Streptococcus   oralis   34:

[-->PO$_4^-$--->6)-α-D-GalpNAc(1-->3)-β-L-Rhap(1-->4)-β-D-Glcp(1-->6)-β-
D-Galf(1-->6)-β–D-GalpNAc(1-->3)-α-D-Galp(1-->]$_n$

Streptococcus   oralis   J22:

[-->PO$_4^-$--->6)-α-D-GalpNAc(1-->3)-β-L-Rhap(1-->4)-β-D-Glcp(1-->6)-β-
                          α-L-Rhap(1--2)-|

D-Galf(1-->6)-β-D-Galp(1-->3)-α-D-GalpNAc-(1-->]$_n$

Streptococcus   oralis   H1[a]:

α-L-Rhap(1-->2)-α-L-Rhap(1-->3)-α-D-Galp(1-->3)-β-D-Galp(1-->4)-β-
D-Glcp(1-->3)-α/β–D-Gal

---

[a] One phosphate is covalently attached to each hexasaccharide unit
but its location is unknown.

FIGURE 3. Structures of carbohydrate receptors recognized in coaggregation among human oral
bacteria.

but each unit contains one phosphate, presumably bound to a carbon atom not
involved in a glycosidic linkage (Cassels, personal communication). Streptococcal
strains 34 and J22 exhibit lactose-, Gal-, or GalNAc-inhibitable coaggregation with
A. naeslundii (A. viscosus) T14V, which appears to recognize the Gal-
NAcβ1→3Gal or the Galβ1→3GalNAc at the reducing end of the respective carbohydrate receptor. The most effective inhibitors of these two coaggregations are
sugars that contain either of the above-mentioned GalNAc-containing disaccha-
rides.

A second kind of surface receptor that lacks N-acetylated sugars has been
found in *S. oralis* H1 (Cassels and London, 1989). This streptococcus bears the
coaggregation receptor for the adhesin on *Capnocytophaga ochracea* ATCC
33596, and the coaggregation is insensitive to GalNAc. The coaggregation is inhibited by lactose and Gal, but L-rhamnose (Rha) is a 16-fold more effective in-

hibitor (Weiss et al., 1987b). Although Rha is a constituent of the above-mentioned two polysaccharides from streptococcal strains 34 and J22, it does not inhibit coaggregations with the respective partners of the strains. Neither *S. oralis* 34 nor *S. oralis* J22 coaggregates with *C. ochracea* ATCC 33596. The internal position of the Rha residues in the hexa- or heptasaccharide units may prevent their recognition by the capnocytophaga adhesin. In contrast, the position of the two Rha residues at the nonreducing end of the *S. oralis* H1 hexasaccharide may present them in the proper configuration to the active site of the capnocytophaga adhesin. This idea of proper configuration of terminal and internal sugar residues of carbohydrate receptors has been discussed in more detail by Karlsson (1989) to explain the use of animal glycolipid and glycosphingolipid molecules as membrane attachment sites for bacteria. Considering the molecular configuration of the functionally active *S. oralis* H1 carbohydrate receptor, the purified hexasaccharide from strain H1 is the most effective coaggregation inhibitor (Cassels and London, 1989). A Cog mutant of strain H1 has no detectable hexasaccharide and does not coaggregate with *C. ochracea* ATCC 33596. Thus, it appears that one kind of streptococcal carbohydrate receptor mediates Rha-inhibitable coaggregation with a gram-negative partner and another kind of receptor mediates GalNAc-inhibitable coaggregation with a gram-positive partner.

Each of the above carbohydrate-bearing streptococci coaggregates with some partners from several genera, which include *Actinomyces, Capnocytophaga, Fusobacterium, Haemophilus, Prevotella* (*Bacteroides,* in part), *Rothia,* and *Veillonella*. Only the Cog mutants of *S. oralis* 34 have been examined for failure to coaggregate with other GalNAc-inhibitable partners. Cog mutants *S. oralis* 102 (Cisar et al., 1985) and *S. oralis* PK1348 (Andersen et al., 1991) do not coaggregate with *Actinomyces* serovar WVA963 strain PK1259, *P. loescheii* PK1295, and *V. parvula* PK1910, all of which show GalNAc-inhibitable coaggregation with the parent, *S. oralis* 34 (Andersen et al., 1991). Mutant *S. oralis* 102 does not react antigenically with antiserum against the carbohydrate receptor and thus appears not to express the coaggregation-mediating carbohydrate (Cisar et al., 1985). Preincubation of the parent, *S. oralis* 34, at a pH of 4.6 with afimbrial adhesin prepared from one of the partners, *P. loescheii* PK1295, blocks the *S. oralis-P. loescheii* coaggregation (London and Allen, 1990). When the prevotella adhesin preparation is adjusted to a pH of 6.8, the parent streptococcus is agglutinated by the aggregated form of the prevotella adhesin. However, it has no effect on cells of mutant strain PK1348 (London and Kolenbrander, unpublished data). Although the Cog mutants no longer participate in GalNAc-inhibitable coaggregations, they retain the characteristics of the parent in coaggregations that are not sugar inhibitable (Andersen et al., 1991). Thus, these streptococcal polysaccharides appear to be specifically involved in mediating GalNAc-inhibitable coaggregations.

One streptococcal adhesin that appears to mediate coaggregation with *A. naeslundii* PK606 was identified in *S. gordonii* PK488 (Kolenbrander and Andersen, 1990). On denaturing gels, it exhibited an M_r of 38,000. Antiserum raised against parent PK488 cells and absorbed with isogenic Cog mutant PK1804 was used in immunoblots. Cell surface preparations from gently sonicated cells of strain PK488 and other streptococci that show the same type of coaggregation with the actinomyces were subjected to denaturing polyacrylamide gel electrophoresis. A protein of a similar size was found in all of them. It was recently shown (Ganeshkumar et al., 1991) that this absorbed antiserum also recognized a 36,000-M_r saliva-binding protein from another streptococcus (Ganeshkumar et al., 1988). Both the 36,000- and the 38,000-M_r proteins have a trypsin-sensitive site, and it is known that this site is near the amino terminus of the 38,000-M_r protein. This may be the first of many examples of oral bacterial surface proteins that mediate more that one adherence function, such as binding to saliva-coated tooth surfaces and coaggregating with other bacteria.

Respective MAbs to adhesins of two species of capnocytophagae, *C. gingivalis* DR2001 (Tempro et al., 1989) and *C. ochracea* ATCC 33596 (Weiss et al., 1990), specifically block coaggregation between each gram-negative capnocytophaga and its gram-positive partner. The adhesins of both capnocytophagae are found in the outer membrane and are 140,000- and 155,000-M_r proteins, respectively. Cog mutants of both capnocytophagae lack the adhesins. MAbs against the *C. gingivalis* adhesin were used to show that between 220 and 280 adhesins are located on each cell and occur singly, in pairs, and in small clusters. Rha-sensitive coaggregations between *C. ochracea* and members of the genera *Streptococcus, Actinomyces,* and *Rothia* were all inhibited by MAbs against the 155,000-M_r adhesin, indicating that the Rha-sensitive adhesin mediates coaggregations with at least three different genera. This kind of specific recognition (i.e., Rha sensitive) among a broad range of coaggregation partners (multigeneric) has been observed with Cog mutants of several other oral bacteria (Kolenbrander, 1988, 1989). In fact, this may be a mechanism that oral bacteria use to conserve their gene pool and maximize their ability to adhere to indigenous bacteria already attached to oral surfaces.

METABOLIC COMMUNICATION

The bacterial population per gram of dental plaque numbers about 10^{11}, and there are about 7×10^8 bacteria per ml of saliva. With this large a community, it would be expected that a considerable exchange of metabolites occurs. Cross-feeding symbionts and food chains should abound, as they do in other ecosystems. Yet, very little is known about the metabolic communication among oral bacteria.

One of the clearest examples is the natural food chain between the lactic acid-producing streptococci and the lactate-utilizing veillonellae. An in vivo study with a gnotobiotic rat model system demonstrated the relationship among metabolism, coaggregation, and colonization of the tooth surface (McBride and van der Hoeven, 1981). Two strains of *S. mutans* were used. Only one strain coaggregated with *V. parvula*. Both streptococcal strains colonized the smooth surface, while the veillonella strain did not. When the coaggregation-positive strain of *S. mutans* was allowed to colonize the teeth before infection with the veillonellae, then the veillonellae attached and colonized the tooth surface. In contrast, no colonization by veillonellae above control values was seen when the coaggregation-negative strain of *S. mutans* was used. Apparently, it was not sufficient to have an available lactate source, since no veillonella colonization occurred with the coaggregation-negative *S. mutans* strain. However, lactate may be required for colonization by veillonellae. Support for this idea comes from an earlier study which showed that veillonellae could not colonize teeth by themselves but could colonize teeth in the presence of lactic acid-producing streptococci (Gibbons et al., 1964). Results from both studies indicated that veillonellae require metabolic and physical contact with other bacteria for successful colonization of the tooth surface.

Another interesting bacterium, *F. nucleatum,* is a predominant inhabitant of subgingival dental plaque regardless of the state of health or disease of the tooth. This organism coaggregates with members of all other 17 genera of bacteria tested to date (Kolenbrander, 1988, 1989). The fusobacteria form corncob configurations with many of their partners, indicating a special morphological arrangement with their partners. The habitat is bathed with gingival crevicular fluid, a serum-derived fluid that is rich in plasma proteins. Fusobacteria themselves and several of their coaggregation partners secrete proteases, which make a ready supply of amino acids that are fermented by the fusobacteria. In addition, fusobacteria ferment three monosaccharides; they transport glucose and galactose by one mechanism and fructose by a second mechanism (Robrish and Thompson, 1990). Both transport systems are driven by amino acid energy sources such as glutamate, lysine, and histidine (Robrish et al., 1987). All three sugars accumulate as the same intracellular glucan independent of the sugar used for growth (Robrish and Thompson, 1989). In the absence of amino acids, the glucan disappears. Most of the resultant glucose is used by the host cell, but some is excreted and potentially available for coaggregation partners (Robrish, personal communication). Indeed, the secretion of nutrients by the fusobacteria may contribute to their recognition by such a wide range of coaggregation partners, each of which also may provide the fusobacteria with an ecological advantage over competitors. Their unique ecological niche among oral bacteria, as illustrated by the fact that the biosynthesis and degradation of intracellular glucan are controlled by energy derived by amino acid fermentation, may give them a strong survival advantage in dental plaque. This kind of metabolic communication may be the reason for the predominance of fusobacteria in both healthy and diseased oral sites.

Such stable nutritional relationships have been observed as reproducible, complex communities of oral bacteria in chemostats (McDermid et al., 1986; McKee et al., 1985). Nine different bacteria representing fusobacteria, streptococci, actinomyces, lactobacilli, neisseriae, bacteroides, and veillonellae were all established as part of a community under glucose-limiting conditions. Morphological observations revealed numerous corncob configurations involving fusobacteria and either streptococci or veillonellae. By adjustment and control of the pH of the chemostat, the bacteria predominating at pH 7.0 versus pH 4.1 were streptococci and veillonellae versus lactobacilli and veillonellae, respectively. Fusobacteria were consistently high in numbers under glucose-limiting conditions but were dramatically reduced under amino acid-limiting conditions. These results with a microbial consortium in a chemostat are consistent with the physiological and biochemical evidence that fusobacteria require an amino acid source to transport sugars and thus to maintain a competitive edge in the habitat.

The subgingival habitat is the site of periodontal diseases, which affect and often destroy the hard and soft tissues surrounding and supporting teeth. Periodontal disease appears to be a mixed bacterial infection. Some of the features of virulence promotion in mixed bacterial infections have been discussed elsewhere (Mayrand, 1985). Identification of a single organism capable of causing periodontal disease has not been reported. Rather, it seems that a succession of facultatively and obligately anaerobic bacteria occurs. The nutritional interdependence of some of the bacteria that are found in periodontally affected sites has been investigated (Grenier and Mayrand, 1986). *Wolinella recta* produced protoheme, and *P. melaninogenica (Bacteroides melaninogenicus)* excreted formate, which stimulates the growth of wolinellae. The growth of *Prevotella* spp. was dependent on hemin, and protoheme satisfied that dependency. A commensal relationship between *P. gingivalis* and *W. recta* was also observed. The presence of *P. gingivalis* in mixed cultures did not stimulate the growth of *W. recta,* but protoheme was essential for the growth of *P. gingivalis.* The fact that wolinellae and porphyromonads coaggregate with fusobacteria, which are abundant in infected sites on teeth, may be central to the overall success of metabolic communication among these potentially periodontopathic bacteria.

Successful communication appears to occur between coaggregating actinomyces and streptococci (Distler et al., 1989). Coaggregated cells consume more glucose and produce more lactic acid than do cells of either of the individual strains cultured separately or a mixture of an actinomyces and a Cog mutant of a streptococcus. The close proximity of two genetically distinct cell types in intergeneric and multigeneric coaggregates is not unlike the close proximity or juxtapositioning of the massive cellular aggregates of metabolic consortia observed in anaerobic biodegradative ecosystems (Shelton and Tiedje, 1984; Zeikus, 1983). The distinction between these two forms of metabolic communication is that the latter do not require cell-to-cell contact but rather involve metabolic communication within an extracellular biopolymer matrix.

NASCENT SURFACES

Temporal Changes in Microbial Composition on a Clean Tooth Surface

Immediately after a tooth is professionally cleaned, its surface becomes coated with a saliva-derived acquired pellicle. It is quickly colonized by streptococci, actinomyces, and a few other gram-positive bacteria, such as *Arachnia, Corynebacterium, Propionibacterium,* and *Rothia* spp. (Nyvad and Kilian, 1987). Gram-negative bacteria, including *Veillonella* and *Haemophilus* spp., constitute less than 2% of the flora in the first 24 h. Apparently, toxic substances are released by the bacteria or the host responds to this accumulation of bacteria for other reasons. Gingivitis, an inflammation of the gingival tissue surrounding the tooth, often results if bacterial plaque is left undisturbed for 48 h or more (Löe et al., 1965).

Numerous surveys of the altered composition of dental plaque associated with a change from health to gingivitis revealed that the population changes from predominantly gram-positive bacteria to predominantly gram-negative bacteria (Löe et al., 1965; Moore et al., 1987; Moore et al., 1982a; Slots et al., 1978; Syed and Loesche, 1978; Theilade et al., 1966). Gingivitis can progress to periodontal disease, which is characterized by bone destruction, loss of tooth attachment, and a high microbial count in the resultant periodontal pocket. Microbial changes accompanying periodontal disease have been reviewed (Moore, 1987). Some gram-negative bacteria, including *F. nucleatum, P. gingivalis, W. recta,* and spirochetes (Dzink et al., 1985; Moore et al., 1985; Tanner et al., 1984), increase in numbers more than others. In addition, some gram-positive bacteria, such as *Eubacterium* and *Lactobacillus* spp., are as numerous as any of the gram-negative bacteria (Moore et al., 1985). These latter two genera are undetectable or are infrequently detected in samples taken from healthy subgingival sites (Moore et al., 1987).

The accumulation of bacteria on a clean tooth over a period of days to months must occur through either bacterial growth of attached cells or accretion of unattached cells or both. The ecological rules of survival of the fittest would seem to favor those organisms that grow more rapidly than others. However, in a lotic environment, rapid growth is not the only requirement for successful colonization, since adherence is essential. Coaggregation between *Candida albicans* and several oral viridans streptococci, including *S. sanguis, S. gordonii,* and *S. oralis,* was shown to be enhanced more than 10-fold when the yeast was starved for 3 h (Jenkinson et al., 1990). Other properties of yeast-streptococcal coaggregations, such as increased coaggregation in the presence of divalent cations and inhibition of coaggregation by heat or protease treatment of the yeast cells but not the streptococcal cells, were similar to those between bacteria only. So, oral colonization by yeasts might be increased by periods of starvation during which coaggregation with indigenous bacteria of dental plaque might stabilize the yeasts in the habitat. This period of adherence and establishment could then be followed by the active growth of accreted cells and the incorporation of these cells into the population. The fact that *C. albicans* exhibits this property is of considerable medical importance, since *C. albicans* is known to colonize traumatized heart valves and is found frequently in the oral cavity of patients suffering from acquired immunodeficiency syndrome. Both yeasts and viridans streptococci have been isolated from damaged heart tissue, and a ready reservoir of these organisms in the oral cavity may be provided at least partly by their accretive abilities in dental plaque.

Saliva-Coated SHA

Bacterial attachment to SHA may involve several mechanisms, such as ionic and hydrophobic forces, high- and low-affinity binding, and lectinlike interactions (Cowan et al., 1987; Gibbons et al., 1983b; Morris and McBride, 1984; Nesbitt et al., 1982). Other interactions, such as salivary immunoglobulin A aggregation of cells (Liljemark et al., 1979) and bacterial recognition of specific salivary proteins (Gibbons and Hay, 1988; Gibbons et al., 1988), may also be involved.

The best-studied systems are those involving the oral streptococci or acti-nomyces, the predominant colonizers of a freshly cleaned tooth surface. In most instances, fimbriae are involved in the adherence of both streptococci (Fives-Taylor, 1982; Gibbons et al., 1983a) and actinomyces (Gibbons et al., 1988; Wheeler and Clark, 1980), but afimbrial adherence mechanisms may also play a role (Lamont et al., 1988). While antibodies against streptococcal fimbriae (Fa-chon-Kalweit et al., 1985) or actinomyces type 1 fimbriae (Clark et al., 1984) block adherence to SHA, purified fimbrial subunits do not exhibit this function. The structural genes for the respective fimbrial subunits of these two organisms were cloned in *Escherichia coli,* but no adherence activity was reported for the cloned proteins (Fives-Taylor et al., 1987; Yeung et al., 1987). These data suggest that the adhesin responsible for binding to SHA is distinct from the structural fimbrial monomers. It does not appear to be a major surface antigen. Rather, it may be a minor component of fimbriae possibly located at the fimbrial tip, as has been seen for adhesins in *P. loescheii* (Weiss et al., 1988b) and *E. coli* (Abraham et al., 1987; Lindberg et al., 1987).

A streptococcus-derived 36-kDa protein, designated SsaB, was cloned in *E. coli* and inhibited the adherence of *S. sanguis* 12na but only poorly inhibited the adherence of its parent, *S. sanguis* 12 (Ganeshkumar et al., 1988). The parent binds to SHA by both pH-sensitive and neuraminidase-sensitive mechanisms, but the mutant recognizes only the pH-sensitive receptor on SHA. Recently, it was shown that the antiserum raised against this cloned protein also recognizes, in immunoblot analysis, other proteins of about 38 kDa in several other streptococcal strains (Ganeshkumar et al., 1991). These other strains are the reference strains for all six streptococcal coaggregation groups. Antiserum to the 38-kDa protein from one of these streptococcal strains blocks coaggregation of the streptococci with the actinomyces (Kolenbrander and Andersen, 1990) and recognizes the cloned protein in immunoblot analysis. This cloned protein and the coaggregation-mediating proteins from the other streptococci are very closely related and con-stitute the first group of proteins with functional activities involved in both cell-to-cell recognition and salivary component recognition.

One of these streptococcal reference strains, *S. gordonii* DL1 (streptococcal coaggregation group 1), binds to SHA (Ciardi et al., 1987). Its coaggregation with *Propionibacterium acnes* PK93 is lactose reversible. *P. acnes* PK93 is unable to bind to SHA but binds very well when the SHA is first exposed to *S. gordonii* DL1. In fact, about equal numbers of newly adherent *P. acnes* PK93 and already adherent *S. gordonii* DL1 are observed. Moreover, the attachment is lactose reversible, releasing only *P. acnes* PK93 after the addition of lactose, while *S. gordonii* DL1 remains bound to the SHA. Collectively, these studies illustrate that sequential attachment to a hard surface is a combination of primary adherence mechanisms of the initial colonizers of SHA followed by coaggregation of sub-sequent colonizers to already attached cells.

Relationship of Coaggregation to Temporal Changes in Microbial Composition

The early colonizers coaggregate with each other; the late colonizers coag-gregate with each other; but the early colonizers of a clean tooth surface do not necessarily coaggregate with bacteria that increase in numbers with advancing stages of periodontal disease (Kolenbrander, 1988, 1989). These statements are in excellent accord both with the results obtained from numerous surveys of the microbial composition attendant to different states of periodontal health and disease (Dzink et al., 1985, Dzink et al., 1988; Loesche and Syed, 1978; Moore et al., 1982a; Moore et al., 1982b; Moore et al., 1983; Moore et al., 1984; Moore et al., 1985; Moore et al., 1987; Slots, 1977; Socransky et al., 1977; Syed and Loesche, 1978; Theilade et al., 1982; Williams et al., 1985) and with the morphological changes in the accreting microbial mass that occur when teeth are not cleaned (Jones, 1972; Listgarten, 1976; Listgarten et al., 1973; Listgarten et al., 1975; Nyvad and Fejerskov, 1987a, 1987b).

One of the interesting morphological arrangements of accreting cells is the corncob configuration (Jones, 1972; Listgarten et al., 1973), which consists of a central rod-shaped cell surrounded by spherical cells. As the microbial population becomes more diverse with respect to time after cleaning of the tooth, corncob arrangements are observed almost entirely at the periphery or advancing edge of accreting plaque (distal to the tooth surface) (Listgarten et al., 1973). For one of these arrangements, the two cell types have been separated by micromanipulation and identified as *S. sanguis* CC5A and *C. matruchotii* (Mouton et al., 1979). Corncob configurations consisting of slender rod-shaped *F. nucleatum* in the center and *S. sanguis* CC5A surrounding it have been reported (Lancy et al., 1983). In fact, *F. nucleatum* (Fig. 4A) forms corncob configurations with all of its coaggregation partners that have been tested so far, including *S. flueggei, P. gingivalis, A. actinomycetemcomitans, V. atypica,* and *S. sanguis* (Fig. 4B, C, D, E, and F, respectively). These configurations are observed when the partner strain is present in a 10- to 50-fold excess over *F. nucleatum.* It appears, therefore, that *F. nucleatum* has the adherence properties necessary to play a central role in the corncob configurations observed on the periphery of dental plaque. Rapid growth

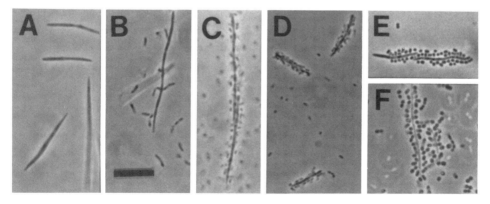

FIGURE 4. Intergeneric coaggregations in the form of corncob configurations. *F. nucleatum* PK1594 (A) forms corncob configurations with its partners *S. flueggei* PK1958 (B), *P. gingivalis,* PK1924 (C), *A. actinomycetemcomitans* Y4 (D), *V. atypica* PK1910 (E), and *S. sanguis* C104 (F) when the partners are present in a 10-fold excess. Bar, 10 μm.

of any one of its partners in vivo could change the cell type ratio and be conducive to corncob formations. Rapid growth is consistent with the clinical observations of periodontal diseases, which are characterized by bursts of activity interspersed with periods of remission (Socransky et al., 1984). Episodes of periodontal destruction may be caused by a succession of assaults by different species; the host responds to each assault with antibodies and other mechanisms of host defense (Williams et al., 1985).

The connection among coaggregation, corncob configurations, and subgingival dental plaque accretion is intriguing, and an attempt to illustrate these phenomena is depicted in Fig. 5. The observation that only streptococci exhibit intrageneric coaggregation was unexpected, but when placed in the context of their predominance in colonizing clean surfaces, it is more understandable. Subsequent colonizers do not need to coaggregate intragenerically. Indeed, fusobacteria, which coaggregate with all other oral genera so far tested, do not exhibit intrageneric coaggregation. As plaque accretes, fusobacteria may form corncob configurations with each new cell type that gains an ecological advantage. It would seem appropriate that these corncobs would predominate at the accreting surface of dental plaque.

The interactions among all the other bacteria in this figure have been observed in the laboratory. All of the coaggregations are consistent with potential advantages to early colonizers interacting with other early colonizers and late colonizers only needing to coaggregate with bacteria that are at the accreting surface. For example, *P. gingivalis*, *S. flueggei*, *Eubacterium nodatum*, and *W. recta* coaggregate with *F. nucleatum* but not with any of the other bacteria. All five of these bacteria are considered periodontopathogens and can contribute to the onset and development of periodontal disease (Moore, 1987). Most only appear when dental plaque is left undisturbed for several days. On the basis of the apparent ubiquitous presence of fusobacteria in plaque samples and their coaggregation range, it is proposed that (i) fusobacteria play a major role in mediating coaggregations between early and late colonizers and (ii) these bridging multigeneric coaggregations are of primary importance in the maturation of dental plaque.

Establishing a stable microbial community in the human oral cavity must involve both adherence and nutritional exchange among the inhabitants. Metabolic end products or secondary metabolites of one species are essential for the

---

FIGURE 5. Diagrammatic representation of proposed temporal relationship of bacterial accretion and multigeneric coaggregation in the formation of dental plaque. The 31 species represented here are *Actinobacillus actinomycetemcomitans*, *Actinomyces naeslundii*, *Actinomyces israelii*, *Capnocytophaga gingivalis*, *Capnocytophaga ochracea*, *Capnocytophaga sputigena*, *Eikenella corrodens*, *Eubacterium nodatum*, *Fusobacterium nucleatum*, *Haemophilus parainfluenzae*, *Peptostreptococcus anaerobius*, *Porphyromonas gingivalis*, *Prevotella denticola*, *Prevotella intermedia*, *Prevotella loescheii*, *Propionibacterium acnes*, *Rothia dentocariosa*, *Selenomonas flueggei*, *Selenomonas infelix*, *Selenomonas noxia*, *Selenomonas sputigena*, *Streptococcus gordonii*, *Streptococcus oralis*, *Streptococcus sanguis*, *Streptococcus* SM, *Veillonella atypica*, *Veillonella dispar*, *Veillonella parvula*, *Wolinella curva*, *Wolinella recta*, and *Wolinella succinogenes*. The thin dark layer at the tooth surface represents the acquired pellicle.

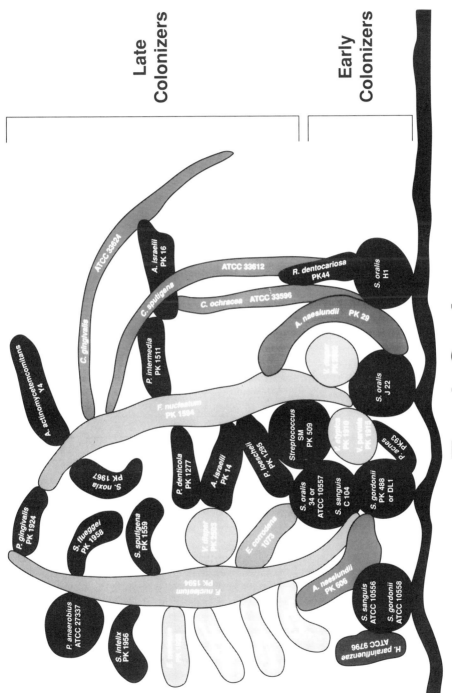

growth of other species. Several kinds of inhabitants may be able to supply the needs of others, and this form of loose, nonspecific interdependence is called protocooperation or nonobligatory mutualism. However, the presence of at least one nutritional partner is obligatory for the successful maintenance of the nutrition-receiving species in the ecological niche. With more than 400 bacterial taxa known to inhabit the oral cavity at various times from birth to adulthood or from oral health to disease, it seems likely that protocooperative interactions account for part of the dynamics of the well-documented bacterial successions in the oral cavity. The role of adherence in bacterial successions seems obvious in that new bacteria in the environment are added sequentially to accreting matrixes. An ecologically advantageous relationship involving both adherence and nutrition should promote the establishment and colonization of new bacteria. The facts (i) that human oral bacteria from 18 genera have been tested to date, (ii) that members of all 18 genera exhibit intergeneric coaggregation, (iii) that 15 of these 18 are the most frequently identified bacteria in subgingival dental plaque samples, (iv) that coaggregating partners occupy the same oral surfaces, and (v) that microscopic observations reveal intergeneric corncob configurations at the periphery of dental plaque offer convincing support for a major role of coaggregation in the development and progression of bacterial communities in the human oral cavity.

ACKNOWLEDGMENTS. I thank Roxanna Andersen for help in preparing the figures and for advice during the writing of this chapter. I thank Christopher Hughes, N. Ganeshkumar, and Angelika Kagermeier for advice and for making unpublished data available. The expertise of my colleagues Fred Cassels, Jack London, Lillian H. Moore, and Stanley Robrish in specific areas discussed here is greatly appreciated.

REFERENCES

Abeygunawardana, C., C. A. Bush, and J. O. Cisar. 1990. Complete structure of the polysaccharide from *Streptococcus sanguis* J22. *Biochemistry* **29:**234–248.

Abraham, S. N., J. D. Goguen, D. Sun, P. Klemm, and E. H. Beachey. 1987. Identification of two ancillary subunits of *Escherichia coli* type 1 fimbriae by using antibodies against synthetic oligopeptides of *fim* gene products. *J. Bacteriol.* **169:**5530–5536.

Andersen, R. N., P. Kolenbrander, C. V. Hughes, and E. Weiss. 1991. Functional relatedness of coaggregation mediators from seven species of oral bacteria. *J. Dent. Res.* **70:**A317.

Cassels, F. J. Personal communication.

Cassels, F. J., H. M. Fales, J. London, R. W. Carlson, and H. van Halbeek. 1990. Structure of a streptococcal adhesin carbohydrate receptor. *J. Biol. Chem.* **265:**14127–14135.

Cassels, F. J., and J. London. 1989. Isolation of a coaggregation-inhibiting cell wall polysaccharide from *Streptococcus sanguis* H1. *J. Bacteriol.* **171:**4019–4025.

Ciardi, J. E., G. F. A. McCray, P. E. Kolenbrander, and A. Lau. 1987. Cell-to-cell interaction of *Streptococcus sanguis* and *Propionibacterium acnes* on saliva-coated hydroxyapatite. *Infect. Immun.* **55:**1441–1446.

Cisar, J. O., M. J. Brennan, and A. L. Sandberg. 1985. Lectin-specific interaction of *Actinomyces* fimbriae with oral streptococci, p. 159–163. *In* S. A. Mergenhagen and B. Rosan (ed.), *Molecular Basis of Oral Microbial Adhesion.* American Society for Microbiology, Washington, D.C.

Cisar, J. O., S. H. Curl, P. E. Kolenbrander, and A. E. Vatter. 1983. Specific absence of type 2 fimbriae on a coaggregation-defective mutant of *Actinomyces viscosus* T14V. *Infect. Immun.* **40:**759–765.

Cisar, J. O., P. E. Kolenbrander, and F. C. McIntire. 1979. Specificity of coaggregation reactions between human oral streptococci and strains of *Actinomyces viscosus* or *Actinomyces naeslundii.* *Infect. Immun.* **24:**742–752.

Clark, W. B., L. L. Bammann, and R. J. Gibbons. 1978. Comparative estimates of bacterial affinities and adsorption sites on hydroxyapatite surfaces. *Infect. Immun.* **19:**846–853.

Clark, W. B., T. T. Wheeler, and J. O. Cisar. 1984. Specific inhibition of adsorption of *Actinomyces viscosus* T14V to saliva-treated hydroxyapatite by antibody against type 1 fimbriae. *Infect. Immun.* **43:**497–501.

Cowan, M. M., K. G. Taylor, and R. J. Doyle. 1987. Energetics of the initial phase of adhesion of *Streptococcus sanguis* to hydroxylapatite. *J. Bacteriol.* **169:**2995–3000.

Crowley, P. J., W. Fischlschweiger, S. E. Coleman, and A. S. Bleiweis. 1987. Intergeneric bacterial coaggregations involving mutans streptococci and oral actinomyces. *Infect. Immun.* **55:**2695–2700.

Distler, W., A. Kagermeier, R. Zajitschek, and A. Kröncke. 1989. Coaggregation enhances the glucose metabolism of mixtures of actinomyces and streptococci. Abstr. 36th ORCA Congr. *Caries Res.* **23:**448.

Dzink, J. L., M. T. Sheenan, and S. S. Socransky. 1990. Proposal of three subspecies of *Fusobacterium nucleatum* Knorr 1922: *Fusobacterium nucleatum* subsp. *nucleatum* subsp. nov., comb. nov.; *Fusobacterium nucleatum* subsp. *polymorphum* subsp. nov., nom. rev., comb. nov.; and *Fusobacterium nucleatum* subsp. *vincentii* subsp. nov., nom. rev., comb. nov. *Int. J. Syst. Bacteriol.* **40:**74–78.

Dzink, J. L., S. S. Socransky, and A. D. Haffajee. 1988. The predominant cultivable microbiota of active and inactive lesions of destructive periodontal diseases. *J. Clin. Periodontol.* **15:**316–323.

Dzink, J. L., A. C. R. Tanner, A. D. Haffajee, and S. S. Socransky. 1985. Gram negative species associated with active destructive periodontal lesions. *J. Clin. Periodontol.* **12:**648–659.

Fachon-Kalweit, S., B. L. Elder, and P. Fives-Taylor. 1985. Antibodies that bind to fimbriae block adhesion of *Streptococcus sanguis* to saliva-coated hydroxyapatite. *Infect. Immun.* **48:**617–624.

Fives-Taylor, P. 1982. Isolation and characterization of a *Streptococcus sanguis* FW213 mutant non-adherent to saliva-coated hydroxyapatite beads, p. 206–209. *In* D. Schlessinger (ed.), *Microbiology—1982.* American Society for Microbiology, Washington, D.C.

Fives-Taylor, P. M., F. L. Macrina, T. J. Pritchard, and S. S. Peene. 1987. Expression of *Streptococcus sanguis* antigens in *Escherichia coli:* cloning of a structural gene for adhesion fimbriae. *Infect. Immun.* **55:**123–128.

Ganeshkumar, N., R. N. Andersen, and P. E. Kolenbrander. 1989. Coaggregation-defective mutants of *Fusobacterium nucleatum*, abstr. D-228, p. 120. Abstr. 89th Annu. Meet. Am. Soc. Microbiol. 1989.

Ganeshkumar, N., P. M. Hannam, P. E. Kolenbrander, and B. C. McBride. 1991. Nucleotide sequence of a gene coding for a saliva-binding protein (SsaB) from *Streptococcus sanguis* 12 and its possible role in coaggregation with actinomyces. *Infect. Immun.* **59:**1093–1099.

Ganeshkumar, N., M. Song, and B. C. McBride. 1988. Cloning of a *Streptococcus sanguis* adhesin which mediates binding to saliva-coated hydroxyapatite. *Infect. Immun.* **56:**1150–1157.

Germaine, G. R., and C. F. Schachtele. 1976. *Streptococcus mutans* dextransucrase: mode of interaction with high-molecular-weight dextran and role in cellular aggregation. *Infect. Immun.* **13:**365–372.

Gibbons, R. J., I. Etherden, and Z. Skobe. 1983a. Association of fimbriae with the hydrophobicity of *Streptococcus sanguis* FC-1 and adherence to salivary pellicles. *Infect. Immun.* **41:**414–417.

Gibbons, R. J., and D. I. Hay. 1988. Human salivary acidic proline-rich proteins and statherin promote the attachment of *Actinomyces viscosus* LY7 to apatitic surfaces. *Infect. Immun.* **56:**439–445.

Gibbons, R. J., D. I. Hay, J. O. Cisar, and W. B. Clark. 1988. Adsorbed salivary proline-rich protein 1 and statherin: receptors for type 1 fimbriae of *Actinomyces viscosus* T14V-J1 on apatitic surfaces. *Infect. Immun.* **56:**2990–2993.

Gibbons, R. J., E. C. Moreno, and I. Etherden. 1983b. Concentration-dependent multiple binding sites on saliva-treated hydroxyapatite for *Streptococcus sanguis*. *Infect. Immun.* **39:**280–289.

Gibbons, R. J., S. S. Socransky, and B. Kapsimalis. 1964. Establishment of human indigenous bacteria in germ-free mice. *J. Bacteriol.* **88:**1316–1323.

Gibbons, R. J., and J. van Houte. 1971. Selective bacterial adherence to oral epithelial surfaces and its role as an ecological determinant. *Infect. Immun.* **3:**567–573.

Grenier, D., and D. Mayrand. 1986. Nutritional relationships between oral bacteria. *Infect. Immun.* **53:**616–620.

Hirsch, P. 1984. Microcolony formation and consortia, p. 373–393. *In* K. C. Marshall (ed.), *Microbial Adhesion and Aggregation*. Dahlem Konferenzen. Springer-Verlag, Berlin.

Hughes, C. V. 1991. Ph.D. thesis. Georgetown University, Washington, D.C.

Hughes, C. V., P. E. Kolenbrander, R. N. Andersen, and L. V. H. Moore. 1988. Coaggregation properties of human oral *Veillonella* spp.: relationship to colonization site and oral ecology. *Appl. Environ. Microbiol.* **54**:1957–1963.

Hughes, C. V., C. A. Roseberry, and P. E. Kolenbrander. 1990. Isolation and characterization of coaggregation-defective mutants of *Veillonella atypica*. *Arch. Oral Biol.* **35**:123S–125S.

Jenkinson, H. F., H. C. Lala, and M. G. Shepherd. 1990. Coaggregation of *Streptococcus sanguis* and other streptococci with *Candida albicans*. *Infect. Immun.* **58**:1429–1436.

Jones, S. J. 1972. A special relationship between spherical and filamentous microorganisms in mature human dental plaque. *Arch. Oral Biol.* **17**:613–616.

Karlsson, K.-A. 1989. Animal glycosphingolipids as membrane attachment sites for bacteria. *Annu. Rev. Biochem.* **58**:309–350.

Kolenbrander, P. E. 1982. Isolation and characterization of coaggregation-defective mutants of *Actinomyces viscosus, Actinomyces naeslundii*, and *Streptococcus sanguis*. *Infect. Immun.* **37**:1200–1208.

Kolenbrander, P. E. 1988. Intergeneric coaggregation among human oral bacteria and ecology of dental plaque. *Annu. Rev. Microbiol.* **42**:627–656.

Kolenbrander, P. E. 1989. Surface recognition among oral bacteria: multigeneric coaggregations and their mediators. *Crit. Rev. Microbiol.* **17**:137–159.

Kolenbrander, P. E., and R. N. Andersen. 1984. Cell-to-cell interactions of *Capnocytophaga* and *Bacteroides* species with other oral bacteria and their potential role in development of plaque. *J. Periodontal Res.* **19**:564–569.

Kolenbrander, P. E., and R. N. Andersen. 1985. Use of coaggregation-defective mutants to study the relationships of cell-to-cell interactions and oral microbial ecology, p. 164–171. *In* S. A. Mergenhagen and B. Rosan (ed.), *Molecular Basis of Oral Microbial Adhesion*. American Society for Microbiology, Washington, D.C.

Kolenbrander, P. E., and R. N. Andersen. 1986. Multigeneric aggregations among oral bacteria: a network of cell-to-cell interactions. *J. Bacteriol.* **168**:851–859.

Kolenbrander, P. E., and R. N. Andersen. 1988. Intergeneric rosettes: sequestered surface recognition among human periodontal bacteria. *Appl. Environ. Microbiol.* **54**:1046–1050.

Kolenbrander, P. E., and R. N. Andersen. 1990. Characterization of *Streptococcus gordonii (S. sanguis)* PK488 adhesin-mediated coaggregation with *Actinomyces naeslundii* PK606. *Infect. Immun.* **58**:3064–3072.

Kolenbrander, P. E., and R. N. Andersen. Unpublished results.

Kolenbrander, P. E., R. N. Andersen, and L. V. Holdeman. 1985. Coaggregation of oral *Bacteroides* species with other bacteria: central role in coaggregation bridges and competitions. *Infect. Immun.* **48**:741–746.

Kolenbrander, P. E., R. N. Andersen, and L. V. H. Moore. 1989. Coaggregation of *Fusobacterium nucleatum, Selenomonas flueggei, Selenomonas infelix, Selenomonas noxia*, and *Selenomonas sputigena* with strains from 11 genera of oral bacteria. *Infect. Immun.* **57**:3194–3203.

Kolenbrander, P. E., R. N. Andersen, and L. V. H. Moore. 1990. Intrageneric coaggregation among strains of human oral bacteria: potential role in primary colonization of the tooth surface. *Appl. Environ. Microbiol.* **56**:3890–3894.

Kolenbrander, P. E., Y. Inouye, and L. V. Holdeman. 1983. New *Actinomyces* and *Streptococcus* coaggregation groups among human oral isolates from the same site. *Infect. Immun.* **41**:501–506.

Kolenbrander, P. E., and B. L. Williams. 1981. Lactose-reversible coaggregation between oral actinomycetes and *Streptococcus sanguis*. *Infect. Immun.* **33**:95–102.

Kolenbrander, P. E., and B. L. Williams. 1983. Prevalence of viridans streptococci exhibiting lactose-inhibitable coaggregation with oral actinomycetes. *Infect. Immun.* **41**:449–452.

Lamont, R. J., and B. Rosan. 1990. Adherence of mutans streptococci to other oral bacteria. *Infect. Immun.* **58**:1738–1743.

Lamont, R. J., B. Rosan, G. M. Murphy, and C. T. Baker. 1988. *Streptococcus sanguis* surface antigens and their interactions with saliva. *Infect. Immun.* **56**:64–70.

Lancy, P., Jr., J. M. DiRienzo, B. Appelbaum, B. Rosan, and S. C. Holt. 1983. Corncob formation between *Fusobacterium nucleatum* and *Streptococcus sanguis*. *Infect. Immun.* **40:**303–309.

Liljemark, W. F., C. G. Bloomquist, and J. C. Ofstehage. 1979. Aggregation and adherence of *Streptococcus sanguis:* role of human salivary immunoglobulin A. *Infect. Immun.* **26:**1104–1110.

Liljemark, W. F., L. J. Fenner, and G. C. Bloomquist. 1986. In vivo colonization of salivary pellicle by *Haemophilus, Actinomyces* and *Streptococcus* species. *Caries Res.* **20:**481–497.

Lindberg, F., B. Lund, L. Johansson, and S. Normark. 1987. Localization of the receptor-binding protein adhesin at the tip of the bacterial pilus. *Nature* (London) **328:**84–87.

Listgarten, M. A. 1976. Structure of the microbial flora associated with periodontal health and disease in man. *J. Periodontol.* **47:**1–18.

Listgarten, M. A., H. Mayo, and M. Amsterdam. 1973. Ultrastructure of the attachment device between coccal and filamentous microorganisms in "corn cob" formation of dental plaque. *Arch. Oral Biol.* **18:**651–656.

Listgarten, M. A., H. E. Mayo, and R. Tremblay. 1975. Development of dental plaque on epoxy resin crowns in man. *J. Periodontol.* **46:**10–25.

Löe, H., E. Theilade, and S. B. Jensen. 1965. Experimental gingivitis in man. *J. Periodontol.* **36:**177–187.

Loesche, W. J., and S. A. Syed. 1978. Bacteriology of human experimental gingivitis: effect of plaque and gingivitis score. *Infect. Immun.* **21:**830–839.

London, J., and J. Allen. 1990. Purification and characterization of a *Bacteroides loescheii* adhesin that interacts with procaryotic and eucaryotic cells. *J. Bacteriol.* **172:**2527–2534.

London, J., A. R. Hand, E. I. Weiss, and J. Allen. 1989. *Bacteroides loescheii* PK1295 cells express two distinct adhesins simultaneously. *Infect. Immun.* **57:**3940–3944.

London, J., and P. E. Kolenbrander. Unpublished data.

Mayrand, D. 1985. Virulence promotion by mixed bacterial infections, p. 281–291. *In* G. G. Jackson and H. Thomas (ed.), *Bayer-Symposium VIII. The Pathogenesis of Bacterial Infections.* Springer-Verlag, Berlin.

McBride, B. C., and J. S. van der Hoeven. 1981. Role of interbacterial adherence in colonization of the oral cavities of gnotobiotic rats infected with *Streptococcus mutans* and *Veillonella alcalescens*. *Infect. Immun.* **33:**467–472.

McCabe, R. M., and J. A. Donkersloot. 1977. Adherence of *Veillonella* species mediated by extracellular glucosyltransferase from *Streptococcus salivarius*. *Infect. Immun.* **18:**726–734.

McDermid, A. S., A. S. McKee, D. C. Ellwood, and P. D. Marsh. 1986. The effect of lowering the pH on the composition and metabolism of a community of nine oral bacteria grown in a chemostat. *J. Gen. Microbiol.* **132:**1205–1214.

McIntire, F. C., L. K. Crosby, A. E. Vatter, J. O. Cisar, M. R. McNeil, C. A. Bush, S. S. Tjoa, and P. V. Fennessey. 1988. A polysaccharide from *Streptococcus sanguis* 34 that inhibits coaggregation of *S. sanguis* 34 with *Actinomyces viscosus* T14V. *J. Bacteriol.* **170:**2229–2235.

McKee, A. S., A. S. McDermid, D. C. Ellwood, and P. D. Marsh. 1985. The establishment of reproducible, complex communities of oral bacteria in the chemostat using defined inocula. *J. Appl. Bacteriol.* **59:**263–275.

Moore, L. V. H., W. E. C. Moore, E. P. Cato, R. M. Smibert, J. A. Burmeister, A. M. Best, and R. R. Ranney. 1987. Bacteriology of human gingivitis. *J. Dent. Res.* **66:**989–995.

Moore, W. E. C. 1987. Microbiology of periodontal disease. *J. Periodontal Res.* **22:**335–341.

Moore, W. E. C., L. V. Holdeman, E. P. Cato, R. M. Smibert, J. A. Burmeister, K. G. Palcanis, and R. R. Ranney. 1985. Comparative bacteriology of juvenile periodontitis. *Infect. Immun.* **48:**507–519.

Moore, W. E. C., L. V. Holdeman, E. P. Cato, R. M. Smibert, J. A. Burmeister, and R. R. Ranney. 1983. Bacteriology of moderate (chronic) periodontitis in mature human adults. *Infect. Immun.* **42:**510–515.

Moore, W. E. C., L. V. Holdeman, E. P. Cato, R. M. Smibert, I. J. Good, J. A. Burmeister, K. G. Palcanis, and R. R. Ranney. 1982a. Bacteriology of experimental gingivitis in young adult humans. *Infect. Immun.* **38:**651–667.

Moore, W. E. C., L. V. Holdeman, R. M. Smibert, E. P. Cato, J. A. Burmeister, K. G. Palcanis, and R. R. Ranney. 1984. Bacteriology of experimental gingivitis in children. *Infect. Immun.* **46:**1–6.

Moore, W. E. C., L. V. Holdeman, R. M. Smibert, D. E. Hash, J. A. Burmeister, and R. R. Ranney. 1982b. Bacteriology of severe periodontitis in young adult humans. *Infect. Immun.* **38**:1137–1148.

Moore, W. E. C., L. V. H. Moore, and E. P. Cato. 1988. You and your flora. *U.S. Fed. Culture Collections Newsl.* **18**:7–22.

Morris, E. J., and B. C. McBride. 1984. Adherence of *Streptococcus sanguis* to saliva-coated hydroxyapatite: evidence for two binding sites. *Infect. Immun.* **43**:656–663.

Mouton, C., H. S. Reynolds, E. A. Gasiecki, and R. J. Genco. 1979. *In vitro* adhesion of tufted oral streptococci to *Bacterionema matruchotii*. *Curr. Microbiol.* **3**:181–186.

Mouton, C., H. S. Reynolds, and R. J. Genco. 1980. Characterization of tufted streptococci isolated from the "corn cob" configuration of human dental plaque. *Infect. Immun.* **27**:235–245.

Nesbitt, W. E., R. J. Doyle, and K. G. Taylor. 1982. Hydrophobic interactions and the adherence of *Streptococcus sanguis* to hydroxylapatite. *Infect. Immun.* **38**:637–644.

Nyvad, B., and O. Fejerskov. 1987a. Scanning electron microscopy of early microbial colonization of human enamel and root surfaces in vivo. *Scand. J. Dent. Res.* **95**:287–296.

Nyvad, B., and O. Fejerskov. 1987b. Transmission electron microscopy of early microbial colonization of human enamel and root surfaces in vivo. *Scand. J. Dent. Res.* **95**:297–307.

Nyvad, B., and M. Kilian. 1987. Microbiology of the early colonization of human enamel and root surfaces in vivo. *Scand. J. Dent. Res.* **95**:369–380.

Nyvad, B., and M. Kilian. 1990. Comparison of the initial streptococcal microflora on dental enamel in caries-active and in caries-inactive individuals. *Caries Res.* **24**:267–272.

Ostrom, C. A., T. Koulourides, F. Hickman, and J. McGhee. 1977. Characterization of an experimental cariogenic plaque in man. *J. Dent. Res.* **56**:550–558.

Poindexter, J. S. 1964. Biological properties and classification of the *Caulobacter* group. *Bacteriol. Rev.* **28**:231–295.

Robrish, S. Personal communication.

Robrish, S. A., C. Oliver, and J. Thompson. 1987. Amino acid-dependent transport of sugars by *Fusobacterium nucleatum* ATCC 10953. *J. Bacteriol.* **169**:3891–3897.

Robrish, S. A., and J. Thompson. 1989. Na$^+$ requirement for glutamate-dependent sugar transport by *Fusobacterium nucleatum* ATCC 10953. *Curr. Microbiol.* **19**:329–334.

Robrish, S. A., and J. Thompson. 1990. Regulation of fructose metabolism and polymer synthesis by *Fusobacterium nucleatum* ATCC 10953. *J. Bacteriol.* **172**:5714–5723.

Schwarz, S., R. P. Ellen, and D. A. Grove. 1987. *Bacteroides gingivalis-Actinomyces viscosus* cohesive interactions as measured by a quantitative binding assay. *Infect. Immun.* **55**:2391–2397.

Selin, Y., and J. L. Johnson. 1981. DNA nucleotide sequence similarities among strains of *Fusobacterium nucleatum*. *J. Dent. Res.* **60**:415.

Shelton, D. R., and J. M. Tiedje. 1984. Isolation and partial characterization of bacteria in an anaerobic consortium that mineralizes 3-chlorobenzoic acid. *Appl. Environ. Microbiol.* **48**:840–848.

Slots, J. 1977. Microflora in the healthy gingival sulcus in man. *Scand. J. Dent. Res.* **85**:247–254.

Slots, J., D. Möenbo, J. Langebaek, and A. Frandsen. 1978. Microbiota of gingivitis in man. *Scand. J. Dent. Res.* **86**:174–181.

Socransky, S. S., A. D. Haffajee, J. M. Goodson, and J. Lindhe. 1984. New concepts of destructive periodontal disease. *J. Clin. Periodontol.* **11**:21–32.

Socransky, S. S., A. D. Manganiello, D. Propas, V. Oram, and J. van Houte. 1977. Bacteriological studies of developing supragingival dental plaque. *J. Periodontal Res.* **12**:90–106.

Staat, R. H., S. D. Langley, and R. J. Doyle. 1980. *Streptococcus mutans* adherence: presumptive evidence for protein-mediated attachment followed by glucan-dependent cellular accumulation. *Infect. Immun.* **27**:675–681.

Stock, J. B., A. M. Stock, and J. M. Mottonen. 1990. Signal transduction in bacteria. *Nature* (London) **344**:395–400.

Syed, S. A., and W. J. Loesche. 1978. Bacteriology of human experimental gingivitis: effect of plaque age. *Infect. Immun.* **21**:821–829.

Tanner, A. C. R., S. S. Socransky, and J. M. Goodson. 1984. Microbiota of periodontal pockets losing crestal alveolar bone. *J. Periodontal Res.* **19**:279–291.

Tempro, P., F. Cassels, R. Siraganian, A. R. Hand, and J. London. 1989. Use of adhesin-specific monoclonal antibodies to identify and localize an adhesin on the surface of *Capnocytophaga gingivalis* DR2001. *Infect. Immun.* **57**:3418–3424.

Theilade, E., J. Theilade, and L. Mikkelsen. 1982. Microbiological studies on early dento-gingival plaque on teeth and Mylar strips in humans. *J. Periodontal Res.* **17**:12–25.

Theilade, E., W. H. Wright, S. B. Jensen, and H. Löe. 1966. Experimental gingivitis in man. II. A longitudinal clinical and bacteriological investigation. *J. Periodontal Res.* **1**:1–13.

van Houte, J., and D. B. Green. 1974. Relationship between the concentration of bacteria in saliva and the colonization of teeth in humans. *Infect. Immun.* **9**:624–630.

Weiss, E. I., I. Eli, B. Shenitzki, and N. Smorodinsky. 1990. Identification of the rhamnose-sensitive adhesin of *Capnocytophaga ochracea* ATCC 33596. *Arch. Oral Biol.* **35**:127S–130S.

Weiss, E. I., P. E. Kolenbrander, J. London, A. R. Hand, and R. N. Andersen. 1987a. Fimbria-associated proteins of *Bacteroides loescheii* PK1295 mediate intergeneric coaggregations. *J. Bacteriol.* **169**:4215–4222.

Weiss, E. I., J. London, P. E. Kolenbrander, R. N. Andersen, C. Fischler, and R. P. Siraganian. 1988a. Characterization of monoclonal antibodies to fimbria-associated adhesins of *Bacteroides loescheii* PK1295. *Infect. Immun.* **56**:219–224.

Weiss, E. I., J. London, P. E. Kolenbrander, A. R. Hand, and R. Siraganian. 1988b. Localization and enumeration of fimbria-associated adhesins of *Bacteroides loescheii*. *J. Bacteriol.* **170**:1123–1128.

Weiss, E. I., J. London, P. E. Kolenbrander, A. S. Kagermeier, and R. N. Andersen. 1987b. Characterization of lectinlike surface components on *Capnocytophaga ochracea* ATCC 33596 that mediate coaggregation with gram-positive oral bacteria. *Infect. Immun.* **55**:1198–1202.

Wheeler, T. T., and W. B. Clark. 1980. Fibril-mediated adherence of *Actinomyces viscosus* to saliva-treated hydroxyapatite. *Infect. Immun.* **28**:577–584.

Williams, B. L., J. L. Ebersole, M. D. Spektor, and R. C. Page. 1985. Assessment of serum antibody patterns and analysis of subgingival microflora of members of a family with a high prevalence of early-onset periodontitis. *Infect. Immun.* **49**:742–750.

Wittenberger, C. L., A. J. Beaman, L. N. Lee, R. M. McCabe, and J. A. Donkersloot. 1977. Possible role of *Streptococcus salivarius* glucosyltransferase in adherence of *Veillonella* to smooth surfaces, p. 417–421. *In* D. Schlessinger (ed.), *Microbiology—1977*. American Society for Microbiology, Washington, D.C.

Yeung, M. K., B. M. Chassy, and J. O. Cisar. 1987. Cloning and expression of a type 1 fimbrial subunit of *Actinomyces viscosus* T14V. *J. Bacteriol.* **169**:1678–1683.

Zeikus, J. G. 1983. Metabolic communication between biodegradative populations in nature. *Symp. Soc. Gen. Microbiol.* **34**:423–462.

Predator-Prey Interactions

Microbial Cell-Cell Interactions
Edited by Martin Dworkin
© 1991 American Society for Microbiology, Washington, DC 20005

Chapter 11

Intercellular Signalling in the *Bdellovibrio* Developmental Cycle

Kendall M. Gray and Edward G. Ruby

INTRODUCTION

The genus *Bdellovibrio* comprises a small group of predatory bacteria, ubiquitous in both terrestrial and aquatic environments, that grow within the periplasm of a variety of gram-negative bacterial prey (see Rittenberg [1982], Ruby [1991], and Thomashow and Rittenberg [1979] for reviews). There are currently three named species of bdellovibrios, with at least one additional recognized species awaiting further taxonomic characterization (Burnham and Conti, 1984). The most extensively studied of these species, *Bdellovibrio bacteriovorus,* will be the primary subject of this chapter.

A feature common to all bdellovibrios is a predatory life cycle consisting of two distinct developmental stages (Fig. 1). During its free-living attack phase, the bdellovibrio is a small (0.2 to 0.5 by 0.5 to 1.4 μm), highly motile, vibrioid cell bearing a single flagellum at one pole. When the attack-phase bdellovibrio encounters a suitable prey cell, it attaches at its nonflagellated pole and penetrates the outer membrane and cell wall of the prey, shedding its own flagellum in the process. This penetration step represents the transition from the attack phase to

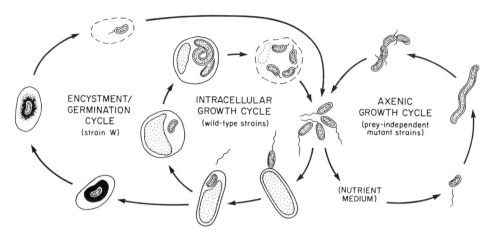

FIGURE 1. The developmental cycles of the bdellovibrios. Every wild-type strain of bdellovibrio currently known exhibits an obligately intracellular growth cycle, although mutants of these strains are capable of axenic growth in complex nutrient medium alone. One strain, *Bdellovibrio* sp. strain W, is unique in its ability occasionally to form an intracellular bdellocyst rather than following the normal filamentous growth cycle.

the growth phase of the bdellovibrio life cycle. Once within the periplasm of its prey, the bdellovibrio biochemically modifies the prey cell's peptidoglycan and lipopolysaccharide layers, rendering it inaccessible to further predation by other attack-phase cells. In this way, the prey cell is formed into an isolated, spherical growth chamber, referred to as a "bdelloplast." Using the cytoplasmic contents of the prey as its growth substrate, the bdellovibrio then elongates into a coiled, aseptate filament that can attain a length of more than 90 attack-phase cell units, depending upon the size of the prey cell (Kessel and Shilo, 1976).

When the prey cell is depleted of some vital nutrient and/or chemical signal, the bdellovibrio filament septates and divides into single-unit cells, each of which grows the polar flagellum characteristic of the attack phase. The bdelloplast is then lysed, releasing the newly differentiated attack-phase bdellovibrios into the surrounding environment to begin the cycle again. For *B. bacteriovorus*, this biphasic developmental cycle takes approximately 3 to 4 h at 30°C, producing an average of four progeny cells when *Escherichia coli* is used as the prey (Thomashow and Rittenberg, 1979).

In contrast to the obligately intracellular life cycle of the wild-type bdellovibrios, there are spontaneous mutations, arising at a frequency of approximately 1 in 10^6 to 10^7 cells, that result in a prey-independent phenotype (Seidler and Starr, 1969). These mutant bdellovibrios continue to follow the same patterns of filamentous elongation, division, and differentiation as the wild-type strains (Burnham et al., 1970; Diedrich et al., 1970), but are either restricted to axenic growth in complex nutrient medium alone ("symbiosis-incompetent" strains) or are capable of facultative growth, either axenically or within a prey cell ("symbiosis-independent" strains) (Varon and Seijffers, 1975).

A second and unique variation on the wild-type life cycle is exhibited by a single strain of an as yet uncharacterized species of *Bdellovibrio* (Burger et al., 1968). In this particular isolate, called strain W, the cells will occasionally differentiate to form resting bodies, or "bdellocysts," following the successful penetration of a prey cell (Tudor and Conti, 1977). Bdellocysts are more resistant than the vegetative cells to heating, sonication, starvation, desiccation, or disruption by nonionic detergents, but are produced at the expense of cellular proliferation, since germination of the cyst invariably results in the liberation of only a single attack-phase bdellovibrio (Fig. 1). The existence of this encysted form makes the bdellovibrios unusual among developmental prokaryotes in that they can be said to possess three distinct differentiated states, consisting of invasive (attack-phase), proliferative (growth-phase), and quiescent (bdellocyst) forms.

Bdellovibrios have evolved several unique physiological capabilities as a consequence of their intracellular lifestyle. Among these are the ability to biosynthetically modify prey cell structures as a means of isolating and conserving their food supply (Ruby and Rittenberg, 1984; Ruby, 1989; Thomashow and Rittenberg, 1978a, 1978b, 1978c), the ability to translocate intact membrane proteins from the prey cell to their own outer membranes (Diedrich et al., 1983; Diedrich et al., 1984; Guerrini et al., 1982; Talley et al., 1987), and the ability to conserve high-energy phosphate bonds already present within the prey cell by taking up nucleic acids (Rittenberg and Langley, 1975; Ruby et al., 1985; Ruby and McCabe, 1986) and possibly fatty acids (Kuenen and Rittenberg, 1975; Nelson and Rittenberg, 1981) in phosphorylated form. As a result of these and still other unique capabilities, bdellovibrios exhibit a growth efficiency in terms of biomass yield per ATP utilized (Y_{ATP}) that is more than twice that of any other bacterium examined (Rittenberg and Hespell, 1975).

Perhaps the most intriguing aspect of the biology of differentiating prokaryotes such as the bdellovibrios, however, is the means by which their developmental cycle is regulated. Given the intimate physical relationship between the bdellovibrio and its prey cell, it is not especially surprising that intercellular signal exchange appears to play a crucial role in the development of bdellovibrios. Furthermore, this predator-prey interaction provides an exceptional instance of cell-cell communication in that only one of the two cells remains viable once the association has been established (Rittenberg and Shilo, 1970), yet it is the nonviable prey cell that appears to serve as the primary source of signal factors. In this chapter, we will review all that is currently known concerning prey-derived signals that regulate the bdellovibrio developmental cycle and attempt to combine the frequently contradictory results of these various studies into a unifying model of intercellular signalling as it relates to bdellovibrio development.

EXAMPLES OF INTERCELLULAR SIGNALLING IN THE *BDELLOVIBRIO* DEVELOPMENTAL CYCLE

Location of Prey

The earliest opportunity for signalling to occur between the bdellovibrio and its prey is actually before they become associated. The greatest problem immediately facing a free-living attack-phase bdellovibrio is the location of a suitable prey bacterium; because of its unusually high respiration rate, the attack-phase cell is estimated to have only a few hours in which to locate its prey before death will occur as a result of the depletion of its limited energetic reserves (Hespell et al., 1974). The bdellovibrio's success in penetrating the prey cell following its initial contact is extremely low, however, suggesting that relatively high prey population densities are required for the survival of bdellovibrios (Varon and Ziegler, 1978). Alternatively, if the attack-phase cells were capable of even limited taxis toward their prey, the chances of a successful encounter, and consequently a successful attack, would be greatly enhanced, thereby reducing the minimum prey cell density required for survival. As a result, there would appear to be a strong selective advantage to a chemotactic "search" mechanism by which the bdellovibrios could more efficiently locate their prey.

Attack-phase bdellovibrios have been demonstrated to exhibit a positive chemotactic response to a variety of respirable substrates including complex media (Straley and Conti, 1974), specific amino acids (LaMarre et al., 1977), and numerous other organic compounds (Straley et al., 1979). In the only publication to address the subject, however, no evidence of chemotaxis by bdellovibrios specifically toward prey cells or to exudates of prey cells was revealed (Straley and Conti, 1977). It should be stressed that numerous experimental alternatives remain to be explored; for example, are nonchemotactic mutants as efficient as the wild-type bdellovibrios in attacking their prey? The data currently available, however, suggest that bdellovibrios do not possess any chemotactic abilities that might lead them specifically to prey cells, or even to populations of other bacteria in general.

Current thinking, therefore, is that the bdellovibrios gain a selective advantage in being attracted to respirable compounds that will delay the onset of their own starvation, even though they cannot proliferate on these compounds. A further advantage to this behavior is that prey bacteria themselves may be chemotactic toward these same compounds, thus indirectly enhancing the bdellovibrios' chances of encountering prey. Once the cells are within the same local environment of a given chemoattractant, however, there is no evidence of any signalling involved in the specific location of prey bacteria by attack-phase bdellovibrios; instead, they would appear to encounter their prey purely by random collision.

Attachment and Penetration

Once the attack-phase bdellovibrio has successfully encountered its prey, the actual mechanism of attachment and penetration appears to be a two-step process that requires continued motility of the bdellovibrio (Dunn et al., 1974). The initial, reversible stage of attachment following collision with the prey is characterized by a loose association in which the bdellovibrio swivels about its long axis from the point of attachment to the prey cell surface. This first stage does not appear to involve the interaction of any specific surface receptors or anitgenic features of the two cells, as attack-phase bdellovibrios will form reversible attachments not only with susceptible prey, but also with nonsuitable prey such as gram-positive bacteria, and even with glass surfaces such as microscope slides or coverslips. The second stage of attachment is an irreversible association and is observed only with bdellovibrios that have first reversibly attached to a suitable prey cell. In this stage, the bdellovibrio forms a tight juncture with the prey cell, which itself begins to rotate about the bdellovibrio's long axis as the attack-phase cell continues to swim. This irreversible second stage of attachment is quickly followed by penetration of the bdellovibrio into the prey cell and subsequent formation of the bdelloplast.

Although a broad range of gram-negative bacteria can serve as prey for bdellovibrios, there remains a certain degree of prey specificity that varies according to the strain of bdellovibrio tested (Taylor et al., 1974) or even the experimental conditions employed (Torrella et al., 1978). This suggests that specific surface receptor interactions might be involved in the recognition of suitable prey by the attack-phase bdellovibrio. Such a possibility is supported by evidence of physical differentiation of the nonflagellated pole of the attack-phase cell envelope; microscopic studies have revealed distinctive ringlike structures with outwardly projecting fibers associated exclusively with the attachment pole of the bdellovibrio (Abram and Davis, 1970; Burnham et al., 1968), although these structures might simply form a nonspecific "holdfast" apparatus (Scherff et al., 1966).

The involvement of specific receptor interactions in prey recognition has been further supported, however, by two additional pieces of evidence. First, a temperature-sensitive bdellovibrio mutant was isolated that is impaired in its ability to form irreversible attachments with a normally susceptible prey strain, although reversible attachment remains unaffected by the mutation (Dunn et al., 1974). Second, a population of prey cells that was maintained in continuous culture with wild-type bdellovibrios gave rise to a mutant strain possessing a predation-resistant phenotype (Varon, 1979). Collectively, these discoveries indicated that mutations in either of the two partners involved in this interaction could specifically block the normal process of attachment and penetration by the bdellovibrio. The question remained, however, as to what specific features were required for the successful attachment of the bdellovibrio to its prey.

Attack-phase bdellovibrios were shown to exhibit an enhanced rate of attachment to "rough" cell wall mutants of susceptible prey (Varon and Shilo, 1969), although the methods employed in these experiments failed to distinguish between simple attachment and actual penetration by the attack-phase cells. Consequently, the observed affinity for these prey cell mutants might simply reflect a decrease in the bdellovibrio's ability to detach from the thicker prey cell envelope following its initial, reversible collision. Furthermore, "smooth" wall mutants of the same species were also preyed upon by the bdellovibrios, suggesting that no specific surface features absolutely crucial to attachment and penetration were affected by these particular mutations.

More recent investigations with *B. bacteriovorus* strains IBFM B-608 and 109D have suggested that it is the core sugars of the prey cell's lipopolysaccharide layer that are specifically required for the irreversible attachment of the attack-phase cell (Chemeris et al., 1984; Schelling and Conti, 1986). Successful attachment and penetration by the bdellovibrios was increasingly inhibited in these experiments by the sequential deletion of residues from the prey cell's lipopolysaccharide core sugars and was sharply reduced by the addition of free sugars to the suspension medium, although this had no apparent effect on the motility of the attack-phase cells. Prey cell outer membrane proteins appeared to play no role at all in the attachment of strain 109D (Schelling and Conti, 1986), raising the possibility that a general lectin-oligosaccharide type of binding interaction might be involved in the formation of the irreversible attachment site between the two cells. Such a description is inherently misleading, however, as there is no strong evidence of any true lectinlike substrate specificity for this particular surface interaction. All that can presently be concluded from this work is that the interaction appears to involve prey surface oligosaccharides of a required length, with no apparent distinction made as to the composition of the sugars.

In contrast to the results obtained with *B. bacteriovorus,* the irreversible attachment of a different species, *Bdellovibrio stolpii,* was found to be completely unaffected by changes in the core sugars of the prey cell's lipopolysaccharide layer. Instead, successful attachment and penetration of the prey by this species appeared to require the presence of specific proteins in the outer membrane of the prey cell (Schelling and Conti, 1986). Consequently, the specific receptors involved in the recognition and penetration of suitable prey by attack-phase bdellovibrios appears to vary, depending upon the species of bdellovibrio being considered.

Because only two strains of *B. bacteriovorus* have been characterized in these studies, it remains uncertain whether equal variability in the mechanisms of prey recognition might not exist within a single species of bdellovibrio as well. Contradictory reports of the ability of various strains of *B. bacteriovorus* to grow on viable versus heat-killed prey (Hespell, 1978; Varon and Shilo, 1968) suggest that such intraspecific variation may well exist, although these results might simply reflect differences in the experimental conditions employed (Torrella et al., 1978). It has also been observed, however, that *B. bacteriovorus* 109J will successfully prey on both viable and heat-killed cells of *E. coli* ML35, but can only attack viable cells of *E. coli* TOE13 (unpublished results). Although there are many possible explanations for this result, it does at least raise the possibility that this particular bdellovibrio strain may possess different "recognition" mechanisms that respond to various surface components, depending upon the antigenic make-up of the prey cell itself.

It seems certain that, at least across species lines, attack-phase bdellovibrios are capable of responding to a variety of cell surface characteristics as a means of identifying suitable prey. This reinforces the idea that bdellovibrios possess only very general, though highly adaptable, abilities of prey recognition that might more accurately be described as simple environmental sensing mechanisms.

The formation of an irreversible attachment to the prey cell must still be considered to entail a true signalling event between the two organisms, however, since it is a necessary prerequisite for the bdellovibrio's production and release of the enzyme activities specifically involved in prey cell penetration and bdelloplast formation (Thomashow and Rittenberg, 1979). This carefully timed developmental response of the bdellovibrio, apparently triggered by its irreversible attachment to the prey cell, clearly indicates that some sort of regulatory signal is received at this stage. Unfortunately, one can only speculate as to whether or not these signals resulting in differential gene expression are the same as those involved in the process of attachment itself. An exciting parallel to the bdellovibrios' attachment and penetration of prey bacteria has recently been found with certain members of the genus *Salmonella*. These intracellular pathogens of eukaryotes have been shown to attach to mammalian epithelial cells by means of a two-step process involving reversible and irreversible stages, with irreversible attachment inducing the synthesis of at least six new bacterial proteins that are essential to the process of penetration (Finlay et al., 1989). Because this induced gene expression appears to require specific surface receptors that are unique to epithelial cells, the similarity of this system to that of the bdellovibrios becomes especially intriguing.

Finally, it is possible that the prey cell's recognition features may play a further role in the process by which the newly differentiated growth-phase bdellovibrio prevents the secondary attack of its bdelloplast by other attack-phase cells. It has been proposed that deacetylation of the prey cell peptidoglycan during bdelloplast formation represents the mechanism of this exclusion, since the glycanase produced by attack-phase cells during penetration cannot cleave the modified bdelloplast wall (Thomashow and Rittenberg, 1978a, 1978b). This view has recently been challenged, however, by the discovery that growth-phase cells of *Bdellovibrio* sp. strain W do not deacetylate the prey cell peptidoglycan, yet they continue to prevent other attack-phase cells from entering the bdelloplast, or even from forming irreversible attachments to its surface (Tudor et al., 1990). Consequently, the exclusion mechanism involved in bdelloplast formation is presumably not limited to deacetylation of the prey cell peptidoglycan alone, and may, in fact, be more specifically related to the outer membrane than to the cell wall itself. It might be speculated, therefore, that other alterations of the bdelloplast envelope serve to inhibit the secondary predation of a given prey cell by limiting the attack-phase bdellovibrios' access to the surface features necessary for the formation of an irreversible attachment.

Initiation of Growth

The initiation of the bdellovibrio growth phase has received perhaps more attention than any other regulatory event of its developmental cycle. Every bdellovibrio strain isolated from nature to date exhibits an absolute requirement for a suitable prey cell in order to initiate growth (Burnham and Conti, 1984), although the exact nature of this prey dependency remains a mystery. Intriguingly, all strains tested have readily yielded prey-independent mutants of one type or another under laboratory selection techniques. The frequency at which these spontaneous mutations occur suggests that only a single point mutation is needed to overcome the obligate prey cell requirement for growth (Seidler and Starr, 1969). The fact that axenically grown prey-independent mutant strains continue to follow the wild-type pattern of filamentous growth, division, and differentiation suggests that the initiation of growth is regulated independently of cell elongation and division and that differentiation into the growth phase normally requires specific chemical or physical cues that are provided by the prey cell during the attack process. This hypothesis has prompted numerous investigations into the possibility of inducing growth in wild-type attack-phase bdellovibrios by exposing them to various components of prey cell extracts.

Reiner and Shilo (1969) were the first to report the existence of a prey-derived signal factor that promoted the extracellular growth of prey-dependent bdellovibrios. This factor, extracted by sonication of a prey cell suspension, was both heat stable and resistant to degradation by RNase, DNase, or pronase. It was proposed to have a molecular weight of greater than 50,000 and was equally distributed between the supernatant fluid and the pellet upon centrifugation at 120,000 $\times$ g. Unfortunately, its activity was sufficient to induce growth in only about 5% of the population tested in each experiment.

Crothers et al. (1972) then reported the extracellular proliferation of normally prey-dependent bdellovibrios in the presence of autoclaved cells of either suitable (gram-negative) or nonsuitable (gram-positive) prey bacteria. Further experiments indicated that extracellular growth of the wild-type bdellovibrios could be induced by the supernatant extract of autoclaved *E. coli* alone, although this growth absolutely required the presence of divalent cations in addition to the "elongation factor" presumed to be released or extracted from the autoclaved prey. A subsequent report revealed that this cation dependency was a requirement of this particular bdellovibrio strain even under intracellular growth conditions, but further stated that sonication extracts of prey cells failed to induce extracellular growth of attack-phase cells, either with or without the addition of free cations (Huang and Starr, 1973). A third paper later reiterated the original observation of extracellular growth in the presence of intact autoclaved prey cells, but provided no new information concerning the reasons for this result (Ross et al., 1974).

Ishiguro (1973) later identified a prey-derived regulatory signal which he called "growth-inducing factor." As with the factor reported by Reiner and Shilo, only a portion of the activity of growth-inducing factor was sedimented at 130,000 $\times$ g, but it differed from both of the previous reports in that it was heat labile, although gentle heating actually stimulated its activity. This factor was originally defined in this study as an activator for the growth and isolation of prey-independent mutant strains, rather than as a signal that is normally required for intracellular growth of the wild-type cells. The frequency at which the bdellovibrios initiated extracellular growth in these experiments (Table 1) supports the assertion that these cells actually represent prey-independent mutant derivatives of the wild-type cells. Because growth-inducing factor continued to be required for axenic growth of these isolates through multiple transfers, however, the distinction of "prey independence" becomes somewhat unclear.

Horowitz et al. (1974) then extended the work begun by Reiner and Shilo, isolating a heat-stable compound that could induce growth in up to 50% of the cells of a given bdellovibrio population. The prey factor that they identified also differed from those of previous reports, however, in that it was sensitive to RNase and pronase and was not sedimented at 140,000 $\times$ g.

Finally, Friedberg (1978) reported the existence of another heat-stable, prey-derived compound that promoted extracellular growth of bdellovibrios with about

TABLE 1
Comparative characteristics of growth-inducing signal factors extracted from prey cells[a]

Reference	Extraction method	Resistance to:				Sedimentation characteristics	Effect of cations	Growth induction efficiency (% pop.)[b]
		Heat	RNase	DNase	Protease			
Reiner and Shilo (1969)	Sonication	+	+	+	+	Partially sedimented at 120,000 × g	NR	≤5%
Crothers et al. (1972)	Autoclaving	+	NR	NR	NR	NR	Requires 2 mM Ca²⁺ or Mg²⁺	NR
Ishiguro (1973)	Sonication	±[c]	NR	NR	NR	Partially sedimented at 130,000 × g	NR	~1 in 10⁸ cells
Horowitz et al. (1974)	Sonication	+	−	+	−	Unsedimented at 140,000 × g	Inhibited by 10 mM Mg²⁺	≤50%
Friedberg (1978)	Sonication	+	+	+	±[d]	Unsedimented at 120,000 × g, sedimented at 220,000 × g	Enhanced by 10 mM Ca²⁺ or Mg²⁺	≤30%

[a] Symbols: growth-inducing activity is (+) or is not (−) resistant to the described treatment; NR, no results reported.
[b] Determined by microscopic observation of cell filamentation, except in the case of Ishiguro (1973), in which growth induction was determined by colony formation on agar plates.
[c] Activity was heat labile, although stimulated by gentle heating (70°C for 10 min).
[d] Signal was inactivated by pronase, but stimulated by treatment with trypsin.

a 30% success rate. This factor remained within the soluble fraction after centrifugation at 120,000 × g, although it could be completely sedimented at 220,000 × g. It was resistant to RNase and DNase, sensitive to pronase, and enhanced in its activity by trypsin digestion or by the presence of divalent cations.

The contradictory nature of these various results (summarized in Table 1) clearly indicates that there is still much to be known concerning the role of prey-derived regulatory factors in the initiation of growth by attack-phase bdellovibrios. The fact that these results have not been reproducible since their original publication further suggests that undetectable and/or uncontrollable variations between the individual preparations of prey cell extract may have been the primary determinants of the results obtained in each case. Intuitively, one would anticipate that the signal to initiate growth, being encountered during the process of attachment and/or penetration of the prey, would be present as part of the prey cell envelope. The active component identified in these various reports was usually either partially or completely sedimented by ultracentrifugation (Table 1), suggesting that at least a portion of the growth-inducing signal exists within the particulate fraction of a prey cell extract. Perhaps the varying degrees of success obtained with these different extracts, as well as the activity's changing location in either soluble or particulate fractions, is simply an effect of the abundance and fragment size, respectively, of prey cell envelope constituents present in each individual extract.

The most recent discoveries concerning the initiation of growth by bdellovibrios were actually obtained from experiments intended to focus upon the physiology of the growth-phase cell following its differentiation from the attack phase (Ruby and Rittenberg, 1983). In these studies, wild-type cells were allowed to attack and penetrate prey bacteria, but were prematurely released from the bdelloplast at various times during their growth cycle by the addition of concentrated preparations of the lytic enzymes normally produced at the end of the growth phase. It was discovered that cells that were released in the first 45 to 60 min after attack, which is before the initiation of net DNA synthesis occurs (Matin and Rittenberg, 1972), would differentiate into the attack phase without any increases in cell number. The fact that this result was reproducible even when the cells were prematurely released into medium that could support growth of prey-independent mutant strains suggests that nutrient conditions alone were not the cause of this differentiation.

The further implication of this result is that penetration of the prey cell and differentiation into the aflagellate growth-phase morphology do not necessarily connote a commitment by the bdellovibrio to increase its cell number. Morphological differentiation into the growth phase is therefore regulated independently of the actual growth and proliferation of these cells. Consequently, it would appear that the prey-derived signals normally required for growth by wild-type bdellovibrios are not encountered during the process of attachment and/or penetration of the prey cell alone; instead, the continued presence of these signals is required at least until the initiation of DNA replication in order for true cellular proliferation to occur.

Elongation and Division

Despite the great amount of interest and work concerning the initiation of growth by attack-phase bdellovibrios, the opposing differentiation from the growth phase into the attack phase has only recently begun to be studied. It has been demonstrated that the duration of the bdellovibrio's growth phase is tightly coupled to the size of the prey cell upon which it feeds (Kessel and Shilo, 1976). Given the close correlation between the amount of prey substrate initially available and the final length of the bdellovibrio filament before division, it was generally assumed that bdellovibrios simply grew to the point of nutrient depletion and that differentiation into the attack phase was initiated in response to conditions of starvation.

In experiments with prematurely released growth-phase cells, however, it was found that release at any stage of the filamentous growth cycle prompted the cells to differentiate into the attack phase immediately upon completion of their previously initiated rounds of DNA replication (Ruby and Rittenberg, 1983). These results suggested that additional prey-derived regulatory signals are required by the bdellovibrio for continued filamentous elongation even after the initiation of growth and DNA replication. This hypothesis was further supported by the observation that the prematurely released bdellovibrios would continue to grow and replicate extracellularly if the nutrient medium in which they were suspended was first supplemented with concentrated prey cell extracts (Fig. 2) (Gray and Ruby, 1989, 1990). Furthermore, prey-independent bdellovibrios exhibited a similar response to prey cell extracts themselves. Synchronously growing cultures of these mutant strains extended the duration of their filamentous growth phase when in

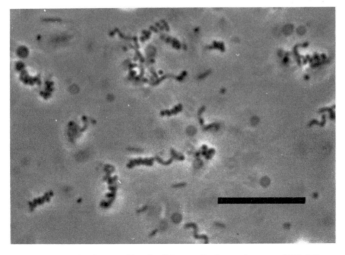

FIGURE 2. Filamentous growth-phase cells of wild-type *B. bacteriovorus* 109J. These cells were prematurely released from the bdelloplast immediately following the initiation of growth and continued to elongate extracellularly in the presence of a cell-free autoclaved extract of the *E. coli* prey for more than 6 h. Bar, 10 μm.

the presence of prey cell extracts, whereas simple nutrient enrichment generally had no such effect (Gray and Ruby, 1990). These results would seem to confirm the hypothesis that there is an additional prey-derived regulatory factor involved in the bdellovibrio developmental cycle that mediates the continuation or termination of the growth phase, rather than its initiation.

The only attempts to isolate or characterize these secondary regulatory signals have been undertaken using prey-independent mutant strains as test organisms (Gray, 1989; Gray and Ruby, 1990). Because the differentiation event of interest in this case is the termination rather than the initiation of growth, synchronous batch cultures of these mutant strains can be used in a relatively simple bioassay in which the duration of the growth phase is measured as a function of the amount of DNA synthesized per filamentous growth cycle. Unfortunately, this bioassay has not proven adaptable for use with the prematurely released wild-type cells because only about 50% of a given population of cells can be induced to remain in the growth phase, even in the presence of high concentrations of prey cell extracts. The remainder of the population differentiates into the attack phase, thereby confusing the results of the assay.

Using this method with prey-independent bdellovibrios, however, a partial characterization of the prey-derived signal compounds responsible for extending the duration of the growth phase has been achieved (Gray and Ruby, 1990). This signal activity was present in extracts of both gram-negative and gram-positive cells, although it appeared to be absent from yeast cells and from the bdellovibrios themselves. The activity remained within the soluble portion of an extract after centrifugation at $120,000 \times g$, was unaffected by heating or by treatment with RNase or DNase, but was inactivated by treatment with proteinase K or chymotrypsin A. The signal activity could not be isolated or significantly purified on the basis of molecular charge, hydrophobicity, or size, although it was restricted to size fractions containing molecules larger than 10,000 Da. Surprisingly, the activity of these signals was retained following digestion with trypsin, although in a range of smaller effective molecular sizes. Consequently, the signal activity appears to reside in soluble proteins or larger peptides of a broad range of native molecular weights that are present in a wide variety of bacteria. These results are all the more surprising given the highly specific nature of the bdellovibrios' response to this activity—that is, its action generally cannot be mimicked or otherwise supplied by complex nutrients alone. As a result, we are left with the paradox of an unusually heterogeneous signal that is common to many kinds of bacteria, but that the bdellovibrios nonetheless are able to perceive as representing a highly specific regulatory cue.

The original hypothesis that a specific prey-derived "signal protein" mediates this differentiation response is not supported by these data. The fact that the signal activity is present in such a wide array of molecular sizes, coupled with its remarkable heat stability, suggests that there is no single protein, or perhaps even class of proteins, responsible for this regulatory effect. The continued activity of the signals following trypsin digestion to a smaller molecular size raises the possibility that differentiation of the growth-phase bdellovibrio is actually regulated by peptides of a specific size range or amino acid composition that are cleaved from the prey cell substrate by the activity of extracellular peptidases produced by the bdellovibrio itself. Growth-phase bdellovibrios are known to produce and export several different protease activities (Engleking and Seidler, 1974; Fackrell and Robinson, 1973; Gloor et al., 1974; Thomashow and Rittenberg, 1978a; Saier, personal communication), and a similar generation of signal factors by means of extracellular proteolysis has recently been proposed for the myxobacteria as well (see chapter 7). This hypothesis has the advantage of explaining the heat stability of the signal activity as well as its presence in such a wide variety of native proteins. It cannot, however, account for the absence of any measurable activity in the extracts of yeast cells or of bdellovibrios themselves (Gray and Ruby, 1990), nor is it clear why continued signal activity of the prey extracts following proteolytic cleavage is apparently specific to trypsin alone, rather than to other endopeptidases as well.

In contrast to these growth-promoting signal factors associated with prey cell extracts, another study has tentatively identified a signal activity produced by the bdellovibrio itself that promotes the division and differentiation of the growth-phase cell (Eksztejn and Varon, 1977). In this study, synchronous cultures of a prey-independent strain derived from *B. bacteriovorus* 109D were used to show that the final length attained by the growth-phase cells before division (corresponding to a single filamentous growth cycle) is greatly affected by the initial cell density of the culture. The observed production of growth-phase filaments with final lengths as short as 2 or as long as 20 cell units suggested that there was no arbitrary maximum or minimum length required for the division of these organisms. This idea was further supported by the fact that aseptate filaments of different lengths invariably possessed fully segregated chromosomes, suggesting that they are competent to divide at any time during the growth cycle.

Attempts to induce the premature division of these bdellovibrios by the addition of chemical agents had no effect on the cells' normal growth and division, as determined by microscopic observation (Eksztejn and Varon, 1977). Alternatively, trichloroacetic acid extracts of late-exponential-phase bdellovibrio cells, or the spent medium from these cultures, were both found to promote an early division and differentiation of the test cells into the attack phase. This result prompted the authors to propose the existence of a growth-terminating "division factor" that is produced by the elongating bdellovibrios themselves.

Detectable amounts of this factor first appeared within the bdellovibrio cells during the middle to late stages of their growth cycle and were released into the culture medium prior to division of the filamentous cells. The observed kinetics of the signal's appearance and subsequent disappearance, both from the bdellovibrio cells and from the surrounding medium, implied that cell division was mediated by the extracellular concentration of this factor and that the signal was consumed or inactivated during the process of division or differentiation. The molecular source of this signal activity was tentatively identified as a small, "apparently cyclic" peptide (Eksztejn and Varon, 1977), although no further report of its isolation or detailed characterization has yet been published.

Unfortunately, the proposed activity of this extracellular division factor provides only a partial explanation for the population density-dependent growth and division of prey-independent bdellovibrios. Because it is the extracellular accumulation of this signal that regulates cell division, the specific timing of its synthesis and release are critical to its activity. Consequently, because the data indicate that the division factor is produced and released only at certain times during the growth cycle, the focus of the problem merely shifts from the question of what triggers division itself to a consideration of what regulates the discontinuous production of the signal to initiate division. Why do dense populations of axenically growing bdellovibrios produce a division factor after only one doubling, while more dilute populations will delay its synthesis for multiple generations of filamentous elongation?

Another difficulty with the proposed division factor model is that it cannot explain why cells that were experimentally induced to divide at an early age continue to grow by a process of binary fission following their differentiation back into the growth phase. Given that the signal required for division is inactivated following the prematurely induced division, one would think that the cells should subsequently elongate as filaments, since they initially divided at a population density below that normally required to induce the production of the division factor. Still other experiments have shown that the growth rate as well as its duration is affected by the addition of bdellovibrio extracts or spent medium (unpublished results). Unfortunately, the possible difference between an early division induced by the activity of regulatory signal factors, as opposed to one resulting from a physiological inhibition of growth in general, cannot be distinguished microscopically, suggesting that the division assay employed in these experiments might itself be inadequate for the accurate identification of a division-inducing signal activity.

Although the evidence presented in favor of an endogenously produced division factor is not especially compelling in itself, the existence of such a regulatory signal remains a distinct possibility. Certainly there is little doubt that the duration of growth by prey-independent bdellovibrios is strongly affected by their population density, and the termination of their first cycle of growth (followed, as it is, by reinitiated growth) cannot be considered to represent a starvation response. These phenomena suggest that the bdellovibrios may well possess some sort of a population density-sensing mechanism which could conceivably act as an inducer of filament division. Although such density-sensing regulatory mechanisms have already been demonstrated for the formation of fruiting bodies by myxobacteria (chapter 7) and for the autoinduction of light production by luminous bacteria (chapter 8), the fact remains that there is currently no strong evidence to support the existence of any such system in the bdellovibrios.

A final point to be made before leaving the subject of bdellovibrio-produced division signals is that the reported absence of a positive, growth-extending activity in bdellovibrio extracts (Gray and Ruby, 1990) most probably was not due to the negative effects of a division factor-like signal. The bdellovibrio extracts used in these experiments adversely affected growth of the test cultures only when used at extremely high concentrations, at which point they inhibited the duration of growth and the growth rate of the cultures as well. Generally, however, these extracts were observed to have no effect on culture growth at all. Furthermore, these bdellovibrio extracts were prepared from washed, stationary cultures, which should already have released any such division signals into the surrounding medium, leaving the cells themselves devoid of such activity.

In addition to the prey-derived signals, several classes of pure compounds have recently been discovered that exert a growth-extending activity similar to that of the prey cell extracts (Gray, 1989). These compounds include the two groups of DNA-binding proteins known collectively as protamines and histones, as well as homopolymers of L-lysine and the protease trypsin. Although these compounds all promote an extended duration of the bdellovibrio growth phase, they clearly differ from the prey-derived signals, as well as from each other, in their biochemical characteristics. The observed activity of protamines and histones could be related to their affinity for binding DNA, although the prey-derived signals themselves could not even be partially purified by DNA-cellulose affinity chromatography (Gray and Ruby, 1990). The strongly basic molecular charge of the DNA-binding proteins and poly-L-lysine did not appear to be critical per se, since these compounds could all be inactivated by proteolytic cleavage, and the addition of equal concentrations of the free amino acid lysine had no effect on the bdellovibrios' growth. The signal activity of trypsin was clearly distinct from these others in that it was entirely dependent upon the continued proteolytic activity of the enzyme. Furthermore, this regulatory effect appeared to be specific to trypsin alone, as no other protease tested, including the structurally similar peptidase chymotrypsin, had any positive effect on the duration of the bdellovibrio growth phase.

Because of these differences, the specific compound regulatory signals appear to be quite distinct from the growth-extending signal activity present in prey cell extracts. This raises the possibility that these compounds, rather than serving as simple analogs to the prey-derived signals, could be operating instead at one or more independent levels of regulatory control.

The determination of what specific factors are involved in regulating the duration of the bdellovibrio growth phase is further confused by the contribution of nutritional factors. Prey-independent strains of *B. bacteriovorus* are clearly affected in their growth and division patterns by the nutrient conditions under which they are cultured, as complex media such as nutrient broth, peptone, tryptone, or yeast extract will all contribute to the degree of filamentation achieved by axenic cultures (Gray and Ruby, 1990). Unlike any of the previously described regulatory signals, however, the extended duration of filamentous growth resulting from increased concentrations of these nutritional factors usually coincides with an increase in the cellular growth rate as well. Furthermore, the bdellovibrios' response to nutrient components is saturated at relatively low substrate concentrations, whereas the extended filamentation in response to prey extracts exhibits a nonsaturating, linear dose dependency, at least to the upper limits of detection by the growth assay (Gray and Ruby, 1990). This result suggests once again that the observed correlation between the size of the prey cell and the final length of the aseptate bdellovibrio filament during its intracellular growth cycle (Kessel and Shilo, 1976) is not solely an effect of nutrient supply, but involves the action of specific prey-derived regulatory signals.

The true nature of the signals that mediate the elongation or division response of growth-phase bdellovibrios is therefore a complex question indeed. The fact that nutrient conditions alone exert a substantial influence on the duration of filamentous growth by prey-independent strains is somewhat problematical to any discussion of intercellular signalling for this differentiation event. The repeated assertion that the compounds present in prey cell extracts represent true regulatory signals, however, rather than simply serving as an additional source of nutrients, is based upon the following lines of evidence.

First, the regulatory activity exhibited by these compounds can be completely inactivated by digestion with proteinase K or chymotrypsin. Such treatment should have no effect on their nutritional contribution to the medium, unless their importance to the bdellovibrios is that of a specific metabolite or cofactor. The heat stability of the signal activity, its presence in a variety of effective molecular weights, and its continued activity following trypsin digestion, however, are not consistent with the latter possibility. Second, the activity of these compounds has been shown to have no effect on the growth rate of the bdellovibrios, even under suboptimal growth conditions (Gray and Ruby, 1990). If the duration of the growth phase is simply a result of the presence or absence of a limiting nutrient or metabolic intermediate, the growth rate might be expected to increase upon the addition of this required component, especially under relatively nutrient-poor growth conditions. Finally, prey-independent bdellovibrio strains invariably reinitiate growth after their first round of division and differentiation and continue to grow for at least two additional generations (Eksztejn and Varon, 1977; Gray and Ruby, 1990). Clearly these cells are not differentiating in response to conditions of nutritional stress; if the medium is inadequate to support continued growth of the cells (thus prompting their differentiation into the attack phase), it should not be competent to support further generations of growth.

The effect of complex nutrient media on the growth of prey-independent bdellovibrios remains a considerable obstacle to our understanding of the role of prey-derived (or other) signals in the regulation of bdellovibrio growth. The available evidence might be summarized by the statement that these organisms are apparently subject to a nutritional response that, while clearly separate, is still only partially distinguishable from their developmental response to the prey-derived regulatory signals.

An important question regarding the relationship between the elongation of wild-type cells and that of their prey-independent mutant derivatives remains to be addressed, however. If bdellovibrios are subject to a nutritional response that can result in increased filamentous elongation, why do the prematurely released wild-type cells fail to continue to grow, even in complex nutrient medium, unless they are supplied with a source of prey cell extracts? One possible response to this question is that, while the mutant strains continue to share many of the regulatory features of the wild-type cells (including the ability to respond to the prey-derived elongation signals) they have also lost at least some of these same requirements. The answer can then be construed in terms of the wild-type cells requiring specific regulatory signals unique to the prey cell prior to any considerations of their responding to nutrient conditions alone. Alternatively, the observed nutritional response of the axenic bdellovibrio cultures could be a feature exclusive to the mutant phenotype itself.

Differentiation and Release of Attack-Phase Progeny

As is apparent from the previous discussion of elongation and division of the growth-phase cell, there is little evidence to suggest that the septation and division of growth-phase bdellovibrios is anything more than a "default" response, resulting from the absence of a specific regulatory cue. The question remains, however, as to what regulates the subsequent differentiation of the individual bdellovibrio cells into the attack phase, or what triggers their release from the bdelloplast.

As demonstrated by premature release experiments with wild-type cells, differentiation of the newly divided growth-phase bdellovibrios does not appear to require any specific extracellular signal factors, but is instead a direct result of the absence of those signals previously involved in promoting an extended growth phase (Ruby and Rittenberg, 1983). Although the onset of flagellar synthesis and bdelloplast lysis following division can occasionally be delayed by intracellular bdellovibrios (Sanchez-Amat and Torrella, 1990), the fact that prey-independent mutant strains invariably differentiate into the attack-phase morphology and exist briefly as motile, single-unit cells before shedding their flagella and reinitiating growth seems to confirm the hypothesis that differentiation is an inevitable consequence of division in these organisms.

Intriguingly enough, however, the observed progression of septation, division, differentiation, and release does not appear to be a sequentially regulated series of events, with each step dependent upon the completion of the previous one. Instead, the "commitment to divide" by the growth-phase bdellovibrio, rather than division itself, appears to be all that is necessary for the subsequent morphological differentiation and release of the attack-phase cells. This somewhat surprising result was obtained from experiments with the antibiotics vancomycin and virginiamycin S, both of which specifically block septum formation, though not elongation, in growing cells. In the presence of these antibiotics, wild-type bdellovibrios will proceed to lyse their bdelloplasts (Varon et al., 1976), and even to elaborate flagella at the correct sites of septum formation (unpublished results), in the complete absence of normal cell septation and division. Furthermore, prey-independent strains treated with vancomycin are apparently capable of reinitiating net DNA synthesis, though at a reduced rate, following the termination of their first round of growth (unpublished results). While this suggests that even the reinitiation of the growth phase itself may be independent of the prior completion of cell division in these organisms, a great deal of work remains to be performed before any specific conclusions can be drawn.

Other Signals and Special Considerations

Intracellular growth at low population densities and low nutrient availability

A largely unexplained phenomenon concerning cell-cell signal interactions within a population of intracellular growth-phase bdellovibrios has been reported with an uncharacterized marine bdellovibrio maintained on a luminous bacterial prey species (Varon et al., 1983). It was observed that the completion of the intracellular growth cycle by this strain is inhibited by the rapid dilution of a population of its bdelloplasts. Regrettably, the growth assay used in these experiments did not distinguish between the suspension of growth by the bdellovibrios (in which case, the growth-phase cells were simply remaining quiescent within their bdelloplasts) and aborted growth (in which case, the bdellovibrios were prematurely differentiating into the attack phase and lysing the prey cell). In either case, normal growth and division of the intracellular bdellovibrios in similarly dilute suspensions was rescued by the addition of polyamines such as spermine, spermidine, or putrescine. The effect of these polyamines did not appear to be related to simple osmotic stabilization of the cell membrane, as there was no similar effect of sucrose, sodium, or magnesium ions when added at a variety of concentrations.

The authors speculate that this developmental response to low population densities is the result of a leakage of one or more essential regulatory components from the bdelloplasts at these higher dilutions. The possibility exists that it is polyamines specifically which are the crucial factors lost from the prey cell in this way. Alternatively, this phenomenon could be restricted to marine bdellovibrios as a group, to the particular strain employed in these experiments, or even to the use of luminous bacteria as prey cells. Whatever the reasons for this inhibition of continued intracellular growth, one might speculate that the renewed growth in response to polyamine additions is functionally related to the reported ability of polyaminic compounds such as poly-L-lysine, protamines, and histones to promote an increased duration of growth by nonmarine, prey-independent bdellovibrios (Gray, 1989).

A more recent study employing a number of different strains of marine bdellovibrios suggests that the suspension of intracellular growth may be a common starvation survival strategy among marine bdellovibrios in general (Sanchez-Amat and Torrella, 1990). In this study, it was found that under conditions of low nutrient concentration and/or low prey cell density the bdellovibrios would interrupt their normal growth cycle by remaining within their bdelloplasts for extended periods of time without lysing the prey cell. Interestingly, even the size of the prey cell itself appeared to have an effect on the bdellovibrios' response, as the use of smaller prey cells resulted in a higher frequency of suspended growth. Microscopic analysis revealed that the growth-phase bdellovibrios within these unlysed bdelloplasts doubled in size and divided to form two normal-sized progeny cells, but failed to elaborate flagella or lyse the prey cell until they were placed in conditions of higher nutrient availability. Sanchez-Amat and Torrella (1990) describe this phenomenon as the formation of a "stable bdelloplast," although the term is somewhat misleading since there is no evidence to indicate that any new or additional modifications are made to the bdelloplast envelope itself. Instead, this suspension of intracellular development would appear to be entirely a behavioral response of the bdellovibrio to the condition of its surrounding nutrient environment.

Although the addition of yeast extract to the medium induced the quiescent bdellovibrios to synthesize flagella and lyse their bdelloplasts, other organic nutrients such as Casamino Acids, purines, and pyrimidines had no such effect, nor did polyaminic compounds such as putrescine or spermine. Consequently, there appears to be a fundamental difference between the regulation of suspended intracellular growth of marine bdellovibrios within "stable" bdelloplasts and that previously reported for rapidly diluted populations of another marine strain. Beyond the fact that each response appears to be regulated differently, however, there is as yet no further understanding of what specific signals (if any) prompt the bdellovibrios' behavior in either case.

Bdellocyst formation and germination

As mentioned in the first section of this review, the taxonomically uncharacterized *Bdellovibrio* sp. strain W is unique in its ability to form cystlike resting bodies within the bdelloplast in lieu of the normal intracellular growth cycle (Burger et al., 1968). Attempts to identify the specific factors that govern this differentiation have revealed a variety of environmental conditions that can influence the relative frequency of encystment as opposed to filamentous growth (Tudor and Conti, 1977). Bdellocyst formation can be induced at least partly in response to conditions of nutrient starvation for the prey cells, although the frequency of encystment is also affected by the numerical density of the attack-phase bdellovibrios, as well as by the ratio of bdellovibrios to prey cells.

Unfortunately, little more than this is currently known about the factors regulating bdellocyst formation. Because a low frequency of encystment invariably occurs under all growth conditions, it remains unclear whether there is any real intercellular signalling involved at all, or if instead this differentiation is simply a feature intrinsic to the developmental programming of this particular bdellovibrio strain. From the limited amount of work that has been performed on prey-independent isolates of strain W, however, it appears that cyst formation does not occur during axenic growth (Tudor, personal communication), suggesting that some feature specific to the intracellular growth environment may play a determinative role in this developmental event.

By contrast, the control of germination of bdellocysts is somewhat better understood than their actual formation. Isolated bdellocysts can be induced to germinate by any of a variety of compounds such as free amino acids or inorganic ions, as well as by heat activation (Tudor and Conti, 1978), as is commonly observed with *Bacillus* endospores. Because germination can be achieved in the complete absence of other bacteria, including prey cells or their exudates, however, this process appears to be a result of an environmental sensing mechanism that is independent of any system of intercellular signalling. Consequently, although undeniably intriguing in its own right, the germination of bdellocysts does not fall within the scope of this particular chapter.

Discrepancies and contradictions within the literature

Finally, some discussion is needed of the contradictory nature of the various reports of developmental regulation found in the literature of bdellovibrio research. Given the sheer number of conflicting results, even for subjects as simple as the bdellovibrios' ability to attack and grow on heat-killed versus viable prey cells (Hespell, 1978; Varon and Shilo, 1968) or their relative rates of DNA and protein synthesis during filamentous growth (Eksztejn and Varon, 1977; Gray and Ruby, 1989), one must ask why there appears to be such a problem in achieving reproducible results with this particular microbial system.

At least some of this variability is a direct consequence of the experimental and taxonomic diversity that continues to exist within the field of bdellovibrio research. There is, unfortunately, no uniform technique applied to the culturing of these organisms, either in the dual-culture maintenance of wild-type cells or in the axenic culturing of prey-independent mutants. Individual laboratories regularly employ their own preferred ratios of attack-phase bdellovibrios to prey cells, utilize different buffers as an attack medium, and use a variety of different complex nutrient media for the axenic maintenance of either the mutant bdellovibrios or the prey organisms. Clearly, such factors can have a profound influence on the observed regulatory and developmental behavior of the bdellovibrios. For example, it has already been mentioned that the phenomenon of prey specificity exhibited by individual bdellovibrio strains is at least partly affected by the experimental conditions employed (Torrella et al., 1978).

Furthermore, the taxonomy of the bdellovibrios is still in a very primitive state (Ruby, 1991). It was almost 10 years after the original description of *Bdellovibrio* as a genus (Stolp and Starr, 1963) that the creation of additional species beyond that of *B. bacteriovorus* was first formally proposed (Seidler et al., 1972). Although those original distinctions were based primarily upon DNA G + C content, subsequent research has indicated a surprising diversity in the basic physiology of bdellovibrios, not only between the different recognized species (Gloor et al., 1974; Odelson et al., 1982; Park and Mahadevan, 1988; Schelling and Conti, 1986) but even among different stock cultures of what were previously considered to be identical strains of a single species (Rayner et al., 1985; Rittenberg, 1972; Talley et al., 1987). There is such taxonomic heterogeneity within the type species of *B. bacteriovorus* itself that the large number of strains currently lumped within that taxon almost certainly represent multiple, distinct species (Schelling and Conti, 1983; Torrella et al., 1978). Consequently, any comparative approach to the bdellovibrio literature will frequently lead to a contradiction in experimental results simply because of differences in the various bdellovibrio strains used in the individual studies.

Finally, and perhaps most importantly, much of the research conducted with bdellovibrios has relied heavily upon the use of prey-independent mutant strains as the experimental organisms. Because there is as yet no genetic characterization of the prey-independent phenotype, however, there are no means of determining how many distinct mutations may give rise to this one selectable trait. A further consideration that generally remains unacknowledged is that, while the prey-independent phenotype may itself arise from a single mutation, subsequent laboratory culture of these mutant strains will inevitably select for secondary, "enhancing" mutations that favor axenic growth. This idea was first formally proposed to account for observed changes in the cell envelope of prey-independent mutant strains (Friedberg and Friedberg, 1977), but can also explain the frequently observed phenomenon of increasing vigor of newly isolated axenic strains during subsequent transfers in pure culture (Ishiguro, 1973, 1974; Varon and Seijffers, 1977; Varon and Shilo, 1976). Similarly, this could explain why the single report of axenic growth of prey-independent bdellovibrio isolates in defined medium was achieved only after repeated transfer of the cells in medium initially supplemented with prey cell extracts (Ishiguro, 1974) and why this result has not been reproducible with other established mutant strains (Rittenberg, personal communication). Such pleiotropic enhancement of the prey-independent phenotype is further demonstrated by the secondarily acquired capability of virtually all such strains to produce membrane pigments as a protection against photo-oxidative damage during their extracellular growth cycle (Friedberg, 1977).

Consequently, given the variety of experimental techniques that have been applied to such an already diverse group of bacteria, the frequent occurrence of discrepant or contradictory results should not be especially surprising. Nonetheless, it invariably becomes necessary to apply such results, however disparate, to the formulation of some sort of general hypothesis that can serve as a unifying model for the biology of these organisms. That somewhat ambitious endeavor will be the focus of the next section of this review.

A MODEL OF INTERCELLULAR SIGNALLING IN THE *BDELLOVIBRIO* LIFE CYCLE

Determinative Cell-Cell Signalling Events

In formulating a model of cell-cell signalling for bdellovibrio development, it is first necessary to catalog the determinative events of the life cycle in sequence, eliminating those that do not appear to involve intercellular signalling. The developmental cycle is first initiated by the free-living attack-phase bdellovibrio locating and colliding with a susceptible prey cell. As previously mentioned, there is as yet no evidence to suggest that anything other than random collision is involved in this stage of the interaction. While the formation of a reversible attachment with the prey cell might be considered to represent the next developmental event, this stage is essentially indistinguishable from the initial collision of the two cells. Furthermore, because reversible attachment can also occur between the bdellovibrio and inanimate surfaces, it is reasonably clear that no form of intercellular communication is involved.

The formation of an irreversible attachment to the prey cell surface is the first event in which the phenomenon of prey cell specificity is manifested. Although this event is clearly mediated by structural features of both organisms, however, there is no strong evidence of actual signal exchange occurring between the two cells. Consequently, this stage of the association may represent a simple surface interaction more analogous to the attachment of a phage particle to its receptor protein than to an example of true regulatory or developmental communication.

The process of penetration of the prey cell may be the first instance of prey-derived signal transduction by the bdellovibrio, since the synthesis and release of the suite of enzyme activities involved in this step appears to be the result of developmentally specific gene expression. It is also during penetration of the prey cell that the first differentiation event of the bdellovibrio life cycle occurs, as the attack-phase cell undergoes the morphological and physiological changes associated with differentiation into the growth phase. Exactly what triggers the physiological responses necessary for bdelloplast formation and cell growth remains unknown; it is possible that these changes are mediated by the same general features required for irreversible attachment itself, since there are no reports of developmental mutants that are capable of irreversible attachment but that are blocked specifically for penetration. As a means of addressing this question further, it would be interesting to know if prey-independent bdellovibrio strains also exhibit developmental expression of the enzyme activities normally associated with penetration and bdelloplast formation even when they are grown axenically.

It is clear, however, that one or more prey-derived signal factors are crucial to this differentiation event of the wild-type developmental cycle. These factors represent the "growth-initiating" signals that have occasionally, though not consistently, been extracted from the prey cells; the normal requirement for these signals can be obviated by the point mutation(s) that results in the prey-independent phenotype. It remains uncertain, however, whether the same or different signals are responsible for triggering the separate events of morphological differentiation of the bdellovibrio cell and production of the various enzyme activities involved in penetration and bdelloplast formation.

As discussed previously, differentiation into the growth-phase morphology does not necessarily connote a commitment by the bdellovibrio to proliferate. Instead, the initiation of cell growth and DNA replication must be considered to be developmentally distinct from differentiation itself, as they absolutely require the continued presence of one or more specific prey-derived regulatory signals. Accordingly, the reinitiation of DNA replication and continued filamentous elongation of the bdellovibrio cell are assumed to be under the control of these same regulatory signal factors.

Finally, division and fragmentation of the growth-phase bdellovibrio following the termination of cell growth and DNA synthesis seem to be the normal response to the complete absence of prey cell signal factors. Subsequent differentiation of the newly divided cells into the attack phase and production of the lytic enzymes responsible for their release from the bdelloplast are apparently programmed along with cell division itself, as even the prey-independent strains will invariably differentiate into the attack phase prior to reinitiating growth.

Thus, of the various developmental stages of the bdellovibrio life cycle, only the two events of differentiation into the growth phase and the initiation of cell growth and DNA replication appear to involve the activity of prey-derived regulatory signals. The possibility that secondary bdellovibrio-produced signal factors may induce the final stages of growth termination and cell division remains an open question that will be discussed later. For now, let us concentrate on the crucial role played by prey-derived signals in the bdellovibrio developmental cycle.

Prey-Derived Initiation and Elongation Signals

A necessary premise for any model of bdellovibrio development is that the initiation of the growth phase is regulated independently of the continued growth that results in filamentous elongation. These two instances of cell-cell signalling are presumed to involve the activity of separate and distinct regulatory signals because, with the exception of a few rare cases, attack-phase cells will not initiate growth extracellularly in response to the presence of prey cell extracts. Once growth is initiated, however, the wild-type bdellovibrios are apparently capable of continued extracellular growth if they remain in the presence of prey cell extracts. If the signal required for continued filamentous elongation were the same as that needed to initiate the growth phase, these extracts should be expected to promote the initiation of extracellular growth by attack-phase cells, which in general they do not.

Given, then, that there are separate prey-derived regulatory signals required either for differentiation into the growth phase (initiation signals) or for the continuation of growth (elongation signals), the attack-phase cell can be said to initiate growth in response to some specific prey-derived regulatory cues encountered during the process of attachment to and/or penetration of the prey cell (Fig. 3, step D). Proliferation of the growth-phase bdellovibrio requires the continued presence of the prey-derived elongation signal factors (Fig. 3, step E), the absence of which will cause the cell to differentiate into the attack phase even before the initiation of DNA replication (Fig. 3, step G). These secondary signals can therefore be said to mediate the developmental decision of continued growth and elongation versus termination of the growth phase. Once the growth phase has been fully initiated, which we now define as including the onset of net DNA synthesis, the bdellovibrio cell appears to be committed to continue growth for at least one complete round of DNA replication before dividing and differentiating back into the attack phase. Subsequent rounds of DNA replication, accompanied by filamentous cell elongation, are presumably mediated by the same elongation signals required for the initial round of DNA replication and cell growth (Fig. 3, step E′ and so on).

The elongation signals that regulate the duration of the growth phase could presumably operate at either of two levels. As a positive cue, they could promote the initiation or reinitiation of DNA replication; alternatively, as a negative signal, they could serve to inhibit the initiation of cell division. The fact that bdellovibrio growth is unbalanced, with the initiation of DNA replication apparently uncoupled from a required cell mass (Gray and Ruby, 1989), suggests that these signals may be operating as positive effectors of DNA synthesis. Because previous studies of the prokaryotic cell cycle have indicated an essential correlation between the completion of DNA replication and the onset of cell division (Képès, 1986), however, the filamentous growth of the bdellovibrio cell suggests that these factors are inhibitors of cell division. The inherent weakness to this second argument is that, although prior completion of DNA replication is generally required for bacterial cell division, there is little evidence to indicate any causal relationship between the termination of DNA synthesis and the onset of cell division. Further-

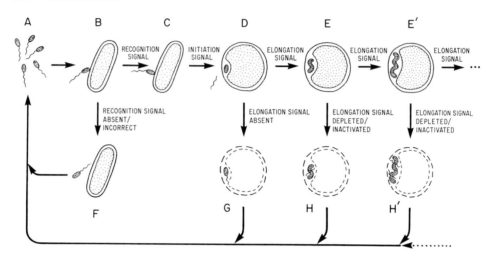

FIGURE 3. A model of the role of prey-derived signal factors in the bdellovibrio growth cycle. The growth cycle of the free-living attack-phase bdellovibrio (A) is initiated following its collision with a susceptible prey cell (B). The formation of an irreversible attachment by the bdellovibrio (C) depends upon the presence of specific surface features associated with the prey cell. In the absence of these recognition signals, no irreversible attachment is formed, and the two cells eventually separate (F). Following irreversible attachment, the bdellovibrio's subsequent penetration into the prey cell, morphological differentiation into the aflagellate growth phase, and formation of the bdelloplast (D) are mediated by one or more uncharacterized initiation signals that are provided by the prey cell itself. These signals may or may not be the same as those required for irreversible attachment; little evidence is currently available on this matter. Following its differentiation into the growth phase, the initiation of DNA replication and cell growth by the bdellovibrio (E) requires the continued presence of one or more elongation signals that are apparently derived from the prey cell in the form of specific soluble proteins or proteinlike compounds. DNA replication is reinitiated in the continued presence of these signal factors, and filamentous growth of the aseptate bdellovibrio cell proceeds without interruption (E', . . .). In the absence of the proper elongation signals (G), or following their presumed depletion or inactivation (H, H'), the growth-phase cell is induced to differentiate and divide (if necessary) prior to lysis of the bdelloplast and the release of the attack-phase progeny into the external environment (A).

more, if the prey-derived elongation signals served only as inhibitors of cell division, one would expect that prematurely released bdellovibrios, having already differentiated into the growth phase, would be capable of at least one round of DNA replication even in the absence of prey cell extracts. Consequently, the best hypothesis is that the prey-derived elongation signals serve to promote the initiation of DNA replication by the growth-phase bdellovibrio. Alternatively, given the remarkably heterogeneous character of these factors, it is possible that they could represent a number of distinct activities that operate at both of the levels described.

The question remains, however, as to why these same signals are unable to induce the initiation of DNA synthesis in attack-phase cells. One possible response to this question is that the signal required by the growth-phase bdellovibrio to initiate DNA replication is somehow unrecognizable to the attack-phase cell. According to this hypothesis, the ability of the bdellovibrio to respond to these signals is dependent upon the physiological changes that accompany morphological differentiation into the growth phase. Consequently, the initiation of DNA replication may be sequentially regulated by the prior appearance of the signals mediating differentiation into the growth phase itself (Fig. 3, steps B through D).

The overall scenario, then, is one in which the growth-phase bdellovibrio, despite having received the signal required to begin its developmental growth phase, is incapable of initiating DNA replication in the absence of specific regulatory cues normally provided by the prey cell. In the presence of the proper elongation signals, however, DNA replication is initiated and then subsequently reinitiated, accompanied by filamentous elongation of the cell. With the depletion of these signals, no new rounds of chromosome replication can be initiated; synchronous division and fragmentation of the filamentous cell then follow the completion of DNA synthesis. The onset of cell division apparently commits the bdellovibrio to differentiate into the attack phase, such that a new initiation signal is required in order to reenter the growth phase.

Such a regulatory system can be construed as the result of a series of adaptations that maximize the bdellovibrios' efficiency in exploiting their intracellular habitat. In simplest terms, bdellovibrios have evolved the ability to employ a set of specific physical and/or chemical cues as indicators both of the presence of a suitable growth environment and of its changing nutritional state. As predatory "specialists," they will differentiate into their proliferative form only in response to signals specifically associated with the attack and penetration of a suitable prey cell. Once within the intracellular environment, the duration of their growth is regulated by the presence of a second set of signals that are also derived from the prey cell. These secondary regulatory signals are soluble components of the prey cell substrate (Gray and Ruby, 1990), thus allowing them to serve as reliable indicators of the remaining cytoplasmic nutrient reserves. Significantly, they are absent from the bdellovibrios themselves, thereby eliminating the risk of a false signal arising from endogenous pools of these same compounds. Similarly, the use of specific signal compounds, rather than the nutrient concentration in general, would also help avoid self-signalling by the bdellovibrio. By contrast, the heterogeneity of the signals enables the bdellovibrios to grow successfully on a broad range of prey species, as well as on variants within the same species. This is of considerable importance to the bdellovibrios' success as a species, given the apparent inefficiency with which the attack-phase cells attack and penetrate their prey (Varon and Zeigler, 1978).

The Role of Endogenous versus Prey-Derived Signals

According to the model as stated, the differentiation of bdellovibrios from the growth phase into the attack phase is regulated solely on the basis of prey-derived regulatory cues. The alternative hypothesis, that differentiation of the growth-phase cell is mediated by endogenous regulatory signals produced by the bdellovibrios, remains to be considered. Although the published evidence supporting the existence of a division factor-like compound is not especially compelling, it is worthwhile to consider the implications of a model in which the growing bdellovibrios can also serve as a source of regulatory signals themselves.

Unfortunately, a regulatory model predicated solely upon the activity of endogenously produced division signals cannot account for the fact that wild-type growth-phase cells prematurely released in the absence of prey cell extracts will differentiate into the attack phase, regardless of their numerical density (Ruby and Rittenberg, 1983). If a bdellovibrio-produced division factor alone were responsible for the differentiation of growth-phase cells into the attack phase, those cells released from the bdelloplast at an early stage of development should continue to grow extracellularly, due to the low concentration of this factor at the time of the cells' release. Moreover, when prey-independent strains are grown axenically on different media, both the rate and duration of growth are depressed in the more dilute medium (Gray and Ruby, 1990). Because the activity of the endogenous division signal is dependent upon cell density alone, however, one would expect the duration of growth to be identical for both cultures, since they both began with equal inocula. Consequently, none of these results is in agreement with the predictions of this model.

However, the contrasting views of developmental regulation by prey-derived signals, as opposed to endogenously produced factors, need not be mutually exclusive. If the hypothesis of a bdellovibrio-produced division factor cannot explain the regulatory behavior of these organisms on its own, it might still prove useful as an addition to the previous model of developmental regulation by prey-derived signals. In such a revised model, differentiation of the growth-phase cell into the attack phase is regulated by means of the counteracting effects of two antagonistic signals: the prey-derived signal activity promotes extended elongation of the growth-phase cell, while the endogenously produced signal factor induces cell division and differentiation into the attack phase.

Ultimately, the addition of this bdellovibrio-produced component to the regulatory model serves only to provide a putative mechanism for the assumed depletion of the prey-derived signal factors. This hypothetical mechanism, however, has the further potential to provide a new explanation for the activity of the various pure-compound signals described earlier. Rather than simply mimicking the growth-promoting effects of the prey-derived elongation factors, these pure-compound signals could operate at a separate level of regulation by inactivating or interfering with the endogenous division-inducing signal factors of the bdellovibrio itself. In this way, these regulatory compounds could extend the duration of the growth phase not by promoting growth directly, but by impairing the bdellovibrio's ability to inactivate or deplete the signals that do promote continued growth.

Problems and Criticisms of the Model

Numerous unanswered questions and alternative possibilities concerning cell-cell signalling in the bdellovibrio developmental cycle remain to be addressed, as do some important weaknesses in the model. An immediate problem concerns the nature of the prey-independent phenotype. According to the model just developed, two distinct prey-derived signals are normally required for the intra-periplasmic growth of bdellovibrios. Because the prey-independent strains appear to have lost the wild-type cells' absolute requirement for both of these prey-derived signal factors, the model would seem to imply that these strains must be double mutants, yet they arise within bdellovibrio populations at a frequency indicative of a single point mutation (Seidler and Starr, 1969). The observed ability of prey-independent strains to respond to the prey-derived elongation signals (Gray and Ruby, 1990) in no way compromises the argument that these signals are no longer a requirement for their growth.

The simplest explanation for this problem is to propose that the bdellovibrios' normal requirement for both the initiation and elongation signals can be alleviated by a single mutation. Even if additional, enhancing mutations inadvertently arise during the cells' subsequent culture in the laboratory, the initial screening of these prey-independent strains requires that they be able to grow and replicate in the complete absence of prey cell signal factors. Because the frequency of their occurrence indicates that only a single mutation is necessary for both morphological differentiation and filamentous cellular growth in axenic culture, we must consider the prey-independent phenotype to be the result of a single mutational event. Consequently, prey-independent cells are postulated to have acquired a single mutation that exhibits pleiotropic regulatory effects by eliminating the requirement for both initiation and elongation signals.

The alternative explanation for this problem is itself a further criticism of the proposed model. If we assume that the proposed distinction between initiation and elongation signal factors is in error and that there is only one regulatory signal derived by the bdellovibrio from its prey, this matter of the prey-independent mutation ceases to be a problem. If this were the case, however, attack-phase bdellovibrios should regularly initiate growth extracellularly when they are suspended in the presence of prey cell extracts, since these would supply the one regulatory signal necessary for their growth. It might be countered that this argument necessarily assumes these signal factors can easily be extracted from the prey cells. The repeated demonstration of extended elongation by both wild-type and prey-independent strains in the presence of any of a variety of such extracts, however, clearly indicates that this activity is easily obtained. Consequently, the proposed distinction between initiation and elongation signals would seem to be reaffirmed by the available evidence.

Perhaps what is needed, however, is a general reevaluation of the data that have been acquired with wild-type bdellovibrios. Is it not possible that the occasional initiation of growth by attack-phase cells in response to prey cell extracts actually represents the same regulatory effect as that observed more frequently (though still not invariably) with the continued growth of prematurely released cells in the presence of similar extracts? If so, we can ascribe the apparent requirement for two distinct signalling events to the activity of a single regulatory factor.

This is a slightly more difficult challenge and requires several different lines of argument for an adequate response. First, the continued growth of the prematurely released wild-type cells has been achieved using several different prey cell extracts prepared by a variety of methods. This reproducibility suggests that these results are not an effect of the special properties of any single extract preparation. This argument is inherently weak in itself, however, as the same might be said for the initiation of extracellular growth by attack-phase bdellovibrios, given the disparate reports of prey cell extracts that have induced that developmental response. When attack-phase cells were exposed to the same extracts that supported continued growth of the prematurely released cells, however, none of the attack-phase cells was ever observed to grow (unpublished results). Neither was there ever any indication of the reinitiation of growth by that proportion of a population of prematurely released cells that differentiated into the attack phase, despite their presence within an environment that supported continued growth of other cells within the same population. Consequently, we maintain that the experimental basis for the proposed distinction between initiation and elongation signals is sound.

A different criticism of the model concerns the early division and subsequent suspension of development by marine strains within "stable" bdelloplasts. The fact that these cells appear to divide after only one doubling in response to environmental cues external to the prey cell itself suggests that the proposed model of developmental control is, at least in the case of marine bdellovibrios, a gross oversimplification. Clearly, there must be at least one additional level of control beyond that of the elongation signal for the regulation of cell division in these organisms; otherwise, they would be expected to undergo a normal cycle of filamentous elongation before their division and subsequent suspension of further development.

An alternative explanation, however, is that this behavioral adaptation of marine bdellovibrios to starvation conditions is in reality more analogous to the encystment of the terrestrial *Bdellovibrio* sp. strain W than it is to a prematurely induced division. During bdellocyst formation by strain W, the prey cell contents are utilized for the de novo synthesis of cyst coat material and for the production of glucose polymers that are stored as inclusion bodies within the bdellovibrio's cytoplasm (Tudor, 1980; Tudor and Bende, 1986). Similarly, in the case of suspended development within stable bdelloplasts, cytoplasmic inclusion bodies also appear within the intracellular bdellovibrios, and the cytoplasmic contents of the prey cell are fully degraded by the time the "premature" division of the bdellovibrios occurs (Sanchez-Amat and Torrella, 1990).

Consequently, the proposed model might still apply in the case of stable bdelloplasts, since the bdellovibrios continue to divide upon depletion of the necessary elongation signal from the prey cell contents. The hypothesis, then, is that the duration of the growth phase remains under the control of prey-derived signals, but the actual mode of growth can be redirected by other, externally derived factors, thus changing from the normal pattern of filamentous elongation to the formation of inclusion bodies (and/or other as yet unidentified products) within the bdellovibrio cells. As with the control of bdellocyst formation, however, the actual signalling mechanisms that trigger such specialized growth responses remain entirely unknown and therefore lie beyond the scope of the model presented here.

Perhaps the greatest problem for the model, however, relates once again to the axenic growth of prey-independent bdellovibrios. As already mentioned, batch cultures of these mutants will follow the wild-type pattern of filamentous elongation for one cycle of growth, but appear to grow for at least two generations thereafter by means of binary fission. Short of once again invoking the hypothetical argument of endogenously produced division signals, the proposed model cannot provide any explanation for this pattern of growth. Why do these strains undergo only a single filamentous growth cycle? Alternatively, why are they induced to divide before reaching a state of complete nutrient exhaustion? The opposite question might also be posed: why do these cells grow filamentously in axenic culture? And how do all of these considerations relate to the largely unexplained effects of complex media on the filamentous growth of these strains?

There are, unfortunately, no answers available for these questions at present. Any discussion of the reasons behind the varying division patterns of prey-independent bdellovibrios, or of the effect of the nutritional environment on their growth morphology, inevitably leads back to the question of what specific features compose the prey-independent phenotype of these strains. Any physiological approach to these questions is thwarted by the fact that these mutant strains generally will not grow in defined media. Ideally, what is needed is a genetic system that will allow us to direct our efforts toward the molecular basis of these observed phenomena. While such a tool has recently become available (Cotter and Thomashow, 1989), we must still await the future for the answers that it will provide.

FUTURE QUESTIONS

In addressing the regulatory aspects of bdellovibrio development, the greatest problems invariably arise from the necessity of working with a two-membered system in which the organism of interest remains inaccessible by virtue of its intracellular existence. Prey-independent mutant strains provide a release from these constraints, but create in their place a host of new problems arising from the need to correlate results obtained with these various mutants to the observed behavior of the wild-type cells. As a result, bdellovibrios have gained the stigma of being a difficult prokaryotic developmental system with which to work. Most of the difficulties currently encountered, however, can and will be resolved with the advent of a genetic system for these organisms.

Unfortunately, the development of a workable genetic system for bdellovibrios has proven to be extremely elusive, due in part to the constraints of their unusual physiology and life style. These organisms have evolved the ability to exploit an environment in which they are typically surrounded by foreign nucleic acids, which they actively utilize as a preferred growth substrate (Hespell and Mertens, 1978; Hespell and Odelson, 1978; Rittenberg and Langley, 1975; Rosson and Rittenberg, 1979, 1981). Consequently, the researcher is faced with the improbable task of achieving the genetic transformation of an organism that is specifically adapted to eliminate foreign nucleic acids from its environment.

On the more positive side, numerous ''bdellophages'' that specifically infect bdellovibrios have been discovered (Althauser et al., 1972; Hashimoto et al., 1970; Roberts et al., 1987; Roberts and Ranu, 1987; Sagi and Levisohn, 1976; Varon and Levisohn, 1972), and at least one of these has been reported to serve as a transducing vector for the wild-type cells (Roberts and Ranu, 1987). As yet, however, no reliable phage-directed genetic technique has been fully developed for the bdellovibrios.

More recently, a mating system allowing the conjugal transfer of DNA to prey-independent bdellovibrio strains has been developed (Cotter and Thomashow, 1989); this technique appears to be reproducible in other laboratories as well, so that a true molecular characterization of these organisms may at last be possible. Furthermore, although there are no reports yet of its successful application to bdellovibrios, there is hope that the method of electroporation may also provide a means of transforming both mutant and wild-type cells.

The development of these techniques allowing the genetic characterization and manipulation of bdellovibrios marks the beginning of many new possibilities for investigations into the underlying mechanisms behind bdellovibrio development. Specifically, we can now begin to elucidate the identity (or identities) of the mutation(s) that gives rise to a prey-independent phenotype, thereby providing the first information of the genetic basis for the obligate prey cell requirement of wild-type bdellovibrios. We can begin to characterize the role of extracellular signalling factors for both of the differentiation events of the wild-type life cycle at the level of developmentally regulated gene expression by the bdellovibrios. We can further analyze the specific molecular and genetic factors required for the initiation of DNA synthesis and/or cell division by growth-phase bdellovibrios. In short, we can begin to test the numerous assumptions of the developmental model presented in this chapter that have been formulated almost solely on the basis of structural and physiological evidence.

It is our hope that this review has convinced the reader that, despite the rather limited and even contradictory data currently available, it is still possible to formulate a workable hypothesis for the role of prey-derived signal factors in the developmental regulation of the bdellovibrio life cycle. Although the model born of these data is, of necessity, highly speculative in its structure, such limitations also reflect the richness of ideas and possibilities that the future holds for bdellovibrio research. This chapter is written with the full expectation that many (if not most) of its premises, ideas, and hypotheses will be rendered obsolete by the results of research conducted in the next few years. We eagerly await these new changes to our perceptions of bdellovibrio development, knowing that these unique organisms, as they have for the 30 years since their initial discovery, will similarly be challenging our perceptions of the limitations and possibilities of cellular life in general.

REFERENCES

Abram, D., and B. K. Davis. 1970. Structural properties and features of parasitic *Bdellovibrio bacteriovorus. J. Bacteriol.* **104:**948–965.

Althauser, M., W. A. Samsonoff, C. Anderson, and S. F. Conti. 1972. Isolation and preliminary characterization of bacteriophage for *Bdellovibrio bacteriovorus. J. Virol.* **10:**516–524.

Burger, A., G. Drews, and R. Ladwig. 1968. Wirtskreis und Infektionscyclus eines neu isolierten *Bdellovibrio bacteriovorus*-Stammes. *Arch. Microbiol.* **61:**261–279.

Burnham, J. C., and S. F. Conti. 1984. Genus *Bdellovibrio* Stolp and Starr 1963, p. 118–124. *In* N. R. Krieg and J. G. Holt (ed.), *Bergey's Manual of Systematic Bacteriology,* vol. 1. Williams & Wilkins, Baltimore.

Burnham, J. C., T. Hashimoto, and S. F. Conti. 1968. Electron microscopic observations on the penetration of *Bdellovibrio bacteriovorus* into gram-negative bacterial hosts. *J. Bacteriol.* **96:**1366–1381.

Burnham, J. C., T. Hashimoto, and S. F. Conti. 1970. Ultrastructure and cell division of a facultatively parasitic strain of *Bdellovibrio bacteriovorus. J. Bacteriol.* **101:**997–1004.

Chemeris, N. A., A. V. Afinogenova, and T. S. Tsarikaeva. 1984. Role of carbohydrate-protein recognition during attachment of *Bdellovibrio* to cells of host bacteria. *Microbiology* **53:**449–451.

Cotter, T. W., and M. F. Thomashow. 1989. Genetic systems for *Bdellovibrio bacteriovorus*, abstr. I-70, p. 229. *Abstr. 89th Annu. Meet. Am. Soc. Microbiol. 1989.*

Crothers, S. F., H. B. Fackrell, J. C.-C. Huang, and J. Robinson. 1972. Relationship between *Bdellovibrio bacteriovorus* 6-5-S and autoclaved host bacteria. *Can. J. Microbiol.* **18:**1941–1948.

Diedrich, D. L., C. P. Curan, and S. F. Conti. 1984. Acquisition of *Escherichia coli* outer membrane proteins by *Bdellovibrio* sp. strain 109D. *J. Bacteriol.* **159:**329–334.

Diedrich, D. L., C. F. Denney, T. Hashimoto, and S. F. Conti. 1970. Facultatively parasitic strain of *Bdellovibrio bacteriovorus. J. Bacteriol.* **101:**989–996.

Diedrich, D. L., C. A. Portnoy, and S. F. Conti. 1983. *Bdellovibrio* possesses a prey-derived OmpF protein in its outer membrane. *Curr. Microbiol.* **8:**51–56.

Dunn, J. E., G. E. Windom, K. L. Hansen, and R. M. Seidler. 1974. Isolation and characterization of temperature-sensitive mutants of host-dependent *Bdellovibrio bacteriovorus* 109D. *J. Bacteriol.* **117:**1341–1349.

Eksztejn, M., and M. Varon. 1977. Elongation and cell division in *Bdellovibrio bacteriovorus. Arch. Microbiol.* **114:**175–181.

Engleking, H. M., and R. J. Seidler. 1974. The involvement of extracellular enzymes in the metabolism of *Bdellovibrio. Arch. Microbiol.* **95:**293–304.

Fackrell, H. B., and J. Robinson. 1973. Purification and characterization of a lytic peptidase produced by *Bdellovibrio bacteriovorus* 6-5-S. *Can. J. Microbiol.* **19:**659–666.

Finlay, B. B., F. Heffron, and S. Falkow. 1989. Epithelial cell surfaces induce *Salmonella* proteins required for bacterial adherence and invasion. *Science* **243:**940–943.

Friedberg, D. 1977. Effect of light on *Bdellovibrio bacteriovorus. J. Bacteriol.* **131:**399–404.

Friedberg, D. 1978. Growth of host-dependent *Bdellovibrio* in host cell free system. *Arch. Microbiol.* **116:**185–190.

Friedberg, I., and D. Friedberg. 1977. Freeze-fracture-etching studies on *Bdellovibrio bacteriovorus* mutants of altered host dependency. *FEMS Microbiol. Lett.* **2:**297–300.

Gloor, L., B. Klubek, and R. J. Seidler. 1974. Molecular heterogeneity of the bdellovibrios: metallo and serine proteases unique to each species. *Arch. Microbiol.* **95:**45–56.

Gray, K. M. 1989. Ph.D. dissertation, p. 78–97. University of Southern California, Los Angeles.

Gray, K. M., and E. G. Ruby. 1989. Unbalanced growth as a normal feature of development of *Bdellovibrio bacteriovorus. Arch. Microbiol.* **152:**420–424.

Gray, K. M., and E. G. Ruby. 1990. Prey-derived signals regulating duration of the developmental growth phase of *Bdellovibrio bacteriovorus. J. Bacteriol.* **172:**4002–4007.

Guerrini, F., V. Romano, M. Valenzi, M. DiGiulio, M. R. Mupo, and M. Sacco. 1982. Molecular parasitism in the *Escherichia coli-Bdellovibrio bacteriovorus* system: translocation of the matrix protein from the host to the parasite outer membrane. *EMBO J.* **1:**1439–1444.

Hashimoto, T., D. L. Diedrich, and S. F. Conti. 1970. Isolation of a bacteriophage for *Bdellovibrio bacteriovorus. J. Virol.* **5:**97–98.

Hespell, R. B. 1978. Intraperiplasmic growth of *Bdellovibrio bacteriovorus* on heat-treated *Escherichia coli. J. Bacteriol.* **33:**1156–1162.

Hespell, R. B., and M. Mertens. 1978. Effects of nucleic acid compounds on viability and cell composition of *Bdellovibrio bacteriovorus* during starvation. *Arch. Microbiol.* **116:**151–159.

Hespell, R. B., and D. A. Odelson. 1978. Metabolism of RNA-ribose by *Bdellovibrio bacteriovorus* during intraperiplasmic growth on *Escherichia coli. J. Bacteriol.* **136:**936–946.

Hespell, R. B., M. F. Thomashow, and S. C. Rittenberg. 1974. Changes in cell composition and viability of *Bdellovibrio bacteriovorus* during starvation. *Arch. Microbiol.* **97:**313–327.

Horowitz, A. T., M. Kessel, and M. Shilo. 1974. Growth cycle of predacious bdellovibrios in a host-free extract system and some properties of the host extract. *J. Bacteriol.* **117:**270–282.

Huang, J. C.-C., and M. P. Starr. 1973. Effects of calcium and magnesium ions and host viability on growth of bdellovibrios. *Antonie van Leeuwenhoek J. Microbiol. Serol.* **39:**151–167.

Ishiguro, E. E. 1973. A growth initiation factor for host-independent derivatives of *Bdellovibrio bacteriovorus. J. Bacteriol.* **115:**243–252.

Ishiguro, E. E. 1974. Minimum nutritional requirements for growth of host-independent derivatives of *Bdellovibrio bacteriovorus* strain 109 Davis. *Can. J. Microbiol.* **20:**263–265.

Képès, F. 1986. The cell cycle of *Escherichia coli* and some of its regulatory systems. *FEMS Microbiol. Rev.* **32:**225–246.

Kessel, M., and M. Shilo. 1976. Relationship of *Bdellovibrio* elongation and fission to host cell size. *J. Bacteriol.* **128:**477–480.

Kuenen, J. G., and S. C. Rittenberg. 1975. Incorporation of long-chain fatty acids of the substrate organism by *Bdellovibrio bacteriovorus* during intraperiplasmic growth. *J. Bacteriol.* **121:**1145–1157.

LaMarre, A. G., S. C. Straley, and S. F. Conti. 1977. Chemotaxis toward amino acids by *Bdellovibrio bacteriovorus. J. Bacteriol.* **131:**201–207.

Matin, A., and S. C. Rittenberg. 1972. Kinetics of deoxyribonucleic acid destruction and synthesis during growth of *Bdellovibrio bacteriovorus* strain 109D on *Pseudomonas putida* and *Escherichia coli. J. Bacteriol.* **111:**664–673.

Nelson, D. R., and S. C. Rittenberg. 1981. Incorporation of substrate cell lipid A components into the lipopolysaccharide of intraperiplasmically grown *Bdellovibrio bacteriovorus. J. Bacteriol.* **147:**860–868.

Odelson, D. A., M. A. Patterson, and R. B. Hespell. 1982. Periplasmic enzymes in *Bdellovibrio bacteriovorus* and *Bdellovibrio stolpii. J. Bacteriol.* **151:**756–763.

Park, J. T., and S. Mahadevan. 1988. Penicillin-binding proteins of bdellovibrios. *J. Bacteriol.* **170:**3750–3751.

Rayner, J. R., W. H. Cover, R. J. Martinez, and S. C. Rittenberg. 1985. *Bdellovibrio bacteriovorus* synthesizes an OmpF-like outer membrane protein during both axenic and intraperiplasmic growth. *J. Bacteriol.* **163:**595–599.

Reiner, A. M., and M. Shilo. 1969. Host-independent growth of *Bdellovibrio bacteriovorus* in microbial extracts. *J. Gen. Microbiol.* **59:**401–410.

Rittenberg, S. C. 1972. Nonidentity of *Bdellovibrio bacteriovorus* strains 109D and 109J. *J. Bacteriol.* **109:**432–433.

Rittenberg, S. C. 1982. Bdellovibrios—intraperiplasmic growth, p. 379–382. *In* R. G. Burns and J. H. Slater (ed.), *Experimental Microbial Ecology.* Blackwell Scientific Publishing, Oxford.

Rittenberg, S. C. 1989. Personal communication.

Rittenberg, S. C., and R. B. Hespell. 1975. Energy efficiency of intraperiplasmic growth of *Bdellovibrio bacteriovorus. J. Bacteriol.* **121:**1158–1165.

Rittenberg, S. C., and D. Langley. 1975. Utilization of nucleoside monophosphates per se for intraperiplasmic growth of *Bdellovibrio bacteriovorus. J. Bacteriol.* **121:**1137–1144.

Rittenberg, S. C., and M. Shilo. 1970. Early host damage in the infection cycle of *Bdellovibrio bacteriovorus. J. Bacteriol.* **102:**149–160.

Roberts, R. C., M. A. Keefer, and R. S. Ranu. 1987. Characterization of *Bdellovibrio bacteriovorus* bdellophage MAC-1. *J. Gen. Microbiol.* **133:**3065–3070.

Roberts, R. C., and R. S. Ranu. 1987. Transfection of *Bdellovibrio bacteriovorus* with bacteriophage MAC-1 DNA. *FEMS Microbiol. Lett.* **43:**207–211.

Ross, E. J., C. F. Robinow, and J. Robinson. 1974. Intracellular growth of *Bdellovibrio bacteriovorus* 6-5-S in heat-killed *Spirillum serpens* VHL. *Can. J. Microbiol.* **20:**847–851.

Rosson, R. A., and S. C. Rittenberg. 1979. Regulated breakdown of *Escherichia coli* deoxyribonucleic acid during intraperiplasmic growth of *Bdellovibrio bacteriovorus* 109J. *J. Bacteriol.* **140:**620–633.

Rosson, R. A., and S. C. Rittenberg. 1981. Pyrimidine metabolism of *Bdellovibrio bacteriovorus* grown intraperiplasmically and axenically. *J. Bacteriol.* **146:**108–116.

Ruby, E. G. 1989. Cell-envelope modifications accompanying intracellular growth of *Bdellovibrio bacteriovorus,* p. 17–34. *In* J. W. Moulder (ed.), *Intracellular Parasitism.* CRC Press, Boca Raton, Fla.

Ruby, E. G. 1991. The genus *Bdellovibrio,* p. 3400–3415. *In* A. Balows, H. G. Trüper, M. Dworkin, W. Harder, and K. H. Schleifer (ed.), *The Prokaryotes,* 2nd ed. Springer-Verlag, New York.

Ruby, E. G., and J. B. McCabe. 1986. An ATP transport system in the intracellular bacterium *Bdellovibrio bacteriovorus* 109J. *J. Bacteriol.* **167:**1066–1070.

Ruby, E. G., J. B. McCabe, and J. I. Barke. 1985. Uptake of intact nucleoside monophosphates by *Bdellovibrio bacteriovorus* 109J. *J. Bacteriol.* **163:**1087–1094.

Ruby, E. G., and S. C. Rittenberg. 1983. Differentiation after premature release of intraperiplasmically growing *Bdellovibrio bacteriovorus. J. Bacteriol.* **154:**32–40.

Ruby, E. G., and S. C. Rittenberg. 1984. Attachment of diaminopimelic acid to bdelloplast peptidoglycan during intraperiplasmic growth of *Bdellovibrio bacteriovorus* 109J. *J. Bacteriol.* **158:**597–602.

Sagi, B., and R. Levisohn. 1976. Isolation of temperate bacteriophage for *Bdellovibrio bacteriovorus. Isr. J. Med. Sci.* **12:**709–710.

Saier, M. 1989. Personal communication.

Sanchez-Amat, A., and F. Torrella. 1990. Formation of stable bdelloplasts as a starvation-survival strategy of marine bdellovibrios. *Appl. Environ. Microbiol.* **56:**2717–2725.

Schelling, M., and S. F. Conti. 1983. Serotyping of bdellovibrios by agglutination and indirect immunofluorescence. *Int. J. Syst. Bacteriol.* **33:**816–821.

Schelling, M., and S. F. Conti. 1986. Host receptor sites involved in the attachment of *Bdellovibrio bacteriovorus* and *Bdellovibrio stolpii. FEMS Microbiol. Lett.* **36:**319–323.

Scherff, R. H., J. E. DeVay, and T. W. Carroll. 1966. Ultrastructure of host-parasite relationships involving reproduction of *Bdellovibrio bacteriovorus* in host bacteria. *Phytopathology* **56:**627–632.

Seidler, R. J., M. Mandel, and J. N. Baptist. 1972. Molecular heterogeneity of the bdellovibrios: evidence of two new species. *J. Bacteriol.* **109:**209–217.

Seidler, R. J., and M. P. Starr. 1969. Isolation and characterization of host-independent bdellovibrios. *J. Bacteriol.* **100:**769–785.

Stolp, H., and M. P. Starr. 1963. *Bdellovibrio bacteriovorus* gen. et sp. n., a predatory, ectoparasitic, and bacteriolytic microorganism. *Antonie van Leeuwenhoek J. Microbiol. Serol.* **29:**217–248.

Straley, S. C., and S. F. Conti. 1974. Chemotaxis in *Bdellovibrio bacteriovorus. J. Bacteriol.* **120:**549–551.

Straley, S. C., and S. F. Conti. 1977. Chemotaxis by *Bdellovibrio bacteriovorus* toward prey. *J. Bacteriol.* **132:**628–640.

Straley, S. C., A. G. LaMarre, L. J. Lawrence, and S. F. Conti. 1979. Chemotaxis of *Bdellovibrio bacteriovorus* toward pure compounds. *J. Bacteriol.* **140:**634–642.

Talley, B. G., R. L. McDade, Jr., and D. L. Diedrich. 1987. Verification of the protein in the outer membrane of *Bdellovibrio bacteriovorus* as the OmpF protein of its *Escherichia coli* prey. *J. Bacteriol.* **169:**694–698.

Taylor, V. I., P. Baumann, J. L. Reichelt, and R. D. Allen. 1974. Isolation, enumeration, and host range of marine bdellovibrios. *Arch. Microbiol.* **98:**101–114.

Thomashow, M. F., and S. C. Rittenberg. 1978a. Intraperiplasmic growth of *Bdellovibrio bacteriovorus* 109J: solubilization of *Escherichia coli* peptidoglycan. *J. Bacteriol.* **135:**998–1007.

Thomashow, M. F., and S. C. Rittenberg. 1978b. Intraperiplasmic growth of *Bdellovibrio bacteriovorus* 109J: N-deacetylation of *Escherichia coli* peptidoglycan amino sugars. *J. Bacteriol.* **135:**1008–1014.

Thomashow, M. F., and S. C. Rittenberg. 1978c. Intraperiplasmic growth of *Bdellovibrio bacteriovorus* 109J: attachment of long-chain fatty acids to *Escherichia coli* peptidoglycan. *J. Bacteriol.* **135:**1015–1023.

Thomashow, M. F., and S. C. Rittenberg. 1979. The intraperiplasmic growth cycle—the life style of the bdellovibrios, p. 115–138. *In* J. H. Parish (ed.), *Developmental Biology of Prokaryotes.* University of California Press, Berkeley.

Torrella, F., R. Guerrero, and R. J. Seidler. 1978. Further taxonomic characterization of the genus *Bdellovibrio. Can. J. Microbiol.* **24:**1387–1394.

Tudor, J. J. 1980. Chemical analysis of the outer cyst wall and inclusion material of *Bdellovibrio* bdellocysts. *Curr. Microbiol.* **4:**251–256.

Tudor, J. J. 1990. Personal communication.

Tudor, J. J., and S. M. Bende. 1986. The outer cyst wall of *Bdellovibrio* bdellocysts is made de novo and not from preformed units from the prey wall. *Curr. Microbiol.* **13:**185–189.

Tudor, J. J., and S. F. Conti. 1977. Characterization of bdellocysts of *Bdellovibrio* sp. *J. Bacteriol.* **131:**314–322.

Tudor, J. J., and S. F. Conti. 1978. Characterization of germination and activation of *Bdellovibrio* bdellocysts. *J. Bacteriol.* **133:**130–138.

Tudor, J. J., M. P. McCann, and I. A. Acrich. 1990. A new model for the penetration of prey cells by bdellovibrios. *J. Bacteriol.* **172:**2421–2426.

Varon, M. 1979. Selection of predation-resistant bacteria in continuous culture. *Nature* (London) **277:**386–388.

Varon, M., C. Cocito, and J. Seijffers. 1976. Effect of virginiamycin on the growth cycle of *Bdellovibrio*. *Antimicrob. Agents Chemother.* **9:**179–188.

Varon, M., M. Fine, and A. Stein. 1983. Effect of polyamines on the intraperiplasmic growth of *Bdellovibrio* at low cell densities. *Arch. Microbiol.* **136:**158–159.

Varon, M., and R. Levisohn. 1972. Three-membered parasitic system: a bacteriophage, *Bdellovibrio bacteriovorus,* and *Escherichia coli. J. Virol.* **9:**519–525.

Varon, M., and J. Seijffers. 1975. Symbiosis-independent and symbiosis-incompetent mutants of *Bdellovibrio bacteriovorus* 109J. *J. Bacteriol.* **124:**1191–1197.

Varon, M., and J. Seijffers. 1977. Osmoregulation in symbiosis-independent mutants of *Bdellovibrio bacteriovorus. Appl. Environ. Microbiol.* **33:**1207–1208.

Varon, M., and M. Shilo. 1968. Interaction of *Bdellovibrio bacteriovorus* and host bacteria. I. Kinetic studies of attachment and invasion of *Escherichia coli* B by *Bdellovibrio bacteriovorus. J. Bacteriol.* **95:**744–753.

Varon, M., and M. Shilo. 1969. Attachment of *Bdellovibrio bacteriovorus* to cell wall mutants of *Salmonella* spp. and *Escherichia coli. J. Bacteriol.* **97:**977–979.

Varon, M., and M. Shilo. 1976. Properties of marine bdellovibrios. *Microb. Ecol.* **2:**284–295.

Varon, M., and B. P. Zeigler. 1978. Bacterial predator-prey interaction at low prey density. *Appl. Environ. Microbiol.* **36:**11–17.

Index